Nucleic acids in chemistry and biology

Nucleic acids in chemistry and biology

Second edition

Edited by

G. MICHAEL BLACKBURN
Department of Chemistry, University of Sheffield

and

MICHAEL J. GAIT
MRC Laboratory of Molecular Biology, Cambridge

Oxford   New York   Tokyo
OXFORD UNIVERSITY PRESS
1996

Oxford University Press, Walton Street, Oxford OX2 6DP

Oxford   New York
Athens   Auckland   Bangkok   Bombay
Calcutta   Cape Town   Dar es Salaam   Delhi
Florence   Hong Kong   Istanbul   Karachi
Kuala Lumpur   Madras   Madrid   Melbourne
Mexico City   Nairobi   Paris   Singapore
Taipei   Tokyo   Toronto

and associated companies in
Berlin   Ibadan

Oxford is a trademark of Oxford University Press

Published in the United States
by Oxford University Press Inc., New York

A catalogue record for this book is available from the British Library

Library of Congress Cataloging-in-Publication Data
Nucleic acids in chemistry and biology/edited by G. Michael
    Blackburn and Michael J. Gait. — 2nd ed.
    Includes index.
    1. Nucleic acids.   I. Blackburn, G. Michael.   II. Gait, Michael J.
QD433.N83   1996   574.87'328—dc20   94-46592
    ISBN 0-19-963534-X (Hbk)
    ISBN 0-19-963533-1 (Pbk)

Typeset by Techset Composition Ltd, Salisbury
Printed in Hong Kong

FOREWORD

Today chemistry and biology are so inextricably meshed that it is hard to imagine a time when they were otherwise. Yet in 1968 as I was finishing my Ph.D. in organic chemistry, most of my colleagues and many of my mentors thought biology was a discipline whose waters were still too murky for a chemist to plunge into. Almost everyone in the Chemistry Department at Sheffield University told me that molecular biology, the subject which I was coming to love, was not even a subject. Happily, the newest appointee to the faculty of Sheffield University was Mike Blackburn. He was a lonely dissenting voice extolling the virtues of biology and encouraging adventurous chemists to graze its wonderful pastures. He was right! When I moved to Harvard University in 1969, for a postdoctoral period with Jack Strominger, I soon became immersed and enthralled by the wonders of molecular biology. At Harvard I worked on tRNAs and spent a whole year deciphering 144 nucleotides of the sequences of two closely related tRNA^{Gly}s by the then-new Sanger method. How things have changed! This same sequence can now be obtained as less than half of a single sequencing reaction! On moving to Cold Spring Harbor in 1972 I moved quickly into DNA and began isolating and characterizing the restriction enzymes that could cut it so effectively. It was nucleic acid sequencing that drove my initial interest in DNA and it was but a short step to studies of mRNA. Our attempts to define a eukaryotic promoter sequence were thwarted initially by the lack of colinearity between adenovirus mRNA and the viral DNA sequence. However, this had the dramatic and rewarding consequence of leading to the discovery of split genes and RNA splicing.

I have always credited my initial training as a chemist for giving me an edge in tackling biological questions. In those early days the text books and many of the practitioners of biology seemed more interested in the phenomenology of biology than in its rigours. Many chemists and physicists, on moving into biology, discovered a rich ground for devising and carrying through experiments, which benefited from the rigour of their early training. Unfortunately, even today where mechanistic questions in biology are more popular, there is often a reluctance to come to grips with the chemistry. I was, therefore, most pleased to discover the first edition of Blackburn and Gait, which captured the essence of nucleic acid chemistry while placing it squarely in a biological context. While the book's proper home is close at hand on my bookshelf, I have found it borrowed far too often. I shall be certain to find a better hiding place for this second edition.

For some years now, Mike Gait has been a valuable colleague as an Executive Editor of *Nucleic Acids Research*. I have come to respect his insight and knowledge on many fronts. It was also pleasing to see that some other former colleagues have contributed to this book. Dieter Soll and I spent much time encouraging and publicizing the use of computers in molecular biology in the early 1980s. Tom RajBhandary taught me most of what I know about tRNAs for he served as my informal mentor during my postdoctoral period. My interactions with Dick Walker, which began through our joint editorial roles at *Nucleic Acids Research*, has now extended into a research collaboration. Again a thorough grounding in chemistry is making possible new research that is of interest and relevance to both fields.

In the past, DNA and RNA were looked upon as fairly passive molecules whose principal function was informational. This changed dramatically with the discovery of ribozymes and it now appears that even DNA can be catalytic. In recent years the discovery of Z-DNA, the G-tetrads at the end of telomeres and the potential for triple helical structures, have all thrown new light on the properties of nucleic acids. The recent finding by Xiaodong Cheng and myself of base flipping, that is illustrated on the cover of this book, shows again the treats that Nature has in store for us. Yet all of these observations, from ribozymes to base flipping, are dependent upon the chemistry of nucleic acids. It is at the chemical level that we need to explain what is going on.

Both those students wishing to know the present state of the field as well as those researchers planning experiments to understand biological phenomenon can gain much from this book. It is a significant contribution to an important and exciting field.

Richard J. Roberts
Director of Research, New England Biolabs

PREFACE TO SECOND EDITION

The first edition of *Nucleic acids in chemistry and biology* in 1990 met the pressing need for a single volume that integrated the chemistry and biology of the nucleic acids in an introductory yet authoritative text. Greatly encouraged by the very favourable response to our book, we have now assembled this second and completely revised edition, written by most of the same team of international experts. In it we have sought to maintain our original objectives, which include a readable style and a strong emphasis on chemical and structural aspects of nucleic acids. At the same time we have introduced a number of significant changes both to its content and to its presentation. These changes have been driven largely by new advances in knowledge or where the passage of time has led to a change in emphasis or interpretation. We have also responded to the many and valued comments that we have received, for example, by the introduction of a new section dealing with the manipulative techniques used in nucleic acid research and in an expansion of the further reading lists appended to each chapter. Our editorial colleagues at Oxford University Press have responded by giving the text a major 'face-lift', extending the use of two-colour presentation and resetting the line diagrams. Many new figures have been introduced, including several full colour plates, in particular to bring to the reader the fullest details of X-ray analysis of protein:nucleic acid structures. At the same time, we have tried to enhance the clarity of presentation even further.

In Chapter 2, we have included a new account of DNA tertiary structure that more accurately reflects the latest opinions concerning sequence-dependent structure modulation and a section on triplex and other unusual structures of nucleic acids. In Chapter 3, we have completely revised the sections on synthesis of oligonucleotide analogues and oligoribonucleotides while Chapter 4 is supplemented with an account of some of the latest drugs produced to interfere with the biosynthesis of nucleic acids. Developments in eukaryotic genome structure are particularly noteworthy in an updated Chapter 5 which has been reshaped to give the topic of DNA recombination more prominence.

Our knowledge of RNA structure and function has provided some of the most exciting recent advances which are described in a recast Chapter 6. This now includes descriptions of new concepts in the folding and tertiary interactions of RNA as well as an expanded section dealing with catalytic RNA. In Chapter 7 we have given extended coverage to the enediyne antibiotics and radical processes by which they achieve remarkable cleavage of both strands of DNA, while in Chapter 8 there is much new material on molecules that bind specifically to G·C sequences in the minor groove of DNA, on RNA intercalators, and on small molecule interactions with multi strand structures.

The greatest area of growth in our knowledge has been in protein : nucleic acid interactions. Chapter 9 has been rewritten in full, with enhanced support for a highly readable text from a range of full colour plates and careful choice of those examples of protein : nucleic acids interactions which best illustrate many fresh insights into this important aspect of molecular recognition. Finally, we have generated a completely new Chapter 10 which deals with the concepts behind essential physical and biological techniques in the nucleic acids which have shaped and continue to influence the development of research in nucleic acids, each one illustrated by an example of a recent application.

In conclusion, we have produced this new edition to meet the needs of a wide range of scientists for a highly accessible account of the chemistry and structure of the nucleic acids. It is firmly set in the context of the exciting developments that continue to stem from biological research, from industrial biotechnology, and out of the effort applied to the conquest on many human diseases.

Sheffield and Cambridge   G.M.B.
November 1995   M.J.G.

PREFACE TO THE FIRST EDITION

Nucleic acids dominate modern molecular science. They have vital roles that are fundamental for the storage and transmission of genetic information within cells. It follows that an accurate and detailed knowledge of their structure and function is of prime importance for molecular scientists of all descriptions. Just as significantly, the genius of biologists and chemists working together has made contemporary research into nucleic acids a rich source of discovery and invention that is dramatically transforming and improving the human condition.

Our own teaching and research experience, shared in discussions with colleagues around the world, has convinced us of the need to fill a significant gap in the modern science library by creating a broad-based yet concise and readable book on nucleic acids. Our single volume is designed to provide a compact, molecular perspective of this great subject. To that end, we have used it unashamedly to emphasize chemical and structural aspects of nucleic acids at all points. In it, we have surveyed a very broad field—up to the point where the frontiers of current studies in nucleic acids are only attainable by reading the latest issues of key journals! In particular, we have strengthened its production by drawing on the talents of an international group of co-authors whose expertise has extended the authority of this book from cover to cover. At the same time, we have tried to keep it selective and simple so as to make it widely accessible to students. This has meant that, of necessity, some sections have focused more on key concepts rather than on fine detail.

We have tried to provide a radically fresh and unified approach. This book builds on a general introduction to the chemistry and biology of the nucleic acids in order to reach out to some of the most significant modern developments of this subject. It is couched in an easily readable style and in a language which, while technically accurate, can yet be grasped quickly by those with a basic scientific background.

We begin with a brief historical perspective designed to point out the significance of later progress. We next provide an outline of the essential features of DNA and RNA structure, highlighting the new subtle insights which have been obtained by detailed analysis of crystals of synthetic oligonucleotides of defined sequence. The next four chapters are the core of the book. The first concentrates on modern chemistry applied to the synthesis of biologically important nucleosides, nucleotides, and oligonucleotides. Then comes a discussion of the biosynthesis of nucleotides, which is given a fresh presentation to emphasize how anti-cancer and anti-viral agents interfere with biosynthetic processes. The core is completed by two chapters which deal with the basic molecular biology of DNA and of RNA, showing how information stored in the form of nucleotide sequence is transmitted into cellular activity. Recent exciting developments in the auto-catalysis of RNA, ribozymes, are especially featured.

Three rather more specialized chapters then focus on the covalent and physical interactions of nucleic acids with small molecules, especially with mutagens and carcinogens and the relevant repair processes, and on their physical interactions with proteins. These important topics are at the forefront of much present research and typify the success of creative symbioses between chemistry and biology. The final chapter contrasts the *in vivo* rearrangements which DNA experiences with the *in vitro* techniques of manipulation of DNA sequences that are the essence of experimentation in cloning and mutagenesis.

This is not a textbook on the molecular biology of nucleic acids. From the outset, we have aimed this book especially at the needs of students and new research workers with a chemical or biochemical background. We hope that molecular biologists and more senior chemists and biochemists alike will find their knowledge of nucleic acids broadened through the special perspectives this book offers.

Sheffield and Cambridge G.M.B.
November 1989 M.J.G.

ACKNOWLEDGEMENTS

Both Mikes express their sincere appreciation for the efforts of all who have supported the production of this book. Principally, our unqualified thanks to those eight expert and understanding co-authors, without whose contributions the revision of this book would not have been possible. Once again, we have taken considerable liberties with their manuscripts, in order to blend all of their contributions into a homogeneous final product. They have responded with equanimity and understanding and have co-operated superbly in the numerous revision processes required for the production of the finished work.

We are very grateful to the many colleagues and fellow scientists who have provided us with valuable comments on the text of the first edition or have read portions of this second edition. They include Tom Brown, Chris Christodoulou, Brian Clarke, Erik de Clercq, Mike Clemens, Bernard Connolly, Geoff Ford, Jane Grasby, Hans Gross, David Hornby, Bill Hunter, Chris Hunter, Paul Kong, Christian Lehmann, David Lilley, Mick McLean, Daniella Rhodes, Gordon Tener, Andrew Travers, Steve West, and Ian Willis. They and others have helped us to rectify many of the minor errors and omissions of the first edition of this book. We remain grateful to Joachim Engels for supplying many of the definitions in the glossary. We particularly thank Lord Todd, Gobind Khorana, and Dan Brown for suggestions and comments on the early part of Chapter 1 which have, we believe, given us a genuine feeling for those key events in the early, seminal areas of nucleic acid studies.

The final production of this book has been supported by many able individuals. We are particularly indebted to Rich Roberts (New England Biolabs), Xiadong Cheng (Cold Spring Harbor), and Bill Scott (LMB, MRC) for access to original graphics for nucleic acids and protein structures on the covers of this volume, and above all to the staff of OUP for establishing even better standards for production of the second edition than for the original work.

Finally, it is inevitable in a book of this breadth that omission, occasional errors, and lapses in the accuracy of interpretation will have escaped the detection of even the most assiduous of proof-readers. We hope that any such mistakes are both minor and minimal and we accept full and exclusive responsibility for them. We shall be grateful to receive your help in their identification for future rectification.

CONTENTS

CONTRIBUTORS

G. MICHAEL BLACKBURN, Chapters 1, 2, 3, and 7
Department of Chemistry, University of Sheffield, Sheffield, S3 7HF, UK

MICHAEL J. GAIT, Chapters 1, 3, and 4
MRC Laboratory of Molecular Biology, Hills Road, Cambridge, CB2 2QH, UK

GORDON C. BARR, Chapter 5
Department of Biochemistry, Medical Sciences Institute, University of Dundee, Dundee, DD1 4HN, UK

ANDREW J. FLAVELL, Chapters 5 and 10
Department of Biochemistry, Medical Sciences Institute, University of Dundee, Dundee, DD1 4HN, UK

TUOMO GLUMOFF, Chapter 9
Centre for Biotechnology, University of Turku, PO Box 123, FIN-20521 Turku, Finland

ADRIAN GOLDMAN, Chapter 9
Centre for Biotechnology, University of Turku, PO Box 123, FIN-20521, Turku, Finland

DAVID NORMAN, Chapter 10
Department of Biochemistry, Medical Sciences Institute, University of Dundee, Dundee, DD1 4HN, UK

UTTAM L. RAJBHANDARY, Chapter 6
Department of Biology, Massachusetts Institute of Technology, 77 Massachusetts Avenue, Cambridge, Massachusetts, 02139, USA

DIETER SÖLL, Chapter 6
Department of Molecular Biophysics and Biochemistry, Yale University, PO Box 208114, New Haven, Connecticut, 06520–8114, USA

RICHARD T. WALKER, Chapter 4
Department of Chemistry, University of Birmingham, PO Box 363, Birmingham, B15 2TT, UK

W. DAVID WILSON, Chapter 8
Department of Chemistry, Georgia State University, Atlanta, Georgia, 30303–3083, USA

NOMENCLATURE

The nomenclature for nucleic acids and their constituents used in this book is derived from the following recommendations:

IUPAC–IUB Joint Commission on Biochemical Nomenclature. Abbreviations and symbols for nucleic acids, polynucleotides and their constituents. Recommendations 1970. (1970) *Biochemistry*, **9**, 4022–7.

IUPAC–IUB Joint Commission on Biochemical Nomenclature. Abbreviations and symbols for the description of conformations of polynucleotide chains. Recommendations 1982. (1983) *Eur. J. Biochem.*, **131**, 9–15.

Definitions and nomenclature of nucleic acid structure parameters. (1989) *EMBO J.*, **8**, 1–4.

Stereodiagrams
The stereo-pair figures used in this book are for parallel viewing, i.e., left diagram to left eye and right to right. They can be viewed either unaided (a little practice helps) or with the help of a simple convex-lens viewer.

INTRODUCTION AND OVERVIEW

1.1 The biological importance of DNA

From the beginning, the study of nucleic acids has drawn together, as though by a powerful unseen force, a galaxy of scientists of the highest ability. Striving to tease apart its secrets, these talented individuals have brought with them a broad range of skills from other disciplines while many of the problems they have encountered have proved to be soluble only by new inventions. Looking at their work, one is constantly made aware that scientists in this field appear to have enjoyed a greater sense of excitement in their work than is given to most. Why?

For over 60 years, such men and women have been fascinated and stimulated by their awareness that the study of nucleic acids is central to a knowledge of life. Let us start by looking at Fred Griffith, who was employed as a scientific civil servant in the British Ministry of Health investigating the nature of epidemics. In 1923, he was able to identify the difference between a virulent, *S*, and a non-virulent, *R*, form of the pneumonia bacterium. Griffith went on to show that this bacterium could be made to undergo a permanent, hereditable change from non-virulent to virulent type. This discovery was a bombshell in bacterial genetics.

Oswald Avery and his group at the Rockefeller Institute in New York set out to identify the molecular mechanism responsible for the change Griffith had discovered, now technically called **bacterial transformation.** They achieved a breakthrough in 1940 when they found that non-virulent *R* pneumococci could be transformed *irreversibly* into a virulent species by treatment with a pure sample of high molecular weight DNA. Avery had purified this DNA from heat-killed bacteria of a virulent strain and showed that it was active at a dilution of 1 part in 10^9.

Avery concluded that '**DNA is responsible for the transforming activity**' and published that analysis in 1944, just three years after Griffith had died in a London air-raid. The staggering implications of Avery's work turned a searchlight on the molecular nature of nucleic acids and it soon became evident that ideas on the chemistry of nucleic acid structure at that time were wholly inadequate to explain such a momentous discovery. As a result, a new wave of scientists directed their attention to DNA and discovered that large parts of the accepted tenets of nucleic acid chemistry had to be set aside before real progress was possible. We need to examine some of the earliest features of that chemistry to appreciate fully the significance of later progress.

1.2 The origins of nucleic acids research

Friedrich Miescher started his research career in Tübingen by looking into the physiology of human lymph cells. In 1868, seeking a more readily available material, he began to study human pus cells which he obtained in abundant supply from the bandages discarded from the local hospital. After defatting the cells with alcohol, he incubated them with a crude preparation of pepsin from pig stomach and so obtained a grey precipitate of pure cell nuclei. Treatment of this with alkali followed by acid gave Miescher a precipitate of a phosphorus-containing substance which he named **nuclein.** He later found this material to be a common constituent of yeast, kidney, liver, testicular, and nucleated red blood cells.

After Miescher moved to Basel in 1872, he found the sperm of Rhine salmon to be a more plentiful source of nuclein. The pure nuclein was a strongly acidic substance which existed in a salt-like combination with a nitrogenous base which Miescher crystallized and called protamine. In fact, his nuclein was really a nucleoprotein and it fell subsequently to Richard Altman in 1889 to obtain the first protein-free material, to which he gave the name **nucleic acid**.

Following William Perkin's invention of mauveine in 1856, the development of aniline dyes had stimulated a systematic study of the colour-staining of biological specimens. Cell nuclei were characteristically stained by basic dyes, and around 1880 Walter Flemming applied that property in his study of the rod-like segments of chromatin (so called because of their colour-staining characteristic) which became visible within the cell nucleus only at certain stages of cell division. Flemming's speculation that the chemical composition of these **chromosomes** was identical with that of Miescher's nuclein was confirmed in 1900 by E. B. Wilson who wrote:

Now chromatin is known to be closely similar to, if not identical with, a substance known as nuclein which analysis shows to be a tolerably definite chemical compound of nucleic acid and albumin. And thus we reach the remarkable conclusion that inheritance may, perhaps, be affected by the physical transmission of a particular compound from parent to offspring.

While this insight was later to be realized in Griffith's 1928 experiments, all of this work was really far ahead of its time. We have to recognize that, at the turn of the century, tests for the purity and identity of substances were relatively primitive. Emil Fischer's classic studies on the chemistry of high molecular weight, polymeric organic molecules were in question until well into the twentieth century. Even in 1920, it was possible to argue that there were only two species of nucleic acids in nature: animal cells were believed to provide **thymus nucleic acid** (DNA), whilst nuclei of plant cells were thought to give **pentose nucleic acid** (RNA).

1.3 Early structural studies on nucleic acids

Accurate molecular studies on nucleic acids essentially date from 1909 when Levene and Jacobs began a reinvestigation of the structure of **nucleotides** at the Rockefeller Institute. Inosinic acid, which Liebig had isolated from beef muscle in 1847, proved to be hypoxanthine-riboside $5'$-phosphate. Guanylic acid, isolated from the nucleoprotein of pancreas glands, was identified as guanine-riboside $5'$-phosphate (Fig. 1.1). Each of these nucleotides was cleaved by alkaline hydrolysis to give phosphate and the corresponding **nucleosides**, inosine and guanosine respectively. Since then, all nucleosides are characterized as the condensation products of a pentose and a nitrogenous base while nucleotides are the phosphate esters of one of the hydroxyl groups of the pentose.

Pentose nucleic acid was available in plentiful supply from yeast and on mild hydrolysis with aqueous pyridine it gave the four pentose-nucleosides adenosine, cytidine, guanosine, and uridine. These were identified as derivatives of the four bases adenine, cytosine, guanine, and uracil (Fig. 1.1).

Thymus nucleic acid, which was readily available from calf tissue, was found to be resistant to alkaline hydrolysis. It was only successfully degraded into deoxynucleosides in 1929 when

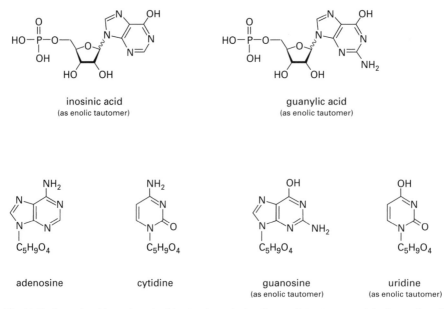

Fig. 1.1 Early nucleoside and nucleotide structures (using the enolic tautomers originally employed). ᨳᨳ denotes unknown stereochemistry at C-1.

Levene adopted enzymes to hydrolyse the deoxyribonucleic acid followed by mild acidic hydrolysis of the deoxynucleotides. He identified its pentose as the hitherto unknown 2-deoxy-D-ribose. These deoxynucleosides involved the four heterocyclic bases, adenine, cytosine, guanine, and thymine, with the latter corresponding to uracil in ribonucleic acid.

Up to 1940, most groups of workers were convinced that hydrolysis of nucleic acids gave the appropriate four bases in **equal relative proportions**. This erroneous conclusion probably resulted from the use of impure nucleic acid or from the use of analytical methods of inadequate accuracy and reliability. It led, naturally enough, to the general acceptance of a **tetranucleotide hypothesis** for the structure of both thymus and yeast nucleic acids, which materially retarded further progress on the molecular structure of nucleic acids.

Several of these tetranucleotide structures were proposed. They all had four nucleosides (one for each of the bases) with an arbitrary location of the two purines and two pyrimidines. They were joined together by four phosphate residues in a variety of ways, among which there was a strong preference for phosphodiester linkages. In 1932, Takahashi showed that yeast nucleic acid contained neither pyrophosphate nor phosphomonoester functions and so disposed of earlier proposals in preference for a neat, cyclic structure which joined the pentoses exclusively using phosphodiester units (Fig. 1.2). It was generally accepted that these bonded 5′- to 3′-positions of adjacent deoxyribonucleosides, but the linkage positions in ribonucleic acid were not known.

One property stuck out like a sore thumb from this picture: the molecular mass of nucleic acids was greatly in excess of that calculated for a tetranucleotide. The best DNA samples were produced by Einar Hammarsten in Stockholm and one of his students, Torjbörn Caspersson, showed that this material was greater in size than protein molecules.

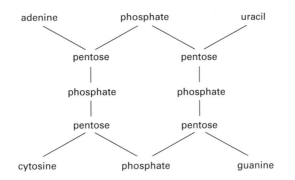

Fig.1.2 The tetranucleotide structure proposed for nucleic acids by Takahashi (1932).

Hammersten's DNA was examined by Rudolf Signer in Bern whose flow-birefringence studies revealed rod-like molecules with a molecular mass of 0.5–1.0×10^6 Daltons (Da). The same material provided Astbury in Leeds with X-ray fibre diffraction measurements that supported Signer's conclusion. Finally, Levene estimated the molecular mass of native DNA at between 200 000 and 1×10^6, based on ultracentrifugation studies.

Scientists compromised. In his Tilden Lecture of 1943, Masson Gulland suggested that the concept of nucleic acid structures of polymerized, uniform tetranucleotides was limited, but he allowed that they could 'form a practical working hypothesis'.

This then was the position in 1944 when Avery published his great work on the transforming activity of bacterial DNA. One can sympathize with Avery's hesitance to press home his case. Levene, in the same Institute, and others were strongly persuaded that the tetranucleotide hypothesis imposed an invariance on the structure of nucleic acids which denied them any role in biological diversity. By contrast, Avery's work showed that DNA was responsible for completely transforming the behaviour of bacteria. It demanded a fresh look at the structure of nucleic acids.

1.4 The discovery of the structure of DNA

From the outset, it was evident that DNA exhibited greater resistance to selective chemical hydrolysis than did RNA. So, the discovery in 1935 that DNA could be cut into **mononucleotides** by an enzyme doped with arsenate was invaluable. Using this procedure, Klein and Thannhauser obtained the four crystalline deoxyribonucleotides whose structures (Fig. 1.3) were later put beyond doubt by total chemical synthesis by Alexander Todd and the Cambridge school he founded in 1944. Todd established the β-configuration and the glycosidic linkage for ribonucleosides in 1951, but found the chemical synthesis of the 2′-deoxyribonucleosides more taxing. The key to success for the Cambridge group was the development of methods of phosphorylation, illustrated in Fig. 1.4 for the preparation of the 3′- and 5′-phosphates of deoxyadenosine (Fig. 1.4).

All the facts were now available to establish the primary structure of DNA as a **linear polynucleotide** in which each deoxyribonucleoside is linked to the next by means of a 3′- to 5′-phosphodiester (Chapter 2, Fig. 2.15). The presence of only diester linkages was essential to

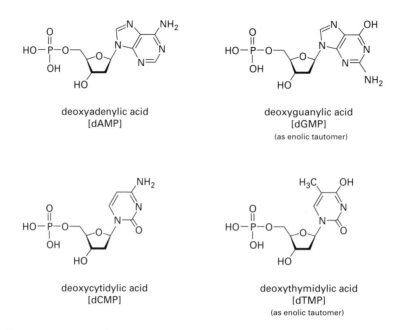

Fig.1.3 Structures of 5′-deoxyribonucleotides (original tautomers for dGMP and dTMP).

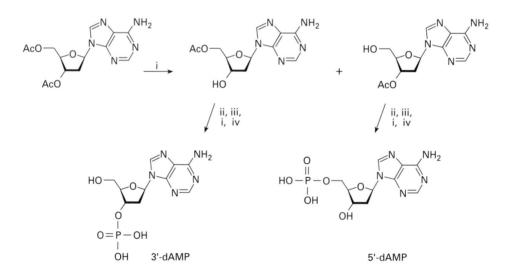

Fig.1.4 Todd's syntheses of deoxyadenosine 3′- and 5′-phosphates (Hayes, D. H., Michelson, A. M., and Todd, A. R. (1955). *J. Chem. Soc.*, 808–15).
Reagents: (i) MeOH, NH₃ (ii) (PhO)₂P(O)OP(H)(O)OCH₂Ph (iii) *N*-chlorosuccinimide (iv) H₂/PdC.

explain the stability of DNA to chemical hydrolysis, since phosphate triesters and monoesters, not to mention pyrophosphates, are more labile. The measured molecular masses for
DNA of about 1×10^6 meant that a single strand of DNA would have some 3000 nucleotides.
Such a size was much greater than that of enzyme molecules, but entirely compatible with
Staudinger's established ideas on macromolecular structure for synthetic and natural polymers. But by the mid-point of the twentieth century, chemists could advance no further with
the primary structure of DNA. Neither of the key requirements for sequence determination
was to hand: there were no methods for obtaining pure samples of DNA with homogeneous
base sequence nor were methods available for the cleavage of DNA strands at a specific base
residue. Consequently, all attention came to focus on the secondary structure of DNA.

Two independent experiments in biophysics showed that DNA possesses an ordered secondary structure. Using a sample of DNA obtained from Hammarsten in 1938, Astbury obtained an X-ray diffraction pattern from stretched, dry fibres of DNA. From the rather
obscure data he deduced '... A spacing of 3.34 Å along the fibre axis corresponds to
that of a close succession of flat or flattish nucleotides standing **out** perpendicularly to the
long axis of the molecule to form a relatively rigid structure.' These conclusions roundly
contradicted the tetranucleotide hypothesis. Some years later, Gulland studied the viscosity
and flow-birefringence of calf thymus DNA and thence postulated the presence of hydrogen
bonds linking the purine–pyrimidine **hydroxyl** groups and some of the amino groups. He suggested that these hydrogen bonds could involve nucleotides either in adjacent chains or within a single chain, but he somewhat hedged his bets between these alternatives.

Sadly, Astbury returned to the investigation of proteins and Gulland died prematurely in a
train derailment in 1947. Both of them left work that was vital for their successors to follow,
but each contribution contained a misconception that was to prove a stumbling block for the
next half-a-dozen years. Thus, Linus Pauling's attempt to create a helical model for DNA
located the pentose-phosphate backbone in its core and the **bases pointing outwards**—as
Astbury had decided. Gulland had subscribed to the wrong tautomeric forms for the heterocyclic bases thymine and guanine, believing them to be **enolic** and having hydroxyl groups.
The importance of the true **keto forms** was only appreciated in 1952.

Erwin Chargaff began to investigate a very different type of order in DNA structure. He
studied the base composition of DNA from a variety of sources using the new technique of
paper chromatography to separate the products of hydrolysis of DNA and employing one of
the first commercial ultraviolet spectrophotometers to quantify their relative abundance. His
data showed that there is a variation in base composition of DNA between species which is
overriden by a universal 1 : 1 ratio of adenine with thymine and of guanine with cytosine. This
meant that the proportion of purines, (A + G), is always equal to the proportion of pyrimidines, (C + T). Although the ratio (G + C)/(A + T) varies from species to species, different
tissues from a single species give DNA of the same composition. Chargaff's results finally dis-
credited the tetranucleotide hypothesis, because that called for equal proportions of all four
bases in DNA.

In 1951, Francis Crick and Jim Watson joined forces in the Cavendish Laboratory in Cambridge to tackle the problem of DNA structure. Both of them were persuaded that the model-
building approach that had led Pauling and Corey to the α-helix structure for peptides should

work just as well for DNA. Almost incredibly, they attempted no other line of direct experimentation but drew on the published and unpublished results of other research teams in order to construct a variety of models, each to be discarded in favour of the next until they created one which satisfied all the facts.

The best X-ray diffraction results were to be found in King's College, London. There, Maurice Wilkins had observed the importance of keeping DNA fibres in a moist state and Rosalind Franklin had found that the X-ray diffraction pattern obtained from such fibres showed the existence of an A-form of DNA at low humidity which changed into a B-form at high humidity. Both forms of DNA were highly crystalline and clearly helical in structure. Consequently, Franklin decided that this behaviour required the phosphate groups to be exposed to water on the **outside** of the helix, with the corollary that the bases were on the **inside** of the helix.

Watson decided that the number of nucleotides in the unit crystallographic cell favoured a double-stranded helix. Crick's physics-trained mind recognized the symmetry implications of the space-group of the A-form diffraction pattern, monoclinic C^2. There had to be local two fold symmetry axes normal to the helix, a feature which called for a double-stranded helix whose two chains must run in opposite directions.

Crick and Watson thus needed merely to solve the final problem: how to construct the core of the helix by packing the bases together in a regular structure. Watson knew about Gulland's conclusions regarding hydrogen bonds joining the DNA bases. This convinced him that the crux of the matter had to be a rule governing hydrogen bonding between bases. Accordingly, Watson experimented with models using the **enolic** tautomeric forms of the bases (Fig. 1.3) and pairing like with like. This structure was quickly rejected by Crick because it had the wrong symmetry for B-DNA. **Self-pairing** had to be rejected because it could not explain Chargaff's 1:1 base ratios, which Crick had perceived were bound to result if you had **complementary base pairing.**

Based on advice from Jerry Donohue in the Cavendish Laboratory, Watson turned to manipulating models of the bases in their **keto forms** and paired adenine with thymine and guanine with cytosine. Almost at once he found a compellingly simple relationship involving two hydrogen bonds for an A·T pair and two or three hydrogen bonds for a G·C pair. The special feature of this base-pairing scheme is that the relative geometry of the bonds joining the bases to the pentoses is virtually identical for the A·T and G·C pairs (Fig. 1.5). It follows that if a purine always pairs with a pyrimidine then an irregular sequence of bases in a single strand

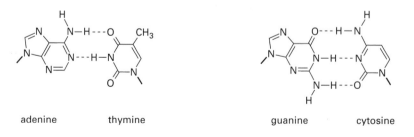

adenine   thymine   guanine   cytosine

Fig. 1.5 Complementary hydrogen bonded base-pairs as proposed by Watson and Crick (thymine and guanine in the revised ketoforms). NB. The G·C structure was later altered to include three hydrogen bonds.

of DNA could nonetheless be paired regularly in the centre of a double helix and without loss of symmetry.

Chargaff's 'rules' were straightaway revealed as an obligatory consequence of a double-helical structure for DNA. Above all, since the base sequence of one chain automatically determines that of its partner, Crick and Watson could easily visualize how one single chain might be the template for creation of a second chain of complementary base sequence.

The structure of the core of DNA had been solved and the whole enterprise fittingly received the ultimate accolade of the scientific establishment when Crick, Watson, and Wilkins shared the Nobel prize for chemistry in 1962, just four years after Rosalind Franklin's early death.

1.5 The advent of molecular biology

It is common to describe the publication of Watson and Crick's paper in *Nature* in April 1953 as the end of the 'classical' period in the study of nucleic acids, up to which time basic discoveries were made by a few gifted academics in an otherwise relatively unexplored field. The excitement aroused by the model of the double helix certainly drew the attention of a much wider scientific audience than before to the importance of nucleic acids, particularly because of the biological implications of the model rather than because of the structure itself. It was immediately apparent that locked into the irregular sequence of nucleotide bases in the DNA of a cell was all the information required to specify the diversity of biological molecules needed to carry out the functions of that cell. The important question now was what was the key, the **genetic code**, through which the sequence of DNA could be translated into protein.

The solution to the coding problem is often attributed to the laboratories in the USA of Marshall Nirenberg and of Severo Ochoa who devised an elegant cell-free system for translating enzymatically synthesized polynucleotides into polypeptides and who by the mid 1960s had established the genetic code for a number of amino acids. In reality, the story of the elucidation of the code involves numerous strands of knowledge obtained from a variety of workers in different laboratories. An essential contribution came from Alexander Dounce in Rochester, New York, who in the early 1950s postulated that RNA, and not DNA, served as a template to direct the synthesis of cellular proteins and that a sequence of three nucleotides might specify a single amino acid. Sydney Brenner and Leslie Barnett in Cambridge later (1961) confirmed the code to be both triplet and non-overlapping. From Robert Holley in Cornell University, New York, and Hans Zachau in Cologne, came the isolation and determination of the sequence of three **transfer RNAs** (tRNA), 'adapter' molecules that each carry an individual amino acid ready for incorporation into protein and which are also responsible for recognizing the triplet code on the **messenger RNA** (mRNA). The mRNA species contain the sequences of individual genes copied from DNA (Chapter 6). Gobind Khorana and his group in Madison, Wisconsin, chemically synthesized all 64 ribotrinucleoside diphosphates and, using a combination of chemistry and enzymology, synthesized a number of polyribonucleotides with repeating di-, tri-, and tetranucleotide sequences. These were used as synthetic mRNA to help identify each triplet in the code. This work was recognized by the award of the Nobel prize for medicine in 1968 jointly to Holley, Khorana, and Nirenberg.

Nucleic acid research in the 1950s and 1960s was preoccupied by the solution to the coding problem and the establishment of the biological roles of tRNA and mRNA. This was not surprising bearing in mind that at that time the smaller size and attainable homogeneity made isolation and purification of RNA a much easier task than it was for DNA. It was clear that in order to approach the fundamental question of what constituted a **gene**—a single hereditable element of DNA that up to then could be defined genetically but not chemically—it was going to be necessary to break down DNA into smaller, more tractable pieces in a specific and predictable way.

The breakthrough came in 1968 when Meselson and Yuan reported the isolation of a **restriction enzyme** from the bacterium *Escherichia coli*. Here at last was an enzyme, a nuclease, that could recognize a defined sequence in a DNA and cut it specifically (Section 9.3). The bacterium used this activity to break down and hence inactive invading (e.g. phage) DNA. It was soon realized that this was a general property of bacteria, and the isolation of other restriction enzymes with different specificities soon followed. But it was not until 1973 that the importance of these enzymes became apparent. At this time, Chang and Cohen at Stanford and Helling and Boyer at the University of California were able to construct in a test tube a biologically functional DNA that combined genetic information from two different sources. This **chimera** was created by cleaving DNA from one source with a restriction enzyme to give a fragment that could then be joined to a carrier DNA, a **plasmid**. The resultant **recombinant DNA** was shown to be able to replicate and express itself in *E. coli*.

This remarkable demonstration of genetic manipulation was to revolutionize biology. It soon became possible to dissect out an individual gene from its source DNA, to amplify it in a bacterium or other organism (**cloning**, Section 10.7), and to study its expression by the synthesis first of RNA and then of protein (Chapters 5 and 6). This single advance by the groups of Cohen and Boyer truly marked the dawn of modern molecular biology.

1.6 The partnership of chemistry and biology

In the 1940s and 1950s the disciplines of chemistry and biology were so separate that it was a rare occurrence for an individual to embrace both. Two young scientists who were just setting out on their careers at that time were exceptional in recognizing the potential of chemistry in the solution of biological problems and both, in their different ways, were to have a substantial and lasting effect in the field of nucleic acids.

One was Frederick Sanger, a product of the Cambridge Biochemistry School, who in the early 1940s set out to determine the sequence of a protein, insulin. This feat had been thought unattainable, since it was widely supposed that proteins were not discrete species with defined primary sequence. Even more remarkably, he went on to develop methods for sequence determination first of RNA and then of DNA (Section 10.6). These methods involved a subtle blend of enzymology and chemistry that few would have thought possible to combine. The results of his efforts have transformed DNA sequencing in only a few years into such a routine procedure that the determination of the sequence of the complete human genome is now regarded as a serious proposition. The award of two Nobel prizes to Sanger (1958 and 1980) hardly seems recognition enough!

The other scientist has already been mentioned in connection with the elucidation of the genetic code. From not long after his post-doctoral studies under George Kenner and Alexander Todd in Cambridge, Gobind Khorana was convinced that chemical synthesis of polynucleotides could make an important contribution to the study of the fundamental process of information flow from DNA to RNA to protein. Having completed the work on the genetic code in the mid 1960s and aware of Holley's recently determined (1965) sequence for an alanine tRNA, he then established a new goal of total synthesis of the corresponding DNA duplex, the gene specifying the tRNA. Like Sanger, he ingeniously devised a combination of nucleic acid chemistry and enzymology to form a general strategy of gene synthesis, which in principle remains unaltered to this day (Section 3.4). Knowledge became available by the early 1970s about the signals required for gene expression and the newly emerging recombinant DNA methods of Cohen and Boyer allowed a second synthetic gene, this time specifying the precursor of a tyrosine suppressor tRNA (Fig. 1.6), to be cloned and shown to be fully functional.

It is ironic that even up to the early 1970s Khorana's gene syntheses were regarded by many biologists to be unlikely to have practical value. Today, synthetic genes are used routinely in the production of proteins. More than this, oligodeoxyribonucleotides, the short pieces of

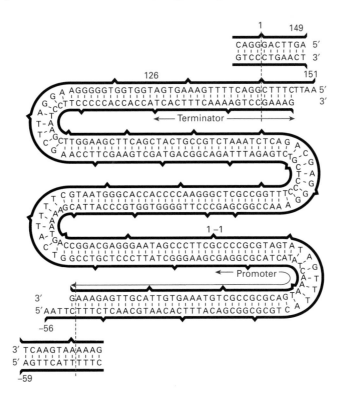

Fig. 1.6 Khorana's totally synthetic DNA corresponding to the tyrosine suppressor transfer RNA gene (from Belagaje, R. *et al.* (1978)). *Chemistry and biology of nucleosides and nucleotides* (ed. R. E. Harmon, R. K. Robins, and L. B. Townsend). Academic Press, New York.

single-stranded DNA for which Khorana developed the first chemical syntheses, have become invaluable general tools in the manipulation of DNA. They are used as primers in DNA sequencing (Section 10.6), as probes in gene detection and isolation (Section 10.10), as mutagenic agents to alter the sequence of DNA (Section 10.8), and more recently as potential therapeutic agents (Section 3.4.6).

The availability of synthetic DNA also provided new impetus in the study of DNA structure. In the early 1970s, new X-ray crystallographic techniques had been developed and applied to solve the structure of the dinucleoside phosphate, ApU, by Rich and co-workers in Cambridge, USA. This was followed by the complete structure of yeast phenylalanine tRNA, determined independently by Rich and by Klug and colleagues in Cambridge, England. For the first time, the complementary base-pairing between two strands could be seen in greater detail than was previously possible from studies of DNA and RNA fibres. ApU formed a double helix by end-to-end packing of molecules, with Watson–Crick pairing clearly in evidence between each strand. The tRNA showed not only Watson–Crick pairs, but also a variety of alternative base-pairs and base-triples, many of which were entirely novel (Chapter 6).

Then in 1978, the structure of synthetic d(pATAT) was solved by Kennard and her group in Cambridge. This tetramer also formed an extended double helix, but excitingly revealed that there was a substantial sequence dependence in its conformation. The angles between neighbouring dA and dT residues were quite different between the A-T sequence and the T-A sequence elements. Soon after, Wang and colleagues discovered that synthetic d(CGCGCG) adopted a totally unpredicted, left-handed Z-conformation. This was soon followed by the demonstration of both a B-DNA helix in a synthetic dodecamer by Dickerson in California and an A-DNA helix in an octamer by Kennard, and finally put paid to the concept that DNA had a rigid, rod-like structure. Clearly, DNA could adopt different conformations dependent on sequence and also on its external environment (Chapter 2). More importantly, an immediate inference could be drawn that conformational differences in DNA (or the potential for their formation) might be recognized by other molecules. Thus, it was not long before synthetic DNA was also being used in the study of DNA binding to carcinogens and drugs (Chapters 7 and 8) and to proteins (Chapter 9).

These spectacular advances were only possible because of the equally dramatic improvements in methods of oligonucleotide synthesis that took place in the late 1970s and early 1980s. The laborious manual work of the early gene synthesis days was replaced by reliable automated DNA synthesis machines, which within hours could assemble sequences well in excess of 100 residues (Section 3.4). Khorana's vision of the importance of synthetic DNA has been fully realized.

1.7 Frontiers in nucleic acids research

The last decade of the twentieth century heralds still further advances in nucleic acids technology. Progress in understanding achieved in previous decades is now being turned into therapeutic and diagnostic products for major human diseases. Furthermore, almost within our grasp is knowledge of the complete DNA sequence of the human genome, information which promises to provide important insights into our genetic make-up yet at the same time presents us with new ethical and moral dilemmas.

A strong revival in the synthesis of nucleoside analogues has led to several therapeutic agents being approved for the treatment of AIDS and HIV infection, perhaps the greatest threat of our generation (Section 4.7.2). More surprisingly, synthetic oligonucleotide analogues have begun to reach the stage of clinical trials for topical and systemic infections as well as for the treatment of some cancers. The development of this 'antisense' technology has been driven by new companies backed by venture capital, but has quickly moved into the domain of the more established pharmaceutical industry. Consequently, many novel oligonucleotide analogues are being prepared and evaluated for their therapeutic utility (Section 3.4.6).

The provision of synthetic RNA has recently become routine (Section 3.5) resulting in major advances in our understanding of catalytic RNA (ribozymes, Section 6.5) and protein–RNA interactions (Section 9.3.7). Ribozyme power is also being harnessed for therapeutic use, the ribozyme being directed to cleave the mRNA product of a harmful gene (Section 6.5.6). Additional new techniques of *in vitro* selection of RNA sequences are extending the potential of RNA to carry out artificial reactions or bind unusual substrates (Section 6.5.7).

Whereas we have learnt how to detect and isolate genes, to sequence them and to clone them, our knowledge about the control of their *in vivo* expression, particularly through the various stages of development, is still very limited. Synthetic antisense oligonucleotides are finding increasing use as sensitive switches to ablate specific gene expression, but even more subtle methods are probably necessary.

Early clinical trials involving **gene therapy**, the replacement in a human being of a harmful gene by a beneficial one, are now under way, but more detailed knowledge of gene control mechanisms will be essential if we are to be able to conquer many of the major genetic diseases or cancers by the gene therapy approach. Part of this knowledge will be gained by determination of the structures of control proteins bound to their target nucleic acids, a subject that is making dramatic advances through use of more powerful X-ray and NMR techniques as well as by use of computer modelling (Chapter 9).

The heady days of the discovery of the double helix and the elucidation of the genetic code are long gone, but in their place have come even more exciting times when many more of us now have the opportunity to answer fundamental questions about genetic structure and function. 'You ain't heard nothin' yet, folks' (Al Jolson, *The Jazz Singer*, July 1927).

Further reading

1.1–1.4

Avery, O. T., MacLeod, C. M., and McCarty, M. (1944). Studies on the chemical nature of the substance inducing transformation of pneumococcal types. *J. Exp. Med.*, **79**, 137–58.

Chargaff, E. (1950). Chemical specificity of nucleic acids and mechanism of their enzymatic degradation. *Experientia*, **6**, 201–9.

Fruton, J. S. (1972). *Molecules and life*, pp. 180–224. Wiley–Interscience, New York.

Judson, H. F. (1979). *The eighth day of creation.* Jonathan Cape, London.

Miescher, F. (1897). *Die Histochemischen und Physiologischen Arbeiten,* Vogel, Leipzig.

Olby, R. (1973). *The path to the double helix.* Macmillan, London.

Portugal, F. H. and Cohen, J. S. (1977). *A century of DNA.* MIT Press, Cambridge, MA.

Watson, J. D. and Crick, F. H. C. (1953). A structure for deoxyribose nucleic acid. *Nature,* **171,** 737–8.

Watson, J. D. (1968). *The double helix.* Athenaeum Press, New York.

1.5

Cohen, S. N. (1975). The manipulation of genes. *Sci. Amer.,* **233**(1), 24–33.

Khorana, H. G. (1965). Polynucleotide synthesis and the genetic code. *Fed. Proc.,* **24,** 1473–87.

Nirenberg, M. W., Matthaei, J. H., Jones, O. W., Martin, R. G., and Barondes, S. H. (1963). Approximation of genetic code *via* cell-free protein synthesis directed by template RNA. *Fed. Proc.,* **22,** 55–61.

Ochoa, S. (1963). Synthetic polynucleotides and the genetic code. *Fed. Proc.,* **22,** 62–74.

1.6

Khorana, H. G. (1979). Total synthesis of a gene. *Science,* **203,** 614–25.

Sanger, F. (1988). Sequences, sequences, and sequences. *Ann. Rev. Biochem.,* **57,** 1–28.

DNA AND RNA
STRUCTURE

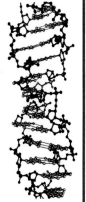

2.1 Structures of components

Nucleic acids are very long, thread-like polymers, made up of a linear array of monomers called **nucleotides**. Different nucleic acids can have from around 80 nucleotides, as in tRNA, to over 10^8 nucleotide pairs in a single eukaryotic chromosome. The unit of size of a nucleic acid is the base-pair (for double-stranded species) or base (for single-stranded species). The abbreviation bp is generally used, as are the larger units Mbp (million base-pairs) and kbp (thousand base-pairs). The chromosome in *E. coli* has 4×10^6 base-pairs, 4 Mbp, which gives it a molecular mass of 3×10^9 Da and a length of 1.5 mm. The size of the (haploid) fruit fly genome is 180 Mbp which, shared between four chromosomes, gives a total length of 56 mm. The genomic DNA of a single human cell has 3900 Mbp and is 990 mm long. How are these extraordinarily long molecules constructed?

2.1.1 Nucleosides and nucleotides

Nucleotides are the phosphate esters of nucleosides and these are components of both ribonucleic acid (RNA) and deoxyribonucleic acid (DNA). RNA is made up of ribonucleotides while the monomers of DNA are 2′-deoxyribonucleotides.

    All nucleotides are constructed from three components: a nitrogen heterocyclic **base**, a pentose **sugar**, and a **phosphate** residue. The major bases are monocyclic **pyrimidines** or bicyclic **purines** (some species of tRNA have tricyclic minor bases such as the Wye (Chapter 3, Fig. 3.17)). The major purines are **adenine (A)** and **guanine (G)** and are found in both DNA and RNA. The major pyrimidines are **cytosine (C)**, **thymine (T)**, and **uracil (U)** (Fig. 2.1).

    In **nucleosides**, the purine or pyrimidine base is joined from a ring nitrogen to carbon-1 of a pentose sugar. In ribonucleic acid, the pentose is **D-ribose** which is locked into a five-mem-

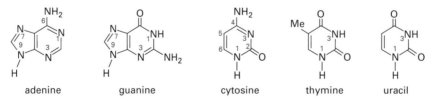

Fig. 2.1 Structures of the five major purine and pyrimidine bases of nucleic acids in their dominant tautomeric forms and with the IUPAC numbering systems for purines and pyrimidines.

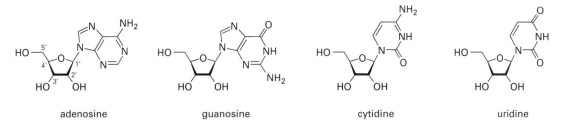

Fig. 2.2 Structures of the four ribonucleosides. The bases retain the same numbering system and the pentose carbons are numbered 1′ through 5′. By convention, the furanose ring is drawn with its ring oxygen at the back and C-2′ and C-3′ at the front. Hydrogen atoms are usually omitted for clarity.

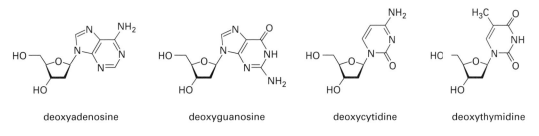

Fig. 2.3 Structures of the four major deoxynucleosides. By convention, only hydrogens bonded to oxygen or nitrogen are depicted.

bered **furanose** ring by the bond from C-1 of the sugar to N-1 of C or U or to N-9 of A or G. This bond is on the same side of the sugar ring as the C-5 hydroxymethyl group and is defined as a β-glycosylic linkage (Fig. 2.2).

In DNA, the pentose is 2-deoxy-D-ribose and the four nucleosides are **deoxyadenosine, deoxyguanosine, deoxycytidine**, and **deoxythymidine** (Fig. 2.3). In DNA, the methylated pyrimidine base thymine takes the place of uracil in RNA, and its nucleoside with deoxyribose is still commonly called thymidine. However, since the discovery of **ribothymidine** as a regular component of tRNA species (Section 6.4.1), it has been preferable to use the name deoxythymidine rather than thymidine. Unless indicated otherwise, it is assumed that nucleosides, nucleotides, and oligonucleotides are derived from D-pentofuranose sugars.

The **phosphate** esters of nucleosides are **nucleotides**, and the simplest of them have one of the hydroxyl groups of the pentose esterified by a single phosphate monoester function. Adenosine 5′-phosphate is a **5′-ribonucleotide** also called adenylic acid and abbreviated to AMP (Fig. 2.4). Similarly, deoxycytidine 3′-phosphate is a **3′-deoxynucleotide**, identified as 3′-dCMP. Nucleotides which have two phosphate monoesters on the same sugar are called **nucleoside bisphosphates** while nucleoside monoesters of pyrophosphoric acid are **nucleoside diphosphates**. By extension, nucleoside esters of tripolyphosphoric acid are **nucleoside triphosphates**, of which the classic example is adenosine 5′-triphosphate (ATP) (Section 3.3.2). Finally, cyclic nucleotides are nucleosides which have two neighbouring hydroxyl groups on the same pentose esterified by a single phosphate as a diester. The most important of these is adenosine 3′,5′-cyclic phosphate (cAMP).

In the most abbreviated nomenclature currently employed, **pN** stands for a 5′-nucleotide, **Np** for a 3′-nucleotide, and **dNp** for a 3′-deoxynucleotide (to be precise, a 2′-deoxynucleoside 3′-phosphate). This shorthand notation is based on the convention that an oligonucleotide chain is drawn horizontally with its 5′-hydroxyl group at the left and its 3′-hydroxyl group at the right-hand end. Thus, pppGpp is the shorthand representation for the 'magic spot' nucleotide, guanosine 3′-diphosphate 5′-triphosphate, while ApG is short for adenylyl-(3′ → 5′)-guanosine, whose 3′ → 5′ internucleotide linkage runs from the nucleoside on the left to that on the right of the phosphate.

2.1.2 Physical properties of nucleosides and nucleotides

Because of their polyionic character, nucleic acids are soluble in water up to about 1 per cent w/v according to size and are precipitated by the addition of alcohol. Their solutions are quite viscous and the long nucleic acid molecules are easily sheared by stirring or by passage through a fine nozzle, such as a hypodermic needle or a fine pipette.

Ionization

The acid–base behaviour of a nucleotide is its most important physical characteristic. It determines its charge, its tautomeric structure, and thus its ability to donate and accept hydrogen bonds, which is the key feature of base : base recognition. The pK_a values for the five bases in the major nucleosides and nucleotides are listed in Table 2.1.

It is clear that all of the bases are uncharged in the physiological range $5 < pH < 9$. The same is true for the pentoses, where the ribose 2′,3′-diol only loses a proton above pH 12 while isolated hydroxyl groups ionize only above pH 15. The nucleotide phosphates lose one proton at pH 1 and a second proton (in the case of monoesters) at pH 7. This pattern of proton equilibria is shown for AMP across the whole pH range (Fig. 2.5).

The three amino bases, A, C, and G, each become protonated on one of the ring nitrogens rather than on the exocyclic amino group since this does not interfere with delocalization of

Table 2.1. pK_a values for bases in nucleosides and nucleotides

Base (site of protonation)		Nucleoside	3′-Nucleotide	5′-Nucleotide
Adenine	(N–1)	3.52	3.70	3.88
Cytosine	(N–3)	4.17	4.43	4.56
Guanine	(N–7)	3.3	(3.5)	(3.6)
Guanine	(N–1)	9.42	9.84	10.00
Thymine	(N–3)	9.93	—	10.47
Uracil	(N–3)	9.38	9.96	10.06

These data relate to 20°C and zero salt concentration. They correspond to *loss* of a proton for $pK_a > 9$ and *capture* of a proton for $pK_a > 5$.

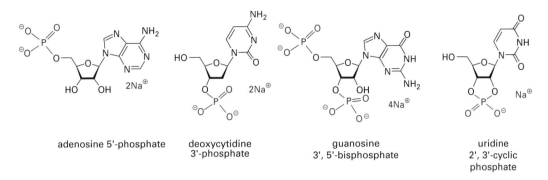

adenosine 5'-phosphate deoxycytidine 3'-phosphate guanosine 3', 5'-bisphosphate uridine 2', 3'-cyclic phosphate

Fig. 2.4 Structures of some common nucleotides. All are presented as their sodium salts in the state of ionization observed at neutral pH.

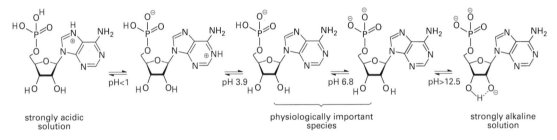

strongly acidic solution physiologically important species strongly alkaline solution

Fig. 2.5 States of protonation of adenosine 5'-phosphate (AMP) from strongly acidic solution (left) to strongly alkaline solution (right).

the NH_2 electron lone pair into the aromatic system. The $C–NH_2$ bonds of A, C, and G are about 1.34 Å long which means that they have 40–50 per cent double bond order, while the C=O bonds of C, G, T, and U have some 85–90 per cent double bond order. It is also noteworthy that the proximity of negative charge on the phosphate residues has a secondary effect, making the ring nitrogens more basic ($\Delta pK_a \approx +0.4$) and the amine protons less acidic ($\Delta pK_a \approx +0.6$).

Tautomerism

A tautomeric equilibrium involves alternative structures which differ only in the location of hydrogen atoms. The choices available to nucleic acid bases are illustrated by the **keto–enol**

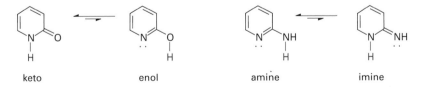

keto enol amine imine

Fig. 2.6 Keto–enol tautomers for 2-pyridone : 2-hydroxypyridine (left) and amine–imine tautomerism for 2-aminopyridine (right).

equilibrium between 2-pyridone and 2-hydroxypyridine and the **amine–imine** equilibrium for 2-aminopyridine (Fig. 2.6). Ultraviolet, NMR, and IR spectroscopies have established that the five major bases exist overwhelmingly (> 99.99 per cent) in the **amino-** and **keto**-tautomeric forms at physiological pH (see Fig. 2.1) and not in the benzene-like **enol** tautomers, in common use before 1950 (Chapter 1, Fig. 1.3).

Hydrogen bonding

The mutual recognition of A by T and of C by G uses hydrogen bonds to establish the fidelity of DNA transcription and translation. The NH groups of the bases are good hydrogen bond donors (**d**), while the sp^2-hybridized electron pairs on the oxygens of the base C=O groups and on the ring nitrogens are much better hydrogen bond acceptors (**a**) than are the oxygens of either the phosphate or the pentose. The **a·d** hydrogen bonds so formed are largely electrostatic in character, with a charge of about $+0.2e$ on the hydrogens and about $-0.2e$ on the oxygens and nitrogens, and they seem to have an average strength of 6–10 kJ mol^{-1}.

The predominant amino–keto tautomer for cytosine has a pattern of hydrogen bond acceptor and donor sites which for $O^2 \cdot N^3 \cdot N^4$ can be expressed as **a·a·d** (Fig. 2.7). Its minor tautomer has a very different pattern: **a·d·a**. In the same way we can establish that the corresponding pattern for the dominant tautomer of dT is **a·d·a** while the pattern for $N^2 \cdot N^1 \cdot O^6$ of dG is **d·d·a** (Fig. 2.7) and that for dA is **(−)·a·d**.

When Jim Watson was engaged in DNA model-building studies in 1952 (Section 1.4), he recognized that the hydrogen bonding capability of an A·T base-pair uses complementarity of **(−)·a·d** to **a·d·a**, while a C·G pair uses the complementarity of **a·a·d** to **d·d·a**. This base-pairing pattern rapidly became known as **Watson–Crick pairing** (Fig. 2.8). There are two hydrogen bonds in an A·T pair and three in a C·G pair (cf. Chapter 1, Fig. 1.5 for the original pairing scheme). The geometry of these pairs has been fully analysed in many structures from dinucleoside phosphates through oligonucleotides to tRNA species, both by the use of X-ray crystallography and, more recently, by NMR spectroscopy.

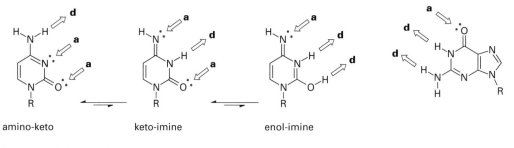

amino-keto   keto-imine   enol-imine

R = deoxyribofuranosyl

Fig. 2.7 Tautomeric equilibria for deoxycytidine showing hydrogen bond acceptor **a** and donor **d** sites as used in nucleic acid base-pairing. The major tautomer for deoxyguanosine is drawn to show its characteristic **d·d·a** hydrogen bond donor–acceptor capacity.

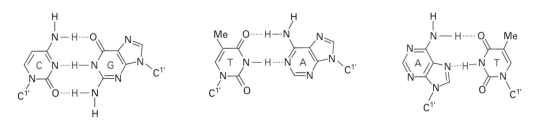

Fig. 2.8 Watson–Crick base-pairing for C·G (left) and T·A (centre). Hoogsteen base-pairing for A·T (right).

In planar base-pairs, the hydrogen bonds join nitrogen and oxygen atoms that are 2.8 Å to 2.95 Å apart. This geometry gives a C-1'– C-1' distance of 10.60 ± 0.15 Å with an angle of $68 \pm 2°$ between the two glycosylic bonds for both the A·T and the C·G base-pairs. As a result of this **isomorphous geometry**, the four base-pair combinations, A·T, T·A, C·G, and G·C, can all be built into the same regular framework of the DNA duplex.

While Watson–Crick base-pairing is the dominant pattern, other pairings have been suggested of which the most significant to have been identified so far are **Hoogsteen pairs** and Crick '**wobble' pairs**. Hoogsteen pairs, illustrated for A·T, are not isomorphous with Watson–Crick pairs because they have an 80° angle between the glycosylic bonds and an 8.6 Å separation of the anomeric carbons (Fig. 2.8). In the case of reversed Hoogsteen pairs and reversed Watson–Crick pairs (not shown), one base is rotated through 180° relative to the other.

Francis Crick proposed the existence of 'wobble' base-pairings to explain the degeneracy of the genetic code (Section 6.6.8). This phenomenon calls for a single base in the 5'-anticodon position of tRNA to be able to recognize either of the pyrimidines or, alternatively, either of the purines as its 3'-codon base partner. Thus a G·U 'wobble' pair has two hydrogen bonds, $G–N^1–H \cdots O^2–U$ and $G–O^6 \cdots H–N^3–U$ and this requires a sideways shift of one base relative to its positions in the regular Watson–Crick geometry (Fig. 2.9). The resulting loss of a hydrogen bond leads to reduced stability which may be offset in part by the improved base stacking (Section 2.3.1) that results from such sideways base displacement.

Base-pairing of these and other non-Watson–Crick patterns is significant in three structural situations. First, the compact structures of tRNAs maximize both base-pairing and base-stacking wherever possible. This has led to the identification of a considerable variety of reversed Hoogsteen and 'wobble' base-pairs as well as of tertiary base-pairs (or base-triplets) (Section 6.3.2). Secondly, where there are triple-stranded helices for DNA and RNA, such as

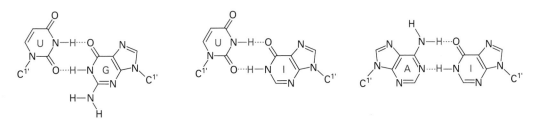

Fig. 2.9 'Wobble' pairings for G·U (left), I·U (centre), and I·A (right).

(poly(dA)·2poly(dT)) and (poly(rG)·2poly(rC)), the second pyrimidine chain binds to the purine in the major groove by Hoogsteen hydrogen bonds and runs parallel to the purine chain (Section 2.4.5). Thirdly, mismatched base-pairs are necessarily identified with anomalous hydrogen bonding and many such patterns have been revealed by X-ray studies on synthetic oligodeoxyribonucleotides (Section 2.3.2).

2.1.3 Spectroscopic properties of nucleosides and nucleotides

Neither the pentose nor the phosphate components of nucleotides show any significant UV absorption above 230 nm. This means that both nucleosides and nucleotides have UV absorption profiles rather similar to those of their constituent bases and absorb strongly with λ_{max} close to 260 nm and molar extinction coefficients of around 10^4 (Table 2.2).

The light absorptions of isolated nucleoside bases given above are measured in solution in high dilution. They undergo marked changes when they are in close proximity to neighbouring bases, as usually shown in ordered secondary structures of oligo- and polynucleotides. In such ordered structures, the bases can stack face-to-face and thus share $\pi–\pi$ electron interactions which profoundly affect the transition dipoles of the bases. Typically, such changes are manifest in a marked reduction in the intensity of UV absorption (by up to 30 per cent) which is known as **hypochromicity** (Section 10.2.1). This phenomenon is reversed on unstacking of the bases.

There are two important applications of this phenomenon. First it is used in the determination of temperature-dependent and pH-dependent changes in base stacking. Secondly, it permits the monitoring of changes in the asymmetric environment of the bases by circular dichroism (CD), or by optical rotatory dispersion (ORD) effects. Both of these techniques are especially valuable for studying helix–coil transitions (Section 2.5.1).

Infrared analysis of nucleic acid components has been less widely used, but the availability of laser Raman and Fourier transform IR methods is making a growing contribution.

Nuclear magnetic resonance has had a dramatic effect on studies of oligonucleotides, largely as a result of a variety of complex spin techniques such as NOESY and COSY for

Table 2.2. Some light absorption characteristics for nucleotides

Compound	$[\alpha_D]$*	pH 1-2		pH > 11	
		λ_{max}/nm	$10^{-4} \times \varepsilon$	λ_{max}/nm	$10^{-4} \times \varepsilon$
Ado 5'-P	−26°	257	1.5	259	1.54
Guo 3'-P	−57°	257	1.22	257	1.13
Cyd 3'-P	+27°	279	1.3	272	0.89
Urd 2'-P	+22°	262	0.99	261	0.73
Thd 5'-P	+7.3°	267	1.0	267[a]	1.0
3',5'-cAMP	−51.3°	256	1.45	260[b]	1.5

[a] pH 7.0; [b] pH 6.0; * specific molar rotation.

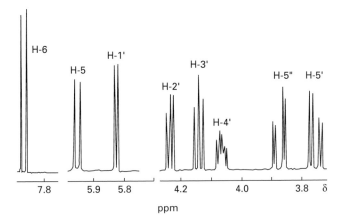

Fig. 2.10 Proton NMR spectrum for cytidine (run in D_2O at 400 MHz).

proton spectra, the use of ^{17}O, ^{18}O, and sulfur substituent effects of ^{31}P NMR, and the analysis of nuclear Overhauser effects (NOE). These provide a useful measure of internuclear distances and with computational analysis can provide solution conformations of oligo-nucleotides (Section 2.2). Nucleosides, nucleotides, and their analogues have relatively simple ^{1}H NMR spectra. The aromatic protons of the pyrimidines and purines resonate at low field ($\delta 7.6$ to $\delta 8.3$ with C^5–H close to $\delta 5.9$). The anomeric hydrogen is a doublet for ribonucleosides and a double-doublet for 2′-deoxynucleosides at $\delta 5.8$–6.4. The pentoses provide a multi-spin system which generally moves from low to high field in the series: H-2′, H-3′, H-4′, H-5′, and H-5″ in the region $\delta 4.3$ to 3.7. Lastly, 2′-deoxynucleosides have H-2′ and H-2″ as an ABMX system near $\delta 2.5$. The 400 MHz spectrum of a simple nucleoside, cytidine (Fig. 2.10), shows how essential two-dimensional spin techniques are in making possible the complete analysis of the spectrum of a large oligomer, which may be equivalent to a dozen such monomer spectra superimposed!

2.1.4 Shapes of nucleotides

Nucleotides have rather compact shapes with several interactions between non-bonded atoms. Their molecular geometry is so closely related to that of the corresponding nucleotide units in oligomers and nucleic acid helices that it was once argued that helix structure is a consequence of the conformational preferences of individual nucleotides. However, the current view is that the sugar-phosphate backbone appears to act as no more than a constraint on the range of conformational space accessible to the base-pairs and that **π–π interactions** between the base-pairs provide the driving force for the different conformations of DNA (Section 2.3.1).

The details of conformational structure are accurately defined by the torsion angles α, β, γ, δ, ε, and ζ in the phosphate backbone, θ_0 to θ_4 in the furanose ring, and χ for the glycosylic bond (Fig. 2.11). Because many of these torsional angles are interdependent, we can more simply describe the shapes of nucleotides in terms of four parameters: the sugar pucker, the *syn–*

anti conformation of the glycosylic bond, the orientation of $C^{4'}$–$C^{5'}$, and the shape of the phosphate ester bonds.

Sugar pucker

The furanose rings are twisted out of plane in order to minimize non-bonded interactions between their substituents. This 'puckering' is described by identifying the major displacement of carbons-2' and -3' from the median plane of $C^{1'}$–$O^{4'}$–$C^{4'}$. Thus, if the *endo* displacement of C-2' is greater than the *exo* displacement of C-3' the conformation is called $C^{2'}$-*endo* and so on (Fig. 2.11). (The *endo* face of the furanose is on the same side as $C^{5'}$ and the base; the *exo* face is on the opposite face to the base.) These sugar puckers are located in the north (**N**) and south (**S**) domains of the pseudorotation cycle of the furanose ring and so spectroscopists frequently use **N** and **S** designations, which also fortuitously reflect the relative shapes of the C–C–C–C bonds in the $C^{2'}$-*endo* and -*exo* forms respectively.

In solution, the **N** and **S** conformations are in rapid equilibrium and are separated by an energy barrier of less than 20 kJ mol^{-1}. The average position of the equilibrium can be estimated from the magnitudes of the ^3J NMR coupling constants linking H$^{1'}$–H$^{2'}$ and H$^{3'}$–H$^{4'}$. This is influenced by (1) the preference of electronegative substituents at $C^{2'}$ and $C^{3'}$ for axial orientation, (2) the orientation of the base (*syn* goes with $C^{2'}$-*endo*), and (3) the formation of an intra-strand hydrogen bond from $O^{2'}$ in one RNA residue to $O^{4'}$ in the next, which favours $C^{3'}$- *endo* pucker.

Syn–anti conformation

The plane of the bases is almost perpendicular to that of the sugars and approximately bisects the $O^{4'}$–$C^{1'}$–$C^{2'}$ angle. This allows the bases to occupy either of two principal orientations. The *anti* conformer has the smaller H-6 (pyrimidine) or H-8 (purine) atom above the sugar

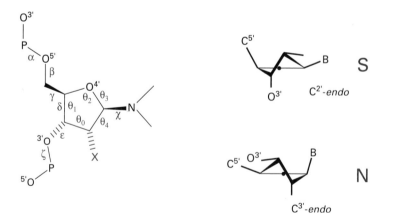

Fig. 2.11 Torsion angle notation (IUPAC) for polynucleotide chains and structures for the $C^{2'}$-*endo* (**S**) and $C^{3'}$-*endo* (**N**) preferred sugar puckers.

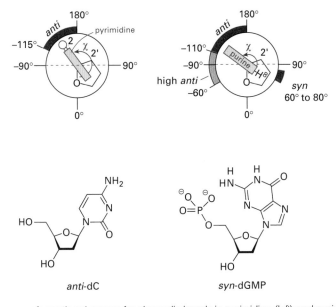

Fig. 2.12 *Anti* and *syn* conformational ranges for glycosylic bonds in pyrimidine (left) and purine (right) nucleosides and drawings of the *anti* conformation for deoxycytidine (lower left) and the *syn* conformation for deoxyguanosine 5′-phosphate (lower right).

ring, while the *syn* conformer has the larger O-2 (pyrimidine) or N-3 (purine) in that position. Pyrimidines occupy a narrow range of *anti* conformations (Fig. 2.12) while purines are found in a wider range of *anti* conformations which can even extend into the high-*anti* range for 8-azapurine nucleosides such as formycin.

One inevitable consequence of this *anti* conformation for the glycosylic bonds is that the backbone chains for A-form and B-form DNA run downward on the *right* of the minor groove and run upward on the *left* of the minor groove, depicted as (↑↓) (see Fig. 9.2).

There is one important exception to the general preference for *anti* forms. Nuclear magnetic resonance, CD, and X-ray analyses all show that guanine prefers the *syn* glycoside in mono-nucleotides, in alternating oligomers like d(CpGpCpG), and in Z-DNA. Theoretical calculations suggest that this effect comes from a favourable electrostatic attraction between the phosphate anion and the C^2-amino group in guanine nucleotides. It results from polarization of one of the nitrogen non-bonding electrons towards the ring. Most unusually, this *syn* conformation can only be built into left-handed helices!

$C^{4'}$–$C^{5'}$ orientation

The conformation of the exocyclic $C^{4'}$–$C^{5'}$ bond determines the position of the 5′-phosphate relative to the sugar ring. The three favoured conformers for this bond (Fig. 2.11) are the classical **synclinal (sc)** and **antiperiplanar (ap)** rotamers. For pyrimidine nucleosides, +**sc** is preferred while for purine nucleosides +**sc** and **ap** are equally populated. However,

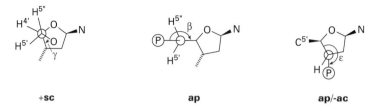

Fig. 2.13 Preferred nucleotide conformations: **+sc** for $C^{4'}$–$C^{5'}$ (left); **ap** for $C^{5'}$–$O^{5'}$ (centre); and **ap/-ac** for $C^{3'}$–$O^{3'}$ (right).

in the nucleotides, the 5'-phosphate reduces the conformational freedom and the dominant conformer for this γ-bond is **+sc** (Figure 2.13). Once again, the demands of Z-DNA have a major effect and the **ap** conformer is found for the *syn* guanine deoxynucleotides.

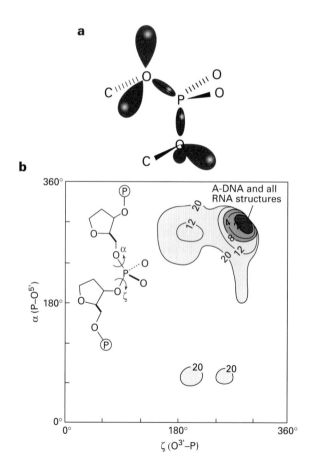

Fig. 2.14 (Upper) Gauche conformation for phosphate diesters showing the antiperiplanar alignment of an occupied non-bonding oxygen orbital with the adjacent P–O bond. (Lower) Contour map for P–O bond rotations calculated for diribose triphosphate (energies in kJ mol^{-1}) (adapted from Govil, G. (1976). *Biopolymers*, **15**, 2303–7. Copyright (1976) John Wiley and Sons, Inc.).

C–O and P–O ester bonds

Phosphate diesters are tetrahedral at phosphorus and show antiperiplanar conformations for the C^5–$O^{5'}$ bond. Similarly, the $C^{3'}$–$O^{3'}$ bond lies in the **antiperiplanar** to **anticlinal** sector. This conformational uniformity has led to the use of the **virtual bond concept** in which the chains $P^{5'}$–$O^{5'}$–$C^{5'}$–$C^{4'}$ and $P^{3'}$–$O^{3'}$–$C^{3'}$–$C^{4'}$ can be analysed as rigid, planar units linked at phosphorus and at $C^{4'}$. Such a simplification has been used to speed up initial calculations of some complex polymeric structures.

Our knowledge of P–O bond conformations comes largely from X-ray structures of tRNA and DNA oligomers. In general, $H^{4'}$–$C^{4'}$–$C^{5'}$–$O^{5'}$–P adopts an extended W-conformation in these structures. A skewed conformation for the C–O–P–O–C system has been observed in structures of simple phosphate diesters such as dimethyl phosphate and also for polynucleotides. This has been described as a **gauche** effect and attributed to the favourable interactions of a non-bonding electron pair on $O^{5'}$ with the P–$O^{3'}$ bond, and vice versa for the P–$O^{5'}$ bond (Fig. 2.14). (This may arise from interaction of the electron lone pair with either phosphorus d orbitals or, more likely, with the P–O antibonding σ orbital. The interaction has been calculated at 30 kJ mol^{-1} more favourable than the extended W-conformation for the C–O–P–O–C system). Other non-bonded interactions dictate that α and ζ both have values close to $+300°$ in helical structures though values of $+60°$ are seen in some dinucleoside phosphate structures.

Other P–O conformations have been observed in non-helical nucleotides while left-handed helices also require changed P–O conformations. These changes take place largely in the rotamers for α. In Z-DNA these are +**sc** for guanines but broadly **antiperiplanar** for cytosines while ζ is +**sc** for cytosines but broadly **synperiplanar** for guanines (Section 2.2.2).

Summary

The building blocks of nucleic acids are nucleotides, which are the phosphate esters of nucleosides. These are formed by condensation of a base and a pentose. In RNA, the pentose is D-ribose and is linked in its furanose form from C-1' to N-9 of a purine, adenine, or guanine, or N-1 of a pyrimidine, cytosine, or uracil. In DNA, 2-deoxy-D-ribose is joined in the same way to the four bases, among which thymine takes the place of uracil. The phosphate esters are strong acids and exist as anions at neutral pH. The 'bases' are, in reality, only very weakly basic and A, C, and G become protonated only below pH 4. The amide NHs in G, T, and U are deprotonated at pHs above 9.

Hydrogen bonds can be formed between the major *amino–keto* tautomers of the bases to link A with T and C with G in Watson–Crick base-pairing. Such hydrogen bonds are largely electrostatic in character. 'Wobble' and Hoogsteen base-pairs offer minor variations to Watson–Crick pairing, and are seen in tRNA structures.

Nucleotides have defined shapes with a general preference for the *anti*-conformers of the glycosylic bond χ, for the $C^{4'}$–$C^{5'}$ bonds γ, and for the two C–O(P) bonds β and ε. The furanose ring is puckered to relieve strain and can adopt either the $C^{2'}$-*endo* or the $C^{3'}$-*endo* conformation, which are in rapid equilibrium at room temperature. Energy barriers between these forms and for the rotamers about the $-O-P-O-$ bonds are sufficiently small to accommodate different helical forms of DNA as dictated by the demands of $\pi-\pi$ base stacking interactions.

2.2 Standard DNA structures

Structural studies on DNA began with the nature of the primary structure of DNA. The classical analysis, completed in mid-century, is easily taken for granted today when we have machines for DNA oligomer synthesis that presuppose the integrity of the 3′-to-5′ phosphodiester linkage. Nonetheless, the classical analysis was the essential key that opened the door to later studies on the regular secondary structure of double-stranded DNA and thereby primed the modern revolution known as molecular biology. **Standard structures** for DNA have generally been determined on heterogeneous duplex material and are thus independent of sequence and apply only to Watson–Crick base-pairing.

2.2.1 Primary structure of DNA

Klein and Thannhauser's work (Section 1.4) established that the primary structure of DNA has each nucleoside joined by a phosphodiester from its 5′-hydroxyl group to the 3′-hydroxyl group of one neighbour and by a second phosphodiester from its 3′-hydroxyl group to the 5′-hydroxyl of its other neighbour. There are no 5′–5′ or 3′–3′ linkages in the regular DNA primary structure (Fig. 2.15). This means that the uniqueness of a given DNA primary structure resides solely in the sequence of its bases.

2.2.2 Secondary structure of DNA

In the first phase of investigation of DNA secondary structure, diffraction studies on heterogeneous DNA fibres identified two distinct conformations for the DNA double helix. At low humidity (and high salt) the favoured form is the highly crystalline A-DNA while at high humidity (and low salt) the dominant structure is B-DNA. We now recognize that there is a wide variety of right-handed double-helical DNA conformations, and this **structural polymorphism** is denoted by the use of the letters A to T as illustrated by A, A′, B, α–B′, β–B′, C, C′, D, E, and T forms of DNA. In broad terms, all of these can be classified in two generically different DNA families: A and B. These are associated with the sugar pucker $C^{3'}$-endo for the A-family and $C^{2'}$-endo (or the equivalent $C^{3'}$-exo) for the B-family. However, as we shall see later (Section 2.3.1), it is the energetics of base-stacking which determines the conformation of the helix and sugar pucker is largely consequential. As we shall see, in B-form DNA the base-pairs sit directly on the helix axis and are nearly perpendicular to it. In A-form DNA the base-pairs are displaced off-axis towards the minor groove and are tilted.

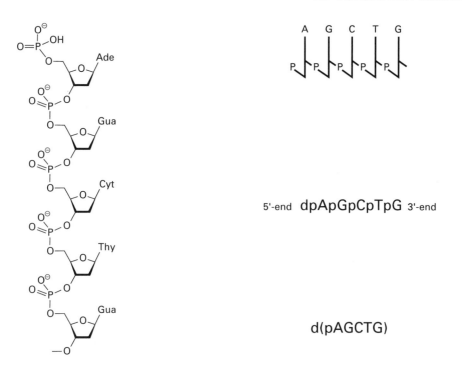

5'-end dpApGpCpTpG 3'-end

d(pAGCTG)

Fig. 2.15 The primary structure of DNA (left) and three of the common shorthand notations: 'Fischer' (upper right), linear alphabetic (centre right), and condensed alphabetic (lower right).

Andrew Wang's unexpected discovery in 1979 that the hexamer d(CGCGCG) adopts a left-handed helical structure, now named Z-DNA, was one of the first dramatic results to stem from the synthesis of oligonucleotides in sufficient quantity for crystallization and X-ray diffraction analysis. Since then, over 100 different oligodeoxynucleotide structures have been solved and these have provided the details on which standard DNA structures are now based. The main features of A-, B-, and Z-DNA are shown in Figs 2.16–2.19 and structural parameters are provided for a range of standard helices in Tables 2.3 and 2.4.

As more highly resolved structures have become available, the idea that these three families of DNA conformations are restricted to standard structures has been whittled away. We now accept that there are local, sequence-dependent modulations of structure which are primarily associated with changes in the orientation of bases. Such changes seek to minimize non- bonded interactions between adjacent bases and maximize base-stacking. They are generally tolerated by the relatively flexible sugar-phosphate backbone. Other studies have explored perturbations in regular helices which result from deliberate mismatching of base-pairs and of lesions caused by chemical modification of bases, such as base methylation and thymine photodimers (Section 7.8.1). In all of these areas, the results derived from X-ray crystallography have been carried into the solution phase by high-resolution NMR analysis, and rationalized by molecular modelling.

Table 2.3 Average helix parameters for the major DNA conformations

Structure type	Helix sense	Residues per turn	Twist per bp $\Omega°$	Displacement bp D_a	Rise per bp/Å	Base tilt $\tau°$	Sugar pucker	Groove width/Å Minor	Groove width/Å Major	Groove depth/Å Minor	Groove depth/Å Major
A–DNA	R	11	32.7	4.5	2.9	20	C3'–endo	11.0	2.7	2.8	13.5
dGGCCGGCC	R	11	32.6	3.6	3.03	12	C3'–endo	9.6	7.9	—	—
B–DNA	R	10	36	−0.2 to −1.8	3.3–3.4	−6	C2'–endo	5.7	11.7	7.5	8.8
dCGCGAATTCGCG	R	9.7	37.1	—	3.34	−1.2	C2'–endo	3.8	11.7	—	—
C–DNA	R	9.33	38.5	−1.0	3.31	−8	C3'–exo	4.8	10.5	7.9	7.5
D–DNA	R	8	45	−1.8	3.03	−16	C3'–exo	1.3	8.9	6.7	5.8
T–DNA	R	8	45	−1.43	3.4	−6	C2'–endo	narrow	wide	deep	shallow
Z–DNA	L	12	−9, −51	−2 to −3	3.7	−7	C3'–endo(syn)	2.0	8.8	13.8	3.7
A–RNA	R	11	32.7	4.4	2.8	16–19	C3'–endo				
A'–RNA	R	11	30	4.4	3.0	10	C3'–endo				

Table 2.4. Comparison of helix parameters for A-DNA and B-DNA crystal structures and for a model Z-DNA helix

1. Base step parameters

Helix	Step	Roll	Tilt	Cup	Slide	Twist	Rise	D_{xy}	Rad_p
B	All	0.6°	0.0°	10.0°	0.4 Å	36.1°	3.36 Å	3.5 Å	9.4 Å
A	All	6.3°	—	—	−1.6 Å	31.1°	2.9 Å	—	9.5 Å
Z	C–G	−5.8°	0.0°	12.5°	5.4 Å	−9.4°	3.92 Å	5.0 Å	6.3 Å
Z	G–C	5.8°	0.0°	−12.5°	−1.1 Å	−50.6°	3.51 Å	6.0 Å	7.3 Å

2. Base-pair parameters

Base	Tip	Inclination	Propeller	Buckle	Shift	Slide	P–P[a]
B	All	0.0°	2.4°	−11.1°	−0.2°	0.8 Å	0.1 Å	8.8–14 Å
A	All	11.0°	12.0°	−8.3°	−2.4°	−4.1 Å	—	11.5–11.9 Å
Z	C	2.9°	−6.2°	−1.3°	−6.2°	3.0 Å	−2.3 Å	13.7 Å
Z	G	−2.9°	−6.2°	−1.3°	6.2°	3.0 Å	2.3 Å	7.7 Å

[a] P–P is the shortest interstrand distance across the minor groove.

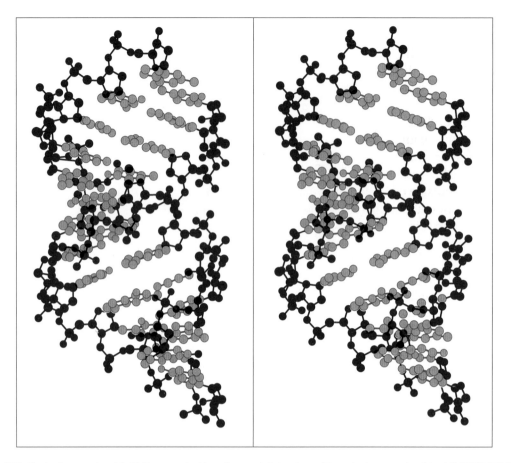

Fig. 2.16 Twelve base-pairs of A-DNA, generated from the crystal structure of the octadeoxynucleotide, d(GGTATACC). The shallow minor groove is frontal at the top and the very deep major groove is frontal at the bottom of the helix. A significant feature of polymorphism in A-DNA structures is the width of opening of the major groove. (Three-dimensional stereo-pair diagram for parallel viewing.)

Finally, our knowledge of higher-order structures, which began with Vinograd's work on DNA supercoiling in 1965, has been extended to studies on DNA cruciform structures, to 'bent' DNA, and to other unusual features of DNA structures.

Regular DNA structures are described by a range of characteristic features. The global parameters of **average rise** D_z **and helix rotation** Ω per base-pair define the pitch of the helix. Sideways tilting of the base-pairs through a **tilt angle** τ permits the separation of the bases along the helix axis D_z to be smaller than the van der Waals distance, 3.4 Å, and so gives a shorter, fatter cylindrical envelope for DNA. The angle τ is positive for A-DNA (positive means a clockwise rotation of the base-pair when viewed end-on and towards the helix axis) but is smaller and negative for B-DNA helices. At the same time, the base-pairs are displaced laterally from the helix axis by a distance D_a. This parameter together with the groove width defines the depth of the major and minor grooves (Table 2.3).

2.2.3 A-DNA

Among the first synthetic oligonucleotides to be crystallized in the late 1970s were d(GGTA-TACC), an iodinated-d(CCGG), and d(GGCCGGCC). They all proved to have A-type DNA structures, similar to the classical A-DNA deduced from fibre analysis at low resolution. Several other oligomers, mostly octamers, also form crystals of the A-structure, but NMR studies suggest that some of these may have the B-form in solution. It is conceivable that crystal packing might especially favour A-DNA for octanucleotides.

The general anatomy of A-DNA follows the Watson–Crick model with antiparallel, right-handed double helices. The sugar rings are parallel to the helix axis and the phosphate backbone is on the outside of a cylinder of about 24 Å diameter (Fig. 2.16).

X-ray diffraction at atomic resolution shows that the bases are displaced 4.5 Å away from the helix axis and this creates a hollow core down the axis which is 3 Å in diameter. There are 11 bases in each turn of 28 Å, which gives a vertical rise of 2.56 Å per base-pair. In order to maintain the normal van der Waals separation of 3.4 Å, the stacked bases are tilted sideways through 20°. The sugar backbone has skewed phosphate ester bonds, and antiperiplanar conformations for the adjacent C−O ester bonds. Finally, the furanose ring has a C3'-*endo* pucker and the glycosylic bond is in the *anti* conformation (Table 2.3). As a result of these features, the major groove of A-DNA is cavernously deep and the minor groove is extremely shallow, as can be appreciated from the three-dimensional picture of the helix (Fig. 2.16). This is further characterized by a 5.4 Å P–P separation between adjacent intra-strand phosphorus atoms.

2.2.4 The B-DNA family

The general features of the B-type structure, obtained from DNA fibres at high relative humidity (95 per cent RH) were first put into sharper focus by X-ray studies on the dodecamer d(CGCGAATTCGCG) and its bromo-derivative at cytosine-9. The B-conformation has now been observed in crystals of several oligomers and standard parameters have been averaged from structures of ten isomorphous deoxyoligonucleotides (Fig. 2.17).

In B-form DNA, the base-pairs sit directly on the helix axis so that the major and minor grooves are of similar depth (Table 2.3). Its bases are stacked predominantly above their neighbours in the same strand and are perpendicular to the helix axis (Table 2.4). The sugars have the $C^{2'}$-*endo* pucker (with some $O^{4'}$-*endo* pucker for the dodecamer), all the glycosides have the *anti* conformation, and most of the other rotamers have normal populations (Table 2.5). Adjacent phosphates in the same chain are a little further apart, P–P $=6.7$ Å, than in A-DNA (Table 2.4).

The interaction of water molecules around a DNA double helix can be very important in stabilizing helix structure, to the extent that water molecules have sometimes been described as the 'fourth component' of DNA structure, after bases, sugars, and phosphates. Just how many water molecules per base-pair can be seen in an X-ray structure depends on quality of structure resolution—and also the optimism of the analyst! In the best structures, up to 14 waters per base-pair have been resolved. For B-DNA, whose structural stability is closely linked to high humidity (Section 2.3.1), highly ordered water molecules can be seen in both

Table 2.5. Average torsion angles (°) for DNA helices

Structure type	α	β	γ	δ	ε	ζ	χ
A–DNA[a]	−50	172	41	79	−146	−78	−154
GGCCGGCC	−75	185	56	91	−166	−75	−149
B-DNA[a]	−41	136	38	139	−133	−157	−102
CGCGAATTCGCG	−63	171	54	123	−169	−108	−117
Z-DNA (C residues)	−137	−139	56	138	−95	80	−159
(G residues)	47	179	−169	99	−104	−69	68
DNA-RNA decamer	−69	175	55	82	−151	−75	−162
A-RNA	−68	178	54	82	−153	−71	−158

[a] Fibres.

major and minor grooves. The broad major groove is 'coated' by a unimolecular layer of water molecules that interact with exposed C=O, N, and NH functions and also extensively solvate the phosphate backbone. The narrow minor groove contains ordered, zig-zag chains of water molecules: two per base-pair. In the A·T region of the Dickerson dodecamer, these waters form a **spine of hydration** with first-shell oxygens buried in the floor of the groove, where they solvate the bases, and alternating second-shell oxygens above and between them near the surface of the groove.

To a first approximation, the difference between the A, B, and other polymorphs of DNA can be described in terms of just two co-ordinates: slide (D_y) and roll (σ). These and other movements of base-pairs are illustrated in Figure 2.18a–c and values of the parameters given in Table 2.4. Clearly A-DNA has high roll and negative slide while B-DNA has little roll and small positive slide. This results in a greater hydrophobic surface area of the bases being exposed in A-DNA per base-pair. From this, it has been argued that B-DNA will have the lesser energy of solvation, explaining its greater stability at high humidity (95 per cent), and that this hydrophobic effect may well tip the balance between the A-form and B-form helices.

Other B-DNA structures have much lower significance. C-DNA is obtained from the lithium salt of natural DNA at rather low humidity. It has 28 bases in three full turns of the helix. D-DNA is observed for alternating A·T regions of DNA and has an overwound helix compared to B-DNA with 8 bp per turn. In phage T2 DNA, where cytosine bases have been replaced by glucosylated 5-hydroxymethylcytosines, the B-conformation observed at high humidity changes into a T-DNA form at low humidity (<60 per cent relative humidity), which also has eightfold symmetry around the helix (see Table 2.3).

2.2.5 Z-DNA

Two of the earliest crystalline oligodeoxyribonucleotides, d(CGCGCG) and d(CGCG), provided structures of a new type of DNA conformer, the left-handed Z-DNA, which has also been

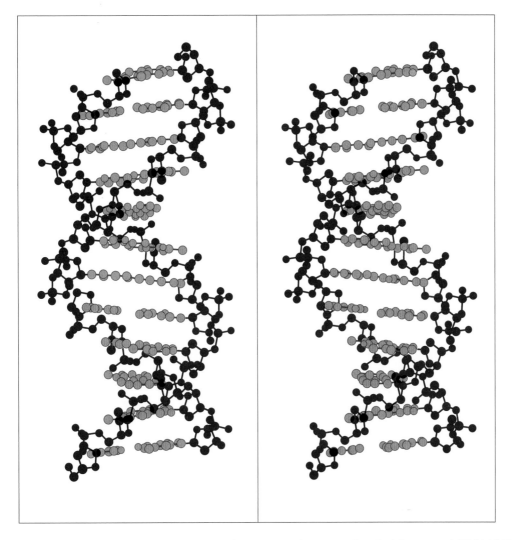

Fig. 2.17 Twelve base-pairs of B-DNA extrapolated from the crystal structure of stacked decamers of d(CCAACGTTGG). The minor groove is frontal at the top and the major groove below. (Three-dimensional stereopair diagrams for parallel viewing.)

found for d(CGCATGCG). Initially it was thought that left-handed DNA had a strict require-ment for alternating purine–pyrimidine sequences. We now know that this condition is neither necessary nor sufficient since left-handed structures have been found for crystals of d(CGATCG) in which cytosines have been modified by C-5 bromination or methylation and have been identified for GTTTG and GACTG sequences by supercoil relaxation studies (Sec-tion 2.3.4).

The Z-helix is also an antiparallel duplex but is a radical departure from A- and B-forms of DNA. It is best typified by an alternating (dG–dC)$_n$ polymer. Its two backbone strands run

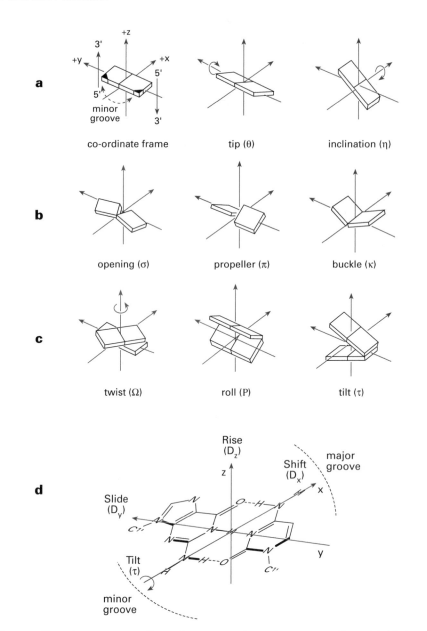

Fig. 2.18 Diagrams illustrating movements of bases in sequence-dependent structures. Rows (a)–(c) show local rotational helix parameters (from the Cambridge DNA nomenclature accord, see R. E. Dickerson *et al.* (1989), *Nucleic Acids Research*, **17**, 1797–803). Within each of the three vertical columns, rotations are in the x, y, and z axes, from *right* to *left* respectively. (**a**) Bases of a pair moving in concert. (**b**) Bases of a pair moving in opposition. (**c**) Steps between two base-pairs. (**d**) Translocational movements of base-pairs relative to the helix axis and to the major and minor grooves.

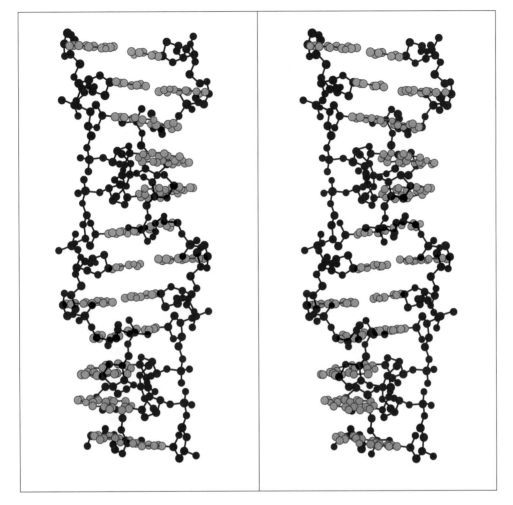

Fig. 2.19 Twelve base-pairs of left-handed, Z-DNA duplex generated from the crystal structure of the hexamer d(CGCGCG). The minor groove is frontal at the top of the helix while the major groove is seen as a convex surface in the front of the lower part of the figure. The zig-zag nature of the phosphate residues is clearly visible, ascending from right to left across the centre of the helix, as is the close approach of the guanosine 5′-phosphate groups across the minor groove. (Three-dimensional stereopair diagram for parallel viewing.)

downward at the left of the minor groove and upward at the right ($\downarrow\uparrow$), and this is the opposite from those of A- and B-DNA ($\uparrow\downarrow$) (NB: the forward direction is defined as the sequence $O^{3'}{\rightarrow}P{\rightarrow}O^{5'}$). In an idealized left-handed duplex, such reversed chain directions would require all the nucleosides to have the *syn*-conformation for their glycosylic bonds. However, this is not possible for the pyrimidines because of the clash between O^2 of the pyrimidine and the sugar furanose ring (Section 2.1.4). So the cytosines take the *anti* conformation and the guanines the *syn* conformation. The name Z-DNA results from this *anti–syn* feature of the glycosylic bonds that **alternates regularly along the backbone**. It causes a local chain reversal that generates a **zig-zag backbone** path and produces a helical repeat consisting of two successive bases (purine-*plus*-pyrimidine), and with an overall chain sense that is the opposite of that of A- and B-DNA (Fig. 2.19). The *syn* conformation of Z-DNA guanines is represented by glycosyl angles χ close to $60°$ while the sugar pucker is $C^{2'}$-*endo* at dC and $C^{3'}$-*endo* at dG residues (Table 2.3).

The switch from B- to Z-DNA conformation appears to be driven by the energetics of $\pi-\pi$ **base stacking**. In Z-DNA, the GpC step is characterized by helical twist $-50.6°$ and base-pair slide -1.1 A. However, for the CpG steps the twist is $-9°$ and the slide is 5.4 Å (Table 2.4; see Fig. 2.18 for an explanation of these terms). These preferences occupy the two extremes of the slide axis and thus appear to be incompatible with a standard right-handed helix, for which helix twist is $36°$ and base-pair slide is 0.4 Å in B-DNA. However, these extremes taken together can be accommodated by a left-handed, Z-type helix. A similar analysis also explains the preference for a Z-helix in the polymer $(dG-dT)_n \cdot (dA-dC)_n$.

The net result of these changes is that the minor groove of Z-DNA is so deep that it actually contains the helix axis whilst the 'major groove' of Z-DNA has become a convex surface on which cytosine-C^5 and guanine-N^7 and -C^8 are exposed (Fig. 2.19 and Table 2.3).

Solution studies on poly(dG-dC) have shown a salt-dependent transition between conformers that can be monitored by circular dichroism (CD) or by ^{31}P NMR (Section 10.2.5). In particular, there is a near inversion in the CD spectrum above 4 M NaCl which has been identified as a change from B- to Z-DNA. It appears that a high salt concentration stabilizes the Z-conformation because it has a much smaller separation between the phosphate anions in opposing strands than is the case for B-DNA, 8 Å as opposed to 11.7 A. A detailed stereochemical examination of this conformational change shows that it calls for an elaborate mechanism and this has posed a problem known as the **chain-sense paradox**: 'How does one reverse the sense of direction of the chains in a B-helix ($\uparrow\downarrow$) to its opposite in a Z-helix ($\downarrow\uparrow$) without unpairing the bases?' Further consideration will be given to this problem later (Section 2.5.4).

The scanning tunnelling microscope has the power to resolve the structure of biological molecules with atomic detail (Section 10.4.2). Much progress has been made with dried samples of duplex DNA, in recording images of DNA under water, and in revealing details of single-stranded poly(dA). Such STM microscopy has provided images of poly(dG-me^{5}dC)·poly(dG-me^{5}dC) in the Z-form. Both the general appearance of the fibres and measurements of helical parameters are in good agreement with models derived from X-ray diffraction data.

Summary

The primary structure of DNA has a string of nucleosides, each joined to its two neighbours through phosphodiester linkages. Each regular 5′-hydroxyl group is linked through a phosphate to a 3′-hydroxyl group, the uniqueness of any primary structure depending only on the sequence of bases in that chain.

A-DNA and B-DNA are the major standard DNA secondary structures with right-handed double helices and Watson–Crick base-pairing.

A-DNA has 11 residues per turn, these bases are tilted 20° to enhance stacking, and they lie 4.5 Å away from the helix axis. This gives a fairly stiff helix, which shows little sequence-dependent variation in structure. The major groove is deep and narrow, the minor groove is broad and shallow.

B-DNA has 10 bases per turn with little tilting of the bases. The wide major groove and narrow minor groove are both of moderate depth and both grooves are well solvated by water molecules. The B-form structure is sufficiently flexible to permit a conformational response in the backbone to particular local base sequences.

Z-DNA is a left-handed double-helical structure stabilized by high concentrations of $MgCl_2$ or NaCl. It is most favoured for alternating G-C sequences. Watson–Crick pairing is maintained but the purines adopt the *syn* glycoside and the $C^{3'}$-*endo* sugar pucker. The phosphate backbone has a zig-zag appearance. The minor groove is narrow and very deep but the major groove has become so shallow that normally inaccessible parts of the bases C and G are exposed. These changes are a consequence of the demands of base stacking which result in large changes in helical twist and base slide parameters for Z-DNA relative to those for A- and B-DNA.

2.3 Real DNA structures

2.3.1 Sequence-dependent modulation of DNA structure

So far we have emphasized the importance of hydrogen bonds in base-pairing and DNA structure and have said little about base stacking. We shall see later that both these two features are important for the energetics and dynamics of DNA helices (Section 2.5) but it is now time to look at the major part played by base stacking in real DNA structures. Two particular hallmarks of B-DNA, in contrast to the A- and Z-forms, are its flexibility and its capacity to make small adjustments in local helix structure in response to particular base sequences.

Different base sequences have their own characteristic signature: they influence groove width, helical twist, curvature, mechanical rigidity, and resistance to bending. It seems probable that these features help proteins to read and recognize one base sequence in preference to another (Chapter 9) possibly only through changes in the positions of the phosphates in the backbone. What do we know about these sequence-dependent structural features?

One surprise to emerge from single-crystal structure analyses of synthetic DNA oligomers has been the breadth of variation of local helix parameters relative to the mean values broadly derived from fibre diffraction analysis and used for the standard A- and B-form DNA structures described earlier. Dickerson has compared eight dodecamer and four decamer B-DNA structures. The mean value of the **helical twist** angle between neighbouring base-pairs is 36.1° but the standard deviation (SD) is 5.9° and the range is from 24° to 51°. Likewise, the mean **rise** per base-pair is 3.36 Å with an SD of 0.46 Å but with a range from 2.5 to 4.4 Å. (NB: because rise is a parameter measured between the C-1′ atoms of adjacent base-pairs, it can be smaller than the thickness of a base-pair if the ends of the two base-pairs bow towards each other. Such **bowing** is also defined as 'positive cup'). Roll angles between successive base-pairs average $+0.6°$ but with an SD of $6.0°$ and a range from $-18°$ to $+16°$. These variations in twist and roll have the effect of substantially re-orienting the potential hydrogen bond acceptors and donors at the edges of the bases along the floor of the DNA grooves, so they may well be a significant component of the sequence-recognition process used by drugs and proteins (Sections 8.3 and 9.3). These and other modes of local changes in the geometry of base-pairs are illustrated in Fig. 2.18.

The major irregularities in the positions of the bases in real DNA structures contrasts with only secondary, small conformational changes in their sugar-phosphate backbones. The main characteristic of these sequence-dependent modulations is **propeller twist**. This results when the bases rotate by some $5°$ to $25°$ relative to their hydrogen-bonded partner around the long axis through C-8 of the purine and C-6 of the pyrimidine (Fig. 2.18b, centre). Sections of oligonucleotides with consecutive A residues, as in d(CGCAAAAAAGCG)·d(CGCTTTTTTGCG), have unusually high propeller twists (approximately $25°$) and these permit the formation of a three-centred hydrogen bonding network in the major groove between adenine-N^6 and two thymine-O^4 residues, the first being the W−C base-pair partner and the second being its $3'$-neighbour, both in the opposing strand. This network of hydrogen bonds gives added rigidity to the duplex and may explain why long runs of adenines are not found in the more sharply curved tracts of chromosomes (Section 2.6.2) yet are found at the ends of nucleosomal DNA, with decreased supercoiling.

Why should the bases twist in this way? The advantage of propeller twist is that it gives improved face-to-face contact between adjacent bases in the same strand and this leads to increased stacking stability in the double-helix. However, there is a penalty! The larger purine bases occupy the centre of the helix so that in alternating purine–pyrimidine sequences they overlap with neighbouring purines in the opposite strand. Consequently, propeller twist causes a clash between such pairs of adjacent purines in opposite strands. For pyrimidine-$(3 \rightarrow 5')$–purine steps, these purine–purine clashes take place in the minor groove where they involve guanine-N^3 and -N^2 and adenine-N^3 atoms. For purine-$(3' \rightarrow 5')$–pyrimidine steps, they take place in the major groove between guanine-O^6 and adenine-N^6 atoms (Fig. 2.20). There are no such clashes for purine–purine and pyrimidine–pyrimidine sequences.

One of the consequences of these effects is that bends may occur at junctions between polyA tracts and mixed-sequence DNA as a result of propeller twist, base-pair inclination, and base stacking differences on two sides of the junction (see below).

Electrostatic interactions between bases

There are two principal types of base–base interaction that drive the local variations in helix parameter described above and in Fig. 2.18a–c. Firstly there are repulsive steric interactions between proximate bases and sugars. They are associated with steric interactions between thymine methyl groups, the guanine amino group, and the configuration of the step pyrimidine–purine (described as YR), purine–pyrimidine (described as RY), and RR/YY. Secondly, there are π–π stacking interactions that are determined by the distribution of π-electron density above and below the planar bases.

Chris Hunter has identified four principal contributions to the energy of π–π interactions between DNA base-pairs.

(1) van der Waals interactions (designated *VDW* and vary as r^{-6}).
(2) Electrostatic interactions between partial atomic charges (designated *atom–atom* and vary as r^{-1}).
(3) Electrostatic interactions between the charge distributions associated with the π-electron density above and below the plane of the bases (designated $\pi\sigma$–$\pi\sigma$ and vary approximately as r^{-5}).
(4) Electrostatic interactions between the charge distributions associated with the π-electron density and the partial atomic charges (designated as *atom–$\pi\sigma$*, this is the cross-term of (2) and (3) and varies as r^{-4}).

He has used these components to calculate the total π–π interaction energies between pairs of stacked bases and applied the results to interpret the source of slide, roll, and helical twist, of propeller twist, and of a range of other conformational preferences that are sequence-dependent. In addition, his calculations correlate very well with experimental observations on polymorphic forms of DNA. The main conclusions can broadly be summarized as follows.

- VDW-steric interactions are seen cross-strand at pyrimidine–purine (YR) and CX/XG steps and can be diminished by reducing propeller twist, reducing helical twist, or by positive slide or positive roll. They are seen as same-strand clashes between the thymine methyl group and the neighbouring 5′-sugar in AX/XT steps which are avoided by introducing negative propeller twist, reducing helical twist, or generating negative slide coupled with negative roll.
- Electrostatic interactions cause positive or negative slide with the sole exception of AA/TT. These slide effects are opposed by the hydrophobic effect which tends to force maximum base overlap and favours a zero-slide B-type conformation.
- Atom–atom interactions are most important for C·G base-pairs where there are large regions of charge and lead to strong conformational preferences for positive slide in CG steps and negative slide in GC steps (see Table 2.4). This leads poly(dCG) to adopt the Z-form left-handed duplex.
- Atom–$\pi\sigma$ interactions lead to sequence-dependent effects which are repulsive in AX/XT, TX/XA, and CX/XG steps where they can be reduced by negative propeller twist, by positive or negative slide, or by introducing buckle.
- $\pi\sigma$–$\pi\sigma$ electrostatic interactions tend to be swamped by other effects and play a relatively minor role in sequence-dependent conformations.

In sequence-dependent structures, propeller twist is most marked for purines on opposing strands in successive base-pairs. The 'purine–purine' clash is much more pronounced for YR steps, where the clash is in the minor groove (Fig. 2.20a), than for RY steps, where the clash is

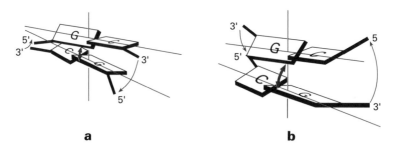

Fig. 2.20 Diagrams illustrating (**a**) clockwise propeller twist for a C → G clash between guanines in the **minor** groove and (**b**) clockwise propeller twist for a G→ C sequence showing purines clashing in the **major** groove.

seen in the major groove (Fig. 2.20b). While its origin was at first thought to result solely from van der Waals interactions, it seems now to be better explained by the total electrostatic interaction picture (see above).

Taken together, these sequence-dependent features suggest that DNA should most easily be unwound and/or unpaired in A-T rich sequences, which have only two hydrogen bonds per base-pair, and in pyrimidine–purine steps. It is noteworthy that the dinucleotide TpA satisfies both of these requirements and has been identified as the base step that serves as a nucleus for DNA unwinding in many enzymatic reactions requiring strand separation.

Calladine's rules

Notwithstanding the apparent success of the above calculations, the evidence from analysis of X-ray structures suggests that base step conformations are influenced by the nature of neighbouring steps. It follows that a better sequence-structure correlation is likely to emerge from examining each step in the context of its flankers: three successive base steps, or a tetrad of four successive base-pairs. However, until a majority of the 136 possible triads has been sampled by analysis of real structures, a set of empirical rules enunciated by Chris Calladine in 1982 will remain useful.

Calladine observed that B-DNA structures respond to minimize the problems of sequence-dependent base clashes in four ways, which he articulated as follows:

- flatten the propeller twist locally for either or both base-pairs
- roll the base-pairs away from their clashing edges
- slide one or both of the base-pairs along their axis to push the purine away from the helix axis
- unwind the helix axis locally to diminish interstrand purine–purine overlap.

The relative motions required to achieve these effects are described by six parameters of which the most significant are ϱ for roll, D_y for slide, and Ω for helix twist. These motions are illustrated for neighbouring G·C base-pairs (Fig. 2.18).

In practice, the structures of crystalline oligomers have exhibited the following six types of conformational modulation which are sequence-dependent and which support these rules:

- the B-DNA helix axis need not be straight but can curve with a radius of 112 Å
- the twist angle, Ω, is not constant at 36° but can vary from 28° to 43°
- propeller twist averages +11° for C·G pairs and +17° for A·T pairs
- base-pairs 'roll' along their long axes to reduce clashing
- sugar pucker varies from $C^{3'}$-*exo* to $O^{4'}$-*endo* to $C^{2'}$-*endo*
- there can be local improved overlap of bases by slide, as in d(TCG) where C-2 moves towards the helix axis to increase stacking with G-3.

The Calladine model is incomplete because it ignores such important factors as electrostatic interactions, hydrogen bonding, and hydration. For example, a major stabilizing influence proposed for the high propeller twist in sequences with consecutive adenines is the existence of cross-strand hydrogen bonding between adenine N-6 in one strand with thymine O-4 of the next base-pair in the opposite strand (see above).

Modulations of B-DNA structure which have been observed in the solid state have to some extent been mirrored by the results of solution studies for d(GCATGC) and d(CTGGATCCAG) obtained by a combination of NMR analysis and restrained molecular dynamics calculations. These oligomers have B-type structures which show clear, sequence-dependent variations in torsion angles and helix parameters. There is strong curvature to the helix axis of the hexamer which results from large positive roll angles at the pyrimidine–purine steps. The decamer has a straight central core but there are bends in the helix axis at the second (TpG) and eighth (CpA) steps which result from positive roll angles and large slide values.

Taken together, these X-ray and NMR analyses give good support for the general conclusion that minor groove clashes at pyrimidine–purine steps are twice as severe as major groove clashes at purine–pyrimidine steps. As a result, it is possible to calculate the behaviour of the helix twist angle, Ω, using sequence data only.

The continuum of right-handed DNA conformations

The simple concept that the standard conformations for right-handed DNA represent discontinuous states, only stable in very different environments, has undergone marked revision. In addition to the range of conformations seen in crystal structures, CD and NMR analyses of solution structures have also undermined that naive picture. In particular, CD studies have shown that there is a continuum of helix conformations in solution that is sequence-dependent while both CD analysis of the complete TFIIIA binding site of 54 bp and the crystal structure of a nonameric fragment from it have identified a conformer that is intermediate between the canonical A- and B-DNA forms. In addition, A- and B-DNA polymorphs can coexist, as seen in the crystal structure of d(GGBrUABrUACC) which contains separate double helices of both the A- and B-DNA forms.

Bending at helix junctions

Bent DNA was first identified as a result of modelling the junction between an A-type and a B-type helix. The best solution to this problem requires a bend of 26° in the helix axis in order to maintain full stacking of the bases. Bent DNA has gained support not only from NMR and CD studies on a DNA·RNA hybrid, [poly(dG)·(rC)$_{11}$–(dC)$_{16}$], but also from studies on regular homopolymers which contain (dA)$_5$·(dT)$_5$ sections occurring in phase in each turn of a 10-fold or 11-fold helix. Moreover, bent DNA containing such dA·dT repeats has been investigated from a variety of natural sources.

It appears that bending of this sort happens at junctions between the stiff [dA·dT] helix and the regular B-helix (see above). In situations where such junctions occur every five bases and in an alternating sense, the net result is a progression of bends which is equivalent to a continuous curve in the DNA.

2.3.2 Mismatched base-pairs

The fidelity of transmission of the genetic code rests on the specific pairings of A·T and C·G bases. Consequently, if changes in shape result from base mismatches, such as A·G, they must be recognized and be repaired by enzymes with high efficiency (Section 7.11).

X-ray analysis of DNA fragments with potential mispairs cannot give any information about the transient occurrence of rare tautomeric forms at the instant of replication. However, it can define the structure of a DNA duplex which incorporates mismatched base-pairs and provide details of the hydrogen bonding scheme, the response of the duplex to the mismatch, the influence of neighbouring sequence on the structure and stability of the mismatch, and the effect of global conformation. All these are intended to provide clues about the ways in which mismatches might be recognized by the proteins that constitute repair systems. High-resolution NMR studies have extended the picture to solution conformations. The different types of base-pair mismatch can be grouped into **transition mismatches**, which pair a purine with the wrong pyrimidine, and **transversion mismatches**, which pair either two purines or two pyrimidines.

Transition mismatches

The G·T base-pair has been observed in crystal structures for A-, B-, and Z-conformations of oligonucleotides. In every case it has been found to be a typical 'wobble' pair having *anti–anti* glycosylic bonds. The structure of the dodecamer, d(CGC**G**AATT**T**GCG). which has two G·T[9] mismatches, can be superimposed on that of the regular dodecamer and shows excellent correspondence of backbone atomic positions.

The A·C pair has been examined in the dodecamer d(CGC**A**AATT**C**GCG) and, once again, the two A[4]·C mismatches are typical 'wobble' pairs, achieved by the protonation of adenine-N^1 (Fig. 2.21). It is notable that there is no significant worsening of base stacking and little perturbation of the helix conformation. However, it appears that no water molecules are bonded to these bases in the minor groove.

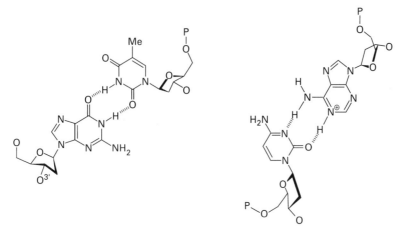

Fig. 2.21 'Wobble' pairs for transition mismatches G·T (left) and A·C (right).

Transversion mismatches

The G·A mismatch is the most thoroughly studied in solution and in the solid state and two different patterns have been found. Crystals of the dodecamer, d(CGC**GA**ATT**A**GCG) have an (*anti*)G·A(*syn*) mismatch with hydrogen bonds from Ade-N^7 to Gua-N^1 and from Ade-N^6 to Gua-O^6 (Fig. 2.21). A similar (*anti*)I·A(*syn*) mismatch has been identified in a related dodecamer structure. Calculations on both of these mismatches suggest that they can be accommodated into a regular B-helix with minimal perturbation.

This work contrasts with both NMR and X-ray studies on d(CCA**A**G**A**TTGG) and NMR work on d(CGA**GA**ATTC**G**CG) which have identified (*anti*)G·A(*anti*) pairings with two hydrogen bonds (Fig. 2.22). The X-ray analysis of the dodecamer (see above) shows a typical B-helix with a broader minor groove and a changed pattern of hydration. This arises in part because the two mismatched G·A pairs are 2.0 Å wider (from C$^{1'}$ to C$^{1'}$) than a conventional Watson–Crick pair.

Insertion–deletion mispairs

When one DNA strand has one nucleotide more than the other, the extra residue can either be accommodated in an intrastrand position or be forced into an extra-strand location. Tridecanucleotides containing an extra A, C, or T residue have been examined in the crystalline solid and solution states. In one case, an extra A has been accommodated into the helix stack while in others a C or A is seen to be extruded into an extrahelical, unstacked location.

In addition to such work on mismatched base-pairs, related investigations have made good progress into structural changes caused by covalent modification of DNA. On the one hand, crystal structures of DNA adducts with cisplatin have characterized its monofunctional linking to guanine sites in a B-DNA helix (Section 7.5.4) and, on the other, NMR studies of O^4-methylthymine residues and of thymine photodimers and psoralen:DNA photoproducts are advancing our understanding of the modifications to DNA structure that result from such

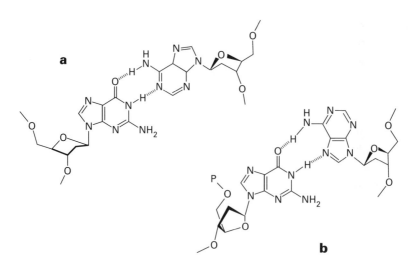

Fig. 2.22 Mismatched G·A pairings (**a**) for the decamer d(CCAAGATTGG) with (*anti*)G·A(*anti*) and (**b**) for the dodecamer d(CGAGAATTCGCG) with (*anti*)G·A(*syn*) conformation.

lesions (Section 7.8.2). It seems likely that the range of patterns of recognition of structural abnormalities may be as wide as the range of enzymes available to repair them!

2.3.3 Unusual DNA structures

Since 1980, there has been a rapid expansion in our awareness of the heterogeneity of DNA structures which has resulted from a widening use of new analytical techniques, notably structure-dependent nuclease action, structure-dependent chemical modification, and physical analysis. Unusual structures are generally sequence-specific, as we have already described for the A–B helix junction (Section 2.3.1). Some of them are also dependent on DNA supercoiling which provides the necessary driving energy for their formation due to the release of torsional strain, as is particularly well defined for cruciform DNA. Consequently, much use has been made of synthetic DNA both in short oligonucleotides and cloned into circular DNA plasmids where the effect of DNA supercoiling can be explored.

Curved DNA

The axial flexibility of DNA is one of the significant factors in DNA–protein interactions (Chapter 9). DNA duplexes up to 150 bp long behave in solution as stiff, though not necessarily straight, rods. By contrast, many large DNA–protein complexes have DNA that is tightly bent. One of the best examples is the bending of DNA in the eukaryotic chromosome where 146 bp of DNA are wrapped around a protein core of histones (Section 9.2.1) to form nearly two complete turns of a left-handed superhelix with a radius of curvature of 43 Å. To achieve this, the major and minor grooves are compressed on the inside of the curve and stretched on its outside. At the same time, the helix axis must change direction.

DNA curvature has also been examined in kinetoplast DNA from trypanosomatids. It provides a source of **open** DNA minicircles whose curvature is sequence dependent rather than being enforced by covalent closure of the circles. Such circles can be examined by electron microscopy and have 360° curvature for about 200 bp. Such kinetoplast DNA has short adenine tracts spaced at 10 bp intervals by general sequence. This fact led to solution studies on synthetic oligomers with repeated sets of four $CA_{5-6}T$ sequences spaced by 2–3 bp. These behave as though they have a 20°–25° bend for each repeat which led to the simple idea that DNA bending is an inherent property of poly(dA) tracts (Fig. 2.23a). In conflict with this idea, poly(dA) tracts in the crystal structures of several oligonucleotides are seen to be straight. What then is the real origin of DNA curvature?

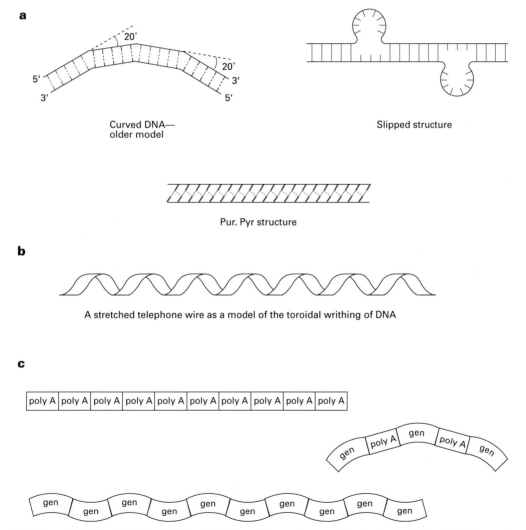

Fig. 2.23 Curved DNA illustrated by straight poly(dA) tracts (upper), consecutive tracts of writhing DNA of general sequence (lower), and curved DNA (right) of alternating segments of linear poly(dA) and curved tracts of general sequence

Richard Dickerson has examined helix bending in a range of B-form crystal structures of oligonucleotides containing poly(dA) tracts and has concluded that poly(dA) tracts are straight and not bent and that regions of A·T base-pairs exhibit a narrow minor groove, large propeller twist, and a spine of hydration in the minor groove. He argues that DNA curvature results from the direct combination of two general features of DNA structure:

- general sequence DNA writhes
- poly(dA) tracts are straight.

Studies on the hydrodynamic properties of DNA show that general-sequence DNA migrates through gels more slowly than expected which is because the DNA helix occupies a cylindrical volume that has a larger diameter than that of a simple B-helix (Section 10.3.3). This phenomenon is a result of **DNA writhing**, which involves a continuously curved distortion of the helix axis to generate a spiral form and is nicely illustrated by the extension of a coiled telephone wire (Fig. 2.23b). It follows that the repeated alternation of straight A-tracts with short sections of general sequence, each having half of a writhing turn, will generate curved DNA (Fig. 2.23c). A detailed structural analysis of this explanation says that curvature of B-DNA involves rolling of base-pairs, compresses the major groove (which corresponds to positive roll), has a sequence-determined continuum in the bending behaviour, and shows anisotropy of flexible bending.

DNA bending

Such intrinsic, sequence-dependent curvature must be distinguished from the bending of DNA which results from the application of an external force. Dickerson has also examined the bending of the DNA helix that occurs in many crystal structures of the B-form. It is associated with the step from a G·C to an A·T base-pair and results from rolling one base-pair over the next along their long axes in a direction that compresses the major groove (Fig. 2.18c). He suggests that this junction is a flexible hinge that is capable of bending or not bending. Such 'facultative bending' responds to the influence of local forces, typically interactions with other macromolecules, for example control proteins or a nucleosome core. By contrast, poly(dA) tracts are known to resist bending in nucleosome reconstitution experiments. It can thus be seen that sequence-dependent variation in **DNA bendability** is an important factor in DNA recognition by proteins.

One important conclusion emerges: DNA has evolved conformationally to interact with other macromolecules. A free, linear DNA helix in solution may, in fact, be the least biologically relevant state of all.

Slipped structures have been postulated to occur at direct repeat sequences, and they have been found upstream of important regulatory sites. The structures described (Fig. 2.23a) are consistent with the pattern of cleavage by single-strand nucleases but otherwise are not well characterized.

Purine–pyrimidine tracts manifest an unusual structure at low temperature with a long-range, sequence-dependent single base shift in base-pairing in the major groove. For the dodecamer d(ACCGGCGCCACA)·d(TGTGGCGCCGGT), the bases in the d(CA)$_n$ tract have high propeller twist $(-32°)$ and are so strongly tilted in the 3′-direction that there is disruption of Watson–Crick pairing in the major groove and formation of interactions with the 5′-neighbour of the complementary base. This alteration propagates along the B-form helix for at least half a turn with a domino-like motion. As a result, the DNA structure is normal when viewed from the minor groove and mismatched when seen from the major groove. Since $(CA)_n$ tracts are involved both in recombination and in transcription, this new recognition pattern has to be considered in the analysis of the various processes involved with reading of genetic information.

Anisomorphic DNA is the description given to DNA conformations associated with direct repair, DR2, sequences at 'joint regions' in viral DNA, which are known to have unusual chemical and physical properties. The two complementary strands have different structures and this leads to structural aberrations at the centre of the tandem sequences that can be seen under conditions of torsional stress induced by negative supercoiling.

Hairpin loops are formed by oligonucleotide single strands which have a segment of inverted complementary sequence. For example, the 16-mer d(CGCGCGTTTTCGCGCG) has a hexamer repeat and its crystal structure shows a hairpin with a loop of four Ts and a Z-DNA hexamer stem (Figure 2.24a). When such inverted sequences are located in a DNA duplex, the conditions exist for formation of a cruciform.

Cruciforms involve intra-strand base-pairing and generate two stems and two hairpin loops from a single unwound duplex region. The inverted sequence repeats are known as **palindromes**, which have a given DNA duplex sequence followed after a short break by the same duplex sequence in the opposite direction. This is illustrated for a segment of the bacterial plasmid pBR322 (Fig. 2.24b), where a palindrome of two undecamer sequences exists.

X-ray, NMR, and sedimentation studies of such stem-loop structures show that the four arms are aligned in pairs to give an oblique X structure with continuity of base stacking and helical axes across the junctions (Section 5.4.2). Also, the loops have an optimum size of from four to six bases. Residues in the loops are sensitive to single-strand nucleases, such as S1 and P1, and especially to chemical reagents such as bromoacetaldehyde, osmium tetroxide, bisulfite, and glyoxal (Sections 7.4 and 7.5). In addition, the junctions are cleavage sites for yeast resolvase and for T4 endonuclease VII.

David Lilley has shown that the formation of two such loops requires the unpairing and unstacking of three or more base-pairs and so will be thermodynamically unstable compared to the corresponding single helix. While there can be some stacking of bases in the loops, the adverse energy of formation of a single cruciform has been calculated to be some 75 kJ mol^{-1}. In experiments on cruciforms using closed circular superhelical DNA, this energy can be provided by the release of strain energy in the form of negative supercoiling (next Section) and is directly related to the length of the arms of the cruciform: the formation of an arm of 10.5 bp unwinds the supercoil by a single turn.

There is also a kinetic barrier to cruciform formation and Lilley has suggested two mechanisms which have clearly distinct physical parameters and may be sequence-dependent. The faster process for cruciform formation, the S-pathway, has $\Delta G^{\ddagger}$ of about 100 kJ mol^{-1} with a small positive entropy of activation. This more common pathway is typified by the behaviour

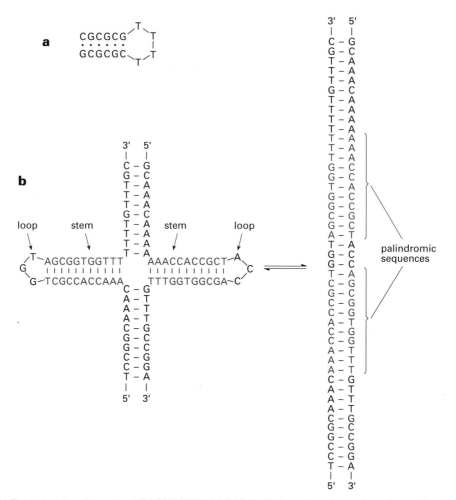

Fig. 2.24 (**a**) The hairpin loop formed by d(CGCGCGTTTCGCGCG). (**b**) Formation of a cruciform from an inverted repeat sequence of the bacterial plasmid pBR322. The inverted palindromic regions, each of 11 bp, are shown in colour.

of plasmid pIRbke8. Following the formation of a relatively small unpaired region, a proto-cruciform intermediate is produced which then grows to equilibrium size by branch migration through the four-way junction (Fig. 2.25). The slower mechanism, the C-pathway, involves the formation of a large bubble followed by its condensation to give the fully-developed cruciform. This behaviour explains the data for the pCollR 315 plasmid whose cruciform kinetics show $\Delta G^{\ddagger}$ about 180 kJ mol^{-1} with a large entropy of activation.

Such extrusion of cruciforms provides the most complete example of the characterization of unusual DNA structures by combined chemical, enzymatic, kinetic, and spectroscopic techniques. However, it is not clear whether cruciforms have any role *in vivo*. One reason may simply be that intracellular superhelical densities may be too low to cause extrusion of inverted repeat sequences. Equally, the kinetics of the process may also be too slow to be of

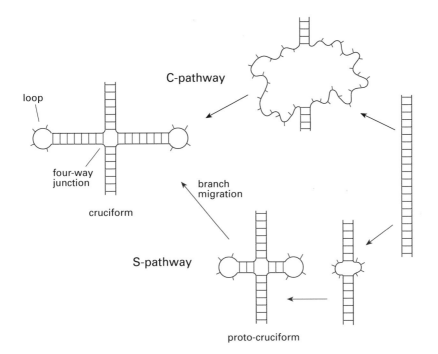

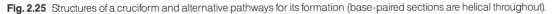

Fig. 2.25 Structures of a cruciform and alternative pathways for its formation (base-paired sections are helical throughout).

physiological significance. However, cruciforms are formally equivalent to Holliday junctions, and these four-way junctions involve two DNA duplexes that are formed during homologous recombination (Section 5.4).

2.3.4 B–Z junctions and B–Z transitions

Segments of left-handed Z-DNA can exist in a single duplex in continuity with segments of right-handed B-DNA. This phenomenon has been observed both *in vitro* and *in vivo*. Because the backbone chains of these polymorphs run in opposite directions, ($\downarrow\uparrow$ and $\uparrow\downarrow$) respectively (Section 2.2.5), there has to be a transitional region between two such segments, and this boundary is known as a **B–Z junction**. While such structures are polymorphic, poorly defined, and are sequence specific, six features can be described:

- B–Z junctions can be as small as 3 bp
- at least one base-pair has neither the B- nor the Z-conformation
- hydrogen bonds between the base-pairs are intact below 50°C
- chemical reagents specific for single-stranded DNA (chloroacetaldehyde, bromoacetaldehyde, and glyoxal) show high reactivity with the junction bases
- junctions are sites for enhanced intercalation for psoralens
- junctions are neither strongly bent nor particularly flexible.

The conformational switch between the right- and left-handed helices is called the B–Z transition. It has a high energy of activation (about 90 kJ mol^{-1}) but is practically independent of temperature ($\Delta G°$ about 0 kJ mol^{-1}). Thus the B–Z transition is co-operative and propagates readily along the helix chains.

Two different models have been suggested to explain the conformational switch that has to occur as a B–Z junction migrates, rather like a bubble, along a double helix. In the first, the bases unpair, guanine flips into the *syn* conformation, the entire deoxycytidine undergoes a conformation switch, and the bases reform their hydrogen bonds. This model appears to be at variance with NMR studies that suggest the bases remain paired because their imino-protons do not become free to exchange with solvent water. The second model suggests that the backbone is stretched until one base-pair has sufficient room to rotate 180° about its glycosyl bonds (**tip**, as shown in Fig. 2.18a), and the bases re-stack. However, one might expect this 'expand–rotate–collapse' process to be impeded by linking bulky molecules to the edge of the base-pairs. Yet bonding *N*-acetoxy-*N*-acetyl-2-aminofluorene to guanine actually facilitates the B–Z transition! Thus, the dynamics of the B–Z transition poses a major conformational problem, and this has sometimes been called the chain-sense paradox.

Ansevin and Wang have suggested an alternative zig-zag model for the left-handed double-helical form of DNA that avoids this paradox and is accessible from B-DNA by simple untwisting. Their **W-DNA** has a Watson–Crick chain sense ($\uparrow\downarrow$) like B-DNA but similar glycosyl geometry to that of Z-DNA. It has reversed sugar puckers, $C^{3'}$-*endo* at cytosine and $C^{2'}$-*endo* at guanine, while in both W- and Z-DNA the minor groove is deep and the major groove broad and very shallow. In addition, this W-model explains (1) the incompatibility of poly(dA-dT)·poly(dA-dT) with a left-handed state, (2) the very slow rate of exchange of hydrogens in the 2-NH$_2$ group of guanine in left-handed DNA, and (3) the incompatibility of a left-handed helix with replacement of O_R oxygens by a methyl group. They argue that Z-DNA has a lower energy than W-DNA and so is adopted in crystals of short oligonucleotides but it may be conformationally inaccessible to longer stretches of DNA in solution.

2.3.5 Circular DNA and supercoiling

The replicative form of bacteriophage ϕX174 DNA was found to be a double-stranded closed circle. It was later shown that bacterial DNA exists as closed circular duplexes, that DNA viruses have either single- or double-helical circular DNA, and that RNA viroids have circular single-stranded RNA as their genomic material. Plasmid DNAs also exist as small, closed circular duplexes.

Topologically unconstrained dsDNA in its linear, relaxed state is either biologically inactive or displays reduced activity in key processes such as recombination, replication, or transcription. It follows that topological changes associated with the constraints of circularization of dsDNA have a profound biological significance. While such circularization can be achieved directly by covalent closure, the same effect can be achieved for eukaryotic DNA as a result of holding DNA loops together by means of a protein scaffold.

The molecular topology of closed circular DNA was described by Vinograd in 1965 and is especially associated with the phenomenon of superhelical DNA, which is also called supercoiled or supertwisted DNA. Vinograd's basic observation was that when a planar, relaxed circle of DNA is strained by changing the pitch of its helical turns, it relieves this torsional strain by winding around itself to form a superhelix whose axis is a diameter of the original circle.

This behaviour is most directly observed by following the sedimentation of negatively supercoiled DNA as the pitch of its helix is changed by intercalating a drug, typically ethidium bromide (Section 8.4). Intercalation is the process of slotting the planar drug molecules between adjacent base-pairs in the helix. For each ethidium molecule intercalated into the helix, there is an increase in **rise** of about 3.4 Å and a linked decrease of about 36° in **twist** (Fig. 2.18c,d). The DNA helix responds first by reducing the number of negative, right-handed supercoils until it is fully relaxed and then by increasing the number of positive, left-handed supercoils. As this happens, the sedimentation coefficient of the DNA first decreases, reaching a minimum when fully relaxed, and then increases as it becomes positively supercoiled. As a control process, the same circular DNA can be nicked in one strand to make it fully relaxed. The result is that it now shows a low sedimentation coefficient at all concentrations of the intercalator species (Fig. 2.26) (Section 10.3.1).

Vinograd showed that the topological state of these covalently closed circles can be defined by three parameters and that the fundamental topological property is **linkage**. The **topological winding number**, T_w, is the number of right-handed helical turns in the relaxed, planar DNA circle and the **writhing number**, W_r, gives the number of left-handed crossovers in the supercoil. The sum of these two is the **linking number**, L_k, which is the number of times one strand of the helix winds around the other (clockwise is +ve) when the circle is constrained to lie in a plane. The simple equation is

$$L_k = T_w + W_r$$

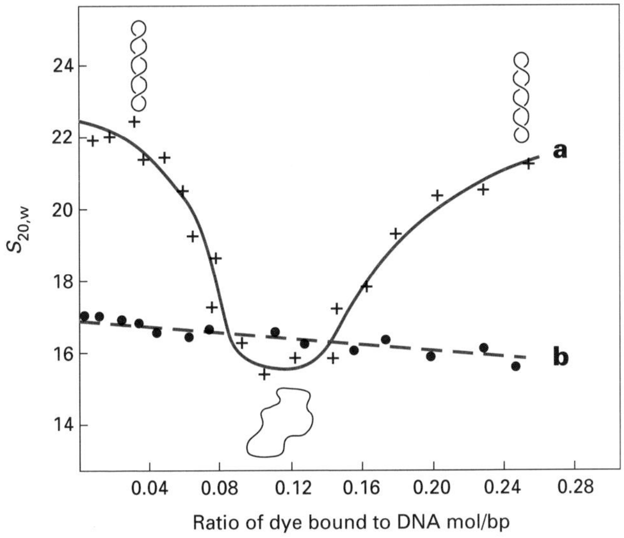

Fig. 2.26 Sedimentation velocity for SV40 DNA as a function of bound ethidium (**a**) for closed circular DNA ($- + - + -$) and (**b**) for nicked circular DNA ($- \cdot - \cdot -$) showing the transition from a negative supercoil (left) through a relaxed circle (centre) to a positive supercoil (right) (adapted from Bauer, W. and Vinograd, J. (1968). *J. Mol. Biol.*, **33**, 141).

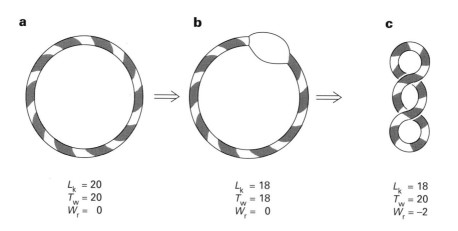

Fig. 2.27 Supercoil formation in closed circular DNA. (**a**) Closed circle of 20 duplex turns (alternate turns in colour). (**b**) circle nicked, underwound two turns, and resealed; (**c**) base-pairing and stacking forces result in the formation of B-helix with two new right-handed helix turns and one compensating right-handed supercoil.

Such behaviour can be illustrated simply (Fig. 2.27) for a relaxed closed circle with 20 helical turns, $T_w = 20$, $L_k = 20$, $W_r = 0$. One strand is now cut, unwound two turns, and resealed to give $L_k = T_w = 18$. This circle is thus underwound by two turns. To restore fully the normal B-DNA base-pairing and base-stacking, the circle needs to gain two right-handed helical turns, $\Delta T_w = +2$, to give $T_w = 20$. Since the DNA circles have remained closed and the linking number stays at 18, the formation of the right-handed helical turns is balanced by the creation of one right-handed supercoil, making $W_r = -2$.

The behaviour of a supercoil can be modelled using a length of rubber tubing. The ends are first held together to form a relaxed, closed circle. If the end in your right hand is given one turn clockwise (right-handed twist) and the other end is given one turn in the opposite sense, the tube will relieve this strain by forming one left-handed supercoil. This is equivalent to unwinding the DNA helix by two turns which generates one positive supercoil (four turns generate two supercoils, and so on). This model shows the relationship: two turns equals one supercoil.

In practice it is sometimes useful to describe the degree of supercoiling using the **superhelical density**, $\sigma = W_r/T_w$, which is close to the number of superhelical turns per 10 bp and is typically around 0.06 for superhelical DNA from cells and virions. The **energy of supercoiling** is a quadratic function of the density of supercoils as described by the equation

$$\Delta G_s = 1050 \times \frac{RT}{N} \times \Delta L_k^2 \quad \text{kJ mol}^{-1}$$

where R is the gas content, T is the absolute temperature, and N is the number of base-pairs.

B–Z transitions are especially important for supercoiling since the conversion of one right-handed B-turn into a left-handed Z-turn causes a change in T_w of -2. This must be complemented by $\Delta W_r = +2$ through the formation of one left-handed superturn.

Enzymology of DNA supercoiling

DNA topoisomers are circular molecules which have identical sequences and differ only in their linking number. A group of enzymes, discovered by Jim Wang, can change that linking number. They fall into two classes: Class I topoisomerases effect integral changes in the linking number, $\Delta L_k = n$, while Class II enzymes interconvert topoisomers with a step rise of $\Delta L_k = \pm 2n$. Topoisomerase I enzymes use a 'nick–swivel–close' mechanism to operate on supercoiled DNA. They break a phosphate diester linkage, hold its ends, and reseal them after allowing exothermic (i.e. passive) free rotation of the other strand. Such enzymes from eukaryotes can operate on either left- or right-handed supercoils while prokaryotic enzymes only work on negative supercoils. The products of topoisomerase I action on plasmid DNA can be observed by gel electrophoresis, and show a ladder of bands, each corresponding to unit change in W_r as the supercoils are unwound, half at a time (Fig. 2.28).

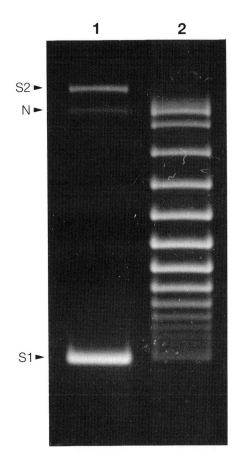

Fig. 2.28 Topoisomers of plasmid pAT153 after incubation with topoisomerase I to produce partial relaxation. Electrophoresis in a 1 per cent agarose gel: Track 1 shows native supercoiled pAT153 (S1), supercoiled dimer (S2), and nicked circular DNA (N); Track 2 shows products of topoisomerase I where ΔL_k up to 14 can be seen clearly (from Lilley, D. M. J. (1986). *Symp. Soc. Gen. Microbiol.*, **39**, 105–17).

By contrast, Class II topoisomerases use a 'double-strand passage' mechanism to effect unit change in the number of supercoils, $\Delta W_r = \pm 2$, and such prokaryotic enzymes can drive the endothermic supercoiling of DNA by coupling the reaction to hydrolysis of ATP. These topoisomerases cleave two phosphate esters to produce an enzyme-bridged gap in both strands. The other DNA duplex is passed through the gap (using energy provided by hydrolysis of ATP), and the gap is resealed. DNA gyrase from *E. coli* is a special example of the Class II enzyme. It is an A_2B_2 tetramer with the energy-free topoisomerase activity of the A subunit being inhibited by quinolone antibiotics such as nalidixic acid. The energy-transducing activity of the B subunit can be inhibited by novobiocin and other coumarin antibiotics. We should point out that such topoisomerases also operate on linear DNA that is torsionally stressed by other processes, most notably at the replication fork in eukaryotic DNA.

Supercoiling is important for a growing range of enzymes as illustrated by two examples. RNA polymerase *in vitro* appears to work ten times faster on supercoiled DNA, $\sigma = 0.06$, than on relaxed DNA, and this phenomenon appears to be related to the enhanced binding of the polymerase to the promoter sequence. Secondly, the *tyrT* promoter in *E. coli* is expressed *in vitro* at least 100 times stronger for supercoiled than for relaxed DNA and this behaviour seems linked to 'preactivation' of the DNA promoter region by negative supercoiling (Section 5.3.13).

Catenated and knotted DNA circles

While Type II topoisomerases usually only effect passage of a duplex from the same molecule through the separated double strands, they can also manipulate a duplex from a second molecule. As a result, two different DNA circles can be interlinked with the formation of a **catenane** (Fig. 2.29). Such catenanes have been identified by electron microscopy and can be artificially generated in high yield from mammalian mitochondria. **Knotted DNA** circles are another unusual topoisomer species which are also formed by intramolecular double-strand passage from an incompletely unwound duplex (Fig. 2.29).

2.3.6 Triple-stranded DNA

Triple helices were first observed for oligoribonucleotides in 1957. A decade later, the same phenomenon was observed for poly(dCT) binding to poly(dGA)·poly(dCT) and for poly(dG) binding to poly(dG)·poly(dC). Oligonucleotides can bind in the major groove of B-form DNA by forming Hoogsteen or reversed Hoogsteen hydrogen bonds using N-7 of the purine bases of the Watson–Crick base-pairs (Fig. 2.30). The resulting base-triplets form the core of a triple helix. In theory, G can form a base-triple with a C·G pair and A with a T·A pair but the only combinations that have isomorphous location of their C-1$'$ atoms are the two triplets TxA·T and C$^+$xG·G, where C$^+$ is the N-3 protonated form of cytosine. This means that the three strands of triple-helical DNA are normally two homopyrimidines and one homopurine. However, despite the backbone distortion that must result from the heteromorphism of other base triplet combinations, oligonucleotides containing G and T, G and A, or G, T, and C have been shown to form helices.

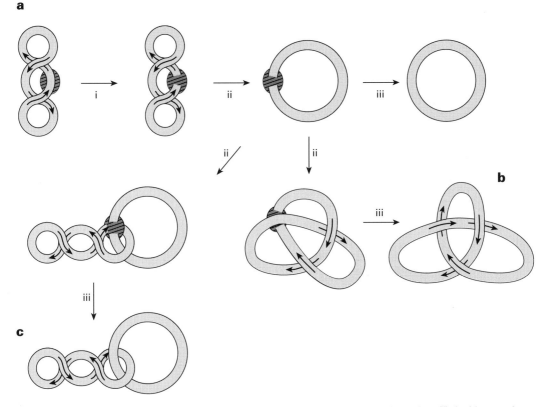

Fig. 2.29 Action of topoisomerase II (red) on singly supercoiled DNA: (i) double-strand opening; (ii) double-strand passage; (iii) resealing to give (**a**) relaxed circle, (**b**) knotted circle, and (**c**) catenated DNA circles.

Intermolecular triple helices are now well characterized for short oligonucleotides binding in the groove of a longer DNA duplex. H-DNA provides an example of an intramolecular triple helix because it has a mirror-repeat sequence relating homopurine and homopyrimidine tracts in a circular double-stranded DNA molecule and triplex formation is driven by supercoiling.

Several studies on third-strand binding to a homopurine·homopyrimidine duplex have established the following features:

- a third homopyrimidine strand binds parallel to the homopurine strand using Hoogsteen hydrogen bonds (i.e. the two homopyrimidine strands are antiparallel)
- a third homopurine strand binds antiparallel to the original homopurine strand using reversed Hoogsteen hydrogen bonds
- the bases in the third strand have the regular *anti* conformation of the glycosylic bond
- synthetic oligodeoxynucleotides having an α-glycosylic linkage also bind as a third strand, parallel for poly(d-α-T) and antiparallel for poly(d-α-TC).

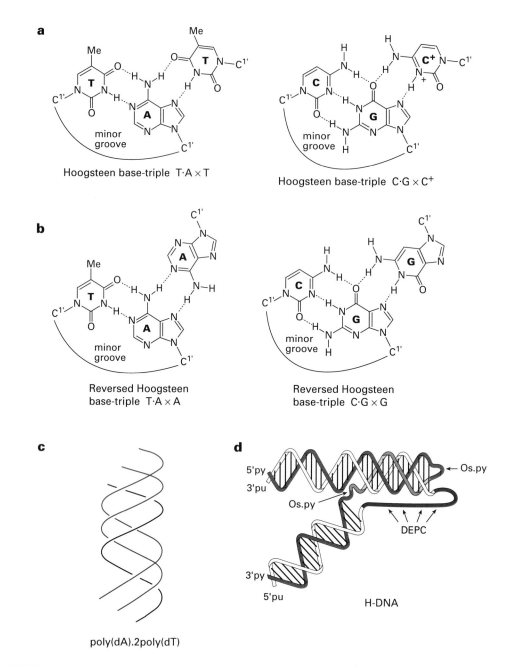

Fig. 2.30 Base-triplets formed by (**a**) Hoogsteen bonding for T·A × T and C·G × C⁺ and (**b**) reversed Hoogsteen bonding for G×G·C and A×A·T. (**c**) Model of triple-helical DNA based on fibre diffraction of poly(dA)·2poly(dT) (kindly provided by courtesy of Professor C. Hélène). (**d**) A schematic representation of H-DNA showing loci for attack by single-strand reagents.

Triple helices are less stable than duplexes. Thermodynamic parameters have been obtained from melting curves, from kinetics, and from the use of differential scanning calorimetry (DSC). With this last technique, values of $\Delta H° = 22 \pm 2$ kJ mol^{-1} and $\Delta S° = 70 \pm 7$ J mol^{-1} K^{-1} have been found for d(C$^+$TTC$^+$C$^+$TC$^+$C$^+$TC$^+$T). The pK_a value of cytidine in isolation is 4.3, but it is higher in oligonucleotides because of their polyanionic phosphate backbone. So it is to be expected that the stability of triple helices is seen to decrease as the pH rises above 5.

Triple helix stability can be enhanced by the use of modified nucleotides. 5-Methylcytosine increases stability at neutral pH, probably by a hydrophobic effect, and 5-bromouracil can usefully replace thymine. Oligoribonucleotides bind more strongly than do deoxyribonucleotides and 2′-O-methylribonucleotides bind even better. Finally, Hélène has shown that the attachment of an intercalating agent to the 5′- (or 3′-end) of the third strand can greatly enhance the stability of the triple-stranded helix.

The major application of triple helices relates to the specificity of the interaction between the single strand and a much larger DNA duplex. This is because homopyrimidines have been identified as potential vehicles for the sequence-specific delivery of agents that can modify DNA and thereby control genes. The DNA of the bacterium *E. coli* has 4.5 Mbp, so the minimum number of base-pairs needed to define a unique sequence in its genome is 11 bp (i.e. $4^{11} = 4\ 194\ 304$ assuming a statistically random distribution of the four bases). The corresponding number for the human genome is about 17 bp. Thus a synthetic 17-mer could be expected to identify and bind to a unique human DNA target and thus deliver a lethal agent to a specific sequence of DNA. In practice, the energetics of mismatched base-triples is complex and depends on nearest neighbours, metal ions, and other parameters. However, a value of about 1.5 kJ mol^{-1} per mismatch seems to fit much of the data and suggests that the specificity of triple helices is at least as good as that of double-helical complexes.

H-DNA

A new polymorph of DNA was discovered in 1985 within a sequence d(A-G)$_{16}$ in the polypurine strand of a recombinant plasmid pEJ4. Its requirement for protons led to the name H-DNA (half of its C residues are protonated, so the transition depends on acid pH as well as on a degree of negative supercoiling). Probes for single-stranded regions of DNA (especially osmium tetroxide:pyridine and nuclease P1 cleavage) were used to identify specific sites and provide experimental support for the model advanced earlier for a triple helical H-DNA (Fig. 2.30d). This has a Watson–Crick duplex which extends to the centre of the (dT-dC)$_n$·(dG-dA)$_n$ tract and the second half of the homopyrimidine tract then folds back on itself, antiparallel to the first half (see above) and winding down the major groove of the helix. The second half of the polypurine tract also folds back, probably in an unstructured single-stranded form. The energetics of nucleation of H-DNA suggest that it requires at least 15 bp for stability and the consequent loss of twist makes H-DNA favoured by negative supercoiling.

Although antibodies have been raised to detect triple-stranded structures, no evidence has yet been found for their existence in cells *in vivo.*

Summary

While there is a continuum of structures linking the A- and B-forms of DNA, there is a discontinuity between B- and Z-DNA. The B–Z transition calls for a major conformational reorganization for long DNA strands.

Irregular structures for DNA range in scale from local conformational perturbations of the flexible B-form, through isomers interchanged by unpairing and re-pairing of bases, to topological isomers such as catenanes and knots which can only be interconverted with single circles by the cleavage of both DNA strands.

Sequence-dependent modulation of DNA structure results largely from propeller twist of base-pairs which optimizes base stacking between adjacent pairs. This results in purine clashes between neighbouring bases in opposite strands which are relieved by local conformational adjustments in the positions of the bases. The energetic requirements of stacking interactions are readily accommodated by changes in the backbone conformation, which are of relatively low energy.

Mismatched base-pairs also cause local structural irregularities which can be accommodated by 'wobble' base-pairing for A·C and G·T pairs or by more extensive changes for G·A pairs.

DNA shows bends at the junctions between A-form and B-form helices. Bending can also occur at Py-C-A-Pu sequences as a result of rolling of base-pairs under the influence of local forces, such as protein binding. Curved DNA is a result of the alternations of linear poly(dA) tracts with intrinsically curved segments of general sequence whose curvature results from DNA writhing.

The topological behaviour of closed circles of DNA involves superhelices whose energy can be used, *inter alia*, to drive the formation of cruciform structures. Topological isomers have a clear biological role and can be interconverted by topoisomerase enzymes.

Triple-stranded DNA can be generated intermolecularly or intramolecularly. It is stable at low pH and has a second polypyrimidine strand wound in the major groove of B-DNA where it forms Hoogsteen base-triples with a complementary polypurine strand.

2.4 Structures of RNA species

As with DNA, studies on RNA structure began with its primary structure. This quest was pursued in parallel with that of DNA, but had to deal with the extra complexity of the 2'-hydroxyl group in ribonucleosides. Today, we also recognize that RNA has greater structural versatility than DNA in the variety of its species, in its diversity of conformations, and in its chemical reactivity. Different natural RNAs can either form long, double-stranded structures or adopt a globular shape composed of short duplex domains connected by single-stranded segments. Watson–Crick base-pairing seems to be the norm, though tRNA structures have provided a rich source of unusual base-pairs and base-triplets (Section 6.3.2). In general, it is now possible to predict double-helical sections by computer analysis of primary sequence data, and this technique has been used extensively to identify secondary structural components of ribosomal RNA and viral RNA species. In this section, we shall focus attention mainly on regular RNA secondary structure.

2.4.1 Primary structure of RNA

The first degradation studies of RNA using mild alkaline hydrolysis gave a mixture of mononucleotides, originally thought to have only four components—one for each base, A, C, G, and U. However, Waldo Cohn used ion-exchange chromatography to separate each of these four into pairs of isomers which were identified as the ribonucleoside 2'- and 3'-phosphates. This duplicity was overcome by use of a phosphodiesterase isolated from spleen tissue which digests RNA from its 5'-end (Section 9.4.3) to give the four 3'-phosphates, Ap, Cp, Gp, and Up, whilst an internal diesterase (snake venom phosphodiesterase was used later) cleaved RNA to the four 5'-phosphates, pA, pC, pG, and pU. It follows that RNA chains are made up of nucleotides which have $3' \rightarrow 5'$-phosphodiester linkages just like DNA (Fig. 2.31).

   The $3' \rightarrow 5'$ linkage in RNA is, in fact, thermodynamically less stable than the 'unnatural' $2' \rightarrow 5'$ linkage, which might therefore have had an evolutionary role. A rare example of such a polymer is produced in vertebrate cells in response to viral infection. Such cells make a glycoprotein called **interferon** which stimulates the production of an oligonucleotide synthetase. This polymerizes ATP to give oligoadenylates with $2' \rightarrow 5'$ phosphodiester linkages and from 3 to 8 nucleotides long. Such $(2' \rightarrow 5')(A)_n$ (Fig. 2.32) then activates an interferon-induced ribonuclease, RNase L, whose function seems to be to break down the viral messenger RNA. (Note also the $2' \rightarrow 5'$ ester linkage is a key feature of self-splicing RNA (Section 6.5).)

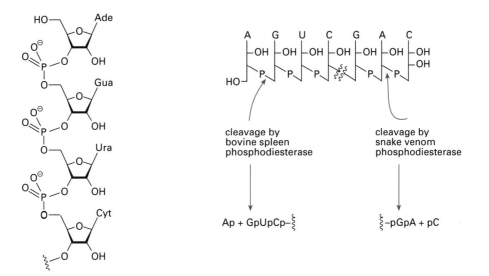

Fig. 2.31 The primary structure of RNA (left) and cleavage patterns with spleen (centre) and snake venom (right) phosphodiesterases.

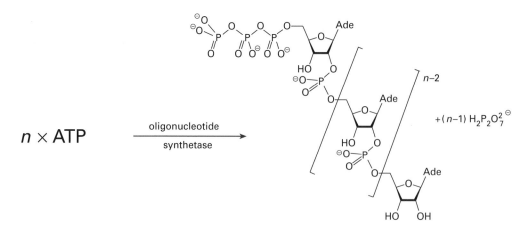

Fig. 2.32 Structure and formation of interferon-induced $(2' \rightarrow 5')(\mathbf{A})_n$.

2.4.2 Secondary structure of RNA: A-RNA and A'-RNA

Two varieties of A-type helices have been observed for fibres of RNA species such as poly(rA)·poly(rU). At low ionic strength, **A-RNA** has 11 bp per turn in a right-handed, anti-parallel double-helix. The sugars adopt a $C^{3'}$-*endo* pucker and the other geometric parameters are all very similar to those for A-DNA (see Tables 2.3 and 2.4). If the salt concentration is raised above 20 per cent, an A'-RNA form is observed which has 12 bp per turn of the duplex. Both structures have typical Watson–Crick base-pairs, which are displaced 4.4 Å from the helix axis and so form a very deep major groove and a rather shallow minor groove.

These features have been confirmed by the analysis of the crystal structure of the oligomer, r(UUAUAUAUAUAUAA). This 14-mer can be treated as three segments of A-helix separated by kinks in the sugar-phosphate backbone, which perturb the major groove dimensions. It is noteworthy that the 2'-hydroxyl groups are prominent at the edges of the relatively open minor groove, where they can be recognized by proteins, and they do not appear to have a structural role (Fig. 2.33). This calls into question the long-held opinion that it is the 2'-OH group that hinders the formation of a B-type helix for RNA. A more recent view of the non-existence of B-RNA is that A·U base-pairs in RNA can freely adopt the positive roll and negative slide that favour an A-form helix (Table 2.4). By contrast, for the equivalent AX/XT steps in DNA, clashing of the thymine methyl group with its 5'-base neighbour prevents positive roll and negative slide so that the B-form helix is preferred.

In one of the first NMR studies of an RNA duplex, Gronenborn and Clore combined two-dimensional NOE analysis with molecular dynamics to identify an A-RNA solution structure for the hexaribonucleotide, 5'-r(GCAUGC)$_2$. It shows sequence-dependent variations in helix parameters, particularly in helix twist and in base-pair roll, slide, and propeller twist. The extent of variation from base to base is much less than for the corresponding DNA hexanucleotide and seems to be dominated by the needs of the structure to achieve very nearly optimal base-stacking. This picture supports experimental studies which indicate that base stacking and hydrogen bonding are equally important as determinants of RNA helix stability.

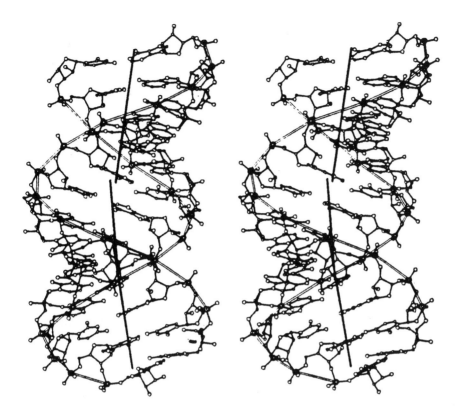

Fig. 2.33 3-D structure of r(UUAUAUAUAUAUAA). This structure can be considered as three segments of A-helix and the three optimal axes are shown.

Antisense RNA is defined as a short RNA transcript that lacks coding capacity, but has a high degree of complementarity to another RNA which enables the two to hybridize. The consequence is that such antisense, or complementary, RNA can act as a repressor of the normal function or expression of the targeted RNA. Such species have been detected in prokaryotic cells with suggested functions concerning RNA-primed replication of plasmid DNA, transcription of bacterial genes, and messenger translation in bacteria and bacteriophages. Quite clearly, such regulation of gene expression depends on the integrity of RNA duplexes.

2.4.3 RNA·DNA duplexes

Helices which have one strand of RNA and one of DNA are very important species in biology.

- They are formed when reverse transcriptase makes a DNA complement to the viral RNA.
- They occur when RNA polymerase transcribes DNA into complementary messenger RNA.
- They are a feature in DNA replication of the short primer sequences in Okazaki fragments (Section 5.4).

- Antisense DNA is a single-stranded oligodeoxynucleotide designed to bind to a short complementary segment of a target nucleic acid (RNA or single-stranded DNA) with the potential for regulation of gene expression.

Such hybrids are formed *in vitro* by annealing together two strands with complementary sequences, such as poly(rA)·poly(dT) and poly(rI)·poly(dC). These two heteroduplexes adopt the A-conformation common to RNA and DNA, the former giving an 11-fold helix typical of A-RNA and the latter a 12-fold helix characteristic of A′-RNA.

A self-complementary decamer r(GCG)d(TATACGC), also generates a hybrid duplex with Watson–Crick base-pairs. It has a helix rotation of 33° with a step-rise of 2.6 Å and $C^{3'}$-*endo* sugar pucker typical of A-DNA and A-RNA (see Table 2.1). Such hybrids appear to be more stable to thermal denaturation than their DNA·DNA counterparts.

The greater stability of RNA·DNA heteroduplexes over DNA·DNA homoduplexes is the basis of the construction of **antisense DNA oligomers**. These are intended to enter the cell where they can pair with, and so inactivate, complementary mRNA sequences. Additional desirable features such as membrane permeability and resistance to enzymic degradation have focused attention on oligodeoxynucleotides either having phosphorothioates, phosphorodithioates, or methylphosphonates replacing the anionic phosphate diester linkages (Section 3.4.6). In most cases, the resulting heteroduplexes have unfortunately proved to have weaker association constants than the natural DNA·RNA duplexes. An exception is the oligo-(2′-O-methyl)-ribonucleotide phosphorothioates which have higher duplex stability than the DNA·RNA parent and are resistant to phosphodiesterase action. However, the potential benefits that can accrue from silencing a particular RNA target are substantial and early successes have stimulated the synthesis and duplex stability of a broad range of antisense oligomers designed to complement an RNA target (Table 2.6).

The subtle differences in conformation between an RNA·DNA hybrid duplex and either DNA·DNA or RNA·RNA duplexes have significance for enzyme action and also for antisense therapy. The therapeutic objective of antisense oligodeoxynucleotides very much depends on their ability to create a duplex with the target RNA and thus make it a substrate for ribonuclease H (Table 2.6; also see Section 9.2.4). Because RNase H cleaves DNA·RNA hybrids but does not cleave the corresponding RNA·RNA duplexes, it can be induced to degrade an endogeneous mRNA species through hybridization with a synthetic antisense oligodeoxyribonucleotide.

Early X-ray structures of crystals of duplexes having DNA and RNA residues in both strands showed them to have pure A-form geometry (see above). The recent advent of efficient chemical synthesis of oligoribonucleotides (Section 3.5) has made possible studies using pure hybrid DNA·RNA species, especially by means of NMR in solutions. The hybrid duplexes d(GTCACATG)·r(CAUGUGAC) and of d(GTGAACCTT)·r(AAGUUCAC) have been analysed by two-dimensional NOE NMR in solution and shown to have neither pure A-form nor pure B-form structure. The sugars of the RNA strands have the regular 3′-*endo* conformation but those in the DNA strand have a novel, intermediate 4′-*endo* conformation. Glycosylic torsion angles in the DNA chain are typical of B-form (near −120°) but those in the RNA chain are typically A-form values (near −140°). Overall, the global structure is that of an A-form helix in

Table 2.6 Properties of antisense oligonucleotides and their analogues

Oligonucleotide type	Duplex stability[a]	Nuclease resistance[b]	RNase H activation[c]
Oligodeoxyribonucleotide (phosphate)	par[a]	— —	yes
Oligodeoxyribonucleotide phosphorothioate	lower	+	yes
Oligodeoxyribonucleotide selenophosphate	lower	+	not known
Oligodeoxyribonucleotide phosphoramidate	lower	+++	no
Oligoribonucleotide (phosphate)	higher	— —	no
Oligoribonucleotide phosphorothioate	higher	+	no
Oligo(2′-*O*-Me)ribonucleotide (phosphate)	higher	+	no
Oligo(2′-*O*-Me)ribonucleotide phosphorothioate	higher	++	no
Oligodeoxyribonucleotide methylphosphonate	lower	+++	no

[a] Compared to DNA·RNA stability under physiological conditions.
[b] Compared to DNA (phosphodiesterase digestion).
[c] Activation of RNase H by the duplex formed between the oligonucleotide and RNA.

which the base-pairs have the small rise and positive inclination typical of an A-form duplex (Fig. 2.18a,d). However, the width of the minor groove appears to be intermediate between A-form and B-form duplexes and such structures have been modelled into the active site of RNase H. The results suggest that additional interactions of the protein with the DNA strand are possible only for this intermediate hybrid duplex conformation but not for an RNA · RNA duplex. So, it seems possible that these subtle changes in nucleotide conformation may explain the selectivity of RNase H for hybrid DNA·RNA duplexes.

2.4.4 RNA bulges, hairpins, and loops

The functional diversity of RNA species reflects the diversity in three-dimensional structure. Several structural elements have been identified that make up folded RNA and their thermodynamic stabilities relative to the unfolded single strand have been evaluated. The folded conformations are largely stabilized by antiparallel double-stranded helical regions, in which intrastrand and interstrand base-stacking and hydrogen bonding provide most of the stabilization. Base-paired regions are separated by regions of unpaired bases, either as various types of loop or as single strands, as illustrated for a 55-nucleotide fragment from R17 virus (Fig. 2.34).

Hairpin loops were first identified as components of tRNA structures (Section 6.3.2) where they contain many bases. In the secondary structure deduced for 16S ribosomal RNA, most of the loops have four unpaired bases and these are known as **tetra-loops** (Section 6.3.1). Smaller tri-loops of three bases can also be formed.

Nuclear magnetic resonance studies on such stable tetra-loop hairpins show that their stems have A-form geometry while the loops have additional, unusual hydrogen bonding

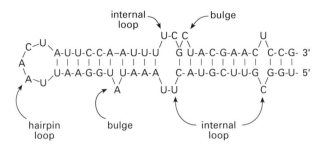

Fig. 2.34 A possible secondary structure for a 55-nucleotide fragment from R17 virus which illustrates hairpin loop, interior loop, and bulge structures. The free energy of this structure has been calculated to have a net $\Delta G°$ of -90 kJ mol^{-1} using appropriate values for base-pairs (Table 2.6) and for loops and bulges.

and base-pair interactions (Section 6.3.1). For example, the GAAA loop has the unusual G·A base-pair and UUYG loops have a reverse wobble U·G base-pair (Figure 6.5). As a result, simple models appear to be inadequate to describe RNA hairpin stem-loop structures. The nonanucleotide r(CGCUUUGCG) forms a stable tri-loop hairpin whose thermodynamic stability has been determined by analysis of T_m curves to be -101 kJ mol^{-1} for $\Delta H°$ and is close to the calculated value (-95 kJ mol^{-1}) for an RNA helix with three C·G base-pairs (see below). Nuclear magnetic resonance analysis shows that the loop has an A-form stem and the chain reversal appears between residues U5 and U6. The three uridine residues in the tri-loop have the 2'-*endo* conformation and show partial base-stacking, notably involving the first U on the 5'-side of the loop. These very high-resolution NMR results give a structure different from those structures computed by restrained molecular dynamics (Section 10.5.1), which shows that further refinement of the computational model is needed.

The hairpin loop is not only an important and stable component of secondary structure, it is a key functional element in a number of well characterized RNA systems. For example, it is required in the RNA TAR region of HIV (human immunodeficiency virus) for transactivation by the tat protein, and several viral coat proteins bind to specific hairpin loop structures.

Bulges are formed when there is an excess of residues on one side of a duplex. For single-base bulges, the extra base can either stack into the duplex, as in the case of an adenine bulge in the coat protein binding site of R17 phage, or be looped out, as shown by NMR studies in uracil bulges in duplexes. Such bulges can provide high affinity sites for intercalators such as ethidium bromide. In general, it appears that a bulge of one or two nucleotides has three effects on structure: (1) it distorts the stacking of bases in the duplex, (2) induces a bend in the RNA, (3) reduces the stability of the helix, and (4) increases the major groove accessibility at base-pairs flanking the bulge.

Internal loops occur where there is non-Watson–Crick apposition of bases. They can involve one or two base-pairs with pyrimidine–pyrimidine opposition (as in Fig. 2.34) or of mismatched purine–purine or pyrimidine–pyrimidine pairs, of which G·A pairs can form a mismatched base-pair compatible with an A-form helix. There are also many examples of larger internal loops. Some of those which are rich in purine residues have been implicated as protein recognition sites. Many of these larger loops show marked resistance to chemical reagents specific for single-strand residues and this, in combination with NMR data, suggests that there is probably a high level of order in such loops, notably of base-stacking and base-triples. One general opinion is that the major differences between loop and stem regions are dynamic rather than structural.

Junctions are regions which connect three or more stems (the connecting region for two stems is an internal loop) and are a common feature of computer-generated secondary structures for large RNAs. A prime example is the four-stem junction in the cloverleaf structure of tRNAs in which stacking continuity between the acceptor and T stems and between the anticodon and D stems is maintained (Section 6.3.2). A junction of three stems forms the hammerhead structure of self-cleaving RNA (Section 6.5.5) and junctions of up to five stems have been observed for 16S RNA.

Thermodynamics of secondary structure elements

The free energy of an RNA conformation has to take account of the contributions of interactions between bases, sugars, phosphates, ions, and solvent. The most reliable parameters are those derived experimentally from the T_m profiles of double-helical regions of RNA and data for each of the 10 nearest-neighbour sequences are given in Table 2.7. They are accurate enough to predict the expected thermodynamic behaviour of any RNA duplex to within about 10 per cent of its experimental value.

Other structural features are less easy to predict. It is clear that stacking interactions are more important than base-pairing so that an odd purine nucleotide 'dangling' at the 3'-end of a stem can contribute some -4 kJ mol^{-1} to the stability of the adjacent duplex. The energies for mispairs or loops are rather less accurate, but always destabilizing, and change with the size of the loop (Table 2.8). Energies of these irregular secondary structures also depend on base composition, for example a single base bulge for uridine costs about $+8$ kJ mol^{-1} and for guanosine about $+14$ kJ mol^{-1}.

Using such data, the prediction of secondary structure is a conceptually simple task that can be handled by a modest computer while the more advanced programmes search suboptimal structures as well as that of lowest free energy. Several examples of such predicted structures are described in Section 6.3.

Interactions between separated regions of secondary structure are defined as tertiary interactions. One example is that of **pseudoknots**, which involve base-pairing between one strand of an internal loop and a distant single-strand region (Section 6.3.3). Pseudoknots can also involve base-pairing between components of two separate hairpin loops and examples with 3–8 bp have been described as a result of NMR analysis. However, the computer prediction of tertiary interactions and base-triples appears to be still beyond the scope of present methodology.

2.4.5 Triple-stranded RNAs

The first triple-stranded nucleic acid was described in 1957 when poly(rU)·poly(rA) was found to form a stable 2 : 1 complex in the presence of magnesium chloride. The extra poly(rU) strand

Table 2.7 Thermodynamic parameters for RNA helix initiation and propagation in 1 M NaCl

Propagation sequence	$\Delta H°$ kJ mol⁻¹	$\Delta S°$ J K⁻¹ mol⁻¹	$\Delta G°$ kJ mol⁻¹	Propagation sequence	$\Delta H°$ kJ mol⁻¹	$\Delta S°$ J K⁻¹ mol⁻¹	$\Delta G°$ kJ mol⁻¹
A·U→ / ↑ A·U→	−27.7	−77.3	−3.8	A·U→ / ↑ G·C→	−55.8	−149	−9.6
U·A→ / ↑ A·U→	−23.9	−65.1	−3.8	U·A→ / ↑ G·C→	−42.8	−110	−8.8
A·U→ / ↑ U·A→	−34.1	−94.9	−4.6	G·C→ / ↑ C·G→	−33.6	−81.5	−8.4
A·U→ / ↑ C·G→	−44.1	−117	−7.6	C·G→ / ↑ G·C→	−59.6	−147	−14.7
U·A→ / ↑ C·G→	−31.9	−80.6	−7.1	G·C→ / ↑ G·C→	−51.2	−125	−12.4
Initiation	(0)	−45.4	14.6				
Symmetry correction (self-complementary)	0	−5.9	1.7	Symmetry correction (non-self-complementary)	0	0	0

Arrows point in a 5′ → 3′ direction to designate the stacking of adjacent base-pairs.
The enthalpy change for helix initiation is assumed to be zero.

Table 2.8 Free energy increments for loops (kJ mol^{-1} in 1 M NaCl, 37°C)

Loop size	Internal loop	Bulge loop	Hairpin loop
1	—	+14	—
2	+4	+22	—
3	+5.4	+25	+31
4	+7.1	+28	+25
5	+8.8	+31	+18.5
6	+10.5	+34.5	+18

is parallel to the poly(rA) strand and forms Hoogsteen base-triples in the major groove of an A-form Watson–Crick helix. Triplexes of 2poly(rA)·poly(rU) can also be formed, while poly(rC) can form a triplex with poly(rG) at pH 6 which has two cytidines per guanine, one of them being protonated to give the (C$^+$xG·C) base-triple also seen for triple-helical DNA (Fig. 2.34). Base-triples are also a very common feature of tRNA structure (Section 6.4).

The importance of added cations to overcome the repulsion between the anionic chains of the Watson–Crick duplex and the polypyrimidine third strand is an essential feature of triple-helix formation. Co^{3+}(NH$_3$)$_6$ and spermine are also effective counter-ions as well as the more usual Mg^{2+}.

Poly(rG), as well as guanosine and GMP, can form structures with four equivalent hydrogen bonded bases in a plane, with all four strands parallel. It is not clear whether this structure has any relevance to RNA folding.

Summary

RNA primary structure is built on the regular 3′→5′ phosphodiester linkage, as for DNA. Oligoadenylates with a 2′→5′ linkage are generated in mammalian cells in response to viral infection.

While RNA species vary in size from 65 nucleotides upwards, their helical structures are restricted to A-form duplexes with 11–12 residues per turn. Sequence-dependent modulation of structure is rather less marked than for B-DNA.

Short strands of RNA can form hairpin loops with a preference for 4 > 3 bases in a loop built on a stem of A-form RNA. Reliable thermodynamic data for the free energies of base stacking, of hairpin loops, of bulge loops, and of interior loops permit the computation of possible stable secondary structures for single-stranded RNA species.

Heteroduplexes with one RNA and one DNA strand can be formed and have higher melting points than the equivalent DNA·DNA homoduplex. RNA·DNA duplexes have a structure that is globally an A-form helix but with modified conformation for the sugars in the DNA strand. This leads to a minor groove that is wider than for A-RNA but narrower than B-DNA, which may explain the selective activity of RNase H for the heteroduplex.

Triple-stranded nucleic acids use Hoogsteen base-pairing to add a polypyrimidine third strand to a polypurine·polypyrimidine duplex. This requires protonation of one cytosine in each base-triplet and the two pyrimidine strands must run antiparallel.

2.5 Dynamics of nucleic acid structures

Any over-emphasis on the stable structures of nucleic acids runs the risk of playing down the dynamic activity of nucleic acids which is intrinsic to their function. Pairing and unpairing, breathing, and winding are integral features of the behaviour of these species.

Established studies on structural transitions of nucleic acids have for a long time used classical physical methods which include light absorption, NMR spectroscopy, ultracentrifugation, viscometry, and X-ray diffraction (Sections 10.2 and 10.3). More recently, these techniques have been augmented by a range of powerful computational methods (Section 10.5). In each case, the choice of experiment is linked to the time-scale and amplitude of the molecular motion under investigation.

2.5.1 Helix–coil transitions of duplexes

Double helices have a lower molecular absorptivity for UV light than would be predicted from the sum of their constituent bases. This **hypochromicity** is usually measured at 259 nm, while C·G base-pairs can also be monitored at 280 nm. It results from coupling of the transition dipoles between neighbouring stacked bases and is larger in amplitude for A·U and A·T pairs than for C·G pairs. As a result, the UV absorption of a DNA duplex **increases** typically by 20–30 per cent when it is denatured. This transition from a helix to an unstacked, strand-separated coil has a strong entropic component and so is temperature dependent. The midpoint of this thermal transition is known as the **melting temperature** (T_m).

Such dissociation of nucleic acid helices in solution to give single-stranded DNA is a function of base composition, sequence, and chain length as well as of temperature, salt concentration, and pH of the solvent. In particular, early observations of the relationship between T_m and base composition for different DNAs showed that A·T pairs are less stable than C·G pairs, a fact which is now expressed in a linear correlation between T_m and the gross composition of a DNA polymer by the equation:

$$T_m = X + 0.41[\%(C + G)] \qquad /^{\circ}C$$

The constant X is dependent on salt concentration and pH and has a value of 69.3°C for 0.3 M sodium ions at pH 7 (Fig. 2.35).

A second consequence is that the steepness of the transition also depends on base sequence. Thus, melting curves for homopolymers have much sharper transitions than those for random-sequence polymers. This is because A·T-rich regions melt first to give unpaired regions which then extend gradually with rising temperature until, finally, even the pure C·G regions have melted (Fig. 2.36). In some cases, the shape of the melting curve can be analysed to identify several components of defined composition melting in series.

Because of end-effects, short homo-oligomers melt at lower temperatures and with broader transitions than longer homopolymers. For example for poly(rA)$_n$·poly(rU)$_n$, the octamer melts at 9°C, the undecamer at 20°C, and long oligomers at 49°C in the same sodium cacodylate buffer at pH 6.9. Consequently, in the design of synthetic, self-complementary duplexes for crystallization and X-ray structure determination, C·G pairs are often placed at the ends of

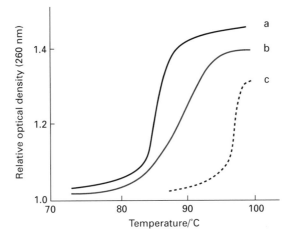

Fig. 2.35 Thermal denaturation of DNAs as a function of base composition (per cent G·C) for three species of bacteria: (**a**) *Pneumococcus* (38 per cent G·C); (**b**) *E. coli* (52 per cent G·C); (**c**) *M. phlei* (66 per cent G·C) (adapted from Marmur, J. and Doty, P. (1959). *Nature*, **183**, 1427–9. Copyright (1959) Macmillan Magazines Ltd).

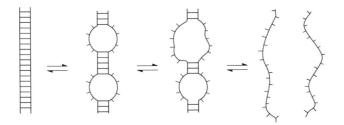

Fig. 2.36 Scheme illustrating the melting of A·T-rich regions (colour) followed by mixed regions, then by C·G-rich regions (black) with rise in temperature (left→right).

hexamers and octamers to stop them 'fraying'. Lastly, the marked dependence of T_m on salt concentration is seen for DNA from *Diplococcus pneumoniae* whose T_m rises from 70°C at 0.01 M KCl to 87°C for 0.1 M KCl and to 98°C at 1.0 M KCl.

Data from many melting profiles have been analysed to give a **stability matrix** for nearest neighbour stacking (Table 2.9). This can be used to predict T_m for a B-DNA polymer of known sequence with a general accuracy of 2–3°C.

The converse of melting is the **renaturation** of two separated complementary strands to form a correctly paired duplex. In practice, the melting curve for denaturation of DNA is reversible only for relatively short oligomers, where the rate determining process is the formation of a nucleation site of about 3 bp followed by rapid zipping-up of the strands and where there is no competition from other impeding processes.

When solutions of unpaired, complementary large nucleic acids are incubated at 10–20°C below their T_m, renaturation takes place over a period of time. For short DNAs of up to several hundred base-pairs, nucleation is rate-limiting at low concentrations and each duplex zips to completion almost instantly (>1000 bp s^{-1}). The nucleation process is bimolecular, so rena-

Table 2.9 Thermal stability matrix for nearest-neighbour stacking in base-paired dinucleotide fragments with B-DNA geometry

5' Neighbour	3' Neighbour			
	A	C	G	T
A	54.50	97.73	58.42	57.02
C	54.71	85.97	72.55	58.42
G	86.44	136.12	85.97	97.73
T	36.73	86.44	54.71	54.50

Numbers give T_m values in °C at 19.5 mM Na$^+$.

turation is concentration dependent with a rate constant around 106 M^{-1} s^{-1}. It is also dependent on the complexity of the single strands. Thus, for the simplest cases of homopolymers and of short heterogeneous oligonucleotides, nucleation sites will usually be fully extended by rapid zipping-up. This gives us an 'all-or-none' model for duplex formation. By contrast, for bacterial DNA each nucleation sequence is present only in very low concentration and the process of finding its correct complement will be slow. Lastly, in the case of eukaryotic DNA, the existence of repeated sequences means that locally viable nucleation sites will form and can be propagated to give relatively stable structures. These will not usually have the two strands in their correct overall register. Because such pairings become more stable as the temperature falls, complete renaturation may take an infinitely long time.

Longer nucleic acid strands are able to generate intrastrand hairpin loops, which optimally have about six bases in the loop and paired sections of variable length. They are formed by rapid, unimolecular processes which can be 100 times faster than the corresponding bimolecular pairing process. Although such hairpins are thermodynamically less stable than a correctly paired duplex, their existence retards the rate of renaturation so that propagation of the duplex is now the rate-limiting process (Fig. 2.37). One notable manifestation of this phenomenon is seen when a solution of melted DNA is quickly quenched to +4°C to give stable denatured DNA.

With longer DNA species, Britten and Kohne have shown that the rate of recombination, which is monitored by UV hypochromicity, can be used to estimate the size of DNA in a homogeneous sample. The time t for renaturation at a given temperature for DNA of single-strand concentration C and total concentration C_0 is related to the rate constant k for the process by an equation which in its simplest form is

$$C/C_0 = (1 + kC_0t)^{-1}$$

In practice, where C/C_0 is 0.5 the value of C_0t is closely related to the complexity of the DNA under investigation.

This **annealing** of two complementary strands has found many applications. For DNA oligomers, it provided a key component of Khorana's chemical synthesis of a gene (Section 3.5). It

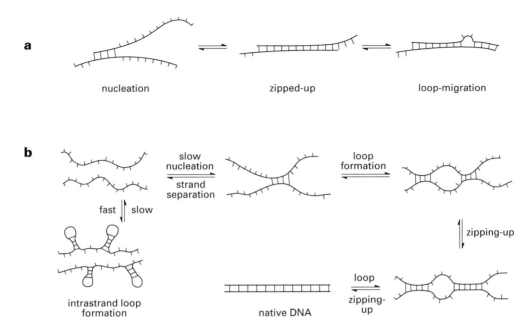

Fig. 2.37 Renaturation processes (**a**) for short oligonucleotide and longer homopolymers and (**b**) for natural DNA strands.

is now an integral feature of the insertion of chemically synthesized DNA into vectors. For RNA·DNA duplexes, it has provided a tool of fundamental importance for gene identification (Section 10.9.9) and is being explored in the applications of antisense DNA.

2.5.2 DNA breathing

Complete separation of two nucleic acid strands in the melting process is a relatively slow, long-range process that is not easily reversible. By contrast, the hydrogen bonds between base-pairs can be disrupted at temperatures well below the melting temperature to give local, short-range separation of the strands. This readily reversible process is known as **breathing.**

The evidence for such dynamic motion comes from chemical reactions which take place at atoms that are completely blocked by normal base-pairing. Those used include tritium exchange studies on hydrogen-bonded protons in base-pairs, the reactivity of formaldehyde with base NH groups, and NMR studies of imino-proton exchange with solvent water. This last technique can be used on a time scale from minutes down to 10 ms which shows that in linear DNA the base-pairs open singly and transiently with a lifetime around 10 ms at 15°C.

Because NMR can distinguish between imino- and amino-proton exchange it can also be used to identify breathing in specific sequences. Some of the most detailed work of this sort has come from studies on tRNA molecules which shows that, with increasing temperature, base-triplets (Fig. 2.30) are destabilized first followed by the ribothymidine helix and then the dihydrouridine helix. Finally the acceptor helix 'melts' after the anticodon helix (Section 6.3.2).

Another possible motion which might be important for the creation of intercalation sites is known as 'soliton excitation'. The concept here is of a stretching vibration of the DNA chain which travels like a wave along the helix axis until, given sufficient energy, it leads to local unstacking of adjacent bases with associated deformation of sugar pucker and other bond conformations.

Such pre-melting behaviour may well relate to the process of drug intercalation, to the association of single-strand specific DNA binding proteins (Section 9.2.2), and to the reaction of small electrophilic reagents with imino and amino groups such as cytosine-N^3 (Section 7.5).

2.5.3 Energetics of the B–Z transition

The isomerization equilibrium between the right-handed B-form and the left-handed Z-form of DNA is determined by three factors:

- chemical structure of the polynucleotide (sequence, modified bases)
- environmental conditions (solvent, pH, temperature, etc.)
- degree of topological stress (supercoiling, cruciform formation).

Many quantitative data have been obtained from spectroscopic, hydrodynamic, and calorimetric studies and linked to theoretical calculations. While these have not yet defined the kinetics or complex mechanisms of the B–Z transition, it is evident that the small transition enthalpies involved lie within the range of the thermal energies available from the environment. So, for example, the intrinsic free energy difference between the Z- and B-forms is close to 2 kJ mol^{-1} for polyd(G-C) base-pairs, only 1 kJ mol^{-1} for polyd(G-m^{5}C) base-pairs, and greater than 5 kJ mol^{-1} for polyd(A-T) base-pairs. It thus appears that local structural fluctuations may be key elements in the mediation of biological regulatory functions through the B–Z transition.

2.5.4 Rapid DNA motions

Rotations of single bonds, either alone or in combination, are responsible for a range of very rapid DNA motions with time-scales down to fractions of a nanosecond. For example, the twisting of base-pairs around the helix axis has a lifetime around 10^{-8} s while crankshaft rotations of the β, α, ζ, and ε C–O–P–O–C bonds (see Fig. 2.11) lead to an oscillation in the position of the phosphorus atom on a millisecond time-scale. Various calculations on the interconversion of 3'-endo and 2'-endo sugar pucker have given low activation energy barriers for their interconversion, in the range 3 to 20 kJ mol^{-1}, showing that these conformers are in rapid, although weighted, equilibrium at 37°C. Lastly, rapid fluctuations in propeller twist can result from oscillations of the glycosylic bond.

Summary

The dynamic motions of nucleic acids range from relatively high-energy processes involving breaking multiple hydrogen bonds to local conformational mobilities of very low activation energy.

Helix–coil transitions result from complete unpairing of the two strands of a duplex with rising temperature. The halfway stage is known as the melting temperature T_m and depends on the C·G content of the duplex, its length and sequence, and the ionic strength of the solution. One C·G pair is worth twice as much energetically as an A·T pair.

The coil–helix transition is fully reversible for short oligonucleotides and homopolymers, but can be an infinitely slow process for very long complementary DNA strands. This intermolecular process has to compete with rapid intramolecular formation of hairpin loops where there is a suitable base sequence.

Local and reversible unpairing of bases is known as breathing and can be monitored by chemical and spectroscopic methods at temperatures below T_m.

The best studied short-range motions are the interconversions of *syn* and *anti* glycosylic linkages and of the 3'-*endo* and 2'-*endo* sugar puckers and the crankshaft rotations of the C–O–P–O–C phosphate esters. These and other motions must be involved in the B–Z transition of left-handed into right-handed DNA but, though the energetics of this process have been measured, its detailed mechanism is still uncertain.

2.6 Higher-order DNA structures

The way in which eukaryotic DNA is packaged in the cell nucleus is one of the wonders of macromolecular structure. In general, higher organisms have more DNA than lower ones (Table 2.10) and this calls for correspondingly greater condensation of the double helix. Human cells contain a total of 7.8×10^9 bp, which corresponds to an extended length of about 2 m. The DNA is packed into 46 cylindrical chromosomes of total length 200 µm,

Table 2.10 Cellular DNA content of various species

Organism	Number of base-pairs	DNA length mm	Number of chromosomes
Escherichia coli	4×10^6	1.4	1
Yeast (*Saccharomyces cerevisiae*)	1.4×10^6	4.6	16
Fruit fly (*Drosophilia melanogaster*)	1.7×10^7	56.0	4
Humans (*Homo sapiens*)	3.9×10^9	990.0	23

Values are provided for haploid genomes.

which gives a net packaging ratio of about 10^4 for such metaphase human chromosomes. The overall process has been broken down into two stages: the formation of nucleosomes and the condensation of nucleosomes into chromatin.

2.6.1 Nucleosome structure

The first stage in the condensation of DNA is the **nucleosome**, whose core has been crystallized by Aaron Klug and John Finch and analysed by X-ray diffraction. The DNA duplex is wrapped around a block of eight histone proteins to give 1.75 turns of a left-handed super-helix (Chapter 9, Fig 9.4a). This process achieves a packing ratio of 7. The number of base-pairs involved in nucleosome structures varies from species to species, being 165 bp for yeast, 183 bp for HeLa cells, 196 bp for rat liver, and 241 bp for sea urchin sperm. Such nucleosomes are joined by linker DNA whose length ranges from 0 bp in neurons to 80 bp in sea urchin sperm but usually averages 30–40 bp. The details of packaging the histone proteins are discussed later (Section 9.2.1).

As the DNA winds around the nucleosome core, the major and minor grooves are compressed on the inside with complementary widening of the grooves on the outside of the curved duplex. Runs of A·T base-pairs, which have an intrinsically narrow minor groove, should be most favourably placed on the inside of the curved segment while runs of G·C base-pairs should be more favourably aligned with minor grooves facing outwards, where they are more accessible to enzyme cleavage. In practice, Drew and Travers measured the periodicities of A·T and C·G base-pairs by cleavage with DNase I and found them to be exactly out of phase and having a periodicity of 10.17 ± 0.5 bp. This result was later confirmed by hydroxyl radical cleavage, which avoids the steric constraints of DNase I.

2.6.2 Chromatin structure

Chromatin is too large and heterogeneous to yield its secrets to X-ray analysis, so electron microscopy is the chosen experimental probe. At intermediate salt concentration (~ 1 mM NaCl), the nucleosomes are revealed as 'beads on a string'. Spherical nucleosomes can be seen with a diameter of 7–10 nm joined by variable-length filaments, often about 14 nm long. If the salt concentration is increased to 0.1 M NaCl, the spacing filaments get shorter and a zig-zag arrangement of nucleosomes is seen in a fibre 10–11 nm wide (Chapter 9, Fig. 9.4b). At even higher salt concentration and in the presence of magnesium, these condense into a 30 nm diameter fibre, called a **solenoid**, which is thought to be either a right-handed or a left-handed helix made up of close-packed nucleosomes with a packing ratio of around 40 (see Fig. 9.4b).

For the further stages in DNA condensation, one of the models proposed suggests that loops of these 30 nm fibres, each containing about 50 solenoid turns and possibly wound in a supercoil, are attached to a central protein core from which they radiate outwards. Organization of these loops around a cylindrical scaffold could give rise to the observed miniband structure of chromosomes, which is some 0.84 µm in diameter and 30 nm in thickness. A

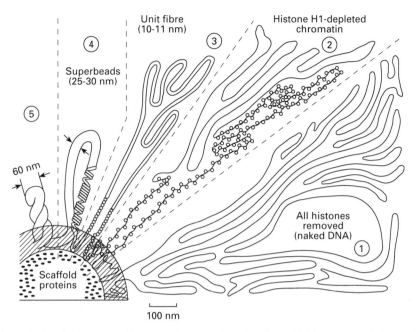

Fig. 2.38 Schematic drawing to illustrate the gradual organization of DNA into highly condensed chromatin. (1) DNA fixed to the protein scaffold; (2) DNA complexed with all histones except H1; (3) aggregation into 100 Å fibre; (4) formation of 'superbeads'; and (5) contraction into 600 Å knob (from Rindt, K.-P. and Nover, L. (1980). *Biol. Zentralblat.*, **99**, 641–73).

continuous helix of loops would then constitute the **chromosome**. These ideas are illustrated in a possible scheme (Fig. 2.38).

It is clear from all of the relevant biological experiments that the single DNA duplex has to be continuously accessible despite all this condensed structure in order for replication to take place. Some of the most exciting electron micrographs of DNA have been obtained from samples where the histones have been digested away leaving only the DNA as a tangled network of interwound superhelices radiating from a central nuclear region where the scaffold proteins remain intact (Fig. 2.39).

Even then, the most condensed packing of nucleic acid is found in the sperm cell. Here a series of arginine-rich proteins called protamines bind to DNA, probably with their α-helices in the major groove of the DNA where they neutralize the phosphate charge and so enable very tight packing of DNA duplexes.

Bacterial DNA is also condensed into a highly organized state (Section 9.2.1). In *E. coli*, the genome has 4400 kbp in a closed circle which is negatively supercoiled. It is condensed around histone-like proteins, HU and H1, to form a **nucleoid** and achieve a compaction of 1000-fold, which is followed by further condensation into supercoil domains. Unlike chromosomal DNA in eukaryotes, there is some additional negative supercoiling in prokaryotes that is not accounted for by protein binding. This is probably a consequence of the activity of the bacterial DNA gyrase, which is capable of actively introducing further negative supercoiling, driven by hydrolysis of ATP. This whole process differs in several respects from assembly of chromatin in eukaryotes:

Fig. 2.39 Electron micrograph of a histone-depleted chromosome showing that the DNA is attached to the scaffold in loops (from Paulson, J. K. and Laemmli, U. K. (1977). *Cell*, **12**, 817–28. Copyright (1977) Cell Press).

- there is no apparent regular repeating structure equivalent to the eukaryotic nucleosome, although short DNA segments of 60–120 bp are organized by means of their interaction with abundant DNA-binding proteins
- there is no prokaryotic equivalent to the solenoid structure
- bacterial DNA seems to be torsionally strained *in vivo* and organized into independently supercoiled domains of about 100 kbp.

The establishment of DNA architecture in the bacterial chromosome has progressed through the analysis of two types of structure. First, the interaction of a dimer of the HU protein from *B. stearothermophilus* shows loops of antiparallel β-sheets inserted into the DNA minor groove (Section 9.2.1) in non-sequence-specific binding. Secondly, a large nucleoprotein complex in bacteria is involved in integration of phage DNA into the host chromosome and is called an **intasome**. This has the phage DNA wrapped as a left-handed supercoil around a complex of proteins including several copies of two DNA-binding proteins, the phage-coded integrase, and the IHF protein (integration host factor). The IHF binds to a specific DNA sequence.

These developments suggest that structural analysis of the bacterial chromosome may well overtake that of eukaryotic systems.

Summary

DNA has to be packaged very efficiently to fit into the cell nucleus. For human chromosomes, the reduction in length is close to 10^4. In the nucleosome, some 180 bp of DNA wind round a core of a histone octamer in 1.75 turns of a left-handed superhelix. These are joined by linker DNA, whose length is about 30–40 bp, and condense into fibres of 10–11 nm diameter.

Models of higher-order structure have such fibres wound into a solenoid of 30 nm diameter and 11 nm pitch, giving a packing ratio of 40. How further stages of condensation give the eukaryotic chromosome is a subject for inspired modelling.

For prokaryotic systems, some of the details of the structure of the nucleoid are now emerging from X-ray analysis.

Further reading

2.1 and 2.2

Arscott, P. G., Lee, G., Bloomfield, V. A., and Evans, D. F. (1989). Scanning tunnelling microscopy of Z-DNA. *Nature*, **339**, 484–6.

Crothers, D. M., Gartenberg, M. R., and Shrader, T. E. (1992). DNA bending in protein–DNA complexes. *Progress in Nucleic Acids and Molecular Biology*, **208**, 118–45.

Dickerson, R. E. (1992). DNA structure from A to Z. *Adv. Enzymol.*, **211**, 67–111.

Dickerson, R. E., Drew, H. R., Connor, B. N., Wing, R. M., Fratini, A. V., and Kapka, M. L. (1982). The anatomy of A-, B-, and Z-DNA. *Science*, **216**, 475–85.

Johnson, B. H. (1992). Generation and detection of Z-DNA. *Progress in Nucleic Acids and Molecular Biology*, **211**, 127–57.

Jurmak, F. A. and McPherson, A. (ed.) (1984). *Biological macromolecules and assemblies*, Vol. 1. John Wiley, New York.

Saenger, W. (1984). *Principles of nucleic acid structure*. Springer-Verlag, New York.

2.3

Blommers, M. J. J., Walters, A. L. I., Haasnoot, C. A. G., Aelen, J. M. A., van der Marel, G. A., van Boom, J. H., and Hilbers, C. W. (1989). Effects of base sequence on the loop folding in DNA hairpins. *Biochemistry*, **28**, 7491–8.

Calladine, C. R., Drew, H. R., and McCall, M. J. (1988). The intrinsic curvature of DNA in solution. *J. Mol. Biol.*, **210**, 127–37.

Dickenson R. E., Goodsell D. S., and Neidle, S. A. (1994). "...The tyrrany of the lattice..." *Proc. Natl Acad. Sci. USA*, **91**, 3579–83.

Drew, H. R. and Travers, A. A. (1984). DNA structural variations in the *E. coli tyrT* promoter. *Cell*, **37**, 491–502.

Hunter, C. A. (1993). Sequence-dependent DNA structure, the role of base stacking interactions. *J. Mol. Biol.*, **230**, 1025–54.

Kennard, O. (1988). In *DNA and its drug complexes, structure and expression* (ed. M. H. Sarma and R. H. Sarma), Vol. 2. Adenine Press, New York.

Lilley, D. M., Sullivan, K. M., Murchie, A. I. H., and Furlong, J. C. (1988). Cruciform extrusion in supercoiled DNA—mechanisms and contextual influence. In *Unusual DNA structures* (ed. R. D. Wells and S. C. Harvey), Springer-Verlag, Heidelberg, pp. 55–72.

Liv, L. F. (1989). DNA topoisomerase poisons as antitumour drugs. *Ann. Rev. Biochem.*, **58**, 351–75.

Palacek, E. (1991). Local supercoil-stabilized DNA structures. *Crit. Revs. Biochem. Molec. Biol.*, **26**, 151–226.

Thuong, N. T. and Hélène, C. (1993). Sequence-specific recognition and modification of double-helical DNA by oligonucleotides. *Angew. Chem. Int. Ed. Engl.*, **32**, 666–90.

Wang, J. C. (1985). DNA topoisomerases. *Ann. Rev. Biochem.*, **54**, 665–97.

Wells, R. D. (1988). Unusual DNA structures. *J. Biol. Chem.*, **263**, 1095–8.

Yoon, C., Privé, G. G., Goodsell, D. S., and Dickersen, R. E. (1988). Structure of an alternating B-DNA helix and its relationship to A- tract DNA. *Proc. Natl. Acad. Sci. USA*, **85**, 6332–6.

2.4

Bates, A. D. and Maxwell, A. (1993). *DNA topology*. IRL Press, Oxford.

Cohen, J. S. (ed.) (1989). *Oligodeoxynucleotides, antisense inhibitors of gene expression*. Macmillan, London.

Clore, G. M., Oschkinat, H., McLaughlin, L. W., Benseler, F., Happ, C. S., Happ, E., and Gronenborn, A. M. (1988). Refinement of structure of the DNA dodecamer 5′d(CGCGPATTCGCG)$_2$ containing a stable purine–thymine base-pair: combined use of nuclear magnetic resonance and restrained molecular dynamics. *Biochemistry*, **27**, 4185–97.

Feroroff, O. Y., Salazar, M., and Reid, B. R. (1993). Structure of a DNA:RNA hybrid duplex. Why RNase H does not cleave pure RNA. *J. Mol. Biol.*, **233**, 509–23.

Freier, S. M., Kierzek, R., Jaeger, J. A., Sugimoto, N., Caruthers, M. H., Neilson, T., and Turner, D. H. (1986). Improved free energy parameters for predictions of RNA duplex stability. *Proc. Natl. Acad. Sci. USA*, **83**, 9373–7.

Happ, S. C., Happ, E., Nilges, M., Gronenborn, A. M., and Clore, G. M. (1988). Refinement of the solution structure of the ribonucleotide 5′r(GCAUGC)$_2$. *Biochemistry*, **27**, 1735–43.

Houba-He'rin, N. and Inouye, M. (1987). Antisense RNA. In *Nucleic acids and molecular biology* (ed. F. Eckstein and D. M. J. Lilley), Vol. 1. Springer-Verlag, Berlin, pp. 210–21.

Kennard, O. and Hunter, W. N. (1991). Single-crystal X-ray diffraction studies of oligonucleotides and oligonucleotide drug complexes. *Angew. Chem. Int. Edn.*, **30**, 1254–77.

Lane, A. M., Ebel, S., and Brown, T. (1993). NMR assignments and solution conformation of the DNA·RNA hybrid d(GCGAACTT)·r(AAGUUCAC). *Eur. J. Biochem.*, **213**, 297–306.

Jaeger, J. A., SantaLucia, J., and Tinoco, I. (1993). Determination of RNA structure and thermodynamics. *Ann. Rev. Biochem.*, **62**, 255–87.

Moser, H. E. and Dervan, P. B. (1987). Sequence-specific cleavage of double-helical DNA by triple-helix formation. *Science*, **238**, 645–50.

Neuhaus, D. and Williamson, M. P. (1989). *The nuclear Overhauser effect in structural and conformation analysis*. Chapters 5 and 12. VCH, Weinheim.

Schimmel, P. R., Söll, D., and Abelson, J. N. (1979). *Transfer RNA: structure, properties, and recognition*. Cold Spring Harbor Laboratory, USA.

van Knippenberg, P. H. and Hilbers, C. W. (ed.) (1986). *Structure and dynamics of RNA*. NATO ASI Series, Plenum, New York.

Uhlmann, E. and Peyman, A. (1990). Antisense oligonucleotides, a new therapeutic principle. *Chem. Rev.*, **90**, 543–84.

2.5

Bates, A. D. and Maxwell, A. (1993). *DNA topology*. IRL Press, Oxford, pp. 85–7.

Breslauer, K. J., Frank, R., Blöcker, H., and Marky, L. A. (1986). Predicting DNA duplex stability from base sequence. *Proc. Natl. Acad. Sci. USA*, **83**, 3746–50.

James, T. L. (1984). Relaxation behaviour of nucleic acids. In *Phosphorus-31 NMR* (ed. D. G. Gorenstein), Academic Press, New York, pp. 349–400.

McCammon, J. A. and Harvey, S. C. (1987). *Dynamics of proteins and nucleic acids*. Cambridge University Press, Cambridge.

Soumpasis, D. M. and Jovin, Th. M. (1987). Energetics of the B–Z transition. In *Nucleic acids and molecular biology* (ed. F. Eckstein and D. M. J. Lilley), Vol. 1. Springer-Verlag, Heidelberg, pp. 85–111.

Travers, A. (1993). *DNA–protein interactions*. Chapman and Hall, London, pp. 158–70.

2.6

Cold Spring Harbor Symposia (1978). Chromatin. *Cold Spring Harbor Symp. Quant. Biol.*, Vol. 42.

Pederson, D. S., Thoma, F., and Simpson, R. T. (1986). Core particles, fibre and transcriptionally active chromatin structure. *Ann. Rev. Cell Biol.*, **2**, 117–47.

Schmid, M. B. (1988). Structure and function of the bacterial chromosome. *Trends Biol. Sci.*, **13**, 131–5.

Selker, E. U. (1990). DNA methylation and chromatin structure: a view from below. *Trends Biol. Sci.*, **15**, 103–7.

Travers, A. A. (1989). DNA conformation and protein binding. *Ann. Rev. Biochem.*, **58**, 427–52.

Travers, A. A. and Klug, A. (1987). The bending of DNA in nucleosomes and its wider implications. *Phil. Trans. Roy. Soc. Lond. B*, **317**, 537–61.

CHEMICAL
SYNTHESIS

3.1 Synthesis of nucleosides

The first nucleoside syntheses were planned to prove the structures of adenosine and the other ribo- and deoxyribonucleosides. Modern syntheses have been aimed at producing nucleoside analogues, frequently for use as inhibitors of nucleic acid metabolism (Section 4.7). In spite of advances in stereospecific synthesis, it remains more economical to produce the major nucleosides by degrading nucleic acids than by total synthesis.

Modified nucleosides are widely distributed naturally. For example, all species of tRNA contain unusual minor bases (Section 6.4.1) and many bacteria and fungi provide rich sources of nucleosides modified in the base, in the sugar, or in both base and sugar residues. Since some of these have been found to show a wide and useful range of biological activity, hundreds if not thousands more nucleoside analogues have been synthesized in pharmaceutical laboratories across the world. In recent times, industrial targets for this work have been anti-viral and anti-cancer agents (Section 4.7). For instance, the arabinose analogues of adenosine and cytidine, *ara*A and *ara*C (Fig. 3.1) are useful as anti-viral and anti-leukemia drugs while 5-iodouridine is valuable for treating *Herpes simplex* infections of the eye.

D-Ribose and other pentoses are relatively inexpensive starting materials which are especially useful in stereochemically controlled syntheses of modified sugars. Three principal strategies for the synthesis of modified nucleosides have been developed. These are illustrated by retrosynthetic analysis (Fig. 3.2). First, **disconnection A** identifies formation of the glycosylic bond by joining the sugar on to a pre-formed base. In practice, this uses the easy displacement of a leaving group from C-1 of an aldose derivative by a nucleophilic nitrogen (or carbon) atom of the heterocyclic base. Secondly, the **double disconnection B** identifies the process of building a heterocyclic base on to a pre-formed nitrogen or carbon substituent at C-1 of the sugar moiety. Thirdly, a **double disconnection C** shows the formation of a purine base onto a preformed imidazole ribonucleoside, an approach to synthesis especially developed by Gordon Shaw. We shall now explore each of these three routes in turn.

3.1.1 Formation of the glycosylic bond

Two general methods have evolved and been refined over the years for the synthesis of the *N*-glycosylic bond in nucleosides. More recently, each has been applied to the preparation of *C*-glycosides in which the sugar residue is joined from C-1 to a **carbon** atom in the base.

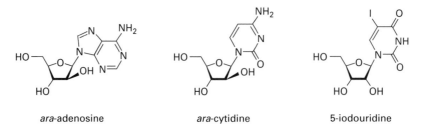

ara-adenosine   *ara*-cytidine   5-iodouridine

Fig. 3.1 Modified nucleosides.

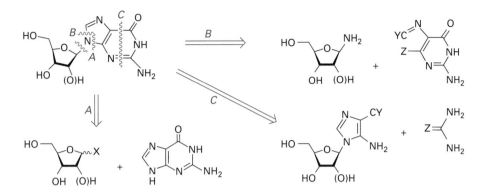

Fig. 3.2 Disconnection analysis of nucleoside synthesis.

Heavy metal salts of bases

Fischer and Helferich and Koenigs and Knorr introduced the use of a heavy metal salt [initially silver(I)] of a purine to catalyse the nucleophilic displacement of a halogen substituent from C-1 of a protected sugar. In a later modification, Davoll and Lowy used mercury(II) salts to improve the yields of products. Typically, chloromercuri-6-benzamidopurine reacts with 2,3,5-tri-*O*-acetyl-D-ribofuranosyl chloride or bromide to give a protected nucleoside from which adenosine is obtained by removal of the protecting groups (Fig. 3.3). These syntheses almost invariably gave the desired regioselectivity, bonding to N-9 of the purine base, and more often than not they gave the desired stereoselectivity, providing predominantly the β-anomer at C-1 of the sugar. The chloromercuri salts of a range of purines can be used, providing that nucleophilic substituents are protected. Thus, amino groups have to be protected by acylation as shown in a synthesis of guanine nucleosides using 2-acetamido-6-chloropurine followed by appropriate hydrolysis (Fig. 3.3).

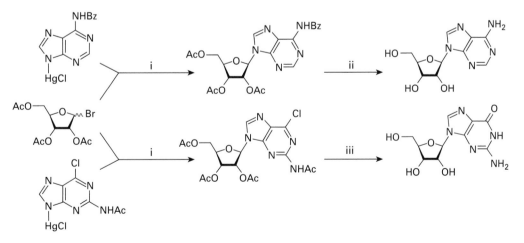

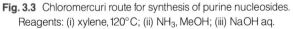

Fig. 3.3 Chloromercuri route for synthesis of purine nucleosides.
Reagents: (i) xylene, 120°C; (ii) NH₃, MeOH; (iii) NaOH aq.

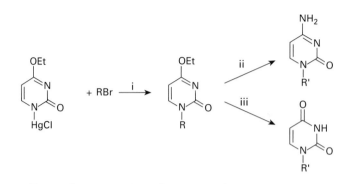

Fig. 3.4 Chloromercuri route for synthesis of pyrimidine nucleosides.
Reagents: (i) xylene, 120°C; (ii) NH$_3$, MeOH; (iii) NaOH aq. R = protected ribofuranosyl;
R′ = 1-β-D-ribofuranosyl.

The chloromercuri derivatives of suitable pyrimidines can be used in much the same way, as illustrated by a synthesis of cytidine from 4-ethoxypyrimidine-2-one (Fig. 3.4). While both of these glycosylations give the desired thermodynamic products at N-9 for purines and N-1 for pyrimidines, the condensation reactions are mechanistically often very much more complex. Thus, there is considerable evidence for pyrimidines that reaction initially gives an O-glycoside or even an O^2,O^4-diglycoside which is then transformed into the desired N-glycoside. For purines, condensation may also take place on to N-7, particularly for bases with a 6-keto substituent (e.g. N^2-acetylguanine and hypoxanthine). Nonetheless, condensations using mercuri derivatives of the heterocyclic bases are still employed with more confidence about the position of substitution of the sugar onto the base and the anomeric configuration than for some more recent methods.

Fusion synthesis of nucleosides

Two disadvantages of the above method are the poor solubility of the mercuri derivatives and the instability of the halogeno-sugar derivative. One early improvement was the combination of 1-acetoxy sugars with Lewis acids such as TiCl$_4$ or SnCl$_4$ as a means of generating the reactive halogeno-sugar *in situ*. That led to the fusion process, in which a melt of the 1-acetoxy sugar and a suitable base *in vacuo*, often with a trace of an acid catalyst, can give acceptable yields of nucleosides. Thus 1,2,3,5-tetra-O-acetyl-D-ribofuranose fused with 2,6,8-trichloropurine or 3-bromo-5-nitro-1,2,4-triazole gives useful yields of the corresponding acylated nucleosides (Fig. 3.5). This method is at its best for weakly basic heterocycles having low melting points.

The quaternization procedure

Hilbert and Johnson noticed that substituted pyrimidines are sufficiently nucleophilic to react directly with halogeno-sugars without any need for electrophilic catalysis. The method,

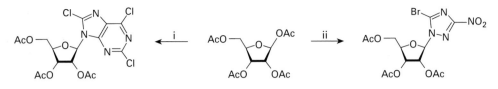

Fig. 3.5 The fusion method of nucleoside synthesis.
Reagents: (i) 2,6,8-trichloropurine, melt at 150°C; (ii) 3-bromo-5-nitro-1,2,4-triazole, melt at 150°C.

which bears their name, involves the alkylation of a 2-alkoxypyrimidine with a halogeno-sugar. The initial product is a quaternary salt which at higher temperatures eliminates an alkyl halide to give an intermediate condensation product. Further chemical modification of substituents on the pyrimidine ring can lead to a range of natural and artificial bases (Fig. 3.6). Such condensations frequently give mixtures of α- and β-anomers although the use of HgBr$_2$ increases the proportion of the β-anomer.

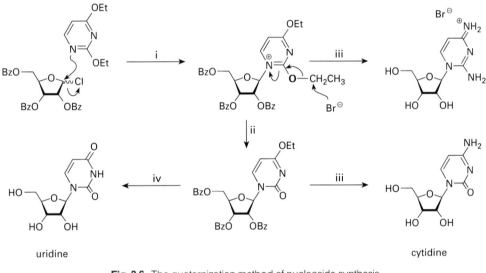

uridine   cytidine

Fig. 3.6 The quaternization method of nucleoside synthesis.
Reagents: (i) CH$_3$CN, 10°C, (ii) CH$_3$CN, reflux; (iii) NH$_3$, MeOH, (iv) NaOH aq.

Silyl base procedure

A major improvement in this method came from the utilization of silylated bases, developed independently by Nishimura, by Birkofer, and by Wittenberg. Such bases have three advantages: (1) they are easily prepared, (2) they react smoothly with sugars in homogeneous solution, and (3) they give intermediate products that can easily be converted into modified bases. The early use of mercuric oxide as a catalyst gave way to Lewis acid catalysts (e.g. SnCl$_4$ or Hg(OAc)$_2$) and they, in turn, have been superseded by the use of silyl esters of strong acids, notably trimethylsilyl triflate, trimethylsilyl nonaflate, or trimethylsilyl perchlorate. Typically, the reaction is carried out in acetonitrile or 1,2-dichloromethane as polar solvent at around $-20°$ to $+50°C$, depending on the reactivity of the components. Some examples are listed (Fig. 3.7). Very often the silylated base is generated *in situ* by the use of bis(trimethylsilyl)acetamide.

This silyl–Hilbert–Johnson method works very well for a large number of nucleoside analogues which have modified bases that are difficult to prepare by other methods. It suffers, as do Koenigs–Knorr procedures, from a lack of precise control of regio- and stereo-selectivity. That is because normally it has the characteristics of an S$_N$1 reaction and depends on the capture of a carbocation at C-1 of the sugar moiety by the most electronegative nitrogen on the base. The problem is well illustrated by the condensation of a 2-azidoarabinose derivative with the trimethylsilyl derivative of 6-chloropurine (Fig. 3.8). Four products have been identified (by UV and NMR spectra) as 9-α (13.7 per cent), 9-β (22.3 per cent), 7-α (10.3 per cent), and 7-β (1.6 per cent) along with other minor species. This lack of regioselectivity is even more marked in the condensation of 2,3,5-tri-*O*-benzoyl-1-*O*-acetyl-D-ribofuranose with a 1,2,4-triazole base when the sugar can become linked to any of the three azole nitrogen atoms.

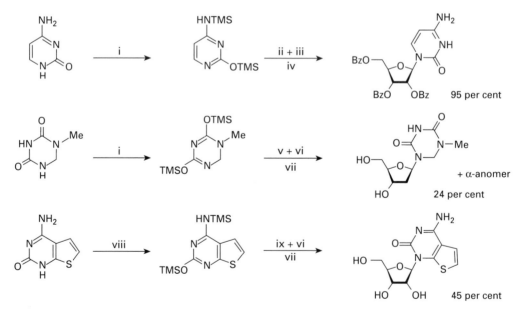

Fig. 3.7 Examples of the silyl base method of nucleoside synthesis.
Reagents: (i) (Me$_3$Si)$_2$NAc; (ii) 1-*O*-acetyl-2,3,5-tri-*O*-benzoyl-D-ribose; (iii) CF$_3$SO$_3$SiMe$_3$; (iv) H$_2$O; (v) 3,5-di-*O*-toluoyl-2-deoxyribofuranosyl chloride; (vi) SnCl$_4$; (vii) NH$_3$, MeOH; (viii) (Me$_3$Si)$_2$NH; (ix) 1,2,3,5-tetra-*O*-acetyl-D-ribose. TMS = Me$_3$Si.

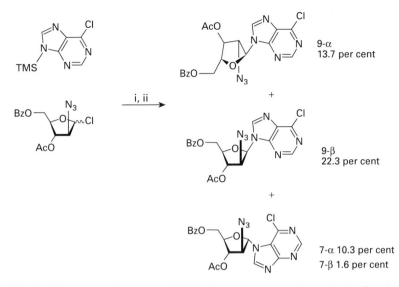

Fig. 3.8 Variable regio- and stereo-specificity in silyl base condensation synthesis of nucleosides. Reagents: (i) $Hg(OAc)_2$; (ii) $ClCH_2CH_2Cl$, 50°C.

Transglycosylation

It is often relatively easy to convert one of the natural nucleosides, typically 2′-deoxythymidine, into a nucleoside with a modified sugar residue, for instance the drug 3′-azido-2′,3′-dideoxythymidine (AZT, Section 4.7.2). At the same time, it can be difficult to achieve the same transformation of 2′-deoxyadenosine into AZdA. In such cases the sugar moiety can be transferred from one base to another by a process known as **transglycosylation**. The reaction is particularly effective for transferring sugars, including quite complex species, from pyrimidines (which are π-deficient heterocycles) to the more basic purines (π-excessive heterocycles) (Fig. 3.9).

This reaction has all the hallmarks of an S_N1 ionization process, as shown both by the intramolecular transfer of a sugar residue from N-7 to N-9 of 6-chloro-1-deazapurine and by the anomerization of β- into α-nucleosides. Transglycosylation is now a favoured method for the preparation of α-anomers of pyrimidine nucleosides from their natural isomers. It also provides a useful synthesis of α-anomers of purine nucleosides (Fig. 3.10). The mixture of α- and β-species which is usually formed can be separated by chromatography. However, only experience is able to predict the thermodynamically favoured regioselectivity of these processes.

In addition to such chemical transglycosylation, both Holy and Hutchinson have made good use of biotransformations of readily available nucleosides into novel derivatives by enzyme catalysed transglycosylation. One can either use a bacterial N-deoxyribofuranosyl transferase (as from *L. leichmanii*) or intact *E. coli* cells immobilized in a gel matrix. Typically 2-deoxy-D-ribose can be transferred from deoxythymidine or deoxyuridine to modified pur-

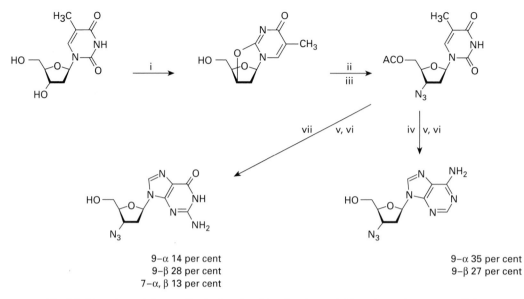

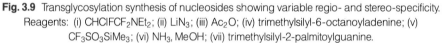

9–α 14 per cent
9–β 28 per cent
7–α, β 13 per cent

9–α 35 per cent
9–β 27 per cent

Fig. 3.9 Transglycosylation synthesis of nucleosides showing variable regio- and stereo-specificity. Reagents: (i) CHClFCF$_2$NEt$_2$; (ii) LiN$_3$; (iii) Ac$_2$O; (iv) trimethylsilyl-6-octanoyladenine; (v) CF$_3$SO$_3$SiMe$_3$; (vi) NH$_3$, MeOH; (vii) trimethylsilyl-2-palmitoylguanine.

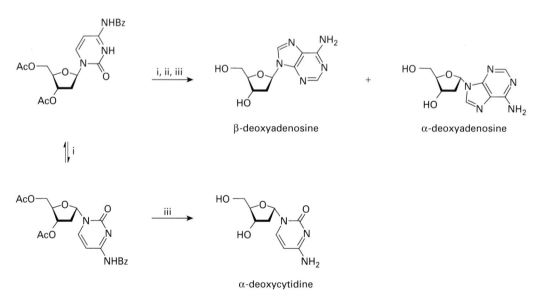

β-deoxyadenosine

α-deoxyadenosine

α-deoxycytidine

Fig. 3.10 Transglycosylation synthesis of α-deoxyribonucleosides. Reagents: (i) (Me$_3$Si)$_2$NAc, CF$_3$SO$_3$SiMe$_3$; (ii) 6-benzoyladenine; (iii) NH$_3$, MeOH.

ines or to substituted imidazoles while the transfer of 2,3-dideoxy-D-ribofuranose is also a practicable proposition. The system is relative efficient, it is highly stereospecific since only β-glycosides are formed, and it can be employed on a gram scale.

Control of anomeric stereochemistry

Sugars with a 2-acyloxy substituent on condensation with a base invariably give N-glycoside products that have the 1,2-*trans*-configuration. This observation led Baker to suggest that neighbouring group participation is responsible. Essentially, the ionization of the leaving group at C-1 of the sugar (from either the α- or the β-face) generates a carbocation which is 'captured' by the carbonyl oxygen of the adjacent acyl group. This bicyclic intermediate is preferentially attacked by the nucleophilic base from the **opposite side** of the furanose ring to the C-2 substituent: hence **Baker's 1,2-*trans*- rule** (Fig. 3.11).

It follows that arabinose and lyxose sugars with a 2-acyloxy substituent give α-anomers while ribose and xylose sugars give β-anomers. In cases where the hydroxyl group at C-2 is protected by a benzyl ether or by an isopropylidene or carbonate group cyclized on to the adjacent 3-hydroxyl group, then neighbouring group participation is not possible. As a result, mixtures of anomers are formed. Similarly, for 2-deoxy-sugars, 2-deoxy-2-fluoro-, or 2-deoxy-2-azido-sugars there is no anomeric control.

There seem to be only a few exceptions to Baker's 1,2-*trans* rule, and these can be explained rationally. For instance, when the acyl group at position-2 is *p*-nitrobenzoyl then an α : β ratio of 1:3 is observed. This has been attributed to the electron-withdrawing effect of the nitro group which impairs participation by the carbonyl oxygen.

An alternative method for the control of anomeric control has involved the use of oxazolidine intermediates in synthesis. These have been employed by a variety of research groups to

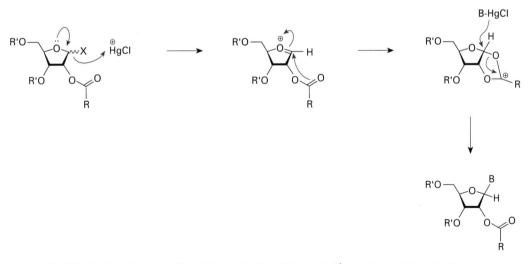

Fig. 3.11 Mechanistic basis of the 1,2-*trans* rule. R = alkyl or aryl; R′ = acyl, benzyl, trimethylsilyl, etc.; B = heterocyclic base (as chloromercuri salt or trimethylsilyl derivative).

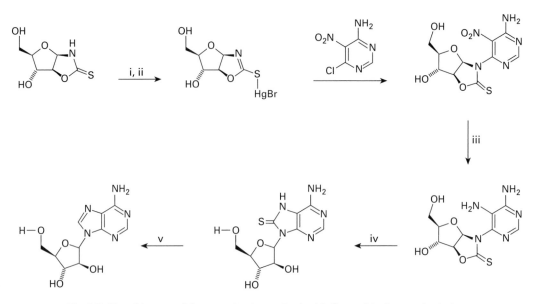

Fig. 3.12 Use of the oxazolidine route for the synthesis of 9-β-D-arabinofuranosyl adenine.
Reagents: (i) NaH, DMF; (ii) HgBr$_2$; (iii) Al/Hg; (iv) heat, H$_2$O; (v) Ra–Ni, H$_2$O, NH$_4$OH.

effect *stereospecific* synthesis of both purine and pyrimidine nucleosides, and they have had rather more general application than methods based on the anomeric effect. The general procedure is well illustrated by a synthesis of 9-β-D-arabinofuranosyladenine from 1-β-D-amino-1-deoxy-arabinofuranose (Fig. 3.12).

A stereoselective synthesis for α-ribonucleosides has been developed which starts from 1-hydroxy sugars. The preferential reaction of the β-anomer (in equilibrium with the α-isomer) with 2-fluoro-1-methylpyridinium tosylate fixes the configuration of the anomeric carbon. Subsequent reaction with a silylated base proceeds by an S$_N$2 displacement with predominant inversion at C-1 to give the α-anomer product (Fig. 3.13)

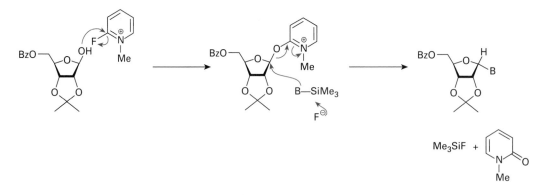

Fig. 3.13 Stereoselective synthesis of α-ribonucleosides.

C-nucleosides

A few C-nucleosides have been made by carbanion displacement reactions at C-1 of a suitably protected sugar, although the high basicity of the carbanion can often lead to an unwanted 1,2-elimination process. A classic example is Brown's synthesis of pseudouridine, a common component of tRNA species (Section 6.4.1), by the reaction of 2,4-bis-(t-butoxy)-5-lithiopyrimidine with 2,4;3,5-bis-O-benzylidene-D-ribose. This gave rather more of the β-pseudouridine (18 per cent) than of the α-anomer (8 per cent) (Fig. 3.14). More often, displacement reactions of carbocations at C-1 of protected sugars are designed to provide intermediates for construction of the base onto the sugar moiety. A useful procedure of this sort is the reaction of 1-fluoropentoses with enol silyl ethers (Fig. 3.14).

For 2-deoxynucleosides, the direct S_N2 displacement of halogen from 1-α-chloro-3,5-di-O-p-toluoyl-D-ribofuranose by the sodium salt of a purine or related heterocycle gives good yields of the β-2'-deoxynucleosides (though the reaction is not always regiospecific) (Fig. 3.15).

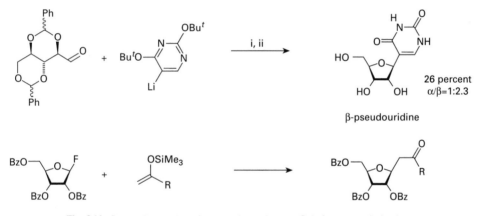

Fig. 3.14 Carbanion and enolate condensations at C-1 of pentose derivatives. Reagents: (i) THF, −78°C; (ii) mild acidic hydrolysis.

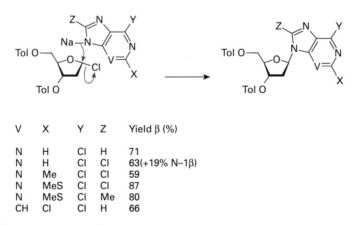

V	X	Y	Z	Yield β (%)
N	H	Cl	H	71
N	H	Cl	Cl	63(+19% N–1β)
N	Me	Cl	Cl	59
N	MeS	Cl	Cl	87
N	MeS	Cl	Me	80
CH	Cl	Cl	H	66

Fig. 3.15 Synthesis of β-2-deoxyribonucleosides by S_N2 displacement of chloride from 1- α-chloro-3,5-di-O-toluoyl-D-2-deoxyribofuranose.

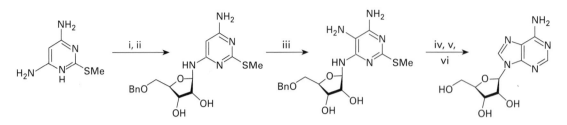

Fig. 3.16   Todd's synthesis of 9-D-ribofuranosyladenine.
Reagents: (i) 5-O-benzyl-2,3,4-tri-O-acetyl-D-ribose; (ii) NH₃; (iii) diazonium coupling or nitrosation
followed by reduction; (iv) thiourea; (v) Raney nickel desulfurization; (vi) H₂/PdC debenzylation.

3.1.2 Building the base on to a C-1 substituent of the sugar

This approach to nucleoside synthesis has three important features. Historically it was later used in Todd's group for a regiospecific synthesis of adenosine (Fig. 3.16). Later, it became the preferred route for the synthesis of *C*-nucleosides and some unusual *N*-nucleosides. Most recently, it has emerged as the most flexible pathway for the synthesis of nucleosides with highly modified sugars linked to normal or to modified bases.

Nucleosides with modified bases

A good example of the use of this route is the synthesis of the fluorescent base Wyosine, which is found in the anticodon loop of some species of tRNA (Section 6.4.1). In this case, the isocyanate function is the foundation for construction of the tricyclic imidazopurine base. The same isocyanate precursor has been used in a synthesis of 5-azacytidine (Fig. 3.17). This nucleoside

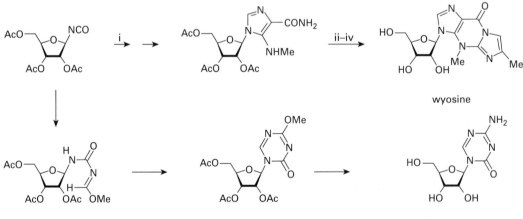

Fig. 3.17   Building the Wye base and 5-azacytosine on to a C-1 isocyanate.
Reagents: (i) three-carbon fragment; (ii) CNBr; (iii) NaOEt, EtOH; (iv) BrCH₂COCH₃.

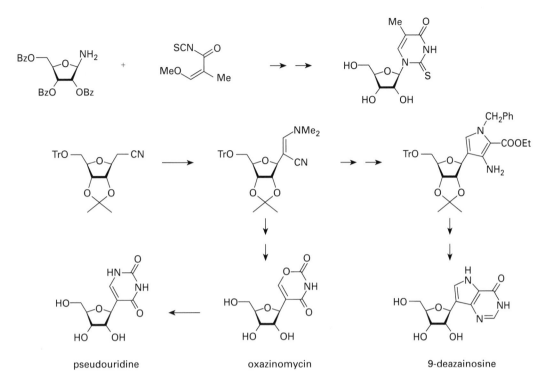

Fig. 3.18 Syntheses of *N*- and *C*-nucleosides by building the base on to the sugar.

is elaborated by a *Streptomyces* species and has been used in the treatment of certain leukaemias.

Syntheses of this type based on 1-amino-1-deoxy-β-D-ribofuranose have the general advantage that the place of attachment of the sugar on to the heterocyclic base is unambiguous and is not determined by the most nucleophilic heteroatom on the base component. Such syntheses have therefore been widely employed for the preparation of the imidazole nucleosides involved in the *de novo* bisynthesis of purine nucleosides (Section 4.1) and modified pyrimidine and purine nucleosides. A typical example of the work of Gordon Shaw in this area is a synthesis of 2-thioribothymidine (Fig. 3.18).

In a similar way, a cyanomethyl group at C-1 of D-ribose supports synthesis of 9-deazainosine, the antibiotic oxazinomycin, and of pseudouridine (Fig. 3.18). Oxazinomycin has both child growth-promoting properties and some anti-tumour activity.

C-nucleosides

With the growing availability of chemical reactions endowed with a high degree of stereochemical selectivity, the synthesis of *C*-nucleosides by this route has moved away from sugars as starting materials. Showdomycin is a product of *Streptomyces showdowensis* and has useful

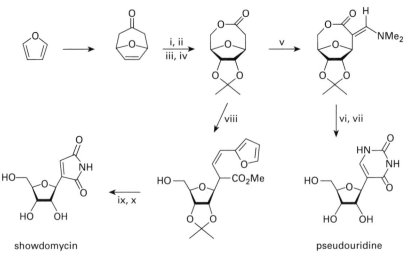

Fig. 3.19 Synthesis of showdomycin and pseudouridine from furan.
Reagents: (i) OsO$_4$, H$_2$O$_2$; (ii) acetone, H$^+$; (iii) CF$_3$CO$_3$H; (iv) resolution; (v) dimethylformamide; (vi) urea; (vii) H$_3$O$^+$; (viii) furfural, MeONa; (ix) ozone; (x) Ph$_3$P=CHCONH$_2$.

cytotoxic and enzyme inhibitory properties. A route starting from a tricyclic precursor can branch to give either showdomycin or pseudouridine in a stereospecific fashion (Fig. 3.19).

Carbocyclic nucleosides

The formal replacement of the 4′-oxygen in the sugar by a methylene group gives a carbocyclic nucleoside. Much of the activity in the synthesis of carbocyclic nucleosides has been carried out in the search for potential anti-tumour and anti-viral agents, especially provoked by the search for agents effective against HIV (Section 4.7). One of the particular values of carbocyclic nucleosides is their greater metabolic stability to the phosphorylase enzymes which cleave the glycosylic bond of normal nucleosides.

The carbocyclic analogue of adenosine was prepared in racemic form by Shealy and Clayson in 1966 and its laevorotatory enantiomer was discovered two years later as a metabolite of *Streptomyces citricolor*, now named aristeromycin. New concepts of carbocyclic nucleosides emerged in 1981 with the isolation of neplanocin A from *Ampullariella regularis* (Fig. 3.20).

Many syntheses use the key 'carbocyclic ribofuranosylamine', which is made from cyclopentadiene in five steps and then built into pyrimidine or purine carbocyclic nucleosides by standard methods. The adaptation of this route for the introduction of a fluorine atom into the 6-position (which may mimic an oxygen lone pair of electrons in binding to a receptor) presents a nice example of the development of such syntheses to highly modified sugars (Fig. 3.20).

The industrial requirement for more efficient and flexible routes to potential pharmaceutical agents has focused attention on *convergent* approaches to synthesis. In such syntheses, an

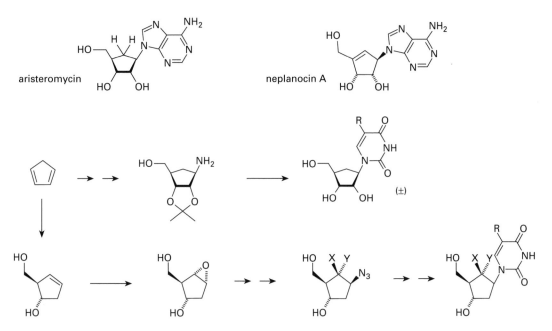

Fig. 3.20 Structures of aristeromycin and neplanocin A (upper). Synthesis of carbocyclic analogues of deoxy- and ribo-uridine (R = H) and thymidine (R = Me) where X and/or Y are H or F (lower).

intact heterocyclic base is coupled directly onto a suitably functionalized carbocyclic sugar moiety. At the same time, the growing awareness that biological activity is normally to be found in only one of the two enantiomers has made synthetic routes to *chiral* carbocyclic nucleosides of paramount importance. One of very many examples of the variety to be found in such syntheses is the use of a chiral epoxide related to 2-deoxy-D-ribose which leads to carbocyclic analogues of deoxyguanosine and deoxythymidine in short, convergent syntheses (Fig. 3.21).

Dioxolane and oxathiolane nucleosides

A recent move in the opposite direction to carbocyclic nucleosides has been the introduction of a second heteroatom into the 'sugar' ring. For example, 1,3-dioxolane nucleosides have been made and found to have useful anti-HIV activity (Section 4.7). The preparation of 2′,3′-dideoxy-3′-oxacytidine by Chu is a good example of stereospecific control in such syntheses (Fig. 3.22). Liotta has synthesized the racemic 1,3-oxathiolane analogue of 5-fluorodeoxycytidine and separated the enantiomers by the action of pig liver esterase on their 5′-butyroyl derivatives. Unexpectedly, he found that it is the unnatural L-(−)-isomer which has both higher anti-viral activity and lower toxicity than the D-(+)-enantiomer (Fig. 3.22).

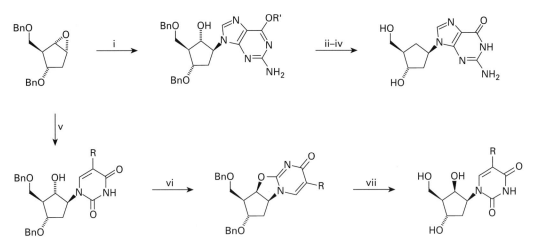

Fig. 3.21 Synthesis of chiral carbocyclic nucleosides of deoxyguanosine (upper), deoxythymidine (R = Me) and deoxyuridine (R = H) (lower) from 3-benzyloxymethyl-4-benzyloxycyclopentene oxide (R' = CH$_2$CH$_2$OMe).
Reagents: (i) 2-amino-6-methoxyethoxypurine; (ii) PhOCSCl then nBu$_3$SnH, AIBN; (iii) H$_2$, Pd/C; (iv) 3M HCl; (v) uracil or thymine, NaH, DMF; (vi) (PhO)$_2$CO; (vii) NaOH, aq. MeOH.

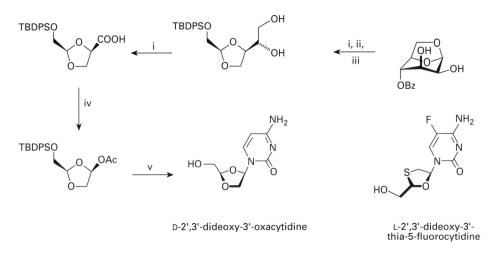

D-2',3'-dideoxy-3'-oxacytidine

L-2',3'-dideoxy- 3'-
thia-5-fluorocytidine

Fig. 3.22 Synthesis of D-2',3'-dideoxy-3'-oxacytidine from 4-O-benzoyl-1,6-anhydro-D-mannose (*upper*) and structure of L-2',3'-dideoxy- 3'-thia-5-fluorocytidine (*lower right*).
Reagents: (i) NaIO$_4$; (ii) NaBH$_4$; (iii) TBDPSCl, pyridine; (iv) Pb(OAc)$_4$; (v) silylated 4- N-acetylcytosine, TMS triflate, 1,2-dichloroethane.

3.1.3 Synthesis of acyclonucleosides

The success of acyclovir for the treatment of genital herpes infections has stimulated much work in this area. In these acyclonucleosides (or seco-nucleosides) the base is usually adenine, guanine, or a related purine base which can be converted into adenine or guanine as a result of metabolic deamination or hydroxylation (i.e. pro-drugs, Section 4.7.2). In principle,

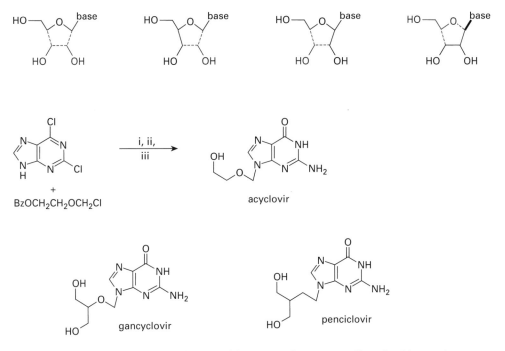

Fig. 3.23 Relationship of various acyclonucleosides to natural prototypes, with exciseable parts in colour (*upper*).
Reagents: (i) Et₃N, dimethylformamide; (ii) NH₃, MeOH; (iii) HNO₂.

four sections of the sugar ring can be 'cut away' and promising biological results have been found in three of these areas. Formally, one can excise (1) C-2′, (2) C-3′, (3) C-2′ + C-3′, and (4) O-4′ + C-4′ + C-5′ (Fig. 3.23). The syntheses of all of these types of acyclonucleoside are invariably based on N-9 alkylation of the desired purine or its precursor, with subsequent manipulation of the necessary protecting groups. Seco-carbocyclic nucleosides have also been found to have useful antiviral activity. One example is penciclovir, which is *N*-9-(4-hydroxy-3-hydroxymethylbutyl)guanine (Fig. 3.23).

Summary

The early need to prove the structure of the major nucleosides led to syntheses based on formation of the glycosylic bond. These have been refined to give good control of regio- and stereo-selectivity. The discovery of *C*-nucleosides resulted in syntheses based on construction of the base on to C-1 of a modified pentose moiety. They have also provided good routes for the preparation of base-modified nucleosides and carbocyclic nucleosides. The search for new anti-viral agents has led to syntheses of acyclonucleoside analogues which lack one or more atoms from the pentofuranose ring.

3.2 Chemistry of esters and anhydrides of phosphorus oxyacids

3.2.1 Phosphate esters

The predominant forms of phosphorus in biology are as orthophosphoric acid, H_3PO_4, its esters, anhydrides, and some amides. Its oxidation state is P(V) and it has a co-ordination number 4 (CN.4). Orthophosphates are tetrahedral at phosphorus. The 'single' P—O bonds use sp^3 hybrid orbitals at phosphorus and are ~1.6 Å long. In triesters the P—O 'double' bond is shorter, ~1.46 Å, and involves a phosphorus d-orbital in a $p_\pi–d_\pi$ hybrid (Fig. 3.24). Because phosphorus has five 3d orbitals, it can participate in such bonding simultaneously to more than one oxygen ligand (the corresponding bonding to neutral nitrogen ligands in phosphoramidates is rather weak so the nitrogen remains moderately basic).

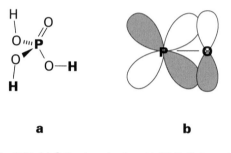

a   **b**

Fig. 3.24 (a) Orthophosphoric acid; **(b)** P—O $d_\pi–p_\pi$ bonding.

Phosphate triesters

Triesters have all three hydrogens of phosphoric acid replaced by alkyl or aryl groups. They are non-ionic, soluble in many organic solvents, and sufficiently stable to be purified by chromatography. The P—O bond is effectively transparent in the UV region and has an IR absorption at 1280 cm^{-1}. When all three ester groups are different, as in butyl ethyl phenyl phosphate (Fig. 3.25a), the phosphorus atom is a centre of asymmetry and optical isomers are possible.

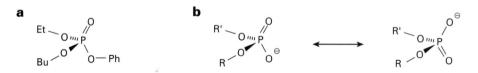

Fig. 3.25 (a) Chiral phosphate triester; **(b)** pro-chiral oxygens in a phosphate diester.

Phosphate diesters

Diesters have two hydrogens replaced by alkyl or aryl groups. The remaining OH ligand is strongly acidic, $pK_a \sim 1.5$. Consequently, phosphate diesters exist as monoanions at pH > 2, and are usually water-soluble. The negative charge is shared equally between the two unsubstituted oxygens (Fig. 3.25b). When the two ester groups are different, the phosphorus atom is a pro-chiral centre and the two unsubstituted oxygens are non-equivalent (i.e. diastereotopic).

Phosphate monoesters

Monoesters have a single alkyl or aryl group and two ionizable OH groups. These have $pK_{a1} \sim 1.6$ and $pK_{a2} \sim 6.6$, so there is an equilibrium in neutral solution (effectively from pH 5 to pH 8) between the monoanion and the dianion. The equivalent oxygens share the negative charge in both mono- and dianions and there is partial double bonding to each. In the monoanion, the hydrogen atom translocates rapidly between the three oxygens making them all equivalent in solution. These three oxygens are pro-pro-chiral. Thus the use of the three isotopes of oxygen, ^{16}O, ^{17}O (=∅), and ^{18}O (=●) has been widely used in stereochemical synthesis and in analysis of substitution reactions of phosphates (Fig. 3.26).

3.2.2 Hydrolysis of phosphate esters

The great stability of phosphate diesters and, to a lesser extent, of monoesters to hydrolysis under physiological conditions is an essential feature of the chemistry of nucleosides and nucleic acids and intrinsic to life itself. Studies of mechanisms for their hydrolysis have had to be carried out at elevated temperatures (up to 100°C) and often at extremes of pH and the data then extrapolated to 37°C and pH 7. Reactivity can also be enhanced by using aryl esters, especially the chromophoric *p*-nitrophenyl esters.

Both C–O and P–O cleavage reactions have been identified. In general, associative, $S_N2(P)$, mechanisms are more common than dissociative ones, $S_N1(P)$, and they are usually linked to inversion of configuration at phosphorus. The associative process is best described by a 5-coordinate species (CN.5) as intermediate or transition state in which ligand positional interchange, often called **pseudorotation**, is usually slower than breakdown of the trigonal bipyramidal species to form products (Fig. 3.27). In dissociative reactions, a planar, 3-co-ordinate

Fig. 3.26 (a) Pro-pro-chiral oxygens in a phosphate monoester; (b) phosphate monoester chiral through having three isotopes of oxgen.

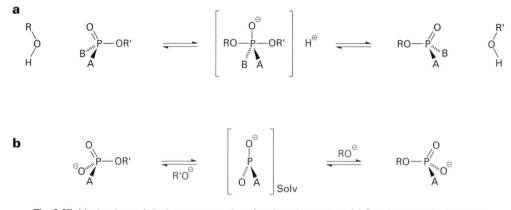

Fig. 3.27 Mechanisms of displacement reactions for phosphate esters. (**a**) Synchronous displacement by an $S_N2(P)$, or 'in line', process via a CN.5 phosphorane; (**b**) stepwise displacement by $S_N1(P)$ process with a transient monomeric metaphosphate (solvated).

species, often described as a monomeric metaphosphate, has the opportunity of capturing the incoming nucleophile on either face and so racemization at phosphorus results (Fig. 3.27).

Hydrolysis of alkyl triesters

Trimethyl phosphate is slowly hydrolysed in alkaline solution in an $S_N2(P)$ process ($k_{OH^-} = 3 \times 10^{-2}$ L mol^{-1} s^{-1} at 100°C). With $H_2^{18}O$ as solvent, no isotope exchange is observed into the P=O group of unreacted ester and no Me^{18}OH is formed. This shows that there is exclusive P–O cleavage. Other 'hard' nucleophiles such as F$^-$ react similarly and, indeed, fluoride catalysis of the transesterification of triesters is a useful process (Fig. 3.28). The intramolecular migration of phosphorus in a triester to a vicinal hydroxyl group is especially easy and must be avoided in the synthesis of oligoribonucleotide and inositol phosphate precursors.

Trimethyl phosphate is hydrolysed extremely slowly in neutral and in acidic conditions ($k_w = 1.6 \times 10^{-7}$ s^{-1} at 44.7°C) with C–O cleavage. Soft nucleophiles, such as RS$^-$, Br$^-$, or I$^-$, also dealkylate phosphate triesters with C–O cleavage. Such reactions are typical S_N2 processes and show a clear preference for dealkylation of Me > Et > R$_2$CH. This characteristic is particularly well exploited in the thiophenolate deprotection of methyl phosphate triesters in oligonucleotide synthesis (Section 3.4.4) (Fig. 3.29).

Alkyl phosphate triesters are also sensitive to β-elimination processes. While such reactions appear to have no role in nucleic acid metabolism, they have been usefully adapted for the selective deprotection of phosphate triesters in oligonucleotide synthesis. The 2-cyano-

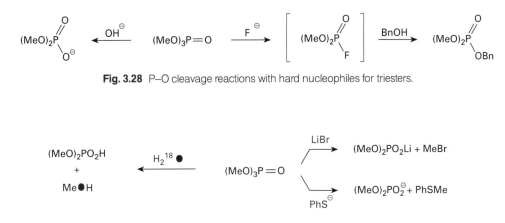

Fig. 3.28 P–O cleavage reactions with hard nucleophiles for triesters.

Fig. 3.29 C–O cleavage reactions with soft nucleophiles for triesters.

ethyl group and its congeners have an acidic β-hydrogen and so are susceptible to cleavage under mildly basic conditions (Section 3.4.4). In a similar fashion, the 2,2,2-trichloroethyl ester group can be eliminated by reduction either using a zinc–copper couple or at a reducing anode using an appropriate electrode potential (Fig. 3.30).

Hydrolysis of aryl triesters

Because aryl phosphates are much more reactive than alkyl phosphate triesters, it is possible to achieve selective, nucleophilic displacement of the phenolic residue in a dialkyl aryl phosphate on account of its better leaving group ability ($pK_a > 5$). One of the best nucleophiles for this purpose is the oximate anion (Fig. 3.31) (Section 3.4.4).

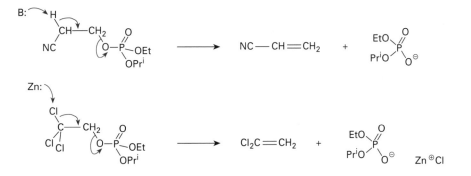

Fig. 3.30 Selective cleavage of alkyl phosphate triesters by β-elimination.

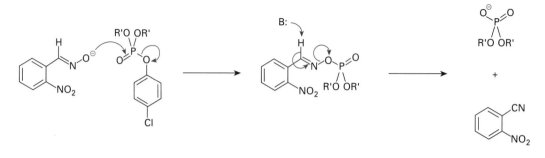

Fig. 3.31 Selective nucleophilic displacement in an aryl triester.

Hydrolysis of phosphate diesters

At pH > 2, phosphate diesters exist as their monoanions, which are stable in boiling water (Table 3.1). Even in strong alkaline conditions, diesters hydrolyse more slowly than triesters; (because of anion–anion repulsion) with predominant C–O cleavage (>90 per cent). In acidic conditions their hydrolysis is rather similar to that of trialkyl phosphates. The diaryl esters are rather more reactive under alkaline conditions, as is to be expected for reaction involving a better leaving group.

This marked stability of the phosphate diester linkage to hydrolysis is a vital feature of the biological role of DNA. It is dramatically changed for esters of 1,2-diols. Here, the vicinal hydroxyl group enormously enhances the rate of hydrolysis of di- and tri-esters and cyclic phosphates of 1,2-diols hydrolyse some 10^7 times faster than do their acyclic or 6- and 7-membered cyclic relatives. This corresponds to a decrease in $\Delta G^{\ddagger}$ of 36 kJ mol^{-1}. About 60 per cent of this acceleration is attributed to relief of strain in the 5-membered cyclic ester, which has a 98° O–P–O angle and an enhanced enthalpy of hydrolysis of -20 kJ mol^{-1}.

The essential observations are that there is acceleration of (1) ring closure, (2) ring opening, and (3) exocyclic P–O bond cleavage (in both phosphate and phosphorane species) (Fig. 3.32).

How can ring strain accelerate both endocyclic and exocyclic substitution at phosphorus? This problem has been investigated by Westheimer and his associated over some 25 years. They have found that the hydrolysis of ethylene phosphate shows incorporation of isotope into P–O bonds from H$_2$^{18}O in both acidic and alkaline conditions, and these reactions must involve an addition–elimination (i.e. an associative) process. It is generally agreed that a transient CN.5 phosphorane intermediate is both stabilized and made kinetically more accessible relative to its acyclic counterpart because of the geometry of the 5-membered ring. It is then reasonable to invoke topoisomerism (in this case the pseudorotation of a trigonal bipyramidal species) to explain most of the phenomena associated with this remarkably enhanced reactivity (Fig. 3.33).

Regardless of the subtleties of the mechanism, this phenomenon is clearly involved in the hydrolysis of RNA by alkali and by ribonucleases. In both cases, the 5-membered 2′,3′-cyclic phosphates of nucleosides are formed by the displacement of the 5′-O-nucleoside residue. The

Table 3.1. Rate constants [log ($10^6 \times k$)] for the hydrolysis of phosphate esters (100°C) and patterns of bond cleavage (colour).

	Conjugate acid (P=OH+) k_{H^+} /s^{-1}	Neutral k_w /s^{-1}	Monoanion k /s^{-1}	Alkaline k_{OH^-} /s^{-1}
(RO)$_3$PO	1 (C–O)	1 (C–O)	—	4.5 (P–O)
(PhO)$_3$PO	1.7 (P–O)	–1.2 (P–O)	—	5.6 (P–O)
(MeO)$_2$PO$_2$H	0.5 (C–O)	0.4	<<–3 (C–O)	–1 (C–O > P–O)
(PhO)$_2$PO$_2$H	0.2 (P–O)	—	–2.5 (P–O)	1.6 (P–O)
(RO)$_2$PO$_2$H	1 (C–O)	–0.3	0.9 (P–O)	<–2
PhOPO$_3$H$_2$	1 (P–O)	1.5	2.5 (P–O)	–0.4 (P–O)
	5.6 (P–O)[a]	—	—	2.7 (P–O)[b]

[a] 30°C; [b] 25°C

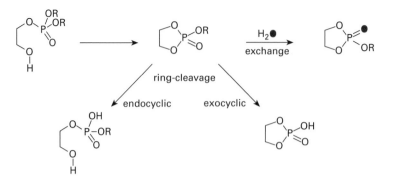

Fig. 3.32 Accelerated P–O cleavages associated with 5-membered ring phosphate esters in acidic and in alkaline solution.

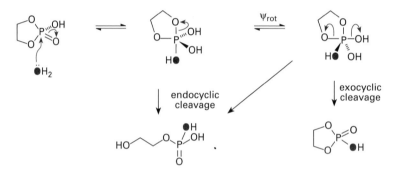

Fig. 3.33 Role of trigonal bipyramidal pseudorotation (ψ_{rot}) in ^{18}O isotope (●) incorporation into ethylene phosphate.

enzymatic reaction is completed by the regioselective ring-opening of the cyclic phosphate to give a 3′-nucleoside phosphate while alkaline hydrolysis leads to a mixture of 2′- and 3′-phosphates (Fig. 3.34). Both reactions exhibit retention of configuration at the phosphorus centre. This is interpreted as a double inversion of stereochemistry as a result of two successive 'in-line' displacement processes. Full details of the mechanism of action of RNase A and RNase T1 have emerged from a combination of X-ray crystallography, site-directed mutagenesis, and mechanistic analysis (Section 9.2.3).

It must be emphasized that this remarkable reactivity is exclusive to 5-membered cyclic phosphate esters and esters of 1,2-diols. That contrasts totally with the stability of esters of 1,3-diols and 6-ring cyclic phosphates. The most notable example is 3′,5′-cAMP, whose key role as the second messenger in cell signalling is dependent on its kinetic stability to non-enzymatic hydrolysis.

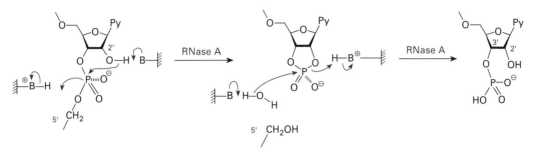

Fig. 3.34 Ribonuclease (and alkaline) hydrolysis of RNA via 2′,3′-cyclic phosphate.

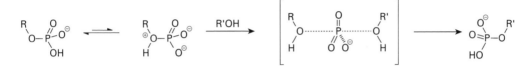

Fig. 3.35 Dissociative hydrolysis of a phosphate monoester monoanion.

Phosphate monoesters

The hydrolysis of monoalkyl phosphates at low pH proceeds via the conjugate acid, $ROP(OH)_3^+$, and is similar in mechanism to that of triesters (Table 3.1). These esters are very resistant to alkaline hydrolysis, as a result of anionic repulsion, but they show an unusually large reactivity for the neutral hydrolysis of the monoanion. This proceeds by P–O cleavage and has all the characteristics of a dissociative process via a CN.3 metaphosphate intermediate. A better description has invoked the idea of an 'exploded transition state' in which there is a very loose association of incoming and outgoing ligands (Fig. 3.35). Similar phenomena have been analysed for spontaneous hydrolyses of acetyl phosphate, creatine phosphate, and ATP (loss of the γ-phosphate), all of which have good leaving groups on a terminal phosphate.

3.2.3 Condensed phosphates

Phosphoric acid can form chains of alternating oxygen–phosphorus linkages which are relatively stable in neutral aqueous solution. The major condensed phosphates of biological importance are pyrophosphoric acid, $(HO)_2P(O)OP(O)(OH)_2$, its esters, and esters of tripolyphosphoric acid. The stability of such species can be related to that of the corresponding phos-

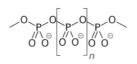

Fig. 3.36 Structure of polyphosphates.

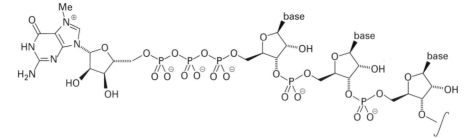

Fig. 3.37 P^1,P^3-Dinucleosidyl triphosphate in the 'cap' structure at the 5′-end of eukaryotic mRNA.

phates after due allowance for (1) the changed stability of the anionic charge and (2) improved leaving group characteristics.

Thus, tetraethyl pyrophosphate is an ethylating agent towards hard nucleophiles and also a phosphorylating agent. Tetraphenyl pyrophosphate is exclusively a phosphorylating agent. P^1,P^2-Dialkyl pyrophosphates have considerable stability towards hydrolysis at ambient pHs where they exist exclusively in the form of the dianion. This feature is very important for the stability of the pyrophosphate link as a structural feature in many co-enzymes, including NADH, FAD, and CoA (Section 3.3). Similarly, P^1,P^3-diesters of tripolyphosphoric acid are stable components of the 'cap' structure of eukaryotic mRNA (Fig. 3.37) (Section 6.4.3) and P^1,P^4-diadenosyl tetraphosphate, Ap$_4$A, is a stable minor nucleotide species found in low concentration in all mammalian tissues (Fig. 3.46).

3.2.4 Synthesis of phosphate esters

The most common approaches to dinucleoside phosphate ester synthesis use phosphorylation reactions in which a 3′-nucleotide component is converted into a reactive phosphorylating species by a condensing agent. One of the major problems is that the more reactive condensing agents not only activate the nucleotide, but may also react with nucleoside bases or with the product, leading to unacceptable loss in yield. The ideal condensing agent should have a high rate of activation of the phosphate species and a low rate of reaction with the hydroxylic component or with the N-protected bases.

Interrelationships of esters of phosphorus oxyacids

The formal relationship between phosphorus halides (X = Cl) and the mono-, di-, and tri-alkyl esters of phosphorus oxyacids is shown schematically (Fig. 3.38). Actual reaction conditions

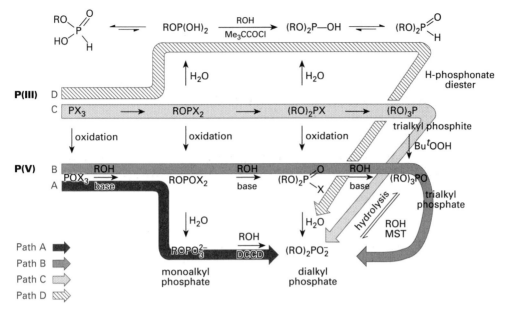

Fig. 3.38 Formal relationship between esters of P(III) and P(V) oxyacids. Path A, the diester route for nucleotide synthesis; Path B, the P(V) triester route; Path C, the P(III) route and Path D, the H-phosphonate route.

have to be controlled carefully to avoid the formation of by-products, especially of alkyl halides. Many of the interconversions shown are best accomplished using nitrogen ligands at phosphorus (X = NPr$_2^i$ or an azole).

Early syntheses worked mainly with mild condensing agents such as dicyclohexylcarbodiimide (DCCD). More powerful reagents, such as the arenesulphonyl chlorides, were introduced next and made more selective by building in steric factors. Since 1980, the demand for faster reactions for oligonucleotide synthesis has switched attention to P(III) chemistry. In a very recent variation, the use of H-phosphonates as a 4-co-ordinate P(III) has built upon the pioneering studies of Todd in the 1950s.

Syntheses via phosphate diesters

In the diester route to oligonucleotides (Fig. 3.38, Path A), the key step is the condensation of a phosphate monoester with an alcohol using DCCD. The reactions are slow, but at room temperature there is no formation of triesters. The mechanism is complex: an initial imidoyl phosphate adduct of DCCD and the 3′-nucleotide is probably converted into condensed phosphates before final formation of the phosphodiester (Fig. 3.39).

Syntheses via phosphate triesters

DCCD was superseded by mesitylenesulphonyl chloride as a faster and more efficient condensing agent. In particular, its greater reactivity enables it to make triesters from dialkyl phos-

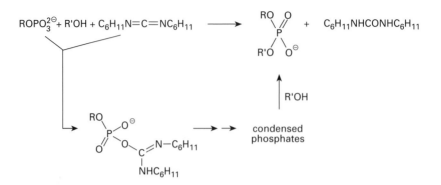

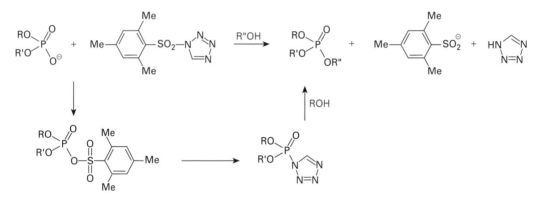

Fig. 3.39 Synthesis of phosphate diesters using dicyclohexylcarbodiimide.

Fig. 3.40 Synthesis of phosphate triesters using mesitylenesulfonyl tetrazolide.

phates and an alcohol, and so it formed the basis of the first triester syntheses of oligonucleotides (Fig. 3.38, Path B). The key step here is the condensation of a suitable nucleotide diester, as $(RO)_2POX$, with a 5′-nucleoside to give a triester, $(RO)_3PO$. To avoid problems arising from the nucleophilicity of chloride anion, the condensing agents now used are mesitylenesulphonyl tetrazolide or nitrotriazolide (Section 3.4.3). A mixed phosphoryl–sulphonyl anhydride is produced initially in which the methyl groups of the mesitylene ring provide steric hindrance to reaction at sulfur and ensure condensation at phosphorus (Fig. 3.40). Subsequent complex reactions, which may also involve condensed phosphate intermediates, lead to the triester product. The final conversion of the triester into the desired diester uses one of the specific cleavages described above (Section 3.2.1).

Syntheses via phosphite triesters

The P(III) triester route to oligonucleotides (Fig. 3.38, Path C) harnesses the intrinsically greater reactivity shown by PCl_3 compared to $POCl_3$ to achieve faster coupling steps. A major breakthrough was achieved by Matteucci and Caruthers, who established the value of alkyl

Fig. 3.41 Synthesis of phosphate triesters via phosphite triesters.
Reagents: (i) R'OH, tetrazole; (ii) R"OH, $Pr^i_2NH^{\ominus}_2$ tetrazolide$^{\ominus}$; (iii) oxidize I_2, H_2O, THF, Et_3N.

phosphoramidites ($X = NPr^i_2$ as stable 3'-derivatives of nucleosides which nonetheless react rapidly and efficiently with nucleoside 5'-hydroxyl groups in the presence of azole catalysts (Section 3.4.3). The resulting product is an unstable phosphite triester which must be oxidized immediately to give the stable phosphate triester in a process that can be cycled up to 100 times on a solid-phase support (Fig. 3.41).

Syntheses via H-phosphonate diesters

The H-phosphonate monoesters of protected 3'-nucleosides are readily prepared using PCl_3 and excess imidazole followed by mild hydrolysis. The intermediates do not require further protection at phosphorus and are rapidly and efficiently activated by a range of condensing agents, such as pivaloyl chloride (Fig. 3.38 Path D). This can be carried out before or after the addition of the second nucleoside if excess condensing agent is avoided. A dinucleoside H-phosphonate is rapidly formed in high yield (Fig. 3.42). The procedure can be repeated many times before a single oxidation step finally converts the H-phosphonate diesters into phosphate diesters. Studies on the mechanism of the condensation indicate that a mixed phosphonate anhydride is a likely intermediate.

Synthesis of phosphate monoesters

Monoesters are usually made either from triesters by selective deprotection or by direct condensation of an alcohol with a reactive phosphorylating agent, usually a polyfunctional species such as $POCl_3$. The first-formed product is immediately hydrolysed to give the desired monoester.

Triester procedures almost always call for selective protection of the nucleoside hydroxyl groups to leave only the reaction centre free. Reagents such as dibenzyl phosphorochloridate (with deprotection via catalytic hydrogenolysis), bis(2,2,2-trichloroethyl) phosphorochloridate (deprotection by zinc reduction), or 1,2-phenylene phosphorochloridate (deprotection by hy-

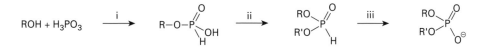

Fig. 3.42 Synthesis of phosphate diesters via H-phosphonates.
Reagents: (i) mesitylenesulphonyl tetrazolide, ROH; (ii) $(CH_3)_3CCOCl$, base, R'OH; (iii) oxidize, I_2/pyridine/H_2O.

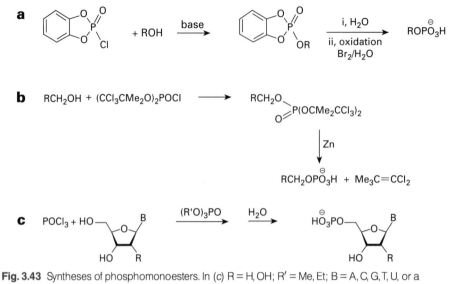

Fig. 3.43 Syntheses of phosphomonoesters. In (*c*) R = H, OH; R′ = Me, Et; B = A, C, G, T, U, or a modified base.

drolysis to the diester then oxidation removal of catechol) give good yields of phosphate monoester (Fig. 3.43a). Phosphoramidates are also useful for this purpose, as illustrated by *bis*(anilino) phosphorochloridate (with deprotection by nitrosation with amyl nitrite).

Several sterically selective phosphorylating species have been designed which condense preferentially with the primary 5′-hydroxyl function in unprotected ribo- and deoxyribonucleosides. These are illustrated by *bis*(2-*t*-butylphenyl) phosphorochloridate (deprotection by acidic hydrolysis) and *bis*(2,2,2-trichloro-1,1-dimethylethyl) phosphorochloridate, which is used in the presence of 4-dimethylaminopyridine and can be cleaved by reduction with zinc or cobalt(I) phthalocyanin (Fig. 3.43b).

For a wide range of sugars and bases, direct phosphorylation at the 5′-position of unprotected nucleosides and 2′-deoxynucleosides has proved especially easy using POCl$_3$ in a trialkyl phosphate solvent system (the Yoshikawa method). This procedure has been refined by the use of aqueous pyridine in acetonitrile solution (the Sowa–Ouchi method) which provides yields greater than 80 per cent with over 90 per cent regioselectivity (Fig. 3.43c). In the same way, the use of PSCl$_3$ leads to phosphorothioates while isotopic oxygen can be readily introduced by the generation of P^{18}OCl$_3$ *in situ*.

Summary

The monoesters and diesters of orthophosphoric acid are relatively unreactive towards hydrolysis and so are stable structural components of nucleotides and nucleic acids. Much of this stability results from the repulsion of nucleophiles by the anionic nature of these esters at neutral pH. A neighbouring vicinal hydroxyl group can have a dramatic effect in lowering this stability, as shown in the relative lability of RNA to hydrolysis by alkali and by ribonucleases. By contrast, non-ionic phosphate triesters and fully esterified condensed phosphates are much more easily hydrolysed.

Most ester syntheses require activation of the phosphate by a condensing reagent which both suppresses its ionization and generates a good leaving group. Some preparations make use of the alkylation of phosphate anions. Phosphate triesters have become the preferred intermediates for many syntheses. That is in part because they can be purified by standard techniques of organic chemistry and in part because methods have been devised for highly selective deprotection to give the desired diesters or monoesters.

3.3 Nucleoside esters of polyphosphates

3.3.1 Structures of nucleoside polyphosphates and co-enzymes

Monoalkyl esters are the most ubiquitous examples of P^1-nucleoside esters of polyphosphates. They include the ribo- and deoxyribo-nucleoside esters of pyrophosphoric acid (NDPs and dNDPs) and of tripolyphosphoric acid (NTPs and dNTPs) (Fig. 3.44). These esters are metabolically labile and participate in a huge range of C–O and P–O cleavage processes. Thiamine pyrophosphate is a co-enzyme which is a metabolically stable monoester of pyrophosphate.

Among the minor nucleoside polyphosphates, the 'magic spot' nucleotides MS1 (ppGpp) and MS2 (pppGpp) are species formed by stringent strains of *Escherichia coli* during amino acid starvation.

Dialkyl esters are biologically significant for di-, tri-, tetra-, and pentapolyphosphoric acids. In every case, the esters are located on the two terminal phosphate residues leaving (as described below) an ionic phosphate at every position which ensures stability to spontaneous hydrolysis. Several of the co-enzymes, such as co-enzyme A, flavine adenine dinucleotide (FAD), and nicotinamide adenine dinucleotide (NAD^+) are stable P^1,P^2-diesters of pyrophosphoric acid (Fig. 3.45a). Their biosynthesis involves the condensation of ATP with a monoalkyl phosphate and the pyrophosphate appears to act generally as a structural unit providing coulombic binding to appropriate enzyme residues.

The recently discovered cyclic ADP ribose (Fig. 3.45b) is a metabolic product formed by the cyclization of NAD^+ (Fig. 3.45a) to close C-1 of the second ribose on to N-1 of the adenine ring.

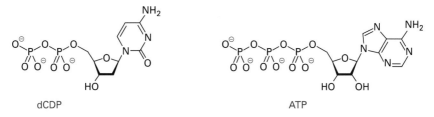

dCDP   ATP

Fig. 3.44 Structures of nucleoside 5′-di- and 5′-triphosphates.

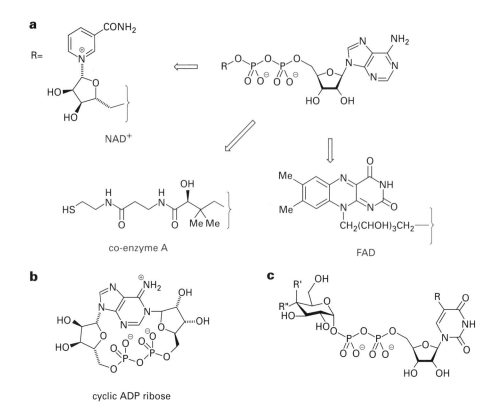

Fig. 3.45 (**a**) Structures of three adenosine coenzymes. (**b**) Structure of cyclic ADP ribose. (**c**) Structures of UDP-hexoses: R′ = H, R′ = OH UDP-glucose; R′ = OH, R″ = H UDP-galactose.

The instability of this nucleotide appears to originate not from its pyrophosphate diester but from the second glycosylic linkage. It has an important role as a second messenger as it appears to be involved in calcium signalling in cells. The active forms of many hexoses are found as pyrophosphate esters of uridine 5′-diphosphate. These include UDP-glucose, UDP-galactose, and UDP-N-acetylglucosamine (Fig. 3.45c). While they are also formed biosynthetically from UTP and the hexose-1-α-phosphate, the pyrophosphate ester is **metabolically labile** and used for catabolic processes involving C-1 of the sugar residue.

The P^1,P^4-dinucleosidyl tetraphosphates, Ap$_4$A and Ap$_4$G, are found in all cells, especially under conditions of metabolic stress (Fig. 3.46). They are produced as a result of the phosphorolysis of aminoacyl adenylates, particularly tryptophanyl and lysyl adenylates, with ATP or

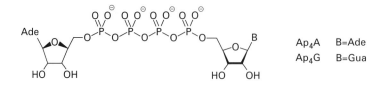

Ap$_4$A B=Ade
Ap$_4$G B=Gua

Fig. 3.46 P^1,P^4-Dinucleoside tetraphosphate structures.

GTP. While these minor nucleotides were discovered by Zamecnik in 1966, their purpose remains uncertain, though they may have a role in the initiation of DNA biosynthesis and their analogues inhibit the aggregation of blood platelets. Somewhat related structures are found in the 'caps' at the 5'-ends of eukaryotic mRNAs (Section 6.4.3), which have a 7-methyl-guanosin-5'-yl residue linked to the 5'-triphosphate (Fig. 3.37). Both of these species, and their analogues, have been targets for synthesis as a means of discovering their biological function.

3.3.2 Synthesis of monoalkyl polyphosphate esters

All of the naturally occurring nucleoside polyphosphates have at least one negative charge on each phosphate residue. This is because uncharged residues in a string of phosphates are readily hydrolysed. As a result, most syntheses have avoided the formation of fully esterified intermediates, though an early synthesis of UTP was achieved (in low yield) by the catalytic hydrogenolysis of its tetrabenzyl ester. Generally, syntheses of monoalkyl esters fall into two classes: they involve C–O bond or P–O–P bond formation.

Poulter has made good use of the alkylating properties of nucleoside 5'-O-tosylates towards pyrophosphate or tripolyphosphate anions and their methylene analogues. This has made possible direct syntheses of nucleoside 5'-diphosphates and -triphosphates and their analogues (Fig. 3.47). A more general procedure depends on the activation of a nucleoside 5'-phosphate or 5'-thiophosphate which is then able to condense with phosphate, pyrophosphate, or a methylenebisphosphonate. Among the condensing agents which have been used widely are DCCD, diphenyl phosphorochloridate, and carbonyl bisimidazole. This procedure is well suited to introduce isotopic oxygen into nucleotides in a non-stereochemically controlled fashion (Fig. 3.48) for subsequent use in positional isotope exchange (PIX) studies.

A great deal of attention has been focused on the stereochemically defined introduction of sulfur and/or oxygen isotopes into nucleotides. These have become prime tools for the investigation of the stereochemistry of enzyme-catalysed phosphoryl transfer processes. For example, the (S_p) isomer of adenosine 5'-O-thiotriphosphate, ATPαS, is readily made from AMPS by the combined action of adenylate kinase and pyruvate kinase (both enzymes can be immobilized on a polymer support for large scale syntheses) and by use of phosphoenol pyruvate with a little ATP to start the cycle. This synthesis illustrates the stereospecificity of of adenylate kinase. The ^{31}P NMR of this product has been used to identify the (R_p) and (S_p) diastereoisomers of dATPαS, which have been synthesized (Fig. 3.49) and separated by ion-exchange

X = O, CH$_2$, CF$_2$

Fig. 3.47 Synthesis of ADP and its analogues by C–O bond formation.

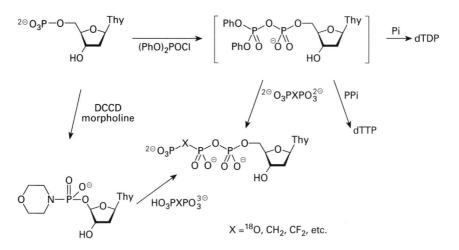

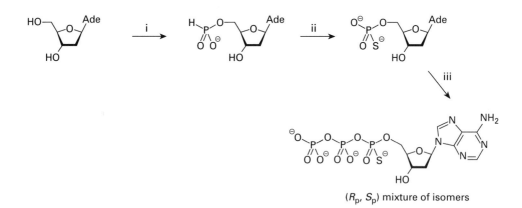

Fig. 3.48 Synthesis of nucleoside diphosphates and triphosphates and analogues by P–O bond formation.

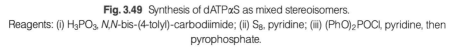

Fig. 3.49 Synthesis of dATPαS as mixed stereoisomers.
Reagents: (i) H_3PO_3, N,N-bis-(4-tolyl)-carbodiimide; (ii) S_8, pyridine; (iii) $(PhO)_2POCl$, pyridine, then pyrophosphate.

chromatography. Such species have been employed *inter alia* to show that DNA polymerase I, T4 RNA ligase, and adenylate cyclase all operate on adenine nucleotides with inversion of configuration at Pα.

For such purposes, ATP has been made with incorporation of either ^{17}O or ^{18}O in just about every possible position in the three phosphate residues. The more useful species for nucleic acid chemistry are the α-phosphate substituted nucleotides. These can be made either by *ab initio* synthesis or by the stereochemically-controlled replacement of sulfur from an α-thio-phosphate residue by isotopic oxygen. This transformation is best carried out by controlled bromine oxidation in ^{17}O- or ^{18}O-enriched water (Fig. 3.50). While this reaction proceeds with inversion of configuration, similar oxidations with *N*-bromosuccinimide or cyanogen bromide have been found to be less stereoselective. In some cases, careful choice of substrates

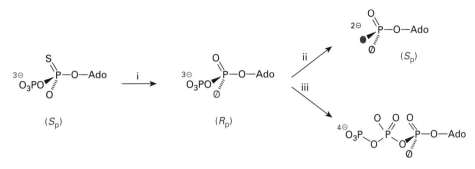

Fig. 3.50 Synthesis of (R_p)-$[\alpha$-$^{17}O]$ ATP and (S_p) AMP.
Reagents: (i) Br_2, $H_2^{17}O$; (ii) snake venom phosphodiesterase, $H_2^{18}\bullet$ (retention);
(iii) pyruvate kinase, Mg^{2+}, K^+, phosphoenolpyruvate.

has been necessary to avoid the apparent migration of the oxygen isotope to a second phosphoryl centre. An alternative procedure, though less widely applied, has used $[^{18}O]$styrene oxide, when the substitution of sulfur by oxygen proceeds with exclusive retention of stereochemistry at phosphorus.

The P^1,P^2-diesters of pyrophosphoric acid are most often made by coupling together two phosphate monoesters using DCCD, by a morpholidate procedure, or by diphenyl phosphorochloridate. A classical example is Khorana's synthesis of co-enzyme A. The same methods have worked well for syntheses of Ap_4A and its analogues, where the use of an excess of activated AMP and limiting pyrophosphate (or one of its analogues) gives acceptable yields of P^1,P^4-dinucleosidyl tetraphosphate or analogue (Fig. 3.51).

For the P^1,P^3-dialkyl triphosphates of the mRNA 'cap' structures, it has proved necessary to devise more sophisticated coupling procedures. This is partly on account of the lability of the glycosylic bond in the 7-MeGuo residues and partly because of the unsymmetrical character of the diester (Fig. 3.52). In general, the major problem encountered in the syntheses of all these species has arisen during purification because there appears to be no good alternative to ion-exchange chromatography, although high-performance silica chromatography has some uses.

Summary

The nucleoside mono- and di-esters of condensed phosphates serve on the one hand as stable structural components of co-enzymes and minor nucleotides and on the other as key catabolic species for the biosynthesis of nucleic acids and for energy-coupling purposes. Most syntheses of these esters can be accomplished by the selective activation of a phosphate monoester followed by coupling to a second phosphate or condensed phosphate. Much use has been made in nucleoside mono-, di- and tri-phosphates of stereochemical labelling of their prochiral oxygens by sulfur and/or isotopic oxygen. These chiral nucleotides have been employed for the analysis of stereochemistry of a wide range of enzyme-catalysed phosphate transfer processes which invariably proceed by single inversion and occasionally by double inversion processes.

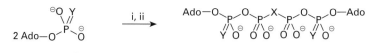

Fig. 3.51 Synthesis of Ap$_4$A and some analogues.
Reagents: (i) (PhO)$_2$POCl, pyridine; (ii) O$_3$PXPO$_3^{4-}$ (X = O, CF$_2$, etc; Y = O or S).

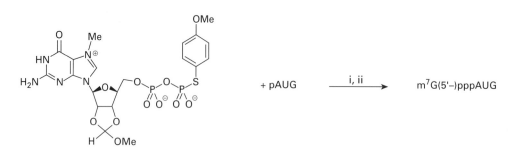

Fig. 3.52 Synthesis of the 'cap' structure of mRNA.
Reagents: (i) Ag$^+$, imidazole; (ii) H$_3$O$^+$.

3.4 Synthesis of oligodeoxyribonucleotides

An oligonucleotide is a single-stranded chain consisting of a number of nucleoside units linked together by phosphodiester bridges. In general the phosphodiesters are formed between a 3'-hydroxyl group of one nucleoside and a 5'-hydroxyl group of another, just as is found in the case of naturally occurring nucleic acids. In the context of nucleic acids the prefix 'oligo' is generally taken to denote a few nucleoside residues, whereas the prefix 'poly' means many. Recently it has become common practice to refer to all chemically synthesized, single-stranded nucleic acid chains of defined length and sequence as oligonucleotides, even if they are well beyond 100 residues in length. The term 'polynucleotide' is used to refer to a synthetic single-stranded nucleic acid of less defined length and sequence obtained by a polymerization reaction (e.g. polycytidylic acid, polyC).

3.4.1 Overall strategy

Nucleic acids are highly sensitive to a wide range of chemical reactions and only the mildest of reaction conditions can be used in assembly of an oligonucleotide chain. The heterocyclic bases are prone to alkylation, oxidation, and reduction (Section 7). The phosphodiester backbone is vulnerable to hydrolysis. In the case of DNA, acidic hydrolysis occurs more readily than alkaline hydrolysis (cf. RNA, Section 3.5), because of the lability of the C^1-N glycosylic bond to acids, especially in the case of purine nucleotides (depurination) (Section 7.1). Such considerations limit the range of chemical reactions in oligodeoxyribonucleotide synthesis to (1) mild alkaline hydrolysis; (2) very mild acidic hydrolysis; (3) mild nucleophilic displacement reactions; (4) base-catalysed elimination reactions; and (5) certain mild redox reactions (e.g. iodine or Ag(I) oxidations, reductive eliminations using zinc).

The key step in synthesis of oligodeoxyribonucleotides is the specific and sequential formation of internucleoside 3′-5′ phosphodiester linkages. The main nucleophilic centres on a 2′-deoxyribonucleoside are the 5′-hydroxyl group, the 3′-hydroxyl group and, in the case of dA, dC, and dG, the base nitrogen atoms. In order to form a specific 3′-5′ linkage between two 2′-deoxyribonucleoside units, the nucleophilic centres not involved in the linkage must be chemically protected. The first (5′)-unit must have a **protecting group** on the 5′-hydroxyl and also on the heterocyclic base, whereas the second (3′)-unit must be protected on the 3′-hydroxyl and on the heterocyclic base. In the example of joining a 5′-dA unit to a 3′-dG unit (Fig. 3.53), R^1 and R^2 protect the 5′-dA unit and R^3 and R^4 protect the 3′-dG unit. One of the two units must now be phosphorylated or phosphitylated on the only available hydroxyl group and then coupled to the other nucleoside unit in a **coupling reaction**. The resultant dinucleoside monophosphate is now fully protected. Commonly the phosphate carries a protecting group, R^5, introduced during the phosphorylation (or phosphitylation) step, such that the internucleoside phosphate becomes a **triester**. To extend the chain, one of the two terminal protecting groups, R^1 or R^3, must be removed selectively to generate a free hydroxyl function to which a new partially protected unit can now be joined.

Where R^1 and R^3 are conventional protecting groups, oligonucleotide synthesis is referred to as **solution-phase**. This has now been largely superseded by the **solid-phase** method, where either R^1 or R^3 is an insoluble polymeric or inorganic support (Section 3.4.4). Whereas extension of the chain in solution-phase synthesis is possible in either the 3′-to-5′ or 5′-to-3′ direction, in solid-phase synthesis the oligonucleotide can only be extended in one direction. The conventional protecting group removed after each coupling step (R^1 or R^3, whichever is not the solid phase) is known as a **temporary** protecting group. R^2, R^4, and R^5 (and in a sense the solid support) are all **permanent** protecting groups. They must retain stable throughout the assembly and be removed at the end of the synthesis in order to generate the final deprotected oligonucleotide.

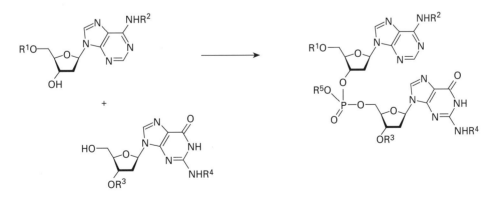

3.53 Joining of a 5′-dA unit to a 3′-dG unit.

Summary

Oligodeoxyribonucleotides must be synthesized using only the mildest reaction conditions. Specific coupling of two deoxyribonucleosides to form an internucleoside 3'-5' phosphodiester linkage occurs if all nucleophilic centres not involved in the linkage are properly protected. In solid-phase synthesis one end of the growing chain is attached to a solid support. A temporary protecting group is removed from the other end after each coupling step.

3.4.2 Protected 2'-deoxyribonucleoside units

The most convenient way to assemble an oligonucleotide is to utilize preformed deoxynucleoside phosphates as basic building units and to couple these sequentially to a terminal nucleoside attached to a solid support. Since the primary 5'-hydroxyl group is a more effective nucleophile than the secondary 3'-hydroxyl, the phosphate is best placed on the 3'-position. To achieve this selectively it is necessary to protect both the heterocyclic amino groups and the 5'-hydroxyl group.

Heterocyclic bases

Good permanent protecting groups for the exocyclic amino groups of adenine, cytosine, and guanine were first developed over 25 years ago by Khorana and co-workers for use in oligonucleotide synthesis. (NB: No protection was found necessary for thymine since it does not carry an exocyclic amino group.) The acyl protecting groups chosen were designed to remain stable for long periods under mildly basic or acidic conditions and during chromatography, but to be removed by treatment with concentrated ammonia at the end of the synthesis. Despite substantial changes that have since arisen in the speed and in the type of chemistry used in assembly of oligonucleotides, these acyl protecting groups are still the most popular today. The benzoyl group is used to protect both adenine and cytosine and the iso-butyryl group to protect guanine (Fig. 3.54).

The protecting groups are introduced by acylation of the parent 2'-deoxynucleoside. In the case of cytosine, selective acylation of the amino groups (as opposed to the sugar hydroxyl

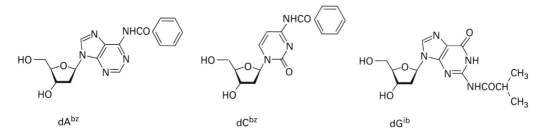

3.54 Common protecting groups for the heterocyclic bases of dA, dC, and dG.

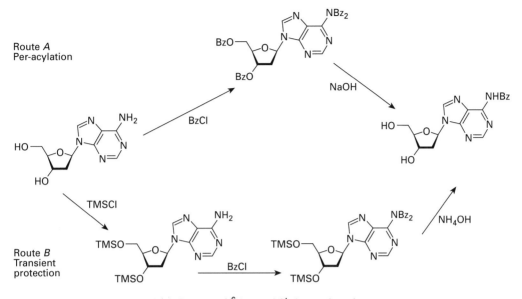

3.55 Routes to N^6-benzoyl-2'-deoxyadenosine.

groups) is possible, but the amino groups of adenine and guanine are too weakly basic for selective reaction. Thus the two most common procedures for acylation are (1) **per-acylation** and (2) **transient protection**. Per-acylation involves use of excess acylating agent to acylate both the hydroxyl and amino functions followed by selective deacylation of the hydroxyl groups. The selectivity is due to the greater stability of amides compared to esters at high pH. For example, Fig. 3.55 (Route A) shows per-acylation of 2'-deoxyadenosine. (NB: The major product of acylation is the N^6,N^6-dibenzoyl derivative. During treatment with alkali one of the N-benzoyl groups is also removed.) In transient protection (Route B), the sugar hydroxyl groups are first silylated with trimethylsilyl chloride (TMSCl). Benzoylation now gives the N^6,N^6-dibenzoyl derivative, which is deacylated to N^6-benzoyl-2'-deoxyadenosine upon treatment with concentrated ammonia solution. Deoxycytidine is benzoylated similarly.

In the case of deoxyguanosine, acylation of the amino group is effected with isobutyric anhydride using either the per-acylation or transient protection route. A complication with deoxyguanosine is that the O^6-position is vulnerable to reaction with certain reagents used in oligonucleotide synthesis. This is particularly so in the case of coupling agents and phosphorylating agents used in the phosphotriester method of synthesis (see below). The O^6-position may be protected using a variety of alkyl or aryl protecting groups, but such protection is not necessary for the now widely used phosphoramidite method (Section 3.4.3).

Recently dA and dG derivatives have been introduced which carry N-phenoxyacetyl or N,N-dimethylaminomethylene protecting groups. These groups are removed under milder ammoniacal conditions than those required for benzoyl and isobutyryl removal. In addition, unwanted depurination is reduced considerably during repetitive acidic treatments required for 5'-deprotection.

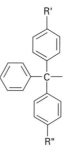

3.56 The triphenylmethyl group and its derivatives. R′ = R″ = H, triphenylmethyl (trityl, Tr); R′ = H, R″ = OCH₃, 4-methoxytriphenylmethyl (monomethoxytrityl, MMTr); R′ = R″ = OCH₃, 4,4′-dimethoxytriphenylmethyl (dimethoxytrityl, DMTr).

5′-Hydroxyl group

By far the most useful protecting group for the 5′-position is the 4,4′-dimethoxytriphenyl-methyl group (dimethoxytrityl, DMTr). This is the most easily cleaved of a family of acid-labile protecting groups (Fig. 3.56), the labilities of which increase as the number of methoxy groups increases.

The DMTr group is introduced on to the 5′-position of the N-acylated deoxynucleoside by reaction with DMTrCl in the presence of a mildly basic catalyst such as pyridine or 4-di-methylaminopyridine. For example, Fig. 3.57 shows 5′-protection of N^6-benzoyl-2′-deoxyade-nosine. The reaction is regioselective for the primary 5′-hydroxyl compared to the secondary 3′-hydroxyl partly because of the bulk of the protecting group. The DMTr group is removed by treatment with acids such as dichloroacetic or trichloroacetic in a non-aqueous solvent, con-ditions just mild enough to prevent unwanted depurination. During deprotection the brightly orange coloured dimethoxytrityl cation is liberated, which can be used as a measure of the amount of deoxynucleoside attached (Section 3.4.4).

Introduction of phosphate

In the original chemistry developed by Khorana (phosphodiester—Section 3.4.3) commer-cially available deoxynucleoside 5′-phosphates were the building blocks. In all other chemis-try developed subsequently, 5′-O-dimethoxytrityl-(N-acylated)-2′-deoxynucleosides are phosphorylated or phosphitylated at the 3′-hydroxyl (Fig. 3.57). In these cases the products of synthesis after assembly of the oligonucleotide chain are phosphate triesters where the in-ternucleoside phosphate carries a protecting group. In **phosphotriester** chemistry [P(V)] the best protecting groups are aryl (usually mono- or di-chlorophenyl derivatives). This is because an aryl phosphodiester is a much more reactive deoxynucleoside building block than an alkyl phosphodiester in a coupling reaction. For example, 5′-O-dimethoxytrityl-N^6-benzoyl-2′-deoxyadenosine gives the corresponding 3′-O-2-(chlorophenyl)phosphodiester by reaction with 2-chlorophenyl phosphoro-bis(triazolide) (Fig. 3.57, Route a). Despite this being a bifunctional phosphorylating agent it acts like a monofunctional one in the absence of any stronger catalyst. In **phosphite-triester** chemistry [P(III)] both aryl and alkyl phosphites are highly reactive species. Here the methyl group or the 2-cyanoethyl group is the protecting group of choice because it can be removed conveniently and selectively at the end of synthesis (Section 3.4.4). Once again a bifunctional reagent is used in a monofunctional capacity, but in order to obtain a sufficiently stable product, a phosphoramidite is prepared (Route *b*).

By contrast, **H-phosphonate** chemistry requires no phosphate protecting group since an assembled oligonucleotide chain containing internucleoside H-phosphonate diester links is relatively inert to the conditions of coupling (Section 3.4.3). In a sense, a proton is the protecting group! A deoxynucleoside 3′-H-phosphonate is simply prepared by the reaction of the deoxynucleoside derivative with phosphorus trichloride and imidazole or triazole plus a basic catalyst such as *N*-methylmorpholine, followed by aqueous work-up (Route *c*).

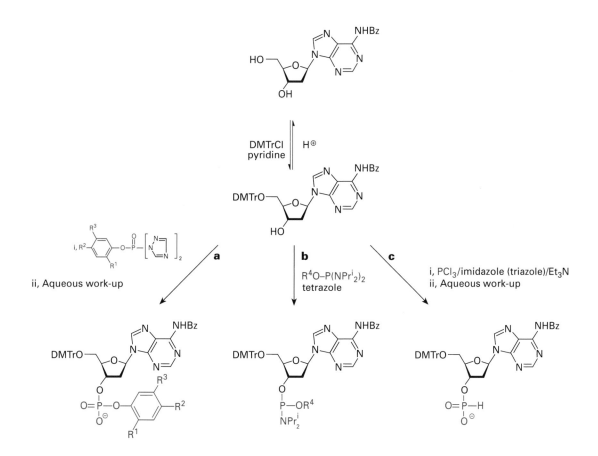

Fig. 3.57 Introduction of a 3′-phosphate by (**a**) phosphorylation, (**b**) phosphitylation, and (**c**) H-phosphonylation. $R^1, R^3 = H, R^2 = Cl$, 4-chlorophenyl; $R^2, R^3 = H, R^1 = Cl$, 2-chlorophenyl; $R^2 = H, R^1, R^3 = Cl$, 2,5-dichlorophenyl; $R^4 =$ methyl or 2-cyanoethyl.

Summary

Acyl groups are used to protect the exocyclic amino groups of adenine, cytosine, and guanine. An acid-labile dimethoxytrityl group is used to protect the 5'-hydroxyl. Phosphorylation or phosphitylation on the 3'-position now gives a suitable building block. In phosphotriester and phosphite-triester syntheses, a phosphoryl-oxygen also carries a protecting group. A proton acts as a virtual protecting group in H-phosphonate synthesis.

3.4.3 Ways of making an internucleotide bond

The develoment of an efficient way of making and internucleotide bond was for many years the most central issue in oligonucleotide synthesis. This problem has been effectively solved by the advent of phosphite triester (phosphoramidite) chemistry and, to some extent, H-phosphonate chemistry. However, an understanding of earlier phosphodiester and phosphotriester chemistry is important (Section 3.2.4).

Phosphodiester

No discussion of oligonucleotide synthesis would be complete without mention of the pioneering gene-synthesis of Khorana and his colleagues in the 1960s and early 1970s. His oligonucleotide synthesis involved coupling a 5'-protected deoxynucleoside derivative with a 3'-protected deoxynucleoside-5'-phosphomonoester (Fig. 3.58). The coupling agent (triisopropylbenzenesulphonyl chloride, TPS) activates the phosphomonoester, via a complex reaction mechanism, to give a powerful phosphorylating agent. Reaction of this with the 3'-hydroxyl group of the 5'-unit gives a dinucleoside phosphodiester. The main drawback is that the product phosphodiester is also vulnerable to phosphorylation by the activated deoxynucleoside phosphomonoester to give a trisubstituted pyrophosphate derivative. An aqueous work-up is necessary to regenerate the desired phosphodiester. Extension of the chain involves removal of the 3'-protecting group using alkali (for R = acetyl) or fluoride ion (for R = t-butyldiphenylsilyl, TBDPS) and coupling with another deoxynucleoside-5'-phosphate derivative. To prepare oligonucleotides beyond five units it is necessary to resort to coupling of preformed **blocks** containing two or more deoxynucleotide residues. Here the reactants contain several unprotected phosphodiesters which undergo considerable side-reactions that substantially reduce yields. These reactions require lengthy procedures for purification of products and substantial efforts in preparation of reactant blocks. Synthesis of an oligonucleotide of 10–15 residues (the effective limit of the method) takes upwards of 3 months.

In the late 1970s, phosphodiester chemistry was successfully applied to solid-phase synthesis. Here the 5'-MMTr group is replaced by linkage to a solid support (Section 3.4.4). Using only monomer units, the time of assembly is reduced to about two weeks and tedious purification of intermediates is avoided. However, the low yields intrinsic to phosphodiester chemistry remain.

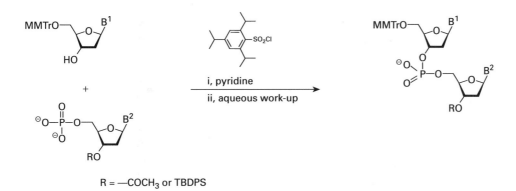

R = —COCH₃ or TBDPS

Fig. 3.58 Formation of an internucleotide bond by the phosphodiester method. B = T, C^{bz}, A^{bz}, or G^{ib}.

Phosphotriester

Although this chemistry was first applied in solution-phase synthesis, it proved particularly successful when applied to solid-phase synthesis in the early 1980s. Here a 5′-O-(chlorophenyl phosphate) (Section 3.4.3) is coupled to a deoxynucleoside attached at its 3′-position to a solid support (Fig. 3.59). The coupling agent (mesitylenesulphonyl 3-nitro-1,2,4-triazolide, MSNT) is similar to that used in phosphodiester synthesis except that 3-nitrotriazolide replaces chloride. The coupling agent activates the deoxynucleoside 3′-phosphodiester and allows reaction with the hydroxyl group of the support-bound deoxynucleoside. The rate of reaction can be enhanced by addition of a nucleophilic catalyst such as N-methylimidazole. This participates in the reaction by forming a more activated phosphorylating intermediate (an N-methylimidazolium phosphodiester), since the N-methylimidazole is a better leaving group. The product is a phosphotriester and accordingly is protected from further reaction with phosphorylating agents. The yield is therefore much better than in the case of a phosphodiester coupling, but phosphotriester chemistry could only be used satisfactorily after the

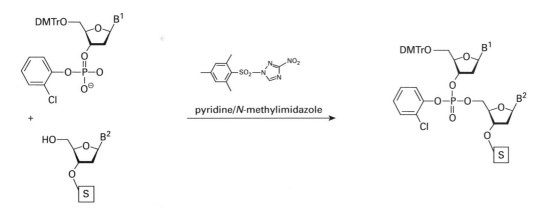

Fig. 3.59 Formation of an internucleotide bond by the solid-phase phosphotriester method.

development of selective reagents for cleavage of the aryl protecting group (Sections 3.2.2 and 3.4.4). To extend the chain, the DMTr group is removed by treatment with acid to liberate the hydroxyl group ready for further coupling. Note that the direction of extension is 3′-to-5′, in contrast to the solid-phase phosphodiester method.

Two side-reactions give rise to limitations. During coupling there is a competitive reaction (about 1 per cent) of sulphonylation of the 5′-hydroxyl group by the coupling agent. This limits the efficiency of phosphotriester coupling to 97–98 per cent and thus also the length of oligo-nucleotide attainable to about 40 residues. More seriously, deoxyguanosine residues are sub-ject to both phosphorylation and nitrotriazole substitution at the O^6-position unless an extra protecting group is used. O^6-Phosphorylation is particularly serious since this is not easily reversible (in contrast to phosphitylation) and leads to chain branching and eventually chain degradation.

The phosphotriester method is particularly useful for large-scale (multi-gram) synthesis of short oligonucleotides. Here the solid support is usually replaced by an acetyl or benzoyl group for solution phase stepwise synthesis or by a soluble polymeric carrier.

Phosphite triester

The development of phosphite triester (now often called phosphoramidite) chemistry by Car-uthers and co-workers in the early 1980s transformed oligonucleotide synthesis from a man-ual or semi-manual procedure carried out by a few specialists into a commercialized process performed using a machine. The crux of this chemistry is a highly efficient coupling reaction between a 5′-hydroxyl group of a support-bound deoxynucleoside and an alkyl 5′-DMTr-(N-acylated)-deoxynucleoside 3′-O-(N,N-diisopropylamino)phosphite (the alkyl group being methyl or 2-cyanoethyl) (Fig. 3.60). In early development of this chemistry, a chlorophosphite was used in place of the N,N-diisopropylaminophosphite, but was found to be too unstable upon storage. By contrast, a phosphoramidite is considerably less reactive and requires proto-nation at nitrogen to make the phosphoramidite into a highly reactive phosphitylating agent. Tetrazole is just sufficiently acidic to do this without causing loss of the DMTr group. The pro-duct of coupling is a dinucleoside phosphite, which must be oxidized with iodine to the phos-photriester before proceeding with chain extension.

The efficiency of coupling is extremely high (>98 per cent) and the only major side reaction is phosphitylation of the O-6 position of guanine. Fortunately, after coupling, treatment with acetic anhydride and N-methylimidazole (introduced originally to cap off any unreacted hy-droxyl groups, see Section 3.4.4) completely reverses this side reaction.

The phosphoramidite method is the procedure of choice for small-scale (microgram to milligram) synthesis) of oligodeoxyribonucleotides up to 150 residues long.

H-Phosphonate

Although the origins of this chemistry lie with Todd and co-workers in the 1950s, it has emerged with high potential in oligonucleotide synthesis only very recently. A deoxynucleo-

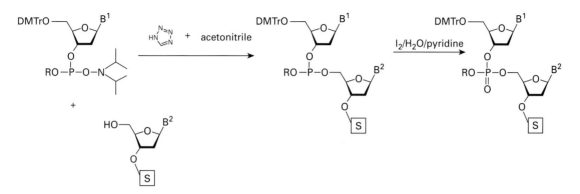

Fig. 3.60 Formation of an internucleotide bond by the solid-phase phosphoramidite method;
R = methyl or 2-cyanoethyl.

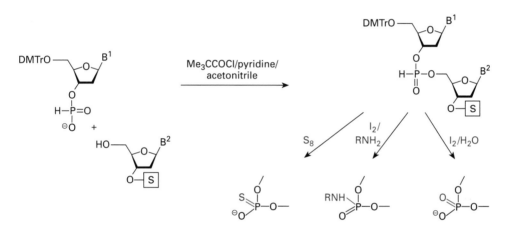

Fig. 3.61 Formation of an internucleotide bond by the solid-phase H-phosphonate method.

side 3′-O-(H-phosphonate) is essentially a tetracoordinated P(III) species, preferring this structure to the tautomeric tricoordinated phosphite monoester. Activation is achieved with a hindered acyl chloride (e.g. pivaloyl chloride), which couples the H-phosphonate to a nucleoside hydroxyl group (Fig. 3.61). The resultant H-phosphonate diester is relatively inert to further phosphitylation, such that the chain may be extended without prior oxidation. Oxidation of all phosphorus centres is carried out simultaneously at the end of the synthesis. An advantage of this chemistry is that oxidation is subject to general base catalysis and this allows nucleophiles other than water to be substituted during oxidation to give a range of oligonucleotide analogues (Section 3.4.6).

Unfortunately, a serious side reaction occurs if an H-phosphonate is premixed with activating agent before coupling. The H-phosphonate rapidly dimerizes to form a symmetrical phosphite anhydride. Subsequent reaction of this with a hydroxyl group gives rise to a branched trinucleoside derivative. The complete elimination of this side reaction, even under optimal conditions, is probably impossible and may account for the marginally lower yields obtained in practice with this route.

Summary

In phosphodiester synthesis, an activated deoxynucleoside 5'-phosphomonoester is reacted with a deoxynucleoside 3'-hydroxyl group. The lack of a protecting group on the resultant phosphodiester gives rise to serious side reactions. Much higher coupling yields are possible in phosphotriester and phosphite triester syntheses, since the triester products are protected from side reactions.

Stable deoxynucleoside 3'-phosphoramidites are powerful phosphitylating agents in the presence of tetrazole and give high coupling yields. Whereas phosphite triesters must be oxidized to phosphotriesters after each coupling step, an H-phosphonate diester is sufficiently stable to allow oxidation of all linkages following the completion of chain assembly.

3.4.4 Solid-phase synthesis

The essence of solid-phase synthesis is the use of a heterogeneous coupling reaction between a deoxynucleotide derivative in solution and another residue bound to an insoluble support. This has the advantage that a large excess of the soluble deoxynucleotide can be used to force the reaction to high yield. The support-bound product dinucleotide can be removed from the excess of reactant mononucleotide simply by filtration and washing. Other reactions can also be carried out heterogeneously and reagents removed similarly. This process is far faster than a conventional separation technique in solution and easily lends itself to mechanization. There are four essential features of solid-phase synthesis.

Attachment of the first deoxynucleoside to the support

Of the many types of support which have been used for solid-phase oligonucleotide synthesis, only controlled pore glass (CPG) and polystyrene have proved to be generally useful. Controlled pore glass beads (CPG) are ideal in being rigid and non-swellable. They are manufactured with different particle sizes and porosities, and they are chemically inert to reactions involved in oligonucleotide synthesis. Currently, 500 and 1000 Å porosities are favoured, the latter for synthesis of chains longer than 80 residues. The silylation reactions involved in functionalization of glass (introduction of reactive sites) are beyond the scope of this chapter. It is sufficient here to note that a long spacer is used to extend the sites away from the surface and ensure accessibility to all reagents. One type of spacer is illustrated (Fig. 3.62). The loading of amino groups on the glass is best kept within a narrow band of 10–50 μmol g^{-1}, below which the reactions become unreproducible and above which they are subject to steric crowding between chains. Highly cross-linked polystyrene beads have the advantage of good

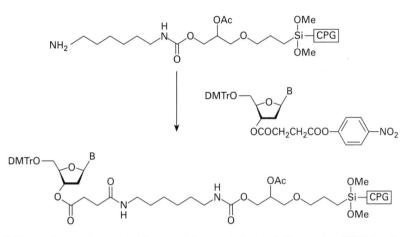

Fig. 3.62 Attachment of a nucleoside to a solid support of controlled pore glass (CPG) functionalized by a long chain alkylamine.

moisture exclusion properties and allow more efficient oligonucleotide synthesis on extremely small scale (40 nmole).

The 3'-terminal deoxynucleoside of the oligonucleotide to be synthesized is attached to the CPG support by conversion of its 5'-O-DMTr-(N-acylated)-derivative into the corresponding 3'-O-(4-nitrophenyl) succinate, which is subsequently reacted with amino groups on the support (Fig. 3.62).

Assembly of oligonucleotide chains

Assembly of the protected oligonucleotide chain is carried out by packing a small column of deoxynucleoside-loaded support and flowing solvents and reagents through in predetermined order. Columns containing only a few mg (40 nmole) or up to several grams (100 µmole) can be used. Assembly is most reproducibly accomplished using a commercial DNA synthesizer, but a manual flow system or even a small sintered glass funnel can be substituted. Machine specification varies considerably, but the basic steps involved in one cycle of nucleotide addition using the popular phosphoramidite chemistry are shown (Fig. 3.63).

Step 1. Detritylation (removal of dimethoxytrityl groups) is accomplished with dichloroacetic or trichloroacetic acid (TCA) in methylene chloride. The orange colour (dimethoxytrityl cation) liberated into solution is compared in intensity with the detritylation of the previous cycle to obtain the **coupling efficiency.**

Step 2. Activation of the phosphoramidite occurs when it is mixed with tetrazole in acetonitrile solution.

Step 3. Addition of activated phosphoramidite to the growing chain.

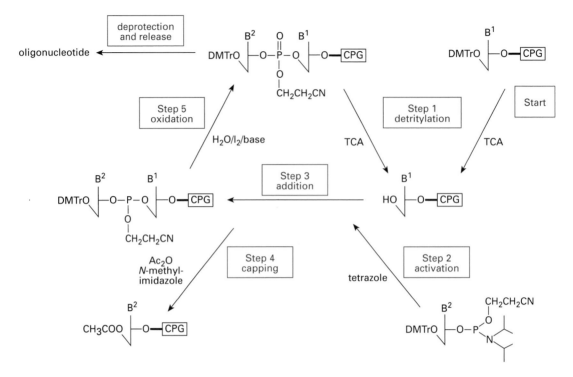

Fig. 3.63 Basic steps in a cycle of nucleotide addition by the phosphoramidite method.

Step 4.   Capping is a safety step introduced to block chains which are somehow not reacted during the coupling reaction and is designed to limit the number of failure sequences (those missing an internal residue). A fortuitous benefit of this step is that phosphitylation of the O-6 position of guanine is reversed.

Step 5.   Oxidation of the intermediate phosphite to the phosphotriester is achieved with iodine and water. Pyridine or 2,6-lutidine is used to neutralize the hydrogen iodide produced.

The cycle is repeated the requisite number of times for the length of oligonucleotide required, with each deoxynucleotide phosphoramidite being added in the desired sequence and building from 3′-to-5′.

Deprotection and removal of oligonucleotide from the support

(1)   The 5′-DMTr group is removed with the same detritrylating agent as used in the assembly cycle.

(2)   The phosphate protecting groups are removed. In the phosphotriester method, an aryl protecting group is selectively displaced using *syn*-2-nitrobenzaldoximate ion or 2-pyridine-carbaldoximate ion. The product undergoes rapid elimination in the presence of water (see Fig. 3.31). In phosphoramidite synthesis, a methyl protecting group is removed with thiophenate ion (generated with thiophenol and triethylamine) which acts by nucleophilic attack on

the methyl group, followed by hydrolysis. Alternatively, a 2-cyanoethyl group is removed by β-elimination using aqueous triethylamine or ammonia (Section 3.2.2).

(3) All heterocyclic base protecting groups are removed with concentrated aqueous ammonia.

(4) The succinate linkage is cleaved with mild aqueous base.

In phosphoramidite chemistry using the 2-cyanoethyl group, steps (2)–(4) are carried out simultaneously using aqueous ammonia. The conditions for ammonia treatment are dependent on the particular heterocyclic base protecting groups used. Conventional benzoyl and isobutyryl groups may be removed at 50° for 5 hours whereas phenoxyacetyl groups are removed within a few hours at room temperature.

Purification of oligonucleotides

The importance of a good separation technique for synthetic oligonucleotides is often neglected. Since the impurities from a large number of reactions accumulate on the support (at least those pertaining to the growing chain), these must all be resolved, preferably in a single chromatographic step. Fortunately, powerful separation methods have been developed for purification of microgram to milligram quantities.

(1) **Polyacrylamide gel electrophoresis** separates oligonucleotides by virtue of their unit charge difference. This method is particularly useful for the purification of long oligonucleotides (>50 residues), but is limited to small scale (up to 1 mg). Single nucleotide resolution is possible to well beyond 100 residues.

(2) **High performance liquid chromatography** is particularly suitable to obtain oligonucleotides of the highest purity. Ion exchange chromatography resolves predominantly by charge difference and is useful both diagnostically and preparatively for oligonucleotides up to about 50 residues. Reversed phase chromatography separates according to hydrophobicity. Here the elution position of a fully deprotected oligonucleotide is hard to predict. More reliable identification of products is achieved by leaving the highly lipophilic 5'-DMTr group intact, such that the oligonucleotide is well resolved from shorter, non-DMTr containing impurities. The DMTr group is removed following chromatography.

Summary

The four essential steps in solid-phase synthesis are: (1) attachment of the first deoxynucleoside to the support, (2) assembly of the oligonucleotide chain, (3) deprotection and removal of the oligonucleotide from the support, and (4) purification.

Assembly of the protected oligonucleotide chain can be carried out on a very small scale with the aid of a machine to add solvents and reagents reproducibly to the solid support, followed by filtration purification.

3.4.5 Duplex DNA and gene synthesis

Classical gene synthesis

The principles of gene assembly were developed 25 years ago by Khorana and his colleagues.

(1) **5′-Phosphorylation.** In order to join the 3′-end of one oligonucleotide to the 5′-end of another, a phosphate group must be attached to one of the ends. This is most easily accomplished at a 5′-end either chemically or enzymatically. The chemical procedure involves reaction of the 5′-hydroxyl group of a protected oligonucleotide whilst still attached to a solid support (following step 1 in Fig. 3.63) with a special phosphoramidite derivative (e.g. $DMTrO(CH_2)_2SO_2(CH_2)_2OP(NPr^i_2)OCH_2CH_2CN$). The DMTr group is removed by acidic treatment and during subsequent ammonia deprotection both the cyanoethyl and hydroxyethyl-sufonylethyl groups are removed to liberate the 5′-phosphate. Alternatively, and in order to introduce a ^{32}P-radiolabel, phosphorylation is carried out enzymatically using T4 polynucleotide kinase (Section 10.7.2) to transfer the γ-phosphate of ATP to the 5′-end of an oligonucleotide.

(2) **Joining of oligonucleotides (ligation).** Figure 3.64 shows schematically the construction of a gene coding for a small bovine protein, caltrin (a protein believed to inhibit calcium transport into spermatozoa). Each synthetic oligonucleotide is denoted by the position of the arrows. These are so arranged such that annealing (heating to 90°C and slow cooling to ambient temperature) of all 10 oligonucleotides simultaneously gives rise to a contiguous section of double-stranded DNA, the sequence of which corresponds to the desired protein sequence (Section 2.5.1). In this example, the oligonucleotides are 24–38 residues long, but chains of 80 residues or more have been used in gene synthesis. Oligonucleotides C2–C9 are previously phosphorylated such that, for example, the 5′-phosphate group of C3 lies adjacent to the 3′-hydroxyl group C1. The duplex is only held together by virtue of the complementary base-

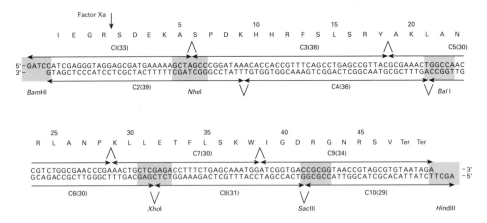

Fig. 3.64 A synthetic gene for bovine caltrin. Oligonucleotides used for the gene assembly are indicated by arrows and caret marks denote points of ligation. The amino acid sequence is shown above. Restriction enzyme recognition sites are shaded [from Heaphy *et al.* (1987). *Protein Engineering*, **1**, 425–31].

pairing between strands. The enzyme T4 DNA ligase (Section 10.7) is now used to join the juxtaposed 5'-phosphate and 3'-hydroxyl groups (the caret marks denote the joins). The overlaps are such that each oligonucleotide acts as a splint for joining of two others.

Note that oligonucleotides C1 and C10 are not phosphorylated. Each end corresponds to a sequence which would be generated by cleavage by a restriction enzyme (Section 9.3.6). Lack of a phosphate group prevents these self-complementary ends from joining to themselves during ligation. The ends are later joined to a vector DNA, previously cleaved by the same two restriction enzymes, to give a closed circular duplex ready for transformation and cloning in *E. coli* (Section 10.7.3).

Other features of the synthetic gene are internal restriction enzyme sites (shaded), which can be introduced artificially merely by judicious choice of codons (Section 6.6.8) specifying the required amino acid sequence. This particular gene is designed without a methionine initiation codon, since the protein is intended to be expressed as a fusion with another vector-encoded protein. This fusion can be cleaved to generate caltrin by treatment with the proteolytic enzyme Factor X_a (an enzyme important in the blood-clotting cascade and whose natural substrate is prothrombin) since the synthetic gene has been designed to include a section encoding the tetrapeptide recognition sequence for this enzyme.

Gene synthesis using the polymerase chain reaction (PCR)

There are numerous procedures for gene synthesis that involve use of PCR (Section 10.7.3). A particularly simple version known as **recursive PCR** has been recently used for the preparation of a 522 bp gene for human lysozyme. Oligonucleotides are synthesized 50–90 residues long but, unlike the classical approach, only their ends have complementarity (Fig. 3.65). Overlaps of 17–20 bp are designed to have annealing temperatures calculated to be in the range 52–56°. A computer search ensures that no two ends are similar in sequence. Recursive PCR is carried out in the presence of all oligonucleotides simultaneously with cycles of heating to 95°, cooling to 56°, and primed DNA repair synthesis at 72° using the four deoxynucleotide triphosphates and the thermostable Vent DNA polymerase derived from *Thermococcus litoralis*. In initial cycles (step 1), each 3'-end is extended using the opposite strand as a template to yield sections of duplex DNA. In further cycles (steps 2–5) one strand of a duplex is displaced by a primer oligonucleotide derived from one strand of a neighbouring duplex. Finally (step 6), a high concentration of the two terminal oligonucleotides drives efficient amplification of the complete duplex. Success is due to the useful characteristics of Vent DNA polymerase which has both a strand displacement activity and an active 3'-5' proofreading activity that reduces the chances of incorrect nucleotide incorporation.

3.4.6 Oligonucleotide analogues

Oligonucleotides have found application as inhibitors of gene expression. For example, they can be designed to be complementary in base sequence to and to hybridize with a section of mRNA or viral RNA (**antisense oligonucleotide**). Depending on the target site, the antisense

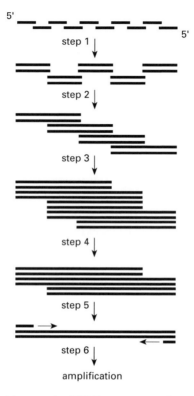

Fig. 3.65 Gene synthesis by recursive PCR. Bars represent oligonucleotides or their extension products after PCR.

oligonucleotide can act to inhibit translation (cap site or AUG initiation site), to block splicing (splice donor or acceptor site) or to induce cleavage by endogenous RNase H which cleaves RNA·DNA hybrids. Alternatively, an oligonucleotide can be designed to bind to duplex DNA, to form a triple helix (**triplex**) (Section 2.3.6), and block transcription. In order to improve resistance to nucleases, a number of oligonucleotide analogues with backbone modifications have been designed. Some of these are now being evaluated clinically as anti-viral and anti-cancer agents but their effectiveness may be limited by their generally poor ability to penetrate the cytoplasm of cells.

Phosphorothioates

Phosphorothioate linkages were first prepared by Fritz Eckstein and his colleagues. They can be made by substituting the oxidation step in either phosphite or H-phosphonate solid-phase synthesis by a sulfurization step using elemental sulfur (Figs. 3.61 and 3.66). In both these cases a mixture of (S_p) and (R_p) isomers is formed. Good chemical routes to chirally pure phosphorothioate oligonucleotides are not yet available except for short oligonucleotides. However, pure (R_p) linkages in DNA can be prepared enzymatically using pure deoxynucleoside $[S_p]$-α-thio triphosphates and a polymerase enzyme, where the reaction goes with inver-

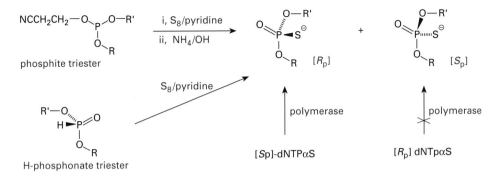

Fig. 3.66 Routes to (R_p) and (S_p) isomers of phosphorothioates.

sion of configuration. Since phosphorothioates are isopolar and isosteric with natural phosphates and since only one isomer is in general utilized as a substrate, nucleotide analogues containing P-S bonds have also been valuable in determining the stereochemical pathway of many nucleic acid recognizing enzymes.

Phosphorodithioates

Phosphorodithioate linkages are isopolar and isosteric phosphate analogues where both nonbridging oxygens are replaced by sulfur. Such linkages are non-chiral and are completely resistant to cleavage by all known nucleases. Caruthers and colleagues have developed good synthetic procedures to phosphorodithioate oligonucleotides, the most convenient of which involves coupling of a 2'-deoxynucleoside 3'-phosphorothioamidite to a support-bound nucleoside 5'-hydroxyl group followed by a sulfurization step (Fig. 3.67). The method allows a phosphorodithioate linkage to be incorporated at any position in an oligonucleotide. The β-thiobenzoylethyl group is removed by β-displacement during ammonia deprotection at the end of synthesis.

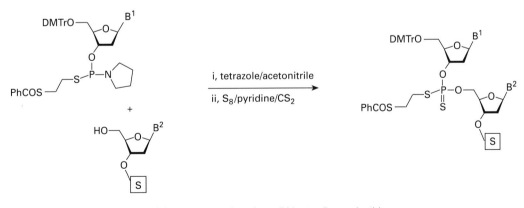

Fig. 3.67 A route to phosphorodithioate oligonucleotides.

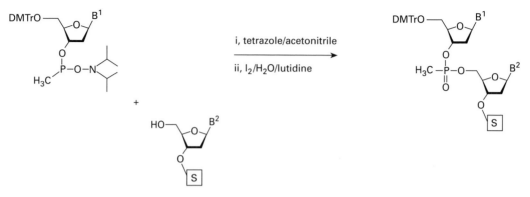

Fig. 3.68 A route to methylphosphonate analogues of oligonucleotides.

Methylphosphonates

These analogues lack the formal negative charge on phosphate and are thus more lipophilic than natural oligonucleotides. Methylphosphonate linkages can be introduced by solid-phase synthesis using 3'-O-methylphosphonamidite deoxynucleosides under similar conditions to standard phosphoramidite synthesis (Fig. 3.68). Once again a mixture of diastereomers generally results. However, it is possible to achieve the synthesis of oligonucleotide methyl-phosphonates of defined stereochemistry at phosphorus.

Peptide nucleic acids

Peptide Nucleic Acids (PNA) are analogues where the phosphate linkage is replaced by a peptide backbone composed of 2-aminoethylglycine units (Fig. 3.67a). PNAs are synthesized by solid-phase peptide synthesis (e.g. using the Boc-protected pentafluorophenyl ester of a thyminyl monomer, Fig. 3.67b). Whereas most analogues lacking a ribose sugar do not interact with natural nucleic acids, PNA have the remarkable property of displacing one strand of duplex DNA to form a triplex containing one DNA and two PNA strands.

a   **b**

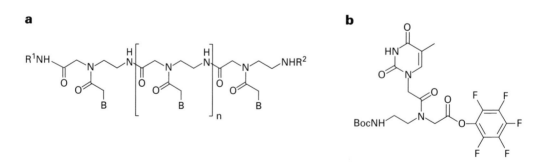

Fig. 3.69 (a) General formula for peptide nucleic acids. B represents any of the four nucleobases. R^1 and R^2 may be amino acid derivatives or other functionalities such as the intercalator acridine. **(b)** A pentafluorophenyl derivative suitable for incorporation of a thyminyl unit into PNA.

Summary

Classical duplex DNA synthesis involves enzymatic phosphorylation of terminal 5'-hydroxyl groups of oligonucleotides and their subsequent joining in a head-to-tail fashion whilst held in a perfect base-paired duplex, each oligonucleotide acting as a splint for joining of two others. Gene assembly can also be accomplished by recursive PCR which involves use of a thermostable DNA polymerase to catalyse the mutual repair of two oligonucleotides annealed at their complementary 3'-ends.

Phosphorothioate oligonucleotides are prepared by sulfurization of intermediate phosphite or H-phosphonate diesters. Phosphorodithioate and methylphosphonate linkages are introduced by solid-phase synthesis using phosphorothioamidite or methylphosphonoamidite monomeric units respectively. Peptide nucleic acids are assembled from units consisting of heterocyclic bases attached to a 2-aminoethylglycine backbone.

3.5 Synthesis of oligoribonucleotides

The development of effective chemical methods for synthesis of oligoribonucleotides has been much slower than for oligodeoxyribonucleotides. This is largely because of the added complications resulting from the 2'-hydroxyl group in a ribonucleoside and the difficulty in finding compatible protecting groups for this and other reactive centres. However, improved techniques for solid-phase oligoribonucleotide synthesis have recently emerged (Section 3.5.2) as well as powerful enzymatic synthesis procedures.

3.5.1 Protected ribonucleoside units

Heterocyclic bases

Acyl protecting groups are required for the exocyclic amino functions of adenine, cytosine, and guanine. The same benzoyl and isobutyryl groups may be used as for the synthesis of oligodeoxyribonucleotides (Section 3.4.2) and these are removed by aqueous ammonia treatment as before. However when 2'-O-silyl protection is used, it is advantageous to employ N-acyl groups which are removable under very mild ammonia conditions in order to avoid partial desilylation that occurs with extended aqueous ammonia treatment. Thus, phenoxyacetyl or dimethylaminomethylene are now commonly used for both guanine and adenine together with benzoyl or isobutyryl for cytosine. If phosphotriester couplings are to be used, extra protection for the O^6-position of guanine and the O^4-position of uracil is necessary, since these sites are particularly prone to phosphorylation or reaction with coupling agents.

Hydroxyl groups

The temporary protecting group commonly used for the 5'-hydroxyl function of ribonucleotide units is the dimethoxytrityl group (DMTr, Section 3.4.2). In this way it is easier for ribonucleotide and deoxyribonucleotide units to be used interchangeably in solid-phase synthesis. Methods for introduction of DMTr on to the 5′-position and subsequent acidic removal are very similar to those used for deoxyribonucleotides. Of paramount importance is the choice of permanent protecting group for the 2′-hydroxyl function. The choice is governed by the sensitivity of RNA to hydrolysis under alkaline conditions (Section 3.2.2). The cleavage mechanism involves generation of a 2′-oxyanion which subsequently attacks the neighbouring internucleotide linkage leading either to migration of the phosphate to the 2′-position or to chain cleavage. This attack is particularly rapid if the phosphate is in triester form. Thus the 2′-protecting group should be retained intact throughout chain assembly and be removed only after deprotection of the internucleotide phosphate. Moreover, any protecting group which upon removal gives rise to oxyanion formation should be avoided.

Two alternative 2′-protecting groups have been found to be compatible with concurrent use of DMTr for the 5′-position. These are the *t*-butyldimethylsilyl (TBDMS) group developed by Ogilvie and colleagues and the 2-fluorophenyl-4-methoxypiperidin-4-yl (Fpmp) group developed by Reese and co-workers (Fig. 3.70a and b).

(1) **The TBDMS group** is stable to acid but is removed by treatment with tetrabutylammonium fluoride. Introduction of TBDMS on to the 2′-position of 5′-*O*-DMTr-protected nucleosides is achieved by use of TBDMS chloride in the presence of imidazole (Fig. 3.71). In some cases, the use of silver salts enhances the rate and 2′-selectivity of this reaction. The 2′- and 3′-silyl isomers must be separated by careful silica gel chromatography avoiding mildly basic conditions which can lead to undesirable 2′- to 3′-migration of the TBDMS group. Phosphitylation is usually carried out by use of (2-cyanoethoxy)-*N*,*N*-diisopropylaminochlorophosphine in the presence of a tertiary base. Alternatively, preparation of the 3′-(H-phosphonate) is analogous to that previously described (Section 3.4.2).

(2) **The Fpmp group** is not strictly orthogonal to the 5′-DMTr group in that both are labile to acid. However at <pH 1 or in the presence of a non-aqueous strong acid (e.g. trichloroacetic acid), the Fpmp group is protonated and thus stable, whereas at pH 2–3 (e.g. 0.01 M HCl) it is unprotonated and cleavable. The introduction of Fpmp and other similar acetals has been significantly simplified by use of a remarkable bifunctional silylating reagent, 1,3-dichloro-1,1,3,3-tetraisopropyldisiloxane, known as the **Markiewicz reagent**. For example, N^6-pivaloyladenosine (obtained by the transient protection route, Section 3.4.2) reacts with the Markiewicz reagent to protect simultaneously both 3′- and 5′-hydroxyl groups (Fig. 3.72). Subsequent introduction of Fpmp is followed by removal of the cyclic silyl group with tetraethylammonium fluoride to give 2′-*O*-Fpmp-N^6-pivaloyladenosine (it should be noted that heterocyclic base protecting groups need not be removable under ammonia conditions as

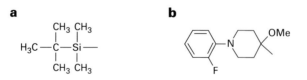

Fig. 3.70 Protecting groups for the 2′-OH group.

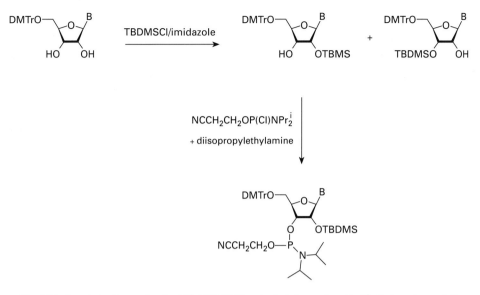

Fig. 3.71 A route for the synthesis of 2′-O-TBDMS protected ribonucleoside 3′-O-phosphoramidite synthons.

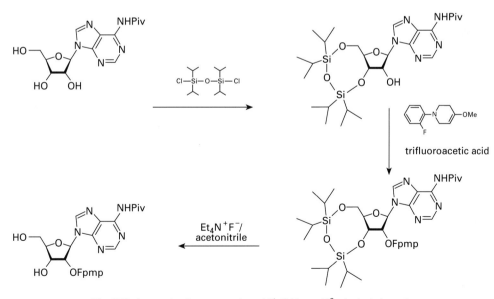

Fig. 3.72 A route for the preparation of 2′-O-Fpmp-N⁶-pivaloyladenosine.

mild as when $2'$- O-silyl protection is used). The $5'$-DMTr group is then introduced in an analogous way to that previously described (Section 3.4.2). Phosphitylation or H-phosphonylation is carried out similarly to the TBDMS route.

Summary

Ribonucleoside building blocks for solid-phase synthesis require a permanent protecting group for the $2'$-hydroxyl that is removed at the end of chain assembly ideally by a mechanism that does not give rise to formation of an oxyanion. It should also be stable to the conditions of removal of the $5'$-DMTr group. The TBDMS and the Fpmp groups both meet these criteria.

3.5.2 Assembly of oligoribonucleotides

The assembly of oligoribonucleotides is entirely analogous to that of oligodeoxyribonucleotides. The only significant difference is the slower rate of some coupling reactions which necessitates either an increase in the time of coupling or use of more powerful activators. This is particularly so when bulky $2'$-protecting groups (such as Fpmp and TBDMS) are used.

A typical coupling reaction using a $3'$-O-phosphoramidite is shown (Fig. 3.73a). Extension of the chain is effected by treatment with acid and further couplings carried out as before (Section 3.4.4). In H-phosphonate synthesis, pivaloyl chloride is once again used as activator. In this case the rate of coupling appears to be less affected by the nature of the $2'$-protecting

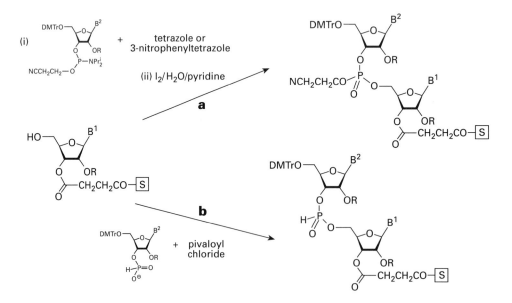

Fig. 3.73 Assembly of oligoribonucleotides by use of (**a**) phosphoramidite coupling reactions and (**b**) H–phosphonate coupling reactions.

group (Fig. 3.73b). After assembly, the phosphonate linkages are oxidized to phosphodiesters using iodine.

Deprotection and purification

Deprotection and removal of oligoribonucleotides from the support follows similar principles to those previously described (Section 3.4.4) but the reagents and conditions depend on the choice of protecting groups.

(1) **5′-Deprotection** of DMTr groups is usually carried out first by acidic treatment while oligonucleotides are still attached to the support.

(2) **Ammonia treatment** simultaneously results in cleavage of the linkage of the oligonucleotide to the support as well as the removal of heterocyclic base protecting groups and 2-cyanoethyl groups from phosphates. Mild methanolic ammonia (phenoxyacetyl protection) or aqueous ammonia (dimethylaminomethylene protection) at room temperature is sufficient for the TBDMS route whereas hot aqueous ammonia is needed for the more stable benzoyl, pivaloyl groups etc used in the Fpmp route.

(3) **2′-Deprotection** is carried out last. Removal of 2′-TBDMS groups is effected with 1 M tetrabutylammonium fluoride in tetrahydrofuran for 16–24 hours. 2′-Fpmp groups are removed by careful treatment at pH 3 with <0.01 M HCl for 16 hours. Optionally, oligoribonucleotides with terminal DMTr and 2′-Fpmp groups intact may be purified by use of reversed phase HPLC and stored in this form to avoid unwanted degradation by ribonucleases. Deprotection is completed thereafter by mild acidic treatment.

Final purification of deprotected oligoribonucleotides is carried out by polyacrylamide gel electrophoresis or by HPLC on ion exchange or reversed phase columns.

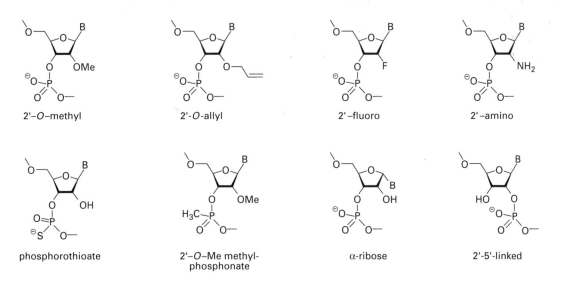

Fig. 3.74 A selection of modified ribonucleotide analogue units.

3.5.3 Oligoribonucleotide analogues

Just as in the case of oligodeoxyribonucleotides (Section 3.4.6), a range of modifications can be introduced on to the base, sugar, or phosphate moieties by preparation of appropriately protected phosphoramidite or H-phosphonate nucleotide synthons and incorporation into oligonucleotides by solid-phase synthesis. A selection of modified ribonucleotide analogues is shown in Fig. 3.74.

Summary

Solid-phase synthesis of oligoribonucleotides follows the same principles as for the synthesis of oligodeoxyribonucleotides but coupling reactions are slower. TBDMS and Fpmp are suitable 2′-protecting groups that are stable to oligonucleotide assembly yet many be removed selectively at the end of synthesis under mild conditions.

3.5.4 Enzymatic synthesis of oligoribonucleotides

Transcription by T7 RNA polymerase

A powerful method of enzymatic oligoribonucleotide synthesis makes use of the RNA polymerase from bacteriophage T7 to copy a synthetic DNA template. The template is prepared from two chemically synthesized oligodeoxyribonucleotides. Upon annealing, a duplex is formed corresponding to base-pairs −17 to +1 of the T7 promoter sequence. Position +1 is the site of initiation of transcription which in natural DNA would be in a fully base-paired duplex. Here instead, the bottom strand carries a single-stranded 5′-extension corresponding to the complement of the desired oligoribonucleotide. Transcription of this template *in vitro* using T7 RNA polymerase and nucleoside triphosphates gives up to 40 μmol of oligoribonucleotide per μmol of template (Fig. 3.75). Since the enzyme can be prepared in large quantities, the scale can be increased to give milligram quantities of oligoribonucleotides.

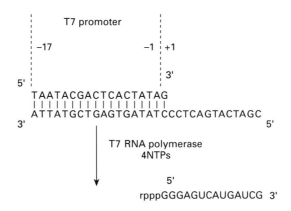

Fig. 3.75 Use of T7 RNA Polymerase to transcribe synthetic DNA templates.

There are two limitations of the method. The exact sequence from $+1$ to $+5$ influences the yield of product considerably. In cases where the sequence is substantially different from the optimum sequence, the enzyme yields a larger proportion of short, abortively initiated chains. Secondly, in some cases a non-template encoded nucleotide is added to the oligonucleotide or it may be one nucleotide shorter than expected. These by-products may need to be resolved by chromatography or gel electrophoresis. Transcription of higher efficiency is obtained by use of a double-stranded DNA template (synthetic or plasmid) which includes a T7 promoter next to the sequence to be transcribed. In order to obtain oligoribonucleotides lacking the terminal triphosphate, transcription of single-stranded or double-stranded templates may be primed by inclusion of a dinucleoside phosphate (e.g. GpG) or an analogue (e.g. 2'-O-MeGpG) such that the dinucleoside is incorporated at the 5'-end of the transcript.

Joining of oligoribonucleotides

The enzyme RNA ligase from bacteriophage T4 catalyses the joining of a 5'-phosphate group of a **donor** molecule (minimum structure pNp) to a 3'-hydroxyl group of an **acceptor** oligonucleotide (minimum structure NpNpN) (Fig. 3.76). The enzyme exhibits a high degree of preference for particular nucleotide sequences, favouring purines in the acceptor and a pyrimidine at the 5'-terminus of the donor, although there are substantial variations depending on the exact sequences of each. In order to prevent other possible joining reactions, the acceptor carries no terminal phosphate whereas the donor is phosphorylated at both ends. The 3'-phosphate of the donor is in essence a protecting group. After joining it can be removed by treatment with alkaline phosphatase to generate a free 3'-hydroxyl group and thus a new potential acceptor. A particularly useful application is the ^{32}P-labelling of RNA where T4 RNA ligase is used to catalyse the addition of (^{32}P)-pCp to the 3'-end of the RNA.

A new and more general method for joining RNA involves the use of the DNA ligase from bacteriophage T4 (normally used to join DNA, Section 10.7.2) to unite two oligoribonucleotides or segments of RNA in the presence of a complementary oligodeoxyribonucleotide splint (Fig. 3.77). Both donor and acceptor oligoribonucleotides can be obtained by T7 RNA polymerase transcription, the donor being prepared, for example, with a GpG primer (this Section) and then phosphorylated by use of ATP and T4 polynucleotide kinase. Advantages of this method include a high selectivity for acceptor oligoribonucleotides of the correct sequence (i.e. incorrect $n + 1$ long acceptor transcripts are not joined) and the lack of a need for 3'-protection of the donor oligonucleotide. The method has proved useful in incorporation of G analogues at the joined site.

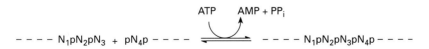

$$- - - -\ N_1pN_2pN_3\ +\ pN_4p\ - - - - \rightleftharpoons - - - -\ N_1pN_2pN_3pN_4p\ - - - -$$

Fig. 3.76 Joining ribonucleotides using RNA ligase

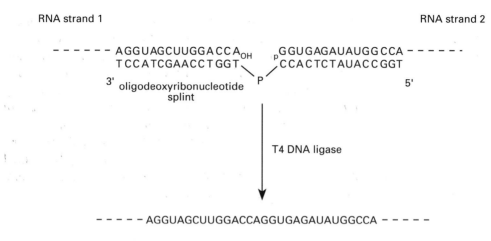

Fig. 3.77 The joining of oligoribonucleotides by use of T4 DNA ligase.

Summary

The RNA polymerase from bacteriophage T7 catalyses efficient transcription from a synthetic T7 promoter and can be used to prepare oligoribonucleotides from single-stranded oligodeoxyribonucleotide templates. The joining of the two oligoribonucleotides can be effected by use of T4 RNA ligase without the need of a template or by the use of T4 DNA ligase in the presence of an oligodeoxyribonucleotide splint.

Further reading

3.1

Hall, R. H. (1971). *Modified nucleosides in nucleic acids*. Columbia University Press, New York.

Hall, R. H. and Dunn, D. B. (1970). Natural occurrence of the modified nucleosides. In *Handbook of biochemistry. Selected data for molecular biology* (ed. H. A. Sober), 2nd edn. CRC Press, Cleveland, Ohio, pp. G99–G105.

Hanessian, S. and Pernet, A. G. (1976). Synthesis of naturally occurring *C*-nucleosides, their analogues, and functionalized *C*-glycosyl precursors. *Adv. Carb. Chem. Biochem.*, **33**, 111–88.

Hobbs, J. B. (1978–1987). In *Organo-phosphorus chemistry. Specialist periodical reports*, Vols. 9–18. Royal Society of Chemistry, London.

Hutchinson, D. W. (1979). *Comprehensive organic chemistry*, Vol. 5. Pergamon Press, Oxford and London, pp. 105–45.

Scheit, K. H. (1983). *Modified nucleotides*. Verlag Chemie, Weinheim, FRG.

Walker, R. T. (1979). *Comprehensive organic chemistry*, Vol. 5, Pergamon Press, Oxford and London, pp. 57–104.

Watanabe, K. (1974). On mechanisms of nucleoside synthesis by condensation reactions. *J. Carbohydrates Nucleosides Nucleotides*, **1**, 1–37.

3.2

Cohn, M. (1982). some properties of the phosphorothioate analogues of adenosine triphosphate as substrates of enzymic reactions. *Acc. Chem. Res.*, **15**, 326–32.

Cox, J. R. and Ramsey, O. B. (1964). Mechanisms of nucleophilic substitution in phosphate esters. *Chem. Revs.*, **64**, 317–52.

Eckstein, F. (1979). Phosphorothioate analogues of nucleotides. *Acc. Chem. Res.*, **12**, 204–10.

Eckstein, F. (1983). Phosphorothioate analogues of nucleotides—Tools for the investigation of biochemical processes. *Angew. Chem. Int. Ed.*, **22**, 423–39.

Frey, P. A. (1982). Stereochemistry of enzymatic reactions of phosphates. *Tetrahedron*, **38**, 1541–67.

Goldwhite, H. (1981). *Introduction to phosphorus chemistry*. Cambridge University Press, Cambridge and London.

Knowles, J. R. (1980). Enzyme-catalyzed phosphoryl transfer reactions. *Ann. Rev. Biochem.*, **49**, 877–919.

Lowe, G. (1983). Chiral[^{16}O, ^{17}O, ^{18}O] phosphate esters. *Acc. Chem. Res.*, **16**, 244–51.

3.3

Khorana, H. G. (1961). *Some recent developments in the chemistry of phosphate esters of biological interest.* Wiley, New York.

Wood, H. S. C. (1979). *Comprehensive organic chemistry,* Vol. 5. Pergamon Press, Oxford and London, pp. 489–548.

3.4

Beaucage, S. L. and Iyer, R. P. (1992). Advances in the synthesis of oligonucleotides by the phosphoramidite approach. *Tetrahedron*, **48**, 2223–311.

Beaucage, S. L. and Iyer, R. P. (1993). The synthesis of modified oligonucleotides by phosphoramidite approach and their applications. *Tetrahedron*, **49**, 6123–94.

Caruthers, M. H. (1985). Gene synthesis machines: DNA chemistry and its uses. *Science*, **230**, 281–5.

Crooke, S. T. and Lebleu, B. (1993). *Antisense research and applications.* CRC Press, Boca Raton, FL.

Eckstein, F. (1985). Nucleoside phosphorothioates. *Ann. Rev. Biochem.*, **54**, 367–402.

Eckstein, F. (1991). *Oligonucleotides and analogues: a practical approach.* IRL Press, Oxford.

Eckstein, F. and Gish, G. (1989). Phosphorothioates in molecular biology. *Trends Biol. Sci.*, **14**, 97–100.

Engels, J. and Uhlmann, E. (1988). Gene synthesis. *Adv. Biochem. Eng./Biotech.*, **37**, 73–127.

Lilley, D. M. J. and Dahlberg, J. E. (1992). DNA structures part A. Synthesis and physical analysis of DNA. *Methods Enzymol.*, **211**, 3–64.

Marshall, W. G. and Caruthers, M. H. (1993). Phosphorodithioate DNA as a potential therapeutic drug. *Science*, **259**, 1564–9.

Miller, P. S. (1991). Oligonucleoside methylphosphonates as antisense reagents. *Biotechnology*, **9**, 358–62.

Miller, P. S., Reddy, P. M., Murakami, A., Blake, K. R., Lin, S.-B., and Agris, C. H. (1986). Solid-phase synthesis of oligodeoxyribonucleoside methylphosphonates. *Biochemistry*, **25**, 5092–7.

Nielsen, P. E., Egholm, M., Berg, R. H., and Buchardt, O. (1991). Sequence-selective recognition of DNA by strand displacement with a thymine-substituted polyamide. *Science*, **254**, 1497–500.

3.5

Gait, M. J., Pritchard, C., and Slim, G. (1991). Oligoribonucleotide synthesis. In *Oligonucleotides and analogues: a practical approach* (ed. F. Eckstein), IRL Press, Oxford, pp. 25–48.

Heidenreich, O., Pieken, W., and Eckstein, F. (1993). Chemically modified RNA: approaches and applications. *FASEB J.*, **7**, 90–6.

Middleton T., Herlihy, W. C., Schimmel, P., and Munro, H. N. (1985). Synthesis and purification of oligoribonucleotides using T4 RNA ligase and reverse-phase chromatography. *Anal. Biochem.*, **144**, 110–17.

Milligan, J. F., Groebe, D. R., Witherell, G. W., and Uhlenbeck, O. C. (1987). Oligoribonucleotide synthesis using T7 RNA polymerase and synthetic DNA templates. *Nucleic Acids Res.*, **15**, 8783–98.

Moore, M. J. and Sharp, P. A. (1992). Site-specific modification of pre-mRNA: the 2'-hydroxyl groups at the splice site. *Science*, **256**, 992–7.

Rao, M. V., Reese, C. B., Schehlmann, V., and Yu, P. S. (1993). Use of 1-(2-fluorophenyl)-4-methoxypiperidin-4-yl (Fpmp) protecting group in the solid-phase synthesis of oligo- and poly-ribonucleotides. *J. Chem. Soc. Perkin 1*, 43–55.

Reese, C. B. (1987). The problem of 2'-protection in rapid oligoribonucleotide synthesis. *Nucleosides and Nucleotides*, **6**, 121–9.

Usman, N. and Cedergren, R. (1992). Exploiting the chemical synthesis of RNA. *Trends Biochem. Sci.*, **17**, 334–9.

Biosynthesis of Nucleotides

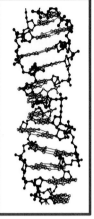

Nucleotides play a key role as the precursors of DNA and RNA, as activated intermediates in many biosynthetic processes, and as metabolic regulators. One particular nucleotide, adenosine 5′-triphosphate (ATP), is an important energy source. For example, a human being turns over 40 kg of ATP per day and during exercise can require 0.5 kg per minute. The biosynthesis of nucleotides involves both constructive (anabolic) and destructive (catabolic) pathways, but the importance of a particular pathway varies substantially between organisms. Moreover, specialized cells sometimes have their own unique metabolic pathways. In this chapter we will concentrate on only the general principles of nucleotide and nucleic acid metabolism and then show how certain steps are prime targets for biosynthetic interference, especially for the design of anti-cancer and anti-viral agents.

4.1 Biosynthesis of purine nucleotides

4.1.1 *De novo* pathways

The key intermediate in the biosynthesis of both pyrimidines and purines is α-D-5-phosphoribosyl-1-pyrophosphate (PRPP) which is formed from α-D-ribose 5-phosphate by a reaction catalysed by the enzyme ribose phosphate pyrophosphokinase (Fig. 4.1). Adenosine 5′-triphosphate acts as the donor of pyrophosphate while ribose 5-phosphate comes mainly from the pentose phosphate pathway.

   In contrast to pyrimidine nucleotide biosynthesis, where a preformed heterocycle is incorporated intact (Section 4.2), in purine nucleotide biosynthesis the purine ring is constructed gradually. The first irreversible step (the **committed step**) is the displacement of pyro-

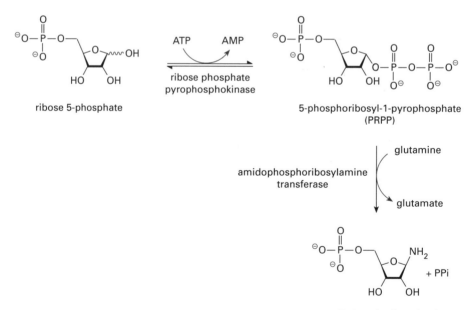

Fig. 4.1 Biosynthesis of 5-phosphoribosylamine.

phosphate at C-1 of PRPP by the side chain amino group of glutamine to give β-D-5-phosphoribosylamine (Fig. 4.1). There is an inversion at C-1 such that the glycosylic bond is now in the β-configuration, the stereochemistry characteristic of all naturally occurring nucleotides. The equilibrium in this reaction is displaced towards the phosphoribosylamine by the hydrolysis of the pyrophosphate co-product.

The five carbon atoms and the remaining three nitrogen atoms of the purine skeleton are derived from no less than six different precursor sources and assembled by nine successive steps (Fig. 4.2). These steps are:

(1) reaction of PRPP with glycine to give glycinamide ribonucleotide;
(2) formylation of the α-amino terminus of the glycine moiety by N^{10}-formyltetrahydrofolate to give α-N-formylglycinamide ribonucleotide;
(3) conversion into the corresponding glycinamidine with a new nitrogen atom derived from glutamine;

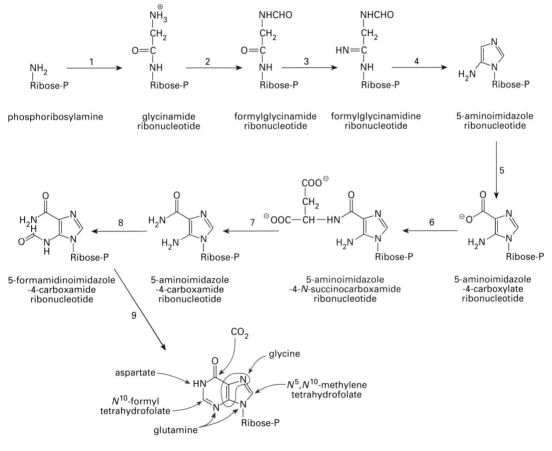

Fig. 4.2 Formation of the purine ring; biosynthesis of IMP.

(4) ring closure to give 5-aminoimidazole ribonucleotide;
(5) carboxylation of the imidazole C-4 (the carbon atom coming from CO_2);
(6) addition of aspartate;
(7) removal of the carbon skeleton of aspartate (as fumarate), leaving behind its amino group to give 5-aminoimidazole-4-carboxamide ribonucleotide;
(8) formylation of the amino group by N^{10}-formyltetrahydrofolate; and
(9) dehydration and ring closure to form inosine 5′-monophosphate (IMP).

Inosine is a nucleoside rarely found in nucleic acids except in the 'wobble' position of some tRNAs (Section 6.6.8). In such cases, the inosine does not come from IMP nor from deamination of adenosine by an adenosine deaminase, but instead the adenine base is removed from adenosine in the preformed tRNA and replaced by hypoxanthine.

IMP is used entirely for the production of the natural purine nucleotides, adenosine 5′-monophosphate (AMP) and guanosine 5′-monophosphate (GMP) (Fig. 4.3). AMP receives its amino group at C-6 from aspartate. GMP is derived in two steps from xanthosine 5′-monophosphate (XMP) with the final amino group being donated by glutamine. In both these pathways, a carbonyl group of an amide is replaced by an amino group to give an amidine. This is a common type of mechanism whereby the amide is phosphorylated by ATP or GTP to its imido-O-phosphoryl ester and then the phosphoryl ester displaced by an amine. The leaving group can be inorganic phosphate, pyrophosphate, or even AMP, while the displacing nucleophile is ammonia, the side chain amide of glutamine, or the α-amino group of aspartate (Fig. 4.4).

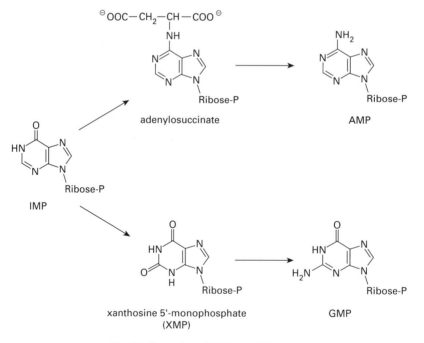

Fig. 4.3 Formation of AMP and GMP from IMP.

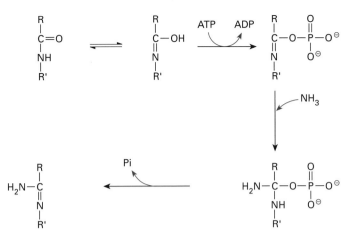

Fig. 4.4 General mechanism for biosynthetic formation of an amidine from an amide.

Steps in the biosynthesis of purine nucleotides furnish good examples of a common control mechanism, namely **feedback inhibition**, where an enzyme catalysing an early step in the pathway is inhibited by the final product of the pathway. For example, the enzyme ribose phosphate pyrophosphokinase (Fig. 4.1) is inhibited by AMP, GMP, and IMP and this inhibition regulates the level of PRPP. Similarly, the enzyme amidophosphoribosyl transferase, which is responsible for catalysing the committed step, is inhibited by a number of purine ribonucleotides including AMP and GMP, which act synergistically. AMP and GMP also inhibit the conversion of IMP into their own immediate precursors, adenylosuccinate and XMP. Another control feature is that GTP is required in the synthesis of AMP, while ATP is required in the synthesis of GMP.

4.1.2 Salvage pathways

Most organisms also use a second pathway of biosynthesis known as **salvage**. This is advantageous since degradation products of nucleic acids can be recycled rather than destroyed, which is much less costly than the energy-demanding reactions of the *de novo* pathways. In some cancer cells or virally infected cells, extra synthesis capacity is required. Here salvage may become the dominant pathway and hence becomes a target for chemotherapeutic inhibitors.

Purine bases, which arise by hydrolytic degradation of nucleotides and nucleic acids, react with PRPP to give the corresponding purine ribonucleotide and pyrophosphate is eliminated (Fig. 4.5). One enzyme, adenine phosphoribosyl transferase, is specific for the reaction with adenine, whereas another enzyme, hypoxanthine-guanine phosphoribosyl transferase (HGPRT) catalyses the formation of IMP and GMP. A deficiency of HGPRT is responsible for the serious Lesch–Nyhan syndrome, which is often characterized by self-mutilation, mental deficiency, and spasticity. Here, elevated concentrations of PRPP give rise to an increase in *de novo* purine nucleotide synthesis and degradation to uric acid (Section 4.5). The purine ana-

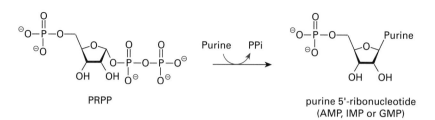

Fig. 4.5 Salvage biosynthesis of purine ribonucleotides.

logue 6-mercaptopurine is used in cancer chemotherapy and particularly against lympho-blastic leukaemia. It is utilised by HGPRT to form the corresponding ribonucleotide, which causes feedback inhibition of amidophosphoribosyl transferase in the synthesis of 5-phos-phoribosylamine from PRPP (see Fig. 4.1) and also prevents IMP being converted into XMP and into adenylosuccinate (see Fig. 4.3).

Another salvage route involves the reaction of a purine (or purine analogue) with ribose 1-phosphate. The reaction is catalysed by a nucleoside phosphorylase (nucleoside phospho-transferase) and the resultant ribonucleoside is then converted into its corresponding 5'-nu-cleotide by a cellular kinase. Similarly a nucleoside phosphotransferase produces deoxy-ribonucleosides from purines and 2-deoxyribose 1-phosphate.

Summary

In biosynthesis of purine nucleotides, the heterocyclic ring is gradually built up from β-D-phosphoribosylamine using six different precursors to give IMP. IMP is then used for the production of AMP and GMP. Several enzymes in the biosynthetic pathway are **inhibited by feedback** in order to regulate the production of purine nucleotides. Purine bases arise by degradation of nucleotides and can be converted into nucleosides by means of **salvage pathways**. Nucleotides are formed from these nucleosides through the action of kinases.

4.2 Biosynthesis of pyrimidine nucleotides

4.2.1 *De novo* pathways

Carbamoyl phosphate is an important intermediate in pyrimidine biosynthesis and is also used in the biosynthesis of urea. Carbamoyl phosphate is formed from glutamine and bicar-bonate in a reaction catalysed by one of two carbamoyl phosphate synthetases (a different one is used in the urea pathway). The reaction requires ATP as an energy source (Fig. 4.6). The committed step is the subsequent formation of N-carbamoyl aspartate from carbamoyl phos-phate and aspartate. This step is subject to feedback inhibition by CTP which is the final pro-duct of the pathway, whereas the synthesis of carbamoyl phosphate is inhibited by UMP. In the next step the pyrimidine ring is formed by cyclization and loss of water followed by a de-hydrogenation to give orotate.

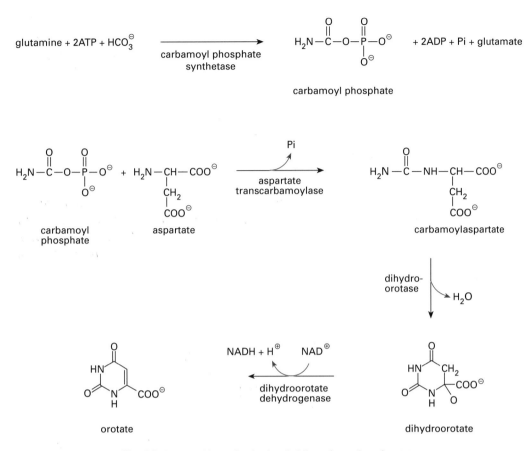

Fig. 4.6 *De novo* biosynthesis of pyrimidines; formation of orotate.

The enzymes involved in the last three steps form a multi-enzyme complex in eukaryotes (but not in prokaryotes) and are located on a single 200 kDa polypeptide chain. A potent inhibitor of the first enzyme, aspartate transcarbamoylase, is *N*-phosphonoacetyl-L-aspartate (PALA) (Fig. 4.7). PALA is an example of a **transition state inhibitor,** which works by mimicking the transition state of a reaction. PALA binds tightly to aspartate transcarbamoylase and has proved to be useful in the production and isolation of the enzyme complex.

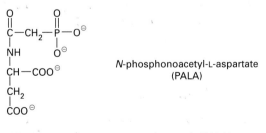

Fig. 4.7 *N*-Phosphonoacetyl-L-aspartate (PALA).

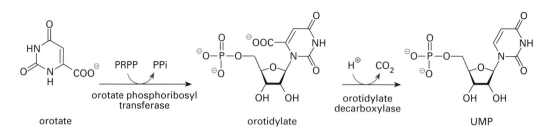

Fig. 4.8 Formation of UMP from orotate.

Orotate then reacts with PRPP to give orotidylate (Fig. 4.8). There is an inversion of configuration at C-1 and a β-nucleotide is formed. The equilibrium of the reaction is once again driven forward by hydrolysis of pyrophosphate. Finally, UMP is produced by decarboxylation. The other pyrimidine nucleotides are derived from UMP after its conversion into UTP (Section 4.3).

4.2.2 Salvage pathways

The enzyme orotate phosphoribosyl transferase, which is involved in the production of orotidylate from orotate, will also utilize a number of other pyrimidines, produced as a result of hydrolysis of DNA or RNA. In a similar way as for the salvage of purines, phosphorylases will catalyse nucleoside formation from a variety of pyrimidines and either ribose 1-phosphate or 2-deoxyribose 1-phosphate. A cellular kinase is also required to convert the nucleoside into its corresponding 5′-nucleotide. Uridine kinase will accept both uridine and cytosine as substrates. Thymidine kinase will accept deoxyuridine as well as deoxythymidine. The fact that many viral thymidine kinases have a lower specificity for their substrates enables a distinction to be made between normal and virally-infected cells and has led to a strategy for viral interference (Section 4.7.2).

Nucleoside transferases will catalyse base-exchange between nucleosides exclusively in the 2′-deoxy series.

Summary

In pyrimidine nucleotide biosynthesis the heterocyclic ring is pre-formed as orotate and reacts with PRPP to give orotidylate. UMP is formed by decarboxylation of orotidylate. Salvage pathways operate in a similar way to that for purines.

4.3 Nucleoside di- and triphosphates

The immediate biosynthetic precursors of the nucleic acids are normally the nucleoside triphosphates, whereas in energy conversions diphosphates are also used. Diphosphates are ob-

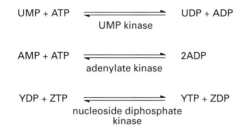

UMP + ATP $\rightleftharpoons$ UDP + ADP
UMP kinase

AMP + ATP $\rightleftharpoons$ 2ADP
adenylate kinase

YDP + ZTP $\rightleftharpoons$ YTP + ZDP
nucleoside diphosphate
kinase

Fig. 4.9 Biosynthesis of nucleoside di- and triphosphates.

tained from the corresponding monophosphates by means of a specific nucleoside monophosphate kinase. Adenylate kinase converts AMP into ADP while UMP kinase converts UMP into UDP. Both enzymes utilize ATP as the phosphoryl donor. Nucleoside triphosphates are interconvertible with diphosphates through nucleoside diphosphate kinase, an enzyme that has a broad specificity. Thus Y and Z (Fig. 4.9) can be any of several purine or pyrimidine ribo- or deoxyribonucleosides.

Cytidine triphosphate (CTP) is formed from UTP by replacement of the carbonyl oxygen atom at C-4 by an amino group. In *Escherichia coli* the donor is ammonia whereas in mammals it is the amide group of glutamine. In both cases ATP is required for the reaction.

4.4 Deoxyribonucleotides

Deoxyribonucleotides are formed by reduction of the corresponding ribonucleotides. The $2'$-hydroxyl group of the ribose is replaced by a hydrogen atom in a reaction that takes place at the level of the ribonucleoside $5'$-diphosphate. The mechanism appears to be simple at first sight but it is in fact extremely complicated. The key enzyme is ribonucleotide reductase (ribonucleoside diphosphate reductase) and the electrons required for reduction of the ribose are transferred from NADPH to sulphydryl groups at the catalytic site of the enzyme. The enzyme from *E. coli* is a prototype for most eukaryotic reductases. A larger subunit (2×86 kDa) binds the NTP substrate, the smaller subunit (2×43 kDa) contains a binuclear iron centre and a tyrosyl radical at residue 122. A mechanism based on all the available data is shown (Fig. 4.10). The reduction of ribonucleoside diphosphates is precisely controlled by allosteric interactions (an allosteric enzyme is one in which the binding of another substance, usually product, alters its kinetic behaviour). Ribonucleotide reductase has two types of allosteric site that bind a number of nucleoside $5'$-triphosphates and lead to a variety of conformations, each with different catalytic properties.

If any dUTP is formed from dUDP, it is rapidly hydrolysed to dUMP by an active dUTPase, which prevents the incorporation of dUTP into DNA. If uracil residues occur in DNA, they are thus likely to have arisen through deamination of cytosines and these mutations are removed by a uracil glycosylase (Section 7.11.2). As a result, deoxythymidine $5'$-triphosphate (dTTP) is the dioxopyrimidine nucleotide incorporated into DNA. First, deoxythymidine $5'$-monophosphate (dTMP) is biosynthesized from dUMP via the enzyme thymidylate synthetase. The methyl group is derived from N^5,N^{10}-methylenetetrahydrofolate, which also acts as an

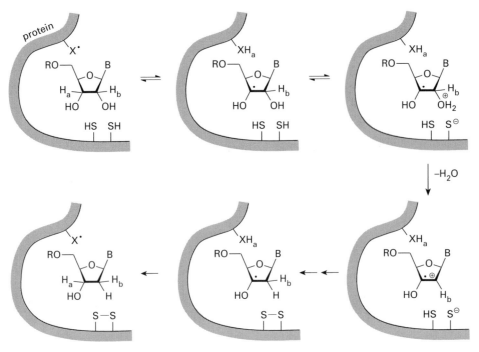

Fig. 4.10   Postulated mechanism for reduction of nucleotides to deoxynucleotides by *E. coli* ribonucleotide reductase (X⋅ = Tyr^{122}−O⋅).

electron donor (Fig. 4.11) and is oxidized to dihydrofolate. Tetrahydrofolate is regenerated by
dihydrofolate reductase using NADPH as the reductant. These two enzymes are excellent targets for cancer chemotherapy because cancer cells have an increased level of DNA synthesis
and thus a heavy requirement for dTMP (see 5-fluorouracil and methotrexate in
Section 4.7.1).

dTMP is next converted into dTTP in two stages by means of a thymidylate kinase and then
a nucleoside diphosphate kinase. In virally infected cells, the viral thymidine kinase (Section
4.3) often also plays the role of a thymidylate kinase.

Summary

Specific nucleoside monophosphate kinases catalyse the formation of nucleoside 5'-
diphosphates from monophosphates. Triphosphates are interconvertible with diphosphates
through the enzyme nucleoside diphosphate kinase. 2'-Deoxyribonucleotides are formed
by reduction of the corresponding ribonucleotides, the reaction taking place at the
diphosphate level. dUTP is hydrolysed to dUMP to prevent its incorporation into DNA.
dUMP is converted into dTTP in three steps.

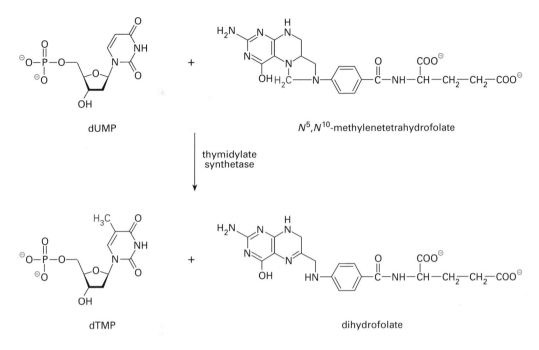

Fig. 4.11 Formation of dTMP from dUMP.

4.5 Catabolism of nucleotides

The degradation of nucleotides is of major importance as a target for drug design. RNA is metabolically much more labile than DNA and is constantly being synthesized and degraded. Degradation occurs initially through the action of ribonucleases and deoxyribonucleases which give oligonucleotides that are further broken down to nucleotides by phosphodiesterases.

Nucleotides are hydrolysed to nucleosides by nucleotidases (and also by phosphatases). Of great importance is the final cleavage of nucleosides by inorganic phosphate to bases and ribose 1-phosphate (or 2-deoxyribose 1-phosphate) catalysed by the widely distributed enzyme purine nucleoside phosphorylase (Fig. 4.12). The ribose phosphate can then be isomerized to ribose 5-phosphate and reused for the synthesis of PRPP. In mammalian tissues, adenosine and deoxyadenosine are resistant to the phosphorylase. AMP is therefore deaminated by adenylate deaminase to IMP and adenosine to inosine by adenosine deaminase, an enzyme which is thought to be present in elevated levels in leukaemic cells. Oxidation of hypoxanthine is catalysed by xanthine oxidase to give xanthine, which is also the deamination product of guanine. Xanthine is further oxidized to uric acid, which in humans is excreted in the urine. Gout is a painful disease caused by excessive production of monosodium urate which is deposited as crystals in the cartilage of joints. Allopurinol, which is an analogue of hypoxanthine, is used to treat gout by acting as a substrate inhibitor of xanthine oxidase. Since the allopurinol becomes tightly bound to the enzyme it is known as a **suicide inhibitor.**

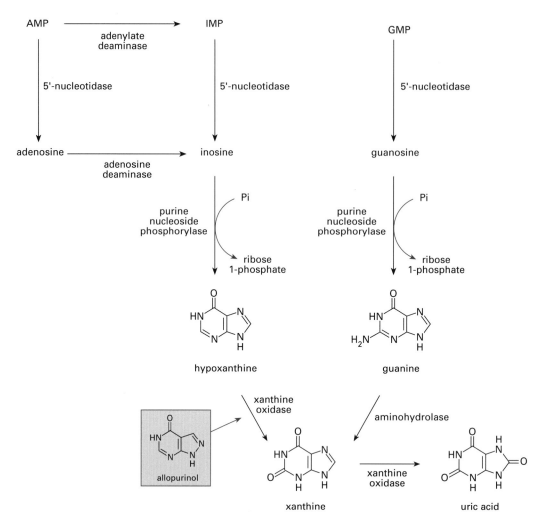

Fig. 4.12 Catabolism of purine nucleotides.

A variety of deaminases can convert cytidine, 2'-deoxycytidine, and dCMP into their corresponding uracil-containing derivatives. All of these products can be broken down to give uracil which, unlike purines, is degraded reductively (Fig. 4.13). Thymine is degraded in an exactly analogous way to uracil.

4.6 Polymerization of nucleotides

While the complex series of reactions involved in the polymerization of nucleotides to form DNA and RNA are described in detail in Chapter 5, we are here concerned primarily with the polymerases as potential targets for chemotherapy since they are the enzymes responsible for polymerization of nucleoside 5'-triphosphates into nucleic acids. In each case there is a re-

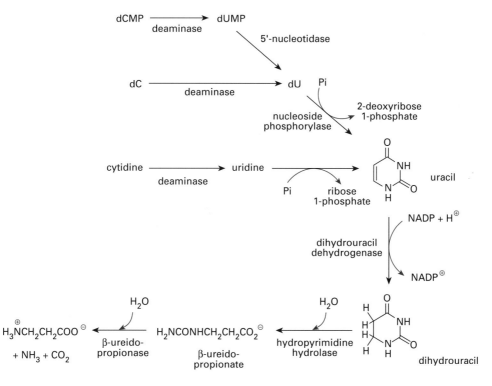

Fig. 4.13 Catabolism of pyrimidine nucleotides.

quirement for a template strand of nucleic acid and an oligoribo- or oligodeoxyribonucleotide primer.

4.6.1 DNA polymerases

In *E. coli* there are three DNA polymerases. DNA polymerase III is the main replicating enzyme and is composed of at least eight different subunits. DNA polymerase I, which was discovered earlier, works primarily as a repair enzyme, but is still indispensable for chromosome replication. The function of DNA polymerase II is not clearly understood. In eukaryotes there are also three DNA polymerases, designated α, β, and γ. DNA polymerase α is involved in cellular replication and is found in the nucleus whereas DNA polymerase γ is exclusively located in the mitochondria. DNA polymerase β is found both in the nucleus and in the cytoplasm but its function is less clear. All these cellular polymerases use DNA as a template and polymerize in a 5′-to-3′ direction. Although these polymerases are potential targets for cancer chemotherapy (e.g. for intercalators, Section 8.4), much greater scope is available for anti-viral therapy since many viruses (e.g. herpesvirus) encode their own DNA polymerases and these often have substrate specificities different from those of the cellular enzymes (Section 4.7.2).

One group of RNA-containing viruses, the retroviruses, replicates via a double-stranded DNA intermediate. Retroviruses are important since many cause cancer and one of them, human immunodeficiency virus (HIV), is responsible for the disease AIDS. The RNA is first transcribed into DNA by an RNA-dependent DNA polymerase, also known as reverse transcriptase (Sections 5.4.7 and 9.2.4). In contrast to the cellular polymerases, these reverse transcriptases are unique to retroviruses and they are also tolerant of a wide range of nucleoside triphosphate analogues which identifies them as targets for chemotherapy (Section 4.7.2).

4.6.2 RNA polymerases

DNA is transcribed into RNA by RNA polymerase. In *E. coli* there is a single RNA polymerase having five subunits. Eukaryotes have three RNA polymerases. RNA polymerase I synthesizes a large precursor that is processed to form ribosomal RNA, RNA polymerase II is responsible for synthesis of the precursors of messenger RNA, and RNA polymerase III synthesizes small RNAs such as 5S RNA and precursors of transfer RNA. All are large and complex enzymes. They use a template of double-stranded DNA and synthesize RNA in a 5′-to-3′ direction.

Several antibiotics are highly potent inhibitors of transcription. Actinomycin D is an intercalator that binds tightly and selectively inhibits ribosomal RNA chain elongation. By contrast, rifampicin interacts directly with one of the subunits of RNA polymerase and inhibits initiation of RNA synthesis. *cis*-Dichlorodiammine-platinum (II) (cisplatin) has strong anti-tumour activity. It cross-links two adjacent guanines present in the same DNA strand at their N^7-positions and interferes with transcription (Section 7.5.4).

Some viruses encode their own RNA-dependent RNA polymerase. These are also potential targets for chemotherapy, since they are generally specific for viral RNA, but as yet no clinically useful inhibitors have emerged.

Summary

Purine nucleotides are catabolized through the action of deaminases to IMP and inosine and thence to xanthine and finally to uric acid. Cytosine-containing nucleotides are deaminated to the corresponding uracil derivative and further degraded reductively.

Nucleotides are polymerized into DNA and RNA through the action of DNA and RNA polymerases respectively. Viral DNA polymerases are often targets for chemotherapy.

4.7 Drug inhibition of nucleic acid biosynthesis

As the end of the twentieth century approaches, most of the serious bacterial diseases exist only in less developed countries. Developed countries still face, however, the scourge of cancer and many viral and parasitic diseases. In the remainder of this chapter we will examine these problems to show the mode of action of some of the compounds presently used clinically and assess the future of rational design for anti-cancer and anti-viral therapy.

Undoubtedly the main reason for the dramatic fall in the spread of bacterial disease is the improved hygiene and better nutrition that have arisen from a higher standard of living. As a result, the natural defences of the body against invasion are given the best possible chance to work. Most bacterial infections are now controlled by vaccination or chemotherapy, but as bacteria are free-living organisms (that is they can exist independently of the organism in which they cause disease), their metabolic processes are usually very different from those of the host and this allows the possibility of selective chemotherapy. It is also much easier to screen compounds for activity or to produce a vaccine against an organism which is under no pressure to change its antigenic determinants and which can easily be grown on a large scale in culture. Thus, bacterial diseases are generally very well controlled.

Many antibacterial agents were discovered by sheer chance (e.g. penicillin). In the case of the cephalosporins and many other compounds, routine screening of extracts from soil samples and other sources gave the first lead. Therefore, it has not proved really necessary in the antibacterial field to have rational design of drugs, since a sufficient number of active compounds has been found in nature and it is relatively easy and cheap to screen for new leads.

The anti-cancer and anti-viral fields are very different. As W. H. Prusoff once said 'a cancer cell is a "happy" host cell'. Unfortunately, it is its neighbours which are invaded by the rapidly dividing cancer cells and become unhappy, and our knowledge of the differences in metabolism between the normal and cancer cell is only at a very rudimentary stage. It is thus unlikely that vaccines of the relatively crude type used to combat bacterial infections could be successful, and specific drug-targeting to a cancer cell is probably required. We need to know more about what signal triggers a cell to proliferate and if possible to interfere with this process. Alternatively, we have to discover the differences in metabolism between a normal cell and a cancer cell and try to exploit them by chemotherapy.

Some leads are being obtained by careful examination of how known anti-cancer compounds work. However, *de novo* design of original compounds in this and in the anti-viral field is altogether another matter, since our knowledge of the criteria necessary to design small molecules which will interact selectively with proteins is very limited. Despite this, anti-viral chemotherapy has made good progress in the past decade and several new anti-viral agents have reached the clinic. Another big success has been vaccines which, for example, have completely eliminated smallpox as well as most of the diseases of childhood, such as tuberculosis, poliomyelitis, and diphtheria, which caused so much suffering and death in previous generations.

Why bother with anti-viral chemotherapy at all if vaccines are so good? There are several answers to this question. First, viruses are usually host and cell-type specific and their metabolic pathways are closely linked to those of their host. For example, hepatitis B virus currently affects 200 million people of whom 40 million will die of liver cirrhosis. It is only recently that it has been possible to grow the virus in the laboratory and produce a vaccine. A second vaccine effective against hepatitis B has been constructed by recombinant DNA techniques (Section 10.7) and is now on the market. However, chemotherapy is still required for patients already infected.

Another problem is that of 'latent virus'. Here, after infection, the viral genome is integrated into the host cell DNA and is thereafter associated with it. For example, most people have been infected by one or more types of herpesvirus and therefore are always candidates for further attack from 'within', leading to symptoms such as cold sores or shingles. Chemotherapy is almost certainly required here. Similarly, after HIV infection, the viral genome is integrated into the genome of the host. Thus while the production of a vaccine may prevent future generations from infection, it can do nothing to cure those already infected.

The most common viral diseases are those caused by influenza virus, the rhinoviruses (common cold), and the enteroviruses (diarrhoea). A vaccine is available against some influenza strains, but the antigenic properties of influenza virus change so rapidly, and there are so many serotypes of rhinovirus and enterovirus, that it is not possible to develop a vaccine which would be effective against anything more than a very small percentage of infections. A similar problem occurs in HIV vaccine development because of the high mutation rate of the HIV genome.

The need for anti-cancer and anti-viral chemotherapy is very much with us and is likely to remain so for some time. We can now take advantage of the understanding of the metabolic pathways of normal, virus-infected, and cancer cells we have gained earlier in this chapter and use them to investigate the role of anti-viral and anti-cancer drugs.

4.7.1 Anti-cancer chemotherapy

Unfortunately our present knowledge is such that it is very difficult to define unique targets in cancer cells for attack by chemotherapy. Cancer cells appear to arise as a result of multiple insults caused by viruses, chemicals, or radiation. While some genes (**proto-oncogenes**) can give rise to **oncogenes** following a retroviral infection, translocation, or mutation, for many cancers such an event, though necessary, is not a sufficient condition. There is on the one hand a genetic component to carcinogenesis, but in addition it is desirable to avoid known carcinogenic and mutagenic agents—hazards such as those which occur in cigarette smoke and radiation, whether it be ultraviolet, X-ray, or arising from radioactive decay (Sections 7.6, 7.8, 7.9).

No cancer cell arising from any of these causes has a qualitatively different metabolism. Many tumours occur with great frequency, but normally pose no problems as they are benign and localized, for example warts. The problems start when such cells spread throughout the body and set up secondary areas of invasive growth. Malignant cells have this ability to **metastasize**.

The main difference between the metabolism of a normal cell and a cancer cell is usually one of rate of replication. Most successful selectivities in chemotherapy are largely based on

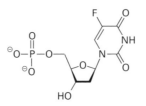

Fig. 4.14 5-Fluoro-2′-deoxyuridine 5′-phosphate.

this phenomenon. Consequently, one side effect of almost all cancer chemotherapy is depression of the immune system, because the cells responsible for the maintenance of this defence against invasion are also rapidly dividing. The selectivities shown for cancer cells in the examples which follow may well be based on slightly different transport properties between cell types, or perhaps a salvage pathway is being used more, or the cell may have a different pH or oxygen tension due to its rapid metabolism.

5-Fluorouracil

There is no complete accord on the mechanism of action of 5-fluorouracil in killing cancer cells. It is possible that it can be incorporated into DNA. It certainly can be incorporated into RNA. However, it is widely agreed that 5-fluoro-2'-deoxyuridine 5'-monophosphate (dFUrdMP) is the active metabolite (Fig. 4.14). This nucleotide is a potent inhibitor of thymidylate synthetase and hence prevents the *de novo* synthesis of dTMP, dTTP, and DNA. The base 5-fluorouracil and its 2'-deoxynucleoside have similar biological effects and since the base is cheaper, it is usually used clinically. 5-Fluorouracil is presumed to be converted into the 5'-nucleotide via the pyrimidine salvage pathway. The mechanism of action and co-factor requirement of thymidylate synthetase have already been discussed (Section 4.4) and decomposition of the ternary complex formed between substrate, co-factor, and enzyme requires the abstraction of a hydrogen atom from C-5 of the uracil ring. The presence of fluorine, whilst small enough to allow complex formation, prevents this reaction because of the stability of the adduct. dFUrdMP is thus is a suicide inhibitor. Much evidence for the presence of such an intermediate has been obtained and a cysteine residue in the enzyme has been identified as the nucleophile which attacks the 6-position of the pyrimidine ring (Fig. 4.15).

Although 5-fluorouracil, and some pro-drugs of it, are widely used in the treatment of common solid tumours, there is no particular reason why it should show any selectivity. It is therefore toxic and causes suppression of the immune system.

Methotrexate

Another way of preventing the *de novo* synthesis of dTTP is to inhibit biosynthesis of the tetrahydrofolate required as the co-factor and donor of the methyl group in the reaction catalysed by thymidylate synthetase (Section 4.4). The dihydrofolate produced in this reaction is normally reduced to tetrahydrofolate by NADPH using the enzyme dihydrofolate

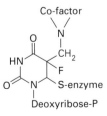

Fig. 4.15 Suicide action of 5-dFUrdMP.

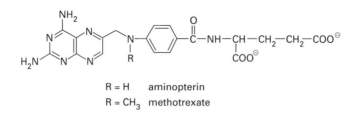

R = H   aminopterin
R = CH₃ methotrexate

Fig. 4.16 Folic acid analogues.

reductase in a process which can be inhibited by folic acid analogues such as **aminopterin** and **methotrexate** (Fig. 4.16).

Methotrexate was introduced as an anti-leukaemia agent ten years before its mode of action was identified, and although many thousands of analogues have been prepared, none has been found to have better properties. This is typical of discoveries in this and in the anti-viral field, even though the original discovery may have been based on only limited or no knowledge of the mode of action of the drug. The selectivity which methotrexate undoubtedly shows for cancer cells appears to be due to preferential uptake of the drug. This highlights a major problem in drug design: even the best drug in the world is of no use if it cannot be delivered intact to the correct site and transported into the target cell.

Cyclophosphamide

Many anti-cancer alkylating agents are also mutagenic and carcinogenic, but are still used in chemotherapy under certain circumstances. One example, cisplatin, is thought to cross-link DNA (Section 7.5.4) and its specificity for tumour cells is not understood. Cyclophosphamide (Fig. 4.17) is a nitrogen mustard and is one of the most clinically useful drugs. It is thought to cross-link DNA and interferes with DNA replication. While the original synthesis was the product of rational drug design, it is now known that the design was based on a false premise and its activity is mediated through an entirely fortuitous mechanism.

Cyclophosphamide itself is inactive as an alkylating agent and it requires enzymatic oxidation in the liver to give the active species (Fig. 4.17). This hydroxylated compound can either degrade enzymatically, leading to relatively inert compounds, or it can undergo an elimination reaction to yield two potent cytotoxic agents: the alkylating agents acrolein and a phosphoramide mustard. Cyclophosphamide is useful because of its selectivity for tumour cells, which may be due to more efficient transport of the active metabolite into or a lack of enzymatic degradation within tumour cells.

In all the above compounds, the eventual target is inhibition of DNA replication, whether by preventing synthesis of precursors or by alkylating the DNA itself. Unfortunately most of the effective drugs are also toxic and patients receiving cancer chemotherapy are unusually susceptible to viral and bacterial infection. There has been little successful rational design so far and also too few targets at which to aim. As a result, pharmaceutical chemists are often reduced to synthesizing analogues of previously recognized active compounds rather than aiming to design entirely novel classes of potentially active compounds.

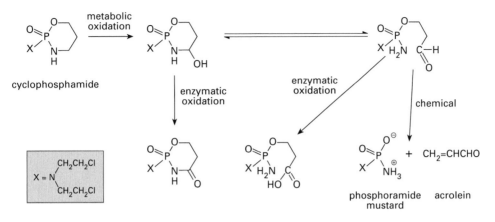

Fig. 4.17 Metabolic activation and deactivation of cyclophosphamide.

Summary

Unique targets for chemotherapy directed against cancer cells are almost impossible to define. Most anti-cancer drugs discovered so far have toxic side effects. Their selectivity for a cancer cell is small and other rapidly proliferating normal cells are also killed. 5-Fluorouracil and methotrexate both probably work by preventing synthesis of dTTP. The former is a thymidylate synthetase inhibitor whilst the latter inhibits *de novo* synthesis of dTTP by inhibition of the enzyme dihydrofolate reductase. Cyclophosphamide requires oxidation by liver enzymes to give a hydroxylated derivative which breaks down to yield a DNA-alkylating agent.

4.7.2 Anti-viral chemotherapy

As our knowledge concerning the methods of viral replication has increased in the past few years, it has become clear that rational targets for anti-viral drugs do indeed exist. The establishment of acyclovir (Fig. 4.18) as an anti-viral agent active against some herpesviruses has shown what is possible in situations where the development of a vaccine poses major problems. However, the facts that most viruses are not related to each other and that the replication cycle of each class is unique appear to leave little chance for the discovery of a wide-activity anti-viral agent comparable to broad-spectrum antibiotics such as the β-lactams.

An equally severe problem is the fact that any drug which affects more than a single viral species is likely to impair a fundamental biochemical pathway. This has the almost inevitable result of host cell toxicity. Because such toxicity is unacceptable, extensive screening is required, even though *in vitro* and animal test systems may be insufficiently related to the human situation to provide really accurate models. As a result, an anti-viral drug can take years to come to market.

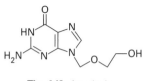

Fig. 4.18 Acyclovir.

Table 4.1. Drugs approved for use against viral infection in humans in 1993

Drug	Virus
Amantadine	Influenza A
Rimantadine	Influenza A (in Russia)
5-Iodo-2'-deoxyuridine	Herpes simplex virus (HSV)
5-Trifluoromethyl-2'-deoxyuridine	HSV
Adenine arabinoside	HSV
5-Ethyl-2'-deoxyuridine	HSV (in Germany)
5-Iodo-2'-deoxycytidine	HSV (in France)
Acyclovir	HSV, Varicella zoster virus (VZV)
(E)-5-(2-Bromovinyl)-2'-deoxyuridine	HSV-1, VZV (in Germany)
3'-Azido-2',3'-dideoxythymidine	Human immunodeficiency virus (HIV)
2',3'-Dideoxyinosine	HIV
2',3'-Dideoxycytidine	HIV
Ribavirin	Respiratory syncytial virus
Gancyclovir	Cytomegalovirus (CMV)
Phosphonoformic acid	CMV

The advent of life-threatening viral infections, particularly AIDS and genital herpes, has brought a dramatic change. When time-consuming screening has to be set against the reality of death as the likely outcome of HIV infection, action can be swift. One remarkable response was the marketing of an anti-AIDS drug, AZT, within a year of the report of its *in vitro* properties.

There are very few anti-viral drugs licensed for use (Table 4.1). Of these, all except three are nucleosides or their analogues and most are effective against herpesvirus. In the following section, we will take a closer look at a few examples of antiviral drugs which act against a number of specific viral targets.

Acyclovir

Before the advent of AZT, this drug was *the* success of ten years of intensive research. It was discovered by chance, it is effective against herpesviruses, and its metabolic conversion into the active form is so bizarre that, with hindsight, it actually gives hope for the possibility of achieving a rational design in the future.

Herpesviruses are a class of double-stranded DNA viruses which cause a variety of diseases in humans: cold sores, eye infections (keratitis), genital sores, chickenpox, shingles, and glandular fever (infectious mononucleosis). One property of all herpesviruses is that they exhibit latency. This means that after a cell as been infected, the virus produced can go into a latent state in the nerve endings from where it can be reactivated by various stimuli (stress, UV light, other viral infections etc.). Since it has been impossible so far to destroy the virus in the latent state (i.e. prevent it from replicating), anti-viral chemotherapy must be directed first against primary infection and then against subsequent recurrent episodes. The symptoms of many herpes infections are relatively severe and cause discomfort on a continuing basis, but acyclovir will only alleviate the symptoms and further episodes of infection are bound to recur.

Why is herpesvirus the target for so many licensed anti-viral nucleosides? It is known that herpesviruses code for many enzymes involved in their own replication and metabolism. The virus is thereby vulnerable because the properties of some of the virally encoded enzymes are slightly different from the corresponding ones in the host cell, although it is by chance that these properties have been targeted. In particular, herpesviruses rely largely on the salvage pathway for the production of dTTP for DNA synthesis and so the virus encodes its own thymidine kinase. Possibly because of the rate at which this enzyme has to work, its specificity is not as great as that of the host cell and it can phosphorylate a wide range of nucleoside analogues which, once activated, can subsequently inhibit viral replication.

Herpesviruses also code for their own DNA polymerase. Again, it is not clear why this is necessary, but the polymerase thus encoded has a different specificity from the cellular polymerases and hence presents a target for selective attack.

Although acyclovir is a purine nucleoside analogue with C-2$'$ and C-3$'$ of the sugar ring missing (Section 3.1.3), it is specifically phosphorylated at the position equivalent to the 5$'$-hydroxyl group by the thymidine kinase of the herpesvirus. Not surprisingly, no metabolism occurs at all in a normal uninfected cell. However, in a virally infected cell, phosphorylated acyclovir is now recognized by the host cell guanylate kinase and is taken to the diphosphate, from which a nucleoside diphosphate kinase produces the 5$'$-triphosphate. This is now a substrate for the herpesvirus-encoded DNA polymerase and it is incorporated into viral DNA. Since the analogue contains no equivalent to the 3$'$-hydroxyl group, it is a chain terminator and thus stops the synthesis of viral DNA (Fig. 4.19).

This is a very convincing *post facto* explanation for the efficacy of acyclovir, but the drug was actually discovered by chance before information concerning the replication of herpesvirus was available. Since the discovery of acyclovir, it has not been possible to capitalize on this chance discovery by the design of novel compounds which are specific substrates for the viral kinase. However, one successful analogue, 6-deoxyacyclovir (Fig. 4.20), can be considered as a pro-drug of acyclovir since it is converted into acyclovir by the enzyme xanthine oxidase (Fig. 4.12), has greater solubility, and so allows higher plasma levels of acyclovir to be achieved.

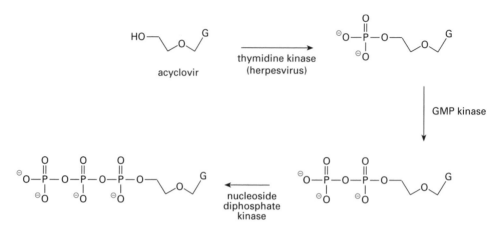

Fig. 4.19 Enzymatic phosphorylation of acyclovir.

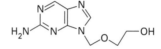

Fig. 4.20 6-Deoxyacyclovir.

5-Substituted-pyrimidine 2′-deoxynucleosides

5-Iodo-2′-deoxyuridine (Fig. 4.21a) was discovered by Prusoff in the 1960s and was the first anti-viral nucleoside drug to be marketed. The mode of action of this nucleoside is still not known over 20 years later, although it is a substrate for both cellular and viral thymidine kinases and it is incorporated into cellular and viral DNA. Its anti-viral properties may be a result either of this incorporation, or of the deoxynucleoside triphosphate being an inhibitor of the viral DNA polymerase, or of some other explanation. It is likely, however, that the toxicity of this drug is a consequence of the fact that both viral and cellular kinases can phosphorylate it and therefore it is further metabolized in infected and in non-infected cells.

A breakthrough came when 5-vinyl-2′-deoxyuridine was shown to be more potent against herpesviruses *in vitro* than the 5-iodo derivative, by many orders of magnitude (Fig. 4.21b).

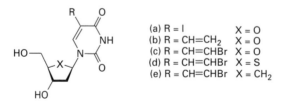

(a) R = I   X = O
(b) R = CH=CH₂   X = O
(c) R = CH=CHBr X = O
(d) R = CH=CHBr X = S
(e) R = CH=CHBr X = CH₂

Fig. 4.21 Anti-viral pyrimidine 2′-deoxynucleosides.

While this compound is very toxic in cell culture, presumably because it is also metabolized by cellular enzymes, in animals it is neither toxic nor does it have anti-viral properties. This is because the nucleoside is a very good substrate for nucleoside phosphorylase, an enzyme which is absent in many tissue culture cell lines. The enzyme degrades it to the heterocyclic base, which has no anti-viral properties. From this example we learn that any analogue in this series must be resistant to nucleoside phosphorylase in order to possess anti-viral activity. Unfortunately we do not know the specificity of this enzyme and so it is difficult to design a nucleoside analogue which is not such a substrate.

The next compound in this series to be discovered (again by chance) was (E)-5-(2-bromovinyl)-2'-deoxyuridine (BVDU; Fig. 4.21c) which was even more effective (ID$_{50}$ 0.001 mg ml^{-1}) against herpes simplex virus (HSV-1) and varicella zoster virus but less so against HSV-2. After the event, this was shown to be a consequence of the nucleoside acting as a substrate for the thymidine kinase of the virus and not for the kinase of the host cell. This viral thymidine kinase is apparently also a thymidylate kinase and produces the 5'-diphosphate, but only in HSV-1-infected cells, since the diphosphate formation does not occur efficiently with the HSV-2-encoded enzyme. Although BVDU is a substrate for nucleoside phosphorylase, it adequately avoids degradation to show useful clinical activity. The base (E)-5-(2-bromovinyl)uracil is not a substrate for pyrimidine-5,6-dihydroreductase, which is the first enzyme in pyrimidine catabolism. Indeed it is an inhibitor of this enzyme and thus the base can actually be salvaged and the 2'-deoxynucleoside regenerated. The triphosphate of BVDU is a substrate for, and an inhibitor of, the virally encoded DNA polymerase. It is still not known which, if either, of these properties is responsible for the anti-viral activity of BVDU. A nucleoside phosphorylase-resistant analogue of BVDU (Fig. 4.21d) has recently been described which has substantially greater activity *in vivo* because it has a much longer half-life in serum. The carbocyclic analogue of BVDU (Fig. 4.21e) has also been found to have significant activity.

Ribavirin

The broad-spectrum anti-viral activity of ribavirin (1-β-D-ribofuranosyl-1,2,4-triazole-3-carboxamide; Fig. 4.22a) was first described in 1972. Since then, it is thought to have been studied in more animals and against more viruses than any other anti-viral agent. It is also apparently active in cell culture against about 85 per cent of all virus species studied and shows little or no cellular toxicity. One is then left to speculate why, after over 20 years, this 'wonder drug' has only been approved for use in the USA in aerosol form against respiratory syncytial virus in young children.

Ribavirin almost certainly does not act in only one way against all viruses. The most abundant form of ribavirin in cells is the 5'-triphosphate (Fig. 4.22b) and this was originally thought to inhibit inosine monophosphate dehydrogenase, which results in a depletion of cellular GTP pools. This in turn means that ribavirin 5'-triphosphate is a more effective competitive inhibitor of the viral-specific RNA polymerase for some viruses. Ribavirin 5'-triphosphate is also known to inhibit the viral-specific mRNA capping enzymes, guanyl

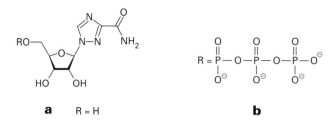

Fig. 4.22 Ribavirin and its 5′-triphosphate.

transferase, and N^7-methyl transfase, so that viral protein synthesis is interrupted. Ribavirin is known to be effective clinically against haemorrhagic fever virus and to cure plant virus infections, and has undergone trials against HIV.

Phosphonoformic acid

Phosphonoacetic acid (PAA, Fig. 4.23a) was discovered to have antiherpetic activity *in vitro* following random screening in 1973. Two years later it was shown to be a selective inhibitor of the virally encoded DNA polymerase, and the related phosphonoformic acid (PFA, Fig. 4.23b) was subsequently found to be an even stronger inhibitor of this enzyme. PFA is also widely used for *in vitro* assays of HIV reverse transcriptase. Both these compounds are analogues of pyrophosphate, a product of the polymerase, and presumably bind to the corresponding site on the enzyme thus preventing replication. One problem with compounds of this sort is that they require no prior activation and therefore one has to rely entirely on the difference in affinity between the virus-encoded and host cell polymerase. Any interaction with the latter will presumably cause toxicity and this may be the reason for some of the problems encountered with these agents.

Retrovirus inhibitors

Much effort has been expended on finding a chemotherapeutic agent which can alleviate the symptoms of AIDS. The human immunodeficiency virus is a member of the lentivirus family (a sub-class of retrovirus) and its reverse transcriptase (Section 4.6.1) was an obvious initial target. A number of 2′-deoxynucleoside 5′-triphosphate analogues were quickly found to act as inhibitors or serve as chain terminators of the enzyme. The compound 3′-azido-2′,3′-dideoxythymidine (AZT; Fig. 4.24a) was found to be the least toxic and is widely used clinically. However, there is a fundamental problem with all compounds of this type. The 5′-triphosphate is necessarily the active species. Since the retrovirus does not encode its own kinase

Fig. 4.23 Antiviral analogues of pyrophosphate.

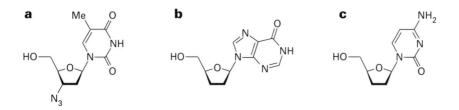

Fig. 4.24 Some anti-HIV nucleosides: (**a**) AZT, (**b**) ddI, (**c**) ddC.

and since it is difficult to get highly anionic nucleotides into cells, one has to rely on the native human cellular thymidine kinase to perform the initial phosphorylation steps.

As we have already seen (Section 4.7.1), the mammalian cellular kinase is highly selective in its requirements. Thus, although many deoxynucleoside triphosphate analogues may be good inhibitors of the viral reverse transcriptase, as long as it is necessary to rely on a specific kinase of the host cell for activation of the analogue, the choice is greatly restricted. A compound like AZT, having only a small modification, is a suitable substrate for the kinase and thus the triphosphate is a substrate for the reverse transcriptase and can cause chain termination. Unfortunately the high dosage levels of AZT required give rise to considerable toxicity. One suggestion is that this is because of an unexplained depletion of the levels of dTTP and dCTP that has been observed. More plausibly, the triphosphate of AZT could be to some extent a substrate for the DNA polymerase of the host cell. Either situation is likely to contribute to bone marrow suppression, which is one of the observed toxic side effects of the drug.

Recently, two more dideoxynucleoside analogues, $2',3'$-dideoxyinosine (ddI; Fig. 4.24b) and $2',3'$-dideoxycytidine (ddC; Fig. 4.24c) have received clinical approval for the treatment of HIV infection, but both suffer from the same problems of toxicity and requirement for phosphorylation by host enzymes as does AZT. Several other nucleoside analogues are in clinical trials in monotherapy or combination studies.

The structure of HIV-1 reverse transcriptase has now been solved by X-ray crystallography (Fig. 9.16). The active form of the enzyme is a heterodimer having one polymerase active site and one RNase H active site (Section 9.2.4). A number of specific inhibitors of HIV-1 reverse transcriptase (and not HIV-2) have been identified (Fig. 4.25) which are not nucleosides and probably do not require metabolism to an active form. One of them, nevirapine (Fig. 4.25c), has been co-crystallized with the transcriptase and its binding site on the enzyme is seen to be a hydrophobic pocket guarded by two tyrosine residues close to the polymerase active site. Binding of the drug is thought directly to inhibit reverse transcription. Although the structures of these non-nucleoside inhibitors appear to be very different, they are all believed to bind in a similar (but not identical) location. They show little toxicity and have a very high anti-HIV activity in cell culture. However, HIV-1 becomes rapidly resistant to these drugs, in most cases due to the selection of strains containing a reverse transcriptase mutated at one or both of the tyrosines, whilst the mutated viral strains retain their infectivity. As the knowledge of the structure of reverse transcriptase improves, so it should be possible to design, using molecular modelling techniques, new inhibitors which interact with enzymatically or structurally important sites on the enzyme, the mutation of which would result in loss of in-

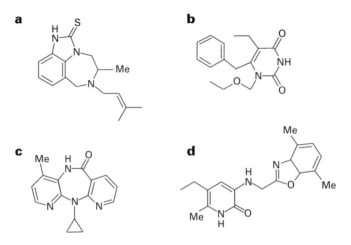

Fig. 4.25 Non-nucleoside HIV reverse transcriptase inhibitors: (**a**) TIBO class, (**b**) HEPT class, (**c**) Nevirapine, and (**d**) 2-Pyridinone class.

fectivity. Some recent clinical trials have shown improved results from combination therapy using two nucleoside analogues (AZT and ddC) together with an inhibitor of the HIV protease enzyme.

Anti-viral drug design

Why should almost all licensed anti-viral drugs be nucleoside analogues? Although such compounds may well be expected to interfere with viral replication and often do have an observed biological effect, they are also very likely to be metabolized by normal cellular enzymes and thus cause toxicity. There are many other viral targets and there is no particular reason why the inhibitor of a viral enzyme should be a nucleoside analogue. Other steps unique to viral replication are, for example, adhesion of the virus to the host cell and subsequent uptake, uncoating of the nucleic acid, maturation and release of the virion.

New classes of compounds are now being found which inhibit some of these steps, particularly in HIV infection. For example, several inhibitors of the HIV aspartyl protease, all peptide substrate analogues, are now in clinical trials. Some phenylisoxazoles are showing promise in the treatment of some RNA virus infections while amantadine, an ion channel inhibitor, is being used for the treatment of influenza virus infection.

It is very difficult to design novel inhibitors until more is known about the details of viral replication at a molecular level. We need to have three-dimensional structures of the enzymes involved and much better models of cell and viral surfaces. Even then, it is one thing to inhibit a function *in vitro*, but quite another to get the inhibitor delivered to the infected cell without degradation, at effective concentration and, in the case of a non-lethal infection, in orally active form.

Much hope for the future has been generated by the recent rational design of a potent inhibitor of influenza virus infection, 4-guanidino-Neu5Ac2en. The inhibitor, which was synthesized based on predictions derived from a knowledge of the structure of the viral-surface sialidase protein, has been found to be 1000 times more effective than amantadine when delivered intranasally to mice or to ferrets.

Summary

In the case of many potential targets for anti-viral chemotherapy, most clinically useful compounds have been directed against virally encoded enzymes which are essential for viral replication. Anti-viral agents directed against non-lethal diseases must be non-toxic to the host cell.

Most of the nucleoside analogues possessing activity against herpesvirus rely upon specific phosphorylation of a herpesvirus-encoded thymidine kinase. In the case of acyclovir, the triphosphate subsequently produced is a substrate for the viral DNA polymerase and is a chain terminator.

Since HIV infection is ultimately lethal, toxic anti-virals have been accepted for its clinical treatment. AZT, ddl, and ddC have shown some effect in alleviating the symptoms of AIDS and, following conversion into the 5′-triphosphate, each compound acts as a chain terminator in the biosynthesis of DNA catalysed by reverse transcriptase.

Few anti-viral compounds have so far been designed on rational principles but, with increasing knowledge of viral replication mechanisms and structures of viral components, it is now becoming possible to design molecules which are effective and non-toxic to the host.

Further reading

4.1–4.6

Nogrady, T. (1985). *Medicinal chemistry—a biochemical approach*. Oxford University Press, Oxford, pp. 343–61.

Stryer, L. (1988). *Biochemistry,* 3rd edn. W. H. Freeman, San Fransisco, pp. 601–26.

Stubbe, J. A. (1989). Protein radical involvement in biological catalysis? *Annu. Rev. Biochem.*, **58**, 257–67.

Zubay, G. (1983). *Biochemistry,* Addison-Wesley, Reading, Mass., pp. 697–735.

4.7

Collier, L. H. and Oxford, J. (ed.) (1980). *Developments in antiviral therapy.* Academic Press, London.

Coulson, C. J. (1988). *Molecular mechanisms of drug action,* Taylor and Francis, London, Chapter 2.

DeClercq, E. (ed.) (1987). *Frontiers in microbiology.* Martinus Nijhoff, Dordrecht.

DeClercq, E. (1992). HIV inhibitors targeted at the reverse transcriptase. *AIDS Research and Human Retroviruses,* **8**, 119–34.

DeClercq, E. and Walker, R. T. (ed.) (1984). *Targets for the design of antiviral agents*. Plenum, New York.

DeClercq, E. and Walker, R. T. (1986). Chemotherapeutic agents for herpesvirus infections. In *Progress in medicinal chemistry* (ed. G. P. Ellis and G. B. West), Vol. 23, pp. 178–218. Elsevier, Amsterdam.

Hertzberg, R. P., Caranfa, M. J., and Hecht, S. M. (1989). On the mechanism of topoisomerase I inhibition by camptothechin: evidence for binding to an enzyme: DNA complex. *Biochemistry*, **28**, 4629–38.

Huryn, D. M. and Okabe, M., (1992). AIDS-driven nucleoside chemistry. *Chem. Rev.*, **92**, 1745–68.

Kensler, T. W. and Cooney, D. A. (1989). Inhibitors of the *de novo* pyrimidine pathway. In *Design of enzyme inhibitors as drugs* (ed. M. Sandler and H. J. Smith), pp. 379–401. Oxford University Press.

Kohlstaedt, L. A., Wang, J., Friedman, J. M., Rice, P. A., and Steitz, T. A. (1992). Crystal structure at 3.5Å resolution of HIV-1 reverse transcriptase complexed with an inhibitor. *Science*, **256**, 1783–90.

Mitsuya, H. and Broder, S. (1987). Strategies for antiviral therapy in AIDS. *Nature*, **325**, 773–8.

Montgomery, J. A. and Bennett, L. I. (1989). Inhibitors of purine biosynthesis. In *Design of enzyme inhibitors as drugs* (ed. M. Sandler and H. J. Smith), pp. 402–34. Oxford University Press.

Neidle, S. and Waring, M. J. (eds.) (1983). Molecular aspects of anti-cancer drug action. *Topics in molecular structural biology*, Vol. 3. Macmillan, London.

Stuart-Harris, C. H. and Oxford, J. (ed.) (1983). *Problems of antiviral therapy.* Academic Press, London.

von Itzstein, M., Wu, W-Y., Kok, G. B., Pegg, M. S., Dyason, J. C., Jin, B., Van Phan, T., Smythe, M. L., White, H. F., Oliver, S. W., Colman, P. M., Varghese, J. N., Ryan, D. M., Woods, J. M., Bethell, R. C., Hotham, V., Cameron, J. M. and Penn, C. R. (1993). Rational design of potent sialidase-based inhibitors of influenza virus replication. *Nature*, **363**, 418–23.

DNA SEQUENCE INFORMATION, REPLICATION, AND REARRANGEMENT

5.1 Gene structure in prokaryotes and phage

5.1.1 The complexity of genetic information in prokaryotes and phage

DNA molecules are the largest of the macromolecules found in a cell. The *E. coli* **chromosome**, for example, consists of a single molecule of double helical DNA about 4 million base-pairs in length. A more useful and common way used to describe this length is to say that the molecule is about 4000 kilobase-pairs (kbp) long. The mass of such a molecule is about 2.6×10^6 kDa and is thus about 10^5 times the mass of an average-sized polypeptide. These sizes have been arrived at in experiments first done by Cairns where DNA was specifically labelled by growing bacteria in media containing [^{3}H]-thymine. Cells were lysed gently so as not to shear DNA and the chromosomes trapped in photographic emulsion. Under the microscope, tracks of silver grains mark the positions of DNA double helices whose contour lengths can be measured. In the case of *E. coli*, the contour length of the chromosome was 1.4 mm and since there is about 0.34 µm between successive base-pairs, the chromosome was estimated to be about 4100 kbp long. A similar size has been deduced recently by the unrelated method of summing the sizes of independent chromosome fragments cloned in cosmids. The *E. coli* chromosome seems to be fairly typical of the length of DNA molecule to be found in other species of bacteria, but it is a tiny molecule compared with those found in the chromosomes of eukaryotes where molecules may be at least 20 to 100 times longer than bacterial DNA molecules (Section 5.3). In phage, DNA molecules are considerably smaller than in bacterial chromosomes. Table 5.1 shows some comparative data.

Before proceeding to consider in detail the nature of DNA in chromosomes, let us stop to consider the implications of DNA size for the variety of DNA sequences possible. At first sight, with only four types of base present, it might seem that there would not be much variety in DNA molecules. However, there are no known restrictions to the order in which bases occur on a polynucleotide chain. Thus, there are 4^2 possible dinucleotides, 4^3 possible trinucleotides and $4^{4\,100\,000}$ possible bacterial chromosomes. As we shall see, it is the sequence of nucleotides in DNA which determines the proteins and enzymes made in a cell and which thus de-

Table 5.1. The size of some genomes

Source of DNA	Size (kb)	Type
E. coli chromosome	3900	Covalently closed circular DNA
Bacillus subtilis chromosome	2000	Covalently closed circular DNA
F plasmid	95.0	Covalently closed circular DNA
pcolE1	6.4	Covalently closed circular DNA
λ phage	48.5	Double-stranded linear DNA
T7 phage	40.0	Double-stranded linear DNA
ϕX174 phage	5.4	Single-stranded circular DNA
M13 phage	6.4	Single-stranded circular DNA
MS2 phage	3.6	Single-stranded linear RNA

termines the form and behaviour of those cells. Of course, not all of the possible sequences will give rise to working proteins let alone to viable bacteria, but nevertheless the scope for genetic variation is enormous. Despite the teeming variety of life on earth, it is tempting to view evolution as having only scratched the surface of the possibilities. What is abundantly clear from DNA sequencing studies is that a great many sequences have been conserved, sometimes strongly so, during the course of evolution over millions of years and we will frequently point out similarities or 'homologies' between sequences, usually with the implication that the homologous sequences serve similar or identical functions.

An unexpected fact to emerge from the experiments of Cairns described above was that the DNA molecules in bacterial chromosomes consist of continuous or 'circular' threads and the same was true of **plasmids**, small extrachromosomal DNA molecules carried by some bacteria. These structures have no free 5'- or 3'-ends and are covalently closed circles of double-stranded DNA (see Section 2.3.5), sometimes referred to as cccDNA. **Bacteriophage** (viruses that infect bacteria) show greater variety in their chromosomal structures. In some phage, the chromosome consists of double-stranded linear DNA molecules. In others, such as λ phage, the DNA is linear and largely double-stranded with short single-stranded and self-complementary ends (or **sticky ends**). Other phage such as M13 and φX174 have single-stranded circular DNA as their chromosome and there are some phage where the chromosome consists of linear RNA, either single-stranded or double-stranded (Table 5.1). Apparently, single-stranded linear DNA chromosomes and double-stranded circular RNA chromosomes do not exist.

The combined but opposing actions of DNA gyrase and topoisomerase (Section 2.3.5) leave small circular molecules of DNA with negative supercoils, and larger circular molecules would seem to have domains lying between anchorage sequences which have more, sometimes less, negative supercoiling than the average for the whole chromosome.

Finally, supercoiled DNA must be folded backwards and forwards many times in bacteria in order to pack 1.4 mm of DNA into a volume equivalent to a cube of side about 0.25 μm. It has been estimated that adjacent double helices would be about 40 Å apart following this packing and it is difficult to see how molecules like RNA polymerase of about 100 Å diameter can rapidly penetrate the **nucleoid**, as this DNA knot is known in bacteria (Section 2.6).

5.1.2 The functional subdivision of DNA sequences

The astonishing success of primitive humans in breeding plants and domesticated animals indicates a strong belief in the idea that living forms contained heritable information. The idea came more into focus from the mid-nineteenth century onwards and the word **gene** was coined to give some substance to the hypothetical units of heredity. With the discovery that DNA causes bacterial transformation (Section 1.1) the way became open to see that genes were made up of DNA and that the linear arrays of genes postulated by geneticists were in molecular terms linear arrays of DNA sequences. At the same time, geneticists such as Beadle and Ephrussi in the 1930s were concluding that the fundamental role of the gene was to produce enzymes and that deliberate or accidental alteration, or **mutation**, of a

gene caused the production of an altered enzyme. Thus was born the **'one gene: one enzyme'
hypothesis**.

Finally, workers such as Crick, Brenner, and Yanofsky established that linear sequences of
nucleotides in DNA produced, through RNA working copies, primary amino acid sequences
in proteins—the **'one gene: one primary sequence' hypothesis**. They were able to show that
special DNA sequences signalled the first, N-terminal amino acid in a polypeptide (i.e. the
beginning of a gene) and that other sequences signalled the last, C-terminal amino acid
(i.e. the end of the gene).

In general, genes are packaged closely along DNA molecules with only a few nucleotides
separating the end of one gene from the beginning of the next. In phage T7, for example, the
DNA contains 50 genes and more than 92 per cent of the 39 936 nucleotide pairs specify or
code for proteins. Overlapping of genes is rare but small overlaps are nevertheless known to
occur, particularly in phage **genomes**, as arrays of covalently connected genes are often
known. Space for packaging DNA inside phage capsids, the protective protein coat of a virus,
is at a premium and the occasional overlap of phage genes may simply reflect the need to
reduce the number of nucleotides between genes to a minimum.

The genome of *E. coli* has been more extensively mapped than any other cellular genome
and more than 1000 genes have been located on it to date. However, while in many parts of
the genome the genes are as closely packed as in phage genomes, there exist a few regions,
sometimes as long as 200 kbp, where no genes have been found. It is possible that the gap is
an accident of the types of gene which have so far been investigated, but there is a strong
feeling that the DNA in these gaps serves some function other than to code for proteins. The
question of apparently redundant DNA is one which will be taken up again when we examine
eukaryotic chromosomes.

In keeping with the finding that *E. coli* chromosomal DNA is circular is the finding that the
genes also form a circular unbranched array. The location of a few of the known genes is
shown (Fig. 5.1). The units of distance on this map are minutes (time), illogical as this may
seem to a physical scientist! The unit arises from the experimental way in which the genetic
mapping was originally carried out. *E. coli* bacteria, variously known as **Hfr** or **donor** or **male**
cells, transfer their chromosomes from a fixed point (0 min) in a regular 'clockwise' direction
to **female** or **recipient** or **F⁻** cells. Transfer proceeds fairly uniformly (Section 5.2.6) and takes
about 100 minutes to complete. A gene such as the one called *trp* in Fig. 5.1 takes 27 minutes
for first appearance in a recipient and is said therefore to be at 27 min on the chromosome.
One minute on the map is equivalent to about 40 kbp. Some day, mapping will have become
sufficiently detailed for us to be able to use kbp as the map unit, as is already the case for the
maps of many phage and viruses.

Nomenclature in bacteria involves use of three lower-case italic letters to designate a gene.
Frequently the letters give a clue to the gene's function. For example, *trp* indicates a gene
coding for an enzyme involved in tryptophan biosynthesis, *lac* indicates a gene involved in
lactose utilization, and *uvr* indicates a gene whose product is involved in conferring resistance of the cell to UV light.

Where more than one gene has been found which affects a given cell function, the genes
are given a further capital letter to distinguish them. Thus, *trpA*, *trpB*, *trpC*, *trpD*, and *trpE*

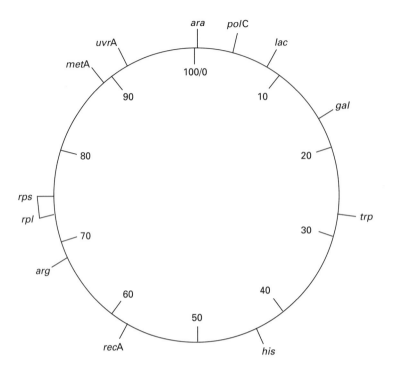

Fig. 5.1 The position of some genes on the *E. coli* chromosome.

genes have been identified each of which codes for the production of a separate polypeptide or a separate enzyme needed for different stages in the biosynthesis of tryptophan. Tables listing the known bacterial genes, their function and their map position are available and are frequently updated.

5.1.3 Clustering of functionally related genes

A consistent finding of genetic mapping is that genes coding for different enzymes involved in one metabolic pathway are frequently clustered or even adjacent on the chromosome. The way in which the five *trp* genes mentioned above are organized in *E. coli* and the functions they serve in tryptophan biosynthesis is shown (Fig. 5.2).

In phage also there is a frequent clustering of genes which have a related function. In phage λ (Fig. 5.3) there is a group of about 20 genes involved in capsid (or phage coat) synthesis subdivided into a group involved in forming the head of the capsid and another group in forming the tail. Some phage genes, unlike bacterial genes, are simply denoted by a single capital letter. Further along the λ chromosome is a group of genes involved in phage DNA recombination, in regulation of phage gene expression, and in phage DNA replication. Finally there are two genes, S and R, coding for proteins which lyse open the phage-infected host to release the newly formed infective particles. However, not all genes of related function are clustered. Instead, such genes may be found scattered throughout the genome.

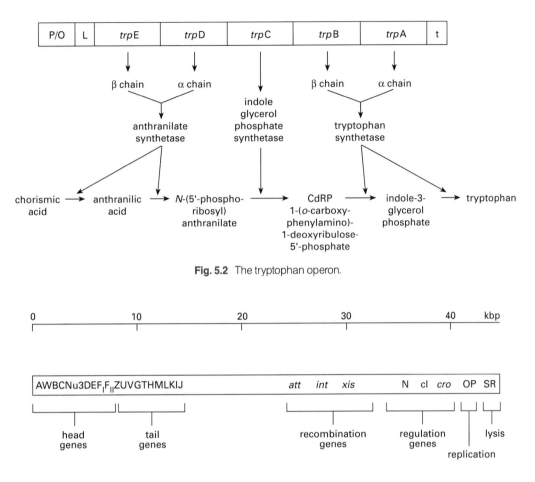

Fig. 5.2 The tryptophan operon.

Fig. 5.3 Clustering of gene functions in the phage λ genome.

5.1.4 Initiation of RNA synthesis

In the first step in expression of a gene as a polypeptide chain, an mRNA working copy of the gene has to be made by RNA polymerase. The enzyme binds to DNA and progresses along it, transcribing the gene in a continuous process to produce mRNA, and finally dissociates from the DNA. The process is thought of in terms of 'flow' such that '**upstream**' refers to DNA sequences before the start of the gene and '**downstream**' refers to those after the end of the gene. **Initiation of transcription** is a very precise process and occurs only at specialized DNA sequences known as **promoters** and situated upstream of genes. The function of a promoter sequence is to be recognized by RNA polymerase and is thus different from the function of a gene sequence, which is to be transcribed and in the end to be translated into an amino acid sequence as a polypeptide.

Promoter sequences have been characterized by the technique known as **footprinting**, illustrated in Fig. 5.4. Purified RNA polymerase is first bound *in vitro* to promoter-containing DNA in which one of the ends has been labelled with ^{32}P. DNA with and without complexed RNA polymerase is next treated with very limiting amounts of an endonuclease, an enzyme which hydrolyses phosphodiester linkages within polynucleotide chains (Section 10.7.2), to give a **nick** in one of the DNA strands. Hydrolysis is nearly random and different individual DNA molecules will suffer nicking at different distances from the labelled end. The population of molecules as a whole will contain representatives which have been nicked at position 1, position 2, position 3, and so on from the labelled end. Next, the DNA is denatured and the single-stranded fragments separated by electrophoresis on gels to produce radioactive bands in which adjacent bands contain molecules differing by one nucleotide in length. With free DNA, the ladder will consist of the complete set of bands, from one nucleotide in length up to a band of complete DNA. However, with DNA complexed to RNA polymerase, those phosphodiester linkages covered by RNA polymerase are protected from hydrolysis by the endonuclease, and hence bands corresponding to the protected region do not appear on the ladder.

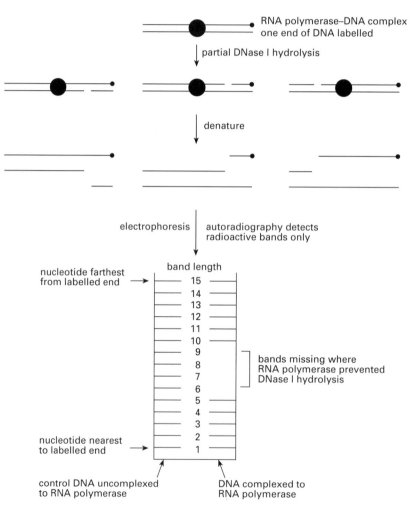

Fig. 5.4 DNA footprinting.

The missing bands are the '**footprint**' left by binding of RNA polymerase. The technique is of general application to other DNA-binding proteins.

Comparison of the footprint (Fig. 5.4 right) and total DNA sequence allows determination of the promoter sequence. Alternatively, complexed and uncomplexed mixtures can be treated to generate four sequencing ladders (Section 10.6) so that the sequence of the promoter foot-print can be read off directly.

RNA polymerase footprints for *E. coli* are about 60 nucleotides long and comparison of a large number of different promoter sequences shows that, while great variety is possible, a number of features are common to all promoters. Secondly, promoter mutations have been isolated which either increase or decrease RNA polymerase binding and hence modify tran-scription from the promoter. Sequencing of these promoter mutations has identified base-pairs which are important for promoter function. The following picture emerges (Fig. 5.5). RNA synthesis begins from a precise nucleotide pair on the DNA, the **transcriptional start**, and is numbered +1 on the sequence. Upstream DNA sequences have negative numbers whereas the gene itself is positively numbered. Most promoters have a sequence TATAAT, sometimes called the **Pribnow box** or the −10 region because it is centred at approximately position −10 on the DNA sequence as numbered above. In different genes, the position of the Pribnow box can vary from the average position by several base-pairs. Also the TATAAT se-quence is not absolutely conserved from promoter to promoter and one or two changes from this so called **consensus sequence** occur in any given promoter. Conservation of the base at each position varies from 45 per cent to nearly 100 per cent and a summary of the consensus would be T(80)A(95)T(45)A(60)A(50)T(96), where the numbers in brackets indicate the per-centage occurrence of the most frequently found base.

Some 16 to 19 nucleotides further upstream and centred approximately at base-pair −35 is another region where sequence similarities occur for different promoters. The consensus se-

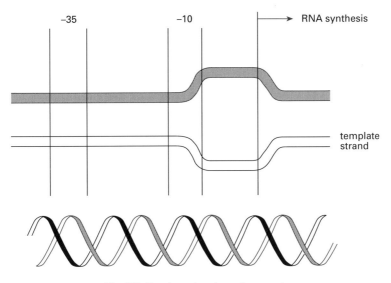

Fig. 5.5 Structure of prokaryotic promoters.

quence can be represented as: T(32)T(84)G(78)A(65)C(54)A(45) for this **−35 region**. The actual intervening sequence does not seem to be important but the distance separating the −10 and −35 regions seems to be critical.

Treatment of promoter–RNA polymerase complexes with chemical reagents which react with specific bases has allowed more precise data to be collected on the sites of protein–DNA interaction. This work has shown that RNA polymerase makes contact with one face of the DNA covering two turns of the double helix from about the −35 region to the −10 region with the −35 region serving as a recognition site for the protein. Two turns along the DNA, the −10 region marks the start of a section of 12 to 14 nucleotides from about −9 to +3 positions where strand separation has been shown to occur by treatment of promoter-polymerase complexes with reagents which specifically attack unpaired bases (Section 2.3.2). This is a region which is usually fairly A·T-rich and since A·T pairs require less energy to separate them than G·C pairs, the region will undergo strand separation with a minimum of input energy (Section 2.5.1).

Another factor which may influence the ease with which this strand separation occurs is the degree of DNA supercoiling since less free energy is needed for the melting in highly supercoiled DNA and, as we will see later (Section 5.1.14), some promoters are sensitive to supercoiling both *in vitro* and *in vivo* (Section 2.3.5).

Finally, RNA synthesis is initiated at position +1 within the melted region and synthesis of the RNA takes place from its 5′-end to its 3′-end. Usually the first nucleotide incorporated is ATP or GTP from a dT or dC residue respectively on the **template strand** of the DNA. The other strand of DNA is called the **coding strand** since the sequence is in the same 5′ → 3′ orientation to the resultant RNA and is identical in base-sequence except T in the DNA is replaced by U in the RNA. The 3′ → 5′ template strand is hence sometimes called the **non-coding strand**.

5.1.5 Termination of RNA synthesis

Once RNA polymerase has initiated mRNA synthesis at a promoter, it moves along the DNA of the gene transcribing the sequence information of the template strand into a complementary RNA until it encounters special termination signals. At these signals, RNA synthesis stops and both enzyme and RNA dissociate from the double-stranded DNA template.

Most prokaryotic termination sites show **palindromic sequences** in the DNA just prior to the termination point and the RNA transcribed has short inverted repeats separated by a small number of nucleotides. Hydrogen bonding between these inverted repeats in the RNA generates **hairpins** with a stem and loop arrangement as shown in Fig. 5.6. Frequently, the palindromic regions are followed by A·T–rich DNA in such a way that the RNA transcribed has a run of U residues, usually about six, hydrogen bonded to dA residues in the template DNA strand. It is thought that these hairpin arrangements in the RNA product cause RNA polymerase to stall in its progress along the DNA template and that because the (oligo-rU)·(oligo-dA) bonding is very weak, RNA breaks free from its DNA template.

The hairpins of terminators have G·C-rich regions which help to stabilize the structure, and mutations have been found which either reduce the efficiency of termination by disrupting base-pairing in the hairpin or increase the efficiency of termination by increasing the

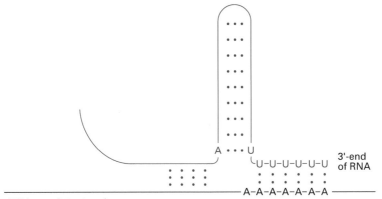

A · · · U
U-U-U-U-U-U   3'-end of RNA

A-A-A-A-A-A-A

DNA template strand

Fig. 5.6 Rho-independent termination of transcription.

stability of the hairpin. Likewise, mutations which reduce or abolish the stretch of U residues in the RNA also reduce or abolish termination.

In addition to these simple terminators, known as **rho-independent terminators**, are the **rho-dependent terminators**. Such terminators also allow the formation of hairpin RNA but the hairpin is not specially G·C-rich and it is not followed by a succession of U residues. Here, termination is thought to be mediated by a special protein called **rho**. One proposal for its mechanism of action is that it binds to the 5'-end of the nascent RNA and travels along behind RNA polymerase (Fig. 5.7). When the polymerase is stalled by the hairpin structure, rho has a chance to catch up. The details of how rho causes termination are not clear but it probably involves the interaction of rho with the β-subunit of RNA polymerase. Possibly the RNA polymerase β-chain is altered in shape by binding of rho such that the enzyme loses its DNA-binding properties. Alternatively, rho may act as a helicase and unwind the RNA from the DNA–RNA complex.

Thus, not only do DNA sequences have to provide information as structural genes, for the synthesis of proteins, they also have to provide the signals for initiation and termination of RNA transcription. We will see shortly that further DNA sequences are involved in regulating gene expression and in DNA replication.

5.1.6 Induction and repression of gene expression

The essential feature of living cells is that almost all their reaction processes are catalysed by enzymes. One way in which a cell can regulate the activities of its metabolic pathways is by alterations in the amounts of the enzymes it contains. Some enzymes are required under all conditions, for example those involved in generating energy, and the genes coding for these enzymes are said to be **constitutively** expressed. Other enzymes are only required when a certain molecule appears in the environment of the cell. Since both protein and the mRNA needed for protein synthesis are very costly to the cell in terms of energy, the efficiency of the cell is vastly improved by synthesizing only those enzymes which are absolutely essential. Thus, some genes are only expressed, or **induced**, when a certain molecule, called the **inducer**,

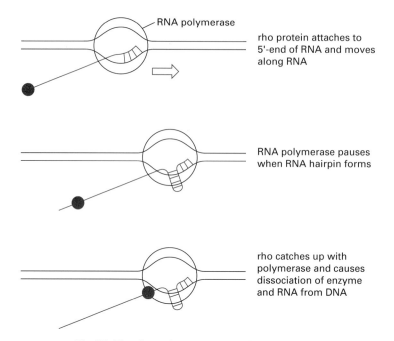

RNA polymerase

rho protein attaches to
5'-end of RNA and moves
along RNA

RNA polymerase pauses
when RNA hairpin forms

rho catches up with
polymerase and causes
dissociation of enzyme
and RNA from DNA

Fig. 5.7 Rho-dependent termination of transcription.

appears in the cell. A well studied case of this which we will look at in some detail is the induction of lactose-metabolizing enzymes when lactose is the only sugar available to provide *E. coli* with essential carbon and energy.

Some genes are normally expressed but are 'switched off' when a compound known as a **repressor** or **co-repressor** appears in the environment of the cell and represses the expression of one or more specific genes. For example, bacteria normally have to synthesize amino acids like tryptophan or histidine from simpler carbon and nitrogen sources through what is often a long series of enzyme-catalysed steps. The presence of preformed amino acids in the environment, however, renders all these enzymes unnecessary and the energy economy of the cell is greatly improved if synthesis of these enzymes is repressed.

Variations in levels of enzymes are also essential in the timing of cell development. This becomes most important in differentiation of eukaryotic cells, but is also found in prokaryotes in periodic processes such as cell division or in the co-ordinated development of phage in infected bacteria. Here, gene expression has to be regulated in a time-related fashion and we find genes divided into **early, middle**, or **late genes** according to when their expression is needed in phage development.

In theory, regulation of gene expression could occur either at the level of translation of mRNA into enzyme, or at the level of transcription of DNA into mRNA, or both. In practice, transcriptional regulation is the most important method in prokaryotes and regulation of promoter activity is the most common way of affecting transcription. However, regulation of transcription at termination sites is also found, particularly in cases of developmentally regulated genes.

Since changes in cell metabolism frequently involve the induction or repression of a whole group of enzymes involved in a pathway rather than induction or repression of one single enzyme, bacteria have evolved ways in which there is co-ordinate induction or repression of all the necessary enzymes. The most important way that bacteria achieve this is to group genes of related function together on the chromosome. When transcription of the group is under the regulation of one single promoter upstream of the first gene in the group and one terminator downstream of the last gene in the group, such a group of genes is called an **operon**. Operon size can vary considerably: operons of two or three genes are common and the operon for histidine biosynthesis in *Salmonella typhimurium* is 11 genes long, while in phage λ there is an operon at least 22 genes in length. Although the mRNA produced from an operon is a single transcript of all the genes which lie between the promoter and the terminator, synthesis of separate polypeptides from this polygenic or **polycistronic messenger** is achieved because of the occurrence within the long mRNA of DNA-coded translational 'start' and 'stop' signals for the ribosome.

The important point to note here is that any process which doubles the frequency with which RNA polymerase initiates at a given promoter will thus double the amount of mRNA for all the genes within the operon and hence will double the amounts of all the proteins coded by all these genes. Therefore we say that there is co-ordinate induction or repression of the genes within the operon.

In the case of genes which are not grouped together in an operon, co-ordinated induction and repression of expression is still possible if the genes share the same regulatory mechanism, usually if their separate promoters share an ability to bind the same regulatory protein. Such a group of co-ordinately regulated genes is known as a **regulon**.

We must now examine one by one the several mechanisms which regulate gene and operon transcription in bacteria. We must also bear in mind that most transcriptional regulation relies on combinations of these mechanisms.

5.1.7 Negative regulation of transcription

The first case we are going to look at is one where initiation of transcription is inhibited when a special protein called a **repressor protein** binds at the promoter and prevents RNA polymerase from binding or prevents it from moving along DNA to transcribe the structural gene. The protein is itself coded by a gene called the **regulator gene**.

The sites where repressor proteins bind to promoters are called **operators** and are designated, for example, as *lacO* and *trpO* in the lactose and tryptophan operons. Operator sequences are about 30 to 40 nucleotide pairs in length and usually overlap with part of the RNA polymerase binding site particularly with the −10 sequences (Fig. 5.8). Their sequences show imperfect dyad symmetry which probably reflects the fact that repressor proteins are symmetrical dimeric or tetrameric proteins.

While bound to operator sequences, repressor protein prevents expression of the gene (Fig. 5.9a). However, combination of an inducer molecule with the repressor protein produces a complex which has lost its ability to bind to the operator. Thus, the promoter is free to bind productively to RNA polymerase and gene expression is induced (Fig. 5.9b).

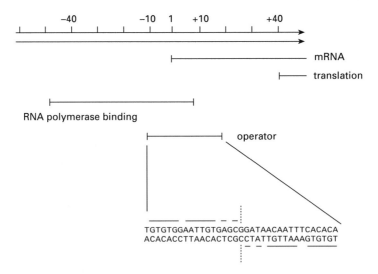

Fig. 5.8 The lactose operon promoter/operator region.

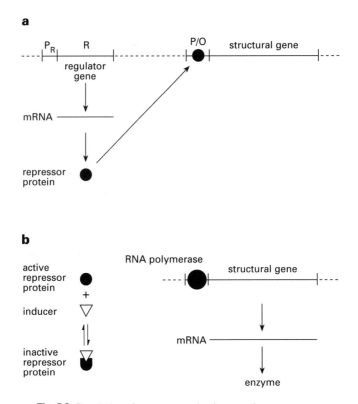

Fig. 5.9 Regulation of gene expression by negative repressors.

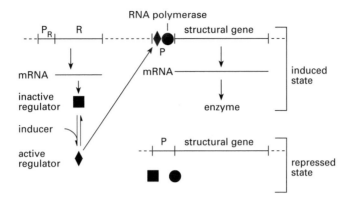

Fig. 5.10 Regulation of gene expression by positive regulators.

5.1.8 Positive regulation of transcription

In the type of regulation just described, the regulator protein acts negatively to inhibit transcription. In other cases, the binding of the regulator protein to the promoter is an essential requirement before transcription can begin. In these positively regulated genes and operons, RNA polymerase cannot bind productively upstream of a structural gene because there is no consensus polymerase-binding sequence present or only a part of one (Section 5.1.4). However, binding of a regulator protein upstream of the gene creates a promoter site where RNA polymerase can bind subsequently. As before, the ability of the positive regulator protein to bind to promoter DNA is affected by the presence or absence of inducer molecules. We illustrate (Fig. 5.10) the situation where the inducer is required to bind to the regulator protein before the latter can bind to the promoter DNA.

5.1.9 Regulation of the lactose operon

The lactose operon consists of three genes (Fig. 5.11). The *lacZ* gene codes for a β-galactosidase which can hydrolyse lactose to glucose and galactose, a *lacY* gene which codes for a lactose

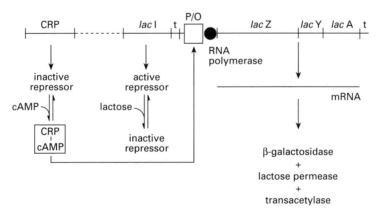

Fig. 5.11 Regulation of *lac* operon expression.

permease that transports the sugar across the cell membrane, and a *lacA* gene whose function is obscure, but which seems to cause acetylation and detoxification of noxious β-galactosides. The operon is induced by lactose, but only if there is no glucose available to the bacteria for metabolism. Lack of glucose raises intracellular cyclic-AMP levels and this converts a positive regulator protein called **CRP (cAMP regulator protein)** into a form which binds to the *lac* operon promoter, *lacP*. RNA polymerase is then able to bind to the promoter, but only provided the operator *lacO* is not occupied by *lac* repressor. This condition will be met if lactose is present. Lactose repressor protein is synthesized by the *lacI* gene which happens to be close to, but not part of, the *lacZYA* operon. Note that repressor proteins are freely diffusible within the cell and the genes coding for them need not be close to the promoters to which they bind.

5.1.10 Attenuation

The expression of a number of operons involved in the biosynthesis of amino acids can be subject to aborted mRNA synthesis called **attenuation**. The *trp* operon of E. *coli* is the best characterized example of this process.

The operon (Fig 5.2) consists of five genes and is subject to negative repression as well as to attenuation. When the cell has plenty of tryptophan, the repressor protein product of the *trpR* gene is converted into an active form which largely prevents transcription from the *trp* promoter upstream of the *trpE* gene. Tryptophan starvation inactivates the TrpR protein and stimulates gene expression some 70-fold.

However, once transcription has started, its completion to the terminator downstream of *trpA* gene is dependent on the severity of the tryptophan starvation and this is sensed when ribosomes try to translate the 5′-end of the newly forming *trp* mRNA. Upstream of the first

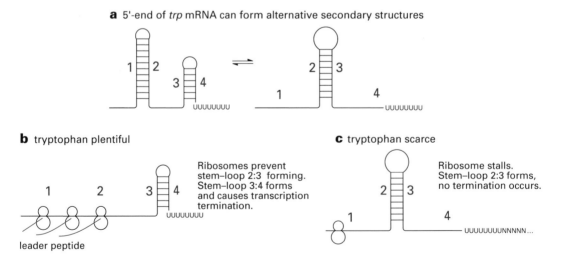

a 5′-end of *trp* mRNA can form alternative secondary structures

b tryptophan plentiful

Ribosomes prevent stem–loop 2:3 forming. Stem–loop 3:4 forms and causes transcription termination.

leader peptide

c tryptophan scarce

Ribosome stalls. Stem–loop 2:3 forms, no termination occurs.

Fig. 5.12 Attenuation of the *trp* operon.

true gene, *trpE*, lies some DNA called *trpL* (**leader peptide**), which codes for RNA with unusual properties. First, the RNA codes for a 13 amino acid peptide (the leader peptide) containing two successive tryptophan residues. Tryptophan has a fairly low frequency of occurrence in proteins so that it is very unusual for two tryptophans to occur side by side. As a result, ribosomes attempting to insert these two tryptophan residues into the leader peptide stall when this amino is in short supply, and this has an effect on the second unusual feature of *trpL* mRNA, a change in its secondary structure. The RNA sequence is such that it can adopt alternative secondary structures as shown (Fig. 5.12a). When tryptophan is plentiful, and the ribosome is able to complete synthesis of the leader peptide (Fig. 5.12b), the RNA forms a stem–loop structure (labelled **3** and **4** in Fig. 5.12a) which is rich in G·C pairs and is followed by a run of eight U residues. This is a typical rho-independent terminator (Section 5.1.5) and mRNA from the rest of the operon is not synthesized. However, when tryptophan is scarce (Fig. 5.12c), the ribosome stalls at the position on *trpL* translation where it is expected to put in the two tryptophan residues. This leaves the RNA sequence (labelled **2** and **3** in Fig. 5.12c) to form an alternative hydrogen bonded stem–loop. Involvement of sequence **3** in pairing with sequence **2** precludes it from taking part in the formation of the termination stem–loop. RNA polymerase thus continues to transcribe the *trpEDCBA* genes, thereby providing sufficient mRNA for the synthesis of much-needed stocks of tryptophan.

Thus, tryptophan starvation increases *trp* operon expression by inactivating TrpR negative repressor protein which increases transcription about 70-fold. It also prevents normal attenuation, which increases transcription another 10-fold. Overall, transcription of the operon can be increased some 700-fold. Similar strategies have also been found in the regulation of operons involved in the biosynthesis of histidine, leucine, threonine, and phenylalanine.

5.1.11 Anti-termination in λ phage development

In the previous section we have seen an example where amino acid availability regulates termination of operon transcription. Here, we are going to study a case where a specific protein regulates gene expression by preventing the normal termination of transcription. The case concerns the development of phage λ in *E. coli* **lysogens**. These are cells which are carrying λ phage DNA in a quiescent form integrated into the bacterial chromosome. The integrated λ DNA does not seem to affect survival of the bacteria until they are exposed to agents like UV light or mutagens which damage DNA and induce the bacterial DNA repair systems. Then a series of enzymic changes is induced which result in the orderly production of infectious λ particles and the eventual lysis and death of the host bacterium. The details of this ordered development are complex. Here, we wish to concentrate on how **anti-terminator proteins** can regulate gene expression in a time-related manner.

In the normal lysogen (Fig. 5.13a), λ genes are prevented from being expressed by a negative repressor protein (product of the λ cI gene) binding at two operators within the two promoters P_L (left promoter) and P_R (right promoter). DNA damage, however, causes inactivation of λcI repressor and transcription of λ lysogen DNA begins from the two promoters as shown (Fig. 5.13b). It stops again promptly at the two rho-dependent termination sites t_{L1} and t_{R1}, but not

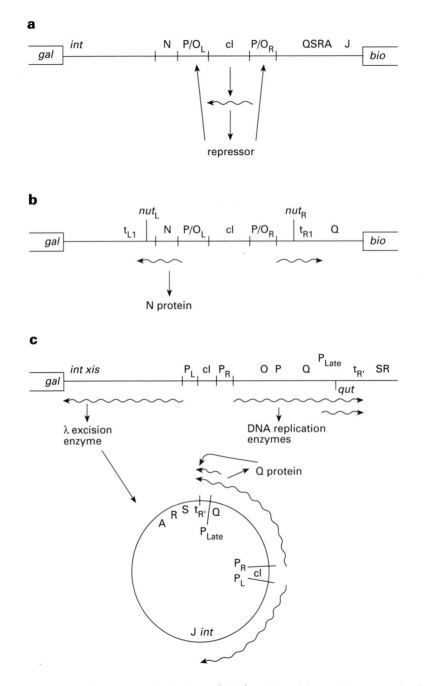

Fig. 5.13 Regulation of gene expression in phage λ. (**a**) λ prophage is inserted between *gal* and *bio* operons of *E. coli*. cl gene expression negatively represses all other λ gene expression (genome sizes not to scale); (**b**) inactivation of λ repressor allows formation of N protein; (**c**) N protein allows transcription of excision and replication enzymes as well as Q terminator. Late genes cannot be expressed because of termination at $t_{R'}$.

before the key gene N has been transcribed. N protein is an anti-terminator protein and, after a time delay in which concentrations of N protein build up, it binds to RNA polymerase and thereby prevents it from terminating. Thus, after a time delay, transcription can proceed past terminator t_{L1} to express genes such as *int* and *xis* involved in excising λ DNA out of the bacterial chromosome as a circular molecule and past t_{R1} to express genes O and P involved in synthesizing multiple copies of phage DNA.

However, transcription of an even later set of genes involved in making the phage coat is prevented by another terminator $t_{R'}$. Another anti-terminator, the product of the λ Q gene, gradually builds up in concentration and can eventually allow RNA polymerase to transcribe past $t_{R'}$ and express the 'late' genes coding for phage structural proteins and for enzymes which lyse open the host bacterium. Thus λ phage development can occur in three sequential stages through the action of two anti-terminator proteins.

The current model for the mechanism by which N protein can cause anti-termination at specific operons is as follows. RNA polymerase transcribes a specific DNA sequence, the *nut* site (*N* utilization), and produces an RNA which, by virtue of its sequence, adopts a specific secondary structure (Section 6.3). N, an RNA-binding protein, specifically binds to this RNA structure and looping of the RNA then allows the RNA–N complex to interact with the downstream RNA polymerase. This is converted into a form having an increased rate of polymerization which allows it to drive through termination signals before they have time to be formed or be recognized. However, the RNA–N–RNA polymerase complex is short-lived and by the time RNA polymerase has reached termination signals further downstream, the complex has broken up and normal termination takes place further downstream. Thus nut_L (Fig. 5.13b) specifically allows N anti-termination at t_{L1} and nut_R specifically allows N anti-termination at t_{R1}. A sequence, *qut* (Fig. 5.13c), acting in a similar way, causes Q binding to RNA transcribed from promoter P_{Late} and specifically allows Q anti-termination at $t_{R'}$. N and Q do not have anti-termination effects at other promoters because the specialized RNA binding structures are not encoded in any other mRNA molecules.

5.1.12 Regulation of gene expression by sigma factors

Recognition of promoter sequences in DNA by RNA polymerase is the responsibility of the σ (sigma) subunit of the enzyme which is present during the formation of initiation complexes. However, there are several versions of σ in bacterial cells and they each recognize a different type of -35 promoter sequence (Section 5.1.4). Cells can therefore influence the type of promoters which are expressed by changing the amounts of the various σ factors synthesized. For example, the normal σ factor of *E. coli* is coded by the *rpoD* gene, but when bacteria experience a heat shock they synthesize an additional new σ factor from the *rpoH* gene which allows expression of a new set of genes with a different type of promoter consensus sequence. These gene products help cells to deal with the heat shock. Similar mechanisms involving other σ factors permit *E. coli* and other bacteria to express selectively genes which allow them to withstand nitrogen starvation or to undergo spore formation or to help them in chemotaxis towards food sources.

5.1.13 Regulation of gene expression by DNA rearrangement

The regulation of transcription of some genes is accomplished by the removal of its promoter to another part of the chromosome when expression is not required and replacement of it when gene expression has to be switched on. One case of this is the synthesis in *Salmonella* of flagellin, the protein subunit of which flagella are made. Which of the two flagellin genes is expressed is determined by the orientation of a particular DNA segment (Section 5.4.5). Similar types of DNA rearrangements have been found to regulate gene expression in phage P1 and phage Mu of *E. coli*.

5.1.14 Regulation of gene expression by changes in DNA topology

If we compare the frequency of transcriptional initiation at a promoter *in vitro* when it is part of a supercoiled, covalently closed circle of DNA and when it is part of a fully relaxed DNA molecule, we sometimes find that initiation occurs much more frequently from the super-coiled DNA. The same is sometimes found *in vivo* when a gene is taken out of its chromosomal context and cloned into a plasmid DNA molecule where its expression varies according to the degree of supercoiling of the plasmid. Presumably, the stress of supercoiling is relieved by se-paration of the DNA strands. Such an essential step of promoter function must considerably aid gene expression.

In bacteria, plasmid DNA is normally negatively supercoiled due to the opposing actions of DNA gyrase, which introduces negative supercoils, and topoisomerase which relaxes the DNA (Section 2.3.5). What happens in the chromosome is much more difficult to assess, but it is believed that here too gyrase and topoisomerase affect the extent of supercoiling. In the cases of genes *proU* and *tonB*, clear evidence has been found that in their normal chromoso-mal locations their expression is sensitive to changes in supercoiling brought about by muta-tions in the genes coding for gyrase, or topoisomerase, or brought about by inhibitors of DNA gyrase. The gene *proU* codes for a betaine transport enzyme whose action protects cells from osmotic stress and *tonB* codes for an iron chelate transport protein.

Summary

Bacterial genomes consist of about 4000 kbp of covalently closed circular DNA, but phage have more varied genomes. DNA sequences are grouped into functional units of heredity called genes, with structural genes coding for enzymes, structural proteins, and regulatory proteins. Genes are expressed by transcription of working copies of DNA sequences into mRNA. This involves binding of RNA polymerase at promoter sequences in DNA and dissociation from the DNA at terminator sequences. Genes with related function are frequently grouped as operons.

Regulation of gene expression in prokaryotes is principally by regulation of transcription, which may be altered by positive regulators, negative regulators, attenuation, DNA rearrangements, or changes in DNA supercoiling, or by combinations of these processes. There is co-ordinate regulation of the genes which make up an operon.

5.2 DNA replication in prokaryotes and phage

5.2.1 Replicons

When a cell divides successfully it must already have created an exactly duplicated set of chromosomes so that both daughter cells carry a set of genes identical to those in the parental cell. Sometimes, in addition to their chromosome, bacteria possess one or more autonomous circular pieces of DNA called **plasmids** which also have to be replicated prior to cell division. Once again, both daughter cells inherit these plasmids. In addition, cells may be infected by phage DNA which replicates independently of other types of DNA. This leads to the production of new infective phage particles, each containing DNA identical to the original infective DNA.

It is useful to be able to refer to each of these independently replicating pieces of DNA as a **replicon**. All replicons have an origin, or **_ori_** sequence, where replication begins and where replication is controlled by interaction with regulators. Once initiated, DNA duplication proceeds to the end of the replicon without the intervention of other regulators. Each regulator is specific for a specific _ori_ sequence, so that different replicons can be replicated independently of one another within one cell.

If DNA molecules which lack _ori_ sequences, or have _ori_ sequences suited only to another species, find their way into a cell, perhaps by phage transduction or by laboratory transformation processes, they cannot be replicated unless they can be incorporated into a suitable replicon by recombination. Even if such molecules are not degraded by deoxyribonucleases, they will eventually be lost from the population of cells when that cell which carries the extra piece of DNA finally dies. Thus, when we attempt to propagate the DNA of one species in cells of another species by genetic engineering, we must provide the foreign DNA with _ori_ sequences suited to the cell type in which we wish it to be propagated.

We might imagine that in a linear or circular replicon, DNA synthesis could begin at any base-pair at random and that this could lead to the formation of a replication eye with the movement outwards of one or both **replication forks** until all the DNA has been duplicated. More than one replication eye might be formed and these, as they grow, might merge until duplication is complete (Fig. 5.14).

However, there is overwhelming evidence that in prokaryotes: (1) there is only one origin per replicon; (2) it is a unique sequence of base-pairs in DNA; and (3) replication forks move outwards from that origin—usually bidirectionally but occasionally unidirectionally. For ex-

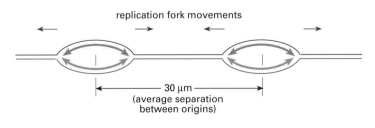

Fig. 5.14 DNA replication eyes.

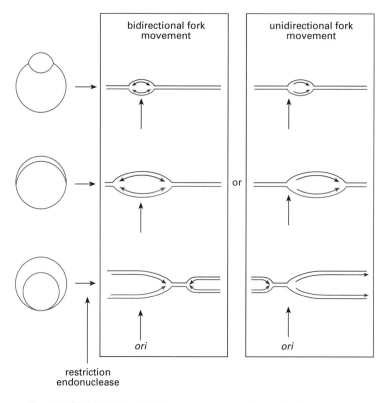

Fig. 5.15 DNA replication initiates at *ori*, a unique site on the chromosome.

ample, examination of a population of replicating plasmid DNA molecules in the electron microscope reveals a collection of theta structures, θ, with a whole range of replication eye lengths (Fig. 5.15). If these molecules are linearized at a unique site, as for example by treatment with a restriction endonuclease which can cut the DNA at only a single site, we find that the linear molecules can be put into a set such as that shown (Fig. 5.15). This analysis shows that all DNA molecules began replication at a site which was at a fixed distance from the unique end. It is also possible to tell whether one or both replication forks are moving with respect to that unique end.

Meselson and Stahl employed heavier ^{15}N-bases to achieve differential labelling of the parental DNA from newly synthesized DNA. They then used density centrifugation (Section 10.3.1) to examine the course of replication. After the first round of replication, they found that both old strands of a double helix are conserved intact—one in each daughter duplex where it is paired with a new strand synthesized to complement the old one. After two rounds of replication, two of the four duplexes each contain one intact 'heavy' strand and two duplexes have only 'light' strands. This type of replication is known as **semi-conservative**.

5.2.2 DNA chain elongation

Bacteria such as *E. coli* and *S. typhimurium* have three distinct DNA polymerases, known as polymerases I, II, and III, but it is DNA polymerase III which is responsible for chain elongation at the replication forks. Structurally, it is a very complex enzyme possessing three activities: (a) 5′ → 3′ chain elongation, (b) 3′ → 5′ exonuclease activity, and (c) 5′ → 3′ exonuclease activity (Section 9.2.4).

Chain elongation (Fig. 5.16) requires a template DNA to be provided. It also requires the four deoxynucleoside 5′-triphosphates and it requires a primer with a 3′-hydroxyl group on to which the polymerase attaches new nucleotide residues according to the hydrogen bonding rules of double-stranded DNA.

It should be noted that the direction of elongation is from 5′-to-3′ only, that the product shows semi-conservative replication (one new strand is laid down against an old strand), and that the sequence of the new strand is complementary to the sequence of the old strand. Thus, there is exact duplication both in amount and in sequence.

No enzyme has ever been found which can synthesize new DNA in the 3′-to-5′ direction and so there is a problem in understanding how a single enzyme with only 5′-to-3′ activity can replicate both of the anti-parallel strands of DNA simultaneously, as was so clearly shown in the experiments of Cairns (Section 5.1.1). The solution to the problem is that only one strand, the **leading strand**, is synthesized continuously in the 5′-to-3′ direction (Fig. 5.17). The other, or **lagging strand**, is synthesized discontinuously. As elongation of the leading strand proceeds, more and more of the lagging strand template becomes single-stranded and short complementary strands are made by DNA polymerase. These small (1 kb to 2 kb) nascent DNA fragments are called **Okazaki fragments** after their discoverer.

Since there must always be a short section of single-stranded DNA on the lagging strand near the replication fork, and since single-stranded DNA is very susceptible to hydrolysis by nucleases, these sections are protected by **single-strand binding proteins** (product of the *ssb* gene). These ssb proteins are later displaced as DNA polymerase moves in to make the new complementary lagging strand.

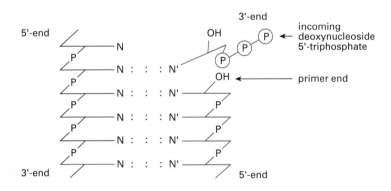

Fig. 5.16 5′ → 3′ Chain elongation by DNA polymerase.

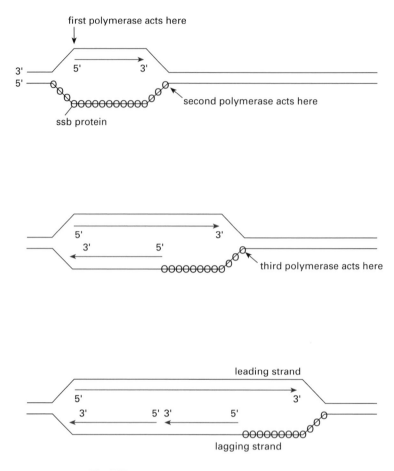

Fig. 5.17 Leading and lagging strand synthesis.

5.2.3 The priming of DNA chain synthesis

Another problem with understanding the action of DNA polymerase is that the enzyme cannot begin DNA chain synthesis *de novo*. There is an absolute requirement for a 3'-hydroxyl priming group paired to a template strand at which the polymerase can add new DNA (Fig. 5.16). In this respect it differs profoundly from RNA polymerase which can initiate new chains *de novo*, for example at promoters. Thus there is a need for a **primer** when leading strand synthesis is initiated at *ori* and also every time synthesis of an Okazaki fragment is initiated on the lagging strand.

The solution to the problem turns out to be that a short section of RNA is first laid down and that the 3'-end of the RNA serves as the primer to which DNA polymerase covalently attaches the first deoxynucleotide. We will return to the priming event at *ori* later (Section 5.2.4) and consider here only the priming of Okazaki fragments. This is carried out by a form of RNA polymerase called DNA primase which is the product of the *dnaG* gene in *E. coli*. Like RNA polymerase, this enzyme begins RNA chain synthesis *de novo* and makes a primer 20 to 30 nucleotides long and complementary in sequence to the template strand. It differs from RNA polymerase, however, in accepting deoxy- as well as ribonucleotides as substrates, so that the primer is a mixture of DNA and RNA.

Once an Okazaki fragment has been primed by DNA primase, DNA polymerase moves along the lagging strand template laying down new DNA until it runs into the 5'-end of the RNA which primed the previous Okazaki fragment (Fig. 5.18). Two things have to be done before lagging strand replication is complete: (1) RNA primer has to be removed and replaced by DNA, (2) individual Okazaki fragments have to be polymerized into a covalently continuous new strand.

The first task falls to DNA polymerase I. This is structurally a much simpler enzyme than DNA polymerase III, but nevertheless it has the same range of enzymic capabilities: a $5' \rightarrow 3'$ polymerase activity, a $3' \rightarrow 5'$ exonuclease 'proofreading' activity, and a $5' \rightarrow 3'$ exonuclease activity. DNA polymerase I is much more abundant than PolIII and is able to find and bind

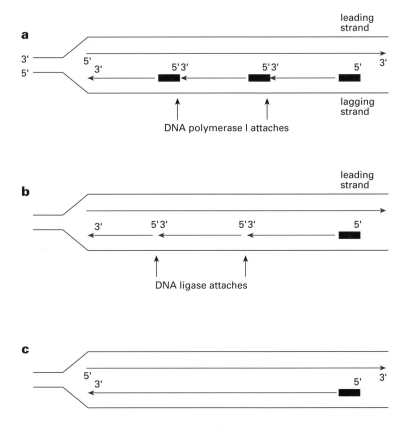

Fig. 5.18 Maturation of Okazaki fragments.

to nicks in DNA, such as the nick remaining between the 3′-DNA end of the Okazaki fragment and the 5′-RNA of the next one (Fig. 5.18a).

DNA polymerase I uses its 5′ → 3′ exonuclease activity to remove the RNA primer and, as it moves along, fill in the gap with new DNA by means of its polymerase activity. Eventually, DNA PolI replaces all the RNA primer by DNA, but it cannot join up the 3′-hydroxyl of the inserted DNA to the 5′-end of the DNA laid down previously by DNA PolIII (Fig. 5.18b). This task is fulfilled by DNA ligase (Fig. 5.18c), the activity of which is described later (Section 10.7.2). Thus, Okazaki fragments are polymerized and new lagging strand DNA becomes covalently continuous just like the new leading strand DNA.

The sequence of events occurring in DNA replication clearly has greater complexity since many more genes are known to be essential for replication than are accounted for in this relatively simple scheme. For example, there are persistent reports that cellular membranes may be involved in some as yet unknown way.

5.2.4 Chain initiation at origins of replication

The most thoroughly studied origin of DNA replication is that belonging to the *E. coli* plasmid pcolE1, described below. However, comparison of the sequence of other origins with that of pcolE1 suggests that similar events occur at all origins (Fig. 5.19).

In pcolE1, replication begins with the formation of an RNA primer 555 nucleotides upstream from the origin (Fig. 5.19a). RNA polymerase, rather than DNA primase, is the enzyme responsible for synthesis of this primer. Transcription actually passes right through the origin with the formation of a DNA–RNA double-stranded structure over the origin. This serves as a substrate for RNase H which hydrolyses RNA when it forms part of a DNA–RNA hybrid and a 3′-hydoxyl priming group is created at *ori* ready for DNA replication to begin (Fig. 5.19b). Without this processing by RNase H, DNA synthesis cannot be initiated and regulation of the initiation, at least in pcolE1, seems to depend on regulation of the processing event.

The regulation is carried out in an unexpected way, not by means of a regulator protein but by means of a regulator RNA. A second species of RNA, called RNA I, is synthesized in the same region as the primer RNA but it is coded from the opposite DNA strand to that used for the primer RNA and it is only 108 nucleotides long. The 3′-hydroxyl group of RNA I lies close to the 5′-end of the primer RNA and the two RNA sequences are, of course, complementary to one another over the 108 nucleotides. They are therefore capable of forming a double-stranded RNA complex at the 5′-end of the primer RNA. This event is thought to affect the secondary structure of the rest of the primer, so that it can no longer serve as a substrate for RNase H.

Thus in the presence of RNA I, the primer is not processed and initiation of DNA replication cannot occur (Fig. 5.19c). As the concentration of RNA I falls, for example as cell volume increases prior to cell division, so there is less RNA I available to complex to the primer RNA and new rounds of plasmid DNA replication are initiated in order to build up plasmid numbers in preparation for cell division.

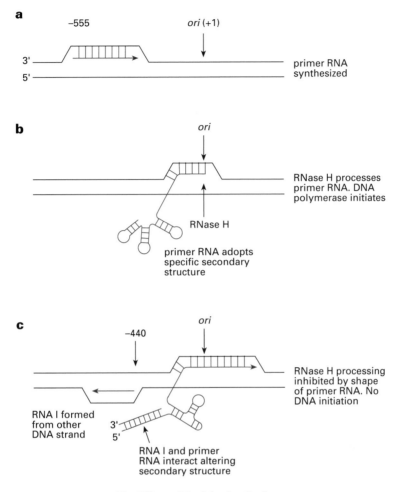

Fig. 5.19 pcolE1 origin of replication.

5.2.5 Accuracy of DNA replication

The accuracy of DNA replication relies on correct hydrogen bonding interaction between an incoming base and a base on the template strand. When we consider that bases can exist (albeit transiently) in tautomeric forms other than the normal ones, we see that, for example, instead of a normal dG·dC pair (Fig. 5.20a), a dT in the template strand could hydrogen bond to the tautomeric form of dG (Fig. 5.20b). Insertion of a dG would be a replication error which would lead to mutation. Based on the expected lifetimes of the tautomeric forms of nucleotides, errors would be expected to occur once every 10^4 or 10^5 nucleotides replicated. In fact, measurements of spontaneous mutation rates show that they are four or five orders of magnitude lower than this expectation; one error occurs roughly every 1000 cycles of bacterial replication, i.e. once in the replication of 10^9 nucleotides.

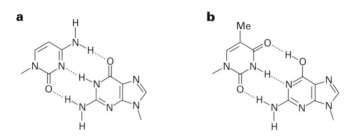

Fig. 5.20 Errors in DNA replication as a result of hydrogen bonding using rare tautomeric forms of bases: (**a**) normal hydrogen-bonding of dG with dC, (**b**) minor tautomeric form of dG bonding with dT.

The reason for this discrepancy seems to be that DNA polymerase exercises a 'proofreading' function. Having inserted a new nucleotide, the enzyme moves forward one nucleotide along the template and, before inserting the next new nucleotide, checks that the last base-pair was formed correctly. An incorrect pair induces a change in shape of the enzyme such that its polymerizing activity is inactivated and its hitherto quiescent $3' \rightarrow 5'$ exonuclease activity is activated. This results in the hydrolytic removal of the offending nucleotide and a return of the enzyme to its original polymerizing form with a second attempt being made to insert the correct nucleotide. The probability that the base is in its rare tautomeric form both at the time of insertion and at the time of proof reading is about 10^{-8} to 10^{-10} which is more in accord with observed mutation rates.

Evidence that $3' \rightarrow 5'$ exonuclease activity is involved in proofreading also comes from the isolation of **mutator** strains of bacteria which have abnormally high rates of mutation and prove to have reduced $3' \rightarrow 5'$ exonuclease activity in their isolated polymerase.

5.2.6 Rolling circle replication

We have seen above how circular replicons can form θ structures during DNA replication, but this mode is not appropriate to the duplication of linear replicons such as linear phage DNA molecules, nor is it appropriate where the duplication of a circular plasmid results in the copy being transferred in a linear fashion from one cell to another. This happens when the *E. coli* **F plasmid** is transferred from an F$^+$ cell to an F$^-$ cell via a sex pilus coded in the donor cell by F plasmid DNA. Rolling circle synthesis of plasmid DNA accompanies this transfer.

First of all, the 95 kbp circular plasmid is nicked at a precise location, *oriT* (origin of transfer synthesis), by a specific endonuclease coded by the *traYZ* genes of F plasmid (Fig. 5.21). Next, the free 5'-end of single-stranded DNA passes down the narrow protein tube of the sex pilus driven by replacement of the lost strand by DNA synthesis in the donor cell. As the 5'-end appears in the F$^-$ recipient cell, a new second strand is synthesized in a series of Okazaki fragments. The template strand in the F$^+$ donor can be regarded as a revolving or rolling circle from which an old strand is peeled off and replaced with a new one. Once a circle has completed one revolution, the endonuclease cuts the old strand free and its 3'-end passes into the F$^-$ cell where it acquires a complementary strand and the linear molecule is circularized.

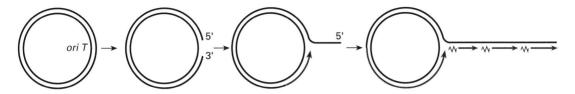

Fig. 5.21 Rolling circle replication during F plasmid transfer synthesis.

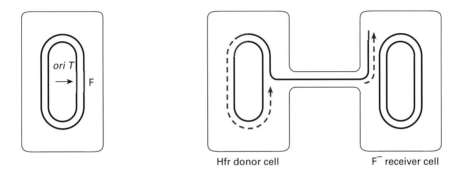

Hfr donor cell   F⁻ receiver cell

Fig. 5.22 Chromosome transfer during mating of Hfr bacteria with F⁻ bacteria.

Thus, the F⁻ cell acquires an F plasmid (and becomes F⁺) and the donor cell retains a copy of the plasmid.

This form of plasmid replication is to be distinguished from that which occurs when F plasmids are duplicated prior to cell division. Here, conventional θ replication takes place which is initiated at an origin quite distinct from the origin of transfer synthesis.

Sometimes F plasmid integrates into the *E. coli* chromosome and, when in this case DNA transfer is initiated from the *oriT* sequence, not only is the F DNA transferred but donor cell chromosomal DNA is also transferred to the F⁻ recipient (Fig. 5.22). The donor cell replaces its chromosome by rolling circle synthesis and the recipient cell acquires a copy of the genes of the donor cell. Donor cells behaving this way have been known as **Hfr** cells because they transfer genetic information with high frequency and they have been extremely important in genetic mapping of *E. coli* (Section 5.1.2).

5.2.7 Rolling circle DNA synthesis during λ replication

λ Phage DNA is linear with short complementary single-stranded ends which are self-cohesive and referred to as *cos* sequences. Within a host cell, this molecule quickly circularizes via its *cos* sequences and DNA ligase of the host cell makes it covalently closed. This circular λ DNA next undergoes a few rounds of θ replication from an *ori* sequence within gene O (Fig. 5.13c), but the major DNA replication occurs through rolling circle mechanisms. Each circle turns many times and many tandem repeats of λ sequence are produced, known as **concatameric DNA** (Fig. 5.23). In order that this DNA can be packaged within the phage heads, an enzyme first makes staggered cuts in the two chains at the *cos* sequences. This re-creates the

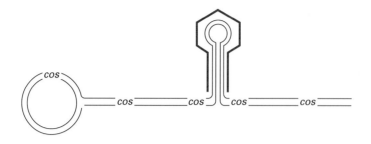

Fig. 5.23 DNA replication in phage λ.

single-stranded ends and then the DNA between the two *cos* sequences is packaged into the phage. The only demands made by the packaging system are that *cos* sequences must be present and that they are about 49 kbp apart on concatameric DNA, a feature which has been exploited in packaging foreign DNA into λ heads *in vitro* and for subsequent efficient entry into *E. coli*.

5.2.8 Replication of single-stranded phage DNA

DNA polymerases can catalyse only the synthesis of sequences complementary to the template they are given. Therefore, sequences identical to the template can be made only by synthesis of a complement of the complement. Thus, in order for copies of single-stranded circular DNA of phage to be made (Fig. 5.24), the infective phage strand, the (+) strand, must be used as a template for the synthesis of a complementary or (−) strand.

The first step is the synthesis of a double-stranded (+−) circular DNA molecule, referred to as the replicative form or **RF DNA**. (+−) RF DNA then undergoes many rounds of rolling circle replication to produce new (+) DNA for packaging into phage coats. In contrast to the type of rolling circle replication described above (Section 5.2.7 and Fig. 5.21), only single-stranded (+) DNA is formed here. As well as serving to make more (+) DNA, the RF DNA is also used for transcribing phage genes into the mRNA required for building the protein parts of the new phage, since only double-stranded DNA can be transcribed into mRNA by RNA polymerase.

Just as in the case of initiation of the synthesis of any new DNA strand, formation of the (−) strand on the infecting phage (+) strand requires priming by RNA to produce the double-stranded RF DNA. This occurs at a specialized region of the (+) DNA where the sequence allows for formation of a stem–loop hydrogen bonded structure. RNA polymerase (in the case of phage M13) or DNA primase (in the case of phage ϕX174) binds to this region of DNA and an RNA primer is produced. The DNA polymerase of the host cell then catalyses the formation of the (−) strand, DNA polymerase I removes the primer and replaces it with DNA. Finally, DNA ligase seals the nicks and DNA gyrase supercoils the RF DNA.

In the case of ϕX174, the subsequent rolling circle stage involves a phage-coded A protein that nicks the supercoiled RF DNA at a particular sequence and binds covalently to the 5′-phosphate end. The (+) single strand then peels off while a new (+) strand is synthesized to replace it. After one turn of the circle, the A protein cuts off the (+) strand and joins the 5′-end

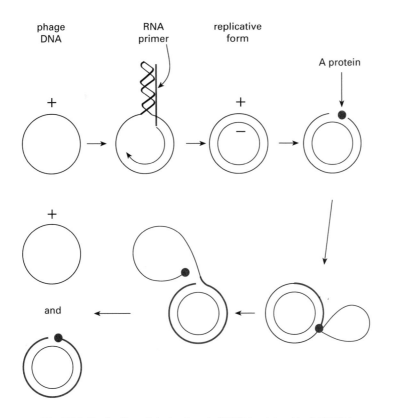

Fig. 5.24 Replication of single-stranded DNA involving (+−) RF DNA.

of the DNA to its 3′-end, thus forming the first single-stranded (+) circle for packaging into new coat proteins. Repeated turns of the RF circle can peel off multiple copies of the single-stranded (+) DNA.

Summary

A replicon is an independently replicating unit of DNA. Replication begins from unique DNA sequences, *ori*, in prokaryotic replicons. Replication eyes proceed bidirectionally, unidirectionally or outwards, with the extension reaction being catalysed in the 5′→3′ direction by DNA polymerase III. High accuracy is achieved by the 3′→5′ exonuclease proofreading activity of DNA polymerases. Initiation of new DNA chains requires formation of RNA primers on to which DNA chains are added. The RNA primer is replaced with DNA by DNA polymerase I and covalent continuity of the DNA strands is effected by DNA ligase.

Circular DNA is most commonly replicated via θ intermediates, but rolling circle replication occurs during plasmid DNA transfer synthesis, during λ phage replication, and during the production of single-stranded phage DNA molecules from double-stranded circular replicative intermediates.

5.3 Genome structure in eukaryotes

Eukaryotic genomes are more complex than those of prokaryotes, since eukaryotes need to construct a larger, more complex cell. There is often an additional requirement of building a complete organism composed of many different cell types, all of which must have been derived from a single cell with a fixed amount of DNA sequence information. It is no surprise therefore to find that eukaryotic genomes are larger than prokaryotic ones. In general, this size increase follows the eukaryotic 'evolutionary tree' but there are some exceptions (Fig. 5.25).

The genome size for some types of organism, such as amphibia, is extremely variable, with sizes varying over a hundred-fold range. It seems highly implausible that one type of amphibian needs much more genetic information than another. Furthermore, the genome size for *Drosophilia melanogaster* (about 2×10^8 base-pairs) seems bigger than is required. If we assume an average size for a gene of 4000 nucleotide pairs, it would mean that *D. melanogaster* contains approximately 50 000 genes (although the calculations are rather approximate). This is roughly 10 times the number which is estimated from saturation mutagenesis experiments on the fly. This dilemma of apparently excessive quantities of DNA present in some organisms has been called the **C value paradox**. In the course of this section we will attempt to find a solution to this paradox by considering what types of DNA sequence are present in eukaryotes and what function they have, if any. It should be mentioned at the outset, however, that much of the DNA found in higher eukaryotes does not encode information which is expressed as protein or stable RNA.

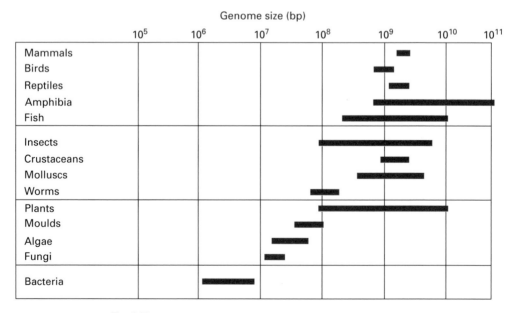

Fig. 5.25 The variation in genome size between different organisms.

5.3.1 Eukaryotic promoters

We will begin our survey with that proportion of DNA which is expressed as RNA and in particular with the DNA sequences which are implicated in the initiation of transcription. Eukaryotic genes are transcribed by one of three different RNA polymerases (excluding the mitochondrial genes). RNA polymerase I transcribes ribosomal RNA exclusively. Since this is the major RNA species in any eukaryotic cell, this polymerase is the predominant enzyme. RNA polymerase II transcribes those RNAs which will form the messenger RNAs encoding all cellular proteins (and some small non-protein-coding RNAs). RNA polymerase III transcribes most small RNAs including tRNAs and 5S RNA. Each RNA polymerase is a multi-subunit complex with some subunits being shared by different polymerases.

Let us look at the promoters which are recognized by eukaryotic RNA polymerases to see if we can find consensus 'boxes' in the same way as we found the Pribnow and −35 boxes in prokaryotes (Section 5.1.4). In the case of RNA polymerase I promoters, no strong consensus emerges between the promoters of different species. Mammals share common sequences between −16 and −21 nucleotides and about −33 and −40 nucleotides upstream of the transcription initiation site, but these are not conserved in amphibia. However, these regions seem to be important to the mammalian genes, since deletion of them seriously impairs initiation. The only sequence shared between all higher eukaryotes studied so far is a G at 16 bases upstream of the initiation site. A point mutation of this base almost completely abolishes transcription of mouse rRNA.

RNA polymerase III is unique in that its promoter region almost always lies downstream of the transcription initiation site. Regions shown to be essential by deletion analysis of three genes are shown in Fig. 5.26. These internal regions are binding sites not for RNA polymerase but rather for a **transcription factor** called **TFIIIA** which interacts with RNA polymerase III so as to promote transcription.

RNA polymerase II is the most interesting of the three polymerases, since it is responsible for the synthesis of all mRNA and ultimately of all protein. Since the thousands of different proteins encoded by eukaryotic DNA have a huge variety of requirements for the timing and location of their expression, there has evolved a correspondingly complex set of regulatory controls on the transcription of the genes which encode them. In practice, this means that there are many types of consensus box to be found upstream of different protein-encoding genes with each type donating a particular developmental specificity of transcription. These

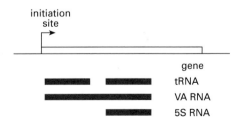

Fig. 5.26 Promoter regions of genes transcribed by RNA polymerase III.

Sequence	Transcription responsive to	Recognized by
5' CTGGAATNTTCTAGA 3' 3' GACCTTANAAGATCT 5'	Heat shock	Heat shock activator protein
5' GGATGTCCATATTAGGACATCT 3' 3' CCTACAGGTATAATCCTGTAGA 5'	Growth factor stimulation	Serum response factor
5' CGNCCCGGNCNC 3' 3' GCNGGGCCNGNG 5'	Cadmium or copper ions	?
5' CCGCCC 3' 3' GGCGGG 5'	Cell type	Protein Sp1
5' GGCCAATCT 3' 3' CCGGTTAGA 5'	Cell type	CAAT binding transcription factor (CTF)
5' ATGAGTCAG 3' 3' TACTCAGTC 5'	Phorbol esters	TPA-modulated *trans*-activating factor (AP1)

Fig. 5.27 Nucleotide sequences of upstream activator/enhancer sequences.

have been variously named **upstream activator sequences** (UASs) or **transcriptional enhancer elements** (Fig. 5.27).

Deletion of these small regions almost abolishes promoter activity. Conversely, addition of, for example, the heat shock consensus box to an upstream region of a gene confers heat shock inducibility upon that gene. Footprinting studies analogous to those used for prokaryotic promoters show that proteins recognize and bind to these consensus boxes. Note also that several of these regions contain dyad symmetry analogous to that seen for restriction enzyme recognition sites (Section 10.7.2).

Many upstream consensus boxes (enhancers) have the property of retaining their transcriptional activating potential when they are either moved or inverted with respect to the initiation site (Fig. 5.28). This is perhaps an artificial distinction created by the experimenter, since enhancers often share consensus boxes with UASs and a UAS can sometimes be converted into an enhancer by the addition of one or two consensus boxes. Boxes with dyad symmetry possess the same sequence in either orientation.

Two major models have been proposed for the interaction between upstream boxes and the RNA polymerase at the initiation site. The first suggests that this interaction occurs by looping out of the DNA between them. The second model suggests that the boxes are entry sites for a protein factor which binds to them and then migrates along the DNA in order to interact with the RNA polymerase (Fig. 5.29). Present evidence strongly favours the first explanation. There is also growing evidence that different proteins can compete for binding to the same

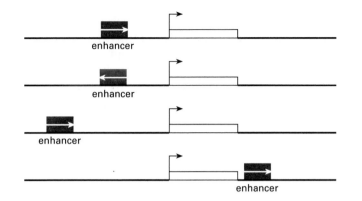

Fig. 5.28 Enhancer function is independent of orientation or distance from the affected gene.

box and prior binding of one protein to a box can inhibit subsequent binding of another protein to an adjacent or overlapping box.

In addition to the varied upstream activator sequences or enhancer elements which have been mentioned above, another type of consensus sequence is often found at RNA polymerase promoters. The consensus sequence is TATAAA. This sequence is usually called the **TATA**

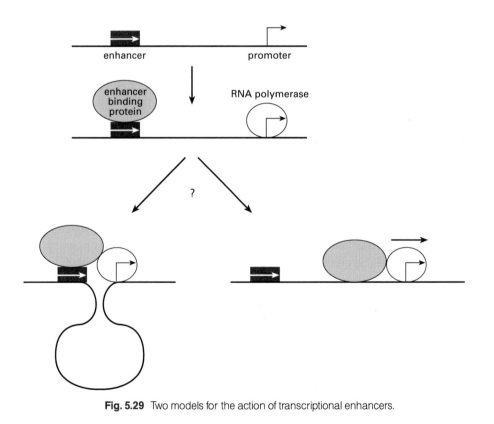

Fig. 5.29 Two models for the action of transcriptional enhancers.

box or **Goldberg/Hogness box** after its discoverers. Such a TATA box is often found at approximately 25 nucleotides upstream of the transcriptional start site. Although it is similar in sequence to the prokaryotic Pribnow box, its function is different. It is required for accurate positioning of the RNA polymerase initiation site. Thus, promoters which possess TATA boxes begin transcription at a unique spot whereas promoters which lack a TATA box yield RNAs with several different, staggered 5′-ends. In such cases TATA boxes do not affect the total level of mRNA production, only the exact structure of the 5′-ends. However, the TATA motif in yeast is more akin to upstream activator sequences than to the Goldberg/Hogness box.

The TATA box is recognized by the transcription factor TFIID. Fascinatingly, TFIID is also involved in the transcriptional initiation of genes lacking TATA boxes. TFIID is itself a multimeric complex of proteins, at least one of which is also found in transcription factors for RNA polymerases I and III.

The transcriptional control regions of some eukaryotic genes can be extremely complex. For instance, the virus SV40 uses a region of about 200 bp to control the transcriptional initiation of its capsid protein genes. At least 30 consensus boxes have been identified in this region and there may be more. Why is there a need for such complexity? We do not really know, but perhaps the necessity for being transcribed in a variety of different cell types, each with its own repertoire of transcription factors, requires such a complicated promoter.

Transcriptional repressors exist in eukaryotes as well as prokaryotes. 'Negative enhancers' or **silencers** can repress transcription at a distance. We assume that such sequences act in an analogous way to UASs and enhancers, only the particular protein factors which bind to them inhibit rather than stimulate transcriptional initiation.

5.3.2 Introns and exons

When we begin to look at the sequences of eukaryotic genes and compare them with those of their prokaryotic counterparts, a major difference immediately becomes apparent. Most eukaryotic genes are discontinuous with respect to the mature RNA or protein encoded by them. A simple example is the β-globin gene of mammals, which was one of the first to be discovered (Fig. 5.30).

The rabbit β-globin gene contains two internal DNA segments which are not found in the mature RNA. These inserts are called **introns** and the regions surrounding them, **exons**. You will see later (Section 6.4) that introns are transcribed by the RNA polymerase but they are removed from the nascent transcript by a processing mechanism called **RNA splicing**. Genes which are transcribed by RNA polymerases I and III also sometimes contain introns.

What is encoded by introns? Do they have any direct function? It is not easy to answer these questions definitely, but in general we believe that introns very often have no functions in the day-to-day expression of those genes which contain them. This has been elegantly demonstrated for an actin gene of yeast by the removal of its sole intron (Section 10.9). No discernible effect on the expression of the actin gene could be seen. Our conclusion must therefore be that, at least in this case, an intron is irrelevant to the expression of a gene.

How much of the eukaryotic genome is comprised of introns? We cannot be sure, but there is often more DNA contained in the introns of genes than is found in their exons. Several ex-

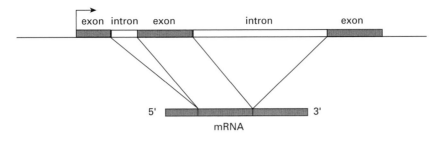

Fig. 5.30 Structure of the β-globin gene which contains intervening sequences (introns) that are not present in the mRNA.

treme examples are known such as the *Antennapaedia* gene of *Drosophila*, where the entire gene stretches across at least 100 000 nucleotide pairs, yet the mRNAs encoded by the gene are a mere 5000 and 3500 nucleotides long.

If many introns have no direct function, do any have indirect functions? Why are they there at all? The answers to these questions probably lie in the ancestry of genes themselves. We know that introns are at least as old as the prokaryote–eukaryote split because they are occasionally found in prokaryotes. This suggests that they are a relic of the process that originally led to the creation of genes. The genes we see today probably derived from the fusion of many smaller modules of genetic information. As an example, many enzymes with widely different functions share a domain which is capable of binding ATP. It is likely that such a segment only evolved once and the many different genes containing it somehow picked up this sequence module during their evolution. Modules of information (exons) may have been 'shuffled' and dealt in many combinations to yield the genes we see today. This model was first proposed by Gilbert and is supported by two lines of evidence. First, exons often encode discrete units of protein structure (often called **domains**) such as the ATP binding site mentioned above or the catalytic site of an enzyme. Secondly, removal of introns from RNA can sometimes occur in the absence of any protein. Indeed it is an inherent property of the nucleic acid (Section 6.5). This amazing property suggests that such RNA splicing evolved before proteins and provides us with a rational explanation for how **exon shuffling** was possible. Presumably, some type of DNA rearrangement must be a prerequisite for a shuffling event, but DNA rearrangements could not be expected to join up sequence modules precisely. There would have to be gaps, many of which would inevitably contain translational stop codons or introduce frame-shifts into the hybrid reading frames. If those gaps had the ability to remove themselves from the RNA containing them by splicing, then these exons would be far more easily shuffled.

There is some evidence to suggest that introns can sometimes be generated by the insertion of DNA segments into genes. These segments are a particular subset of transposable genetic elements (see Section 5.4.7). It is not clear whether these elements have acquired splicing signals near their ends relatively recently or they have always had such abilities. This is a very important question to answer because it could resolve whether some transposable elements evolved from introns or *vice versa*.

Currently, we believe that introns have always occurred in genes simply because they are a part of the way genes evolved in the first place, namely by shuffling of modular units of genetic information (exons) at the DNA level combined with removal of the intervening sequences (introns) at the RNA level.

5.3.3 Gene families

The β-globin gene is not the only globin gene in a mammal. All mammals possess α-globin genes and myoglobin genes. Several genes related to the α- and β-globin genes are only expressed in the fetus or embryo. We call the collection of all these related genes a **gene family**. Often the members of a gene family are all present at the same chromosomal location. It is postulated that each gene family originally evolved from an individual gene. A plausible mechanism which involves recombination, called **unequal crossing-over**, exists to explain how such duplications may occur (Fig. 5.31 and Section 5.4.3). Strong evidence to suggest that this happens in *Homo sapiens* comes from sequence analysis of the human embryo-specific, β-globin-like genes Gγ and Aγ. The DNA sequences of these two genes and of the regions surrounding them are extremely similar, suggesting that a region bigger than the original precursor gene became duplicated relatively recently (in evolutionary terms). Furthermore, the duplicated regions are flanked by pieces of repetitive DNA. The likely mechanism of duplication involved homologous recombination between these repetitive segments. Normally, recombination exchanges a DNA duplex for a closely related sequence. However, if the two DNAs do not align themselves properly, unequal crossing-over transfers a piece of one chromosome to the other, thus generating a duplication (Fig. 5.31).

An extreme example of β-globin gene duplication can be found in the domestic goat. Here, at least four consecutive duplications have given rise to the 12 genes which share homology with β-globin (Fig. 5.32).

Thus, genes may be duplicated many times over in the higher eukaryotes. Does this process fulfil a useful function? We believe it does, since in several instances the duplicated genes encode proteins with different specificities. For instance (Fig. 5.32), some of the β-globin-like genes are expressed only during embryonic developmental stages and the protein encoded by them possesses an advantageous oxygen-binding capacity which, in fetal haemoglobin, enables it efficiently to extract oxygen from the maternal haemoglobin. This is a specific example of a much more general phenomenon of great evolutionary significance. If a gene is duplicated, then the new copy is under no selective evolutionary pressure to carry out its old function, since one copy is perfectly capable of doing this. Therefore, a new copy is able to experiment with its sequence by random mutation through the course of time until it finds a novel function which is useful to the host.

Do all duplicated genes eventually find a use? No, some pick up nonsense or missense mutations and become 'dead' genes or **pseudogenes**. Three genes in the goat β-globin-like cluster are pseudogenes. Thus, genes have a tendency to multiply in higher eukaryotes, thereby contributing both to the large amounts of DNA seen in these organisms and to the evolution of their hosts by the acquisition of novel functions.

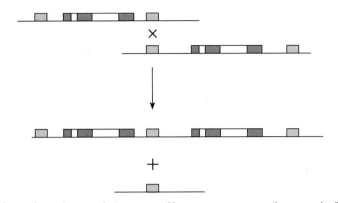

Fig. 5.31 Unequal crossing-over between repetitive sequences generating gene duplications and deletions.

Perhaps the most fascinating gene family discovered thus far in the vertebrates is the immunoglobulin superfamily. A broad class of at least 100 genes which encode proteins that bind to cell surfaces probably are all derived from a single ancestral gene. Most (and perhaps all) of these proteins are involved in recognition of other cells or invading entities. Presumably, this diversification evolved coincidentally with the structural evolution of higher eukaryotes in order better to define and regulate the activity of different cell types in the organism.

Another consequence of multiplication of genes is the ability to produce large amounts of the gene products. Thus, the genes which encode very abundantly expressed RNAs, such as tRNAs, 5S RNA, ribosomal RNA, and histone RNAs, are repeated many times in higher

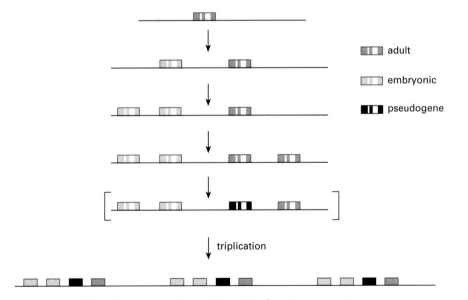

Fig. 5.32 The probable evolution of the β-globin locus of the goat.

Table 5.2. Approximate copy number of genes encoding abundantly expressed RNA or protein

	tRNA	5S RNA	rRNA	Histone H2A
Saccharomyces cerevisiae	250	150	150	2
Drosophila melanogaster	1 000	150	150	100
Xenopus laevis	1 000	25 000	500	50
Homo sapiens	1 000	2 000	300	15

eukaryotes. The approximate copy number of these genes for several organisms is listed in Table 5.2.

5.3.4 Highly repetitive interspersed DNAs

There are two known classes of DNAs which are present in hundreds of thousands or even millions of copies which are interspersed amongst the rest of the genomes of eukaryotes. Short interspersed elements (or **SINES**) are comprised of short stretches of DNA (a few hundred base-pairs long). These are virtually all pseudogenes which are derived from genes for various small RNAs. Each organism has its own characteristic family (or families) of SINE elements. The major element in mammals is the **Alu family**, named after the restriction enzyme Alu I which generates a characteristic fragment seen in digests of human DNA. It is possible that Alu elements neither fulfil any major function for their host nor do they significantly harm their bearer. These sequences have been termed parasitic or **'selfish' DNAs** in that their only function is their propagation through the genomes they inhabit. At present, we are uncertain of whether such sequences are truly parasitic. Perhaps some as yet undiscovered property which helps the host is awaiting discovery.

How do such sequences become amplified? We believe that they are members of a family of transposable genetic elements (Section 5.4.7) which employ **reverse transcription** (conversion of RNA into DNA) to create new genomic copies of themselves. A possible mechanism for the creation of a new SINE element is shown in Fig. 5.33).

A second class of interspersed repeats which is related to the SINE family is found in higher eukaryotes. These long interspersed elements (or **LINES**) are several thousands of nucleotides long. We believe that they are also of no use to the organisms containing them and that they multiply by a fundamentally similar mechanism to that used by SINES. However, LINES encode the reverse transcriptase which is involved in this process and it is even possible that LINE reverse transcriptase is the enzyme which aids the amplification of SINES (which are far too small to encode such a protein).

5.3.5 Satellite DNA

The last major repetitive component of eukaryotic DNAs which we will consider is satellite DNA. This DNA is composed of many thousands or even millions of repeats of short se-

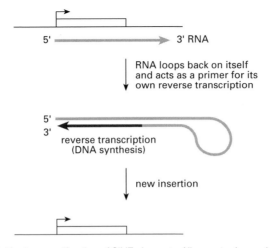

Fig. 5.33 A model for the amplification of SINE elements. All repeats share a 'core' sequence which resembles the *chi* site of *E. coli*.

quences. Satellite repeats are organized in huge tandem arrays. Because such arrays are so homogeneous in sequence, they have a characteristic G–C content. This is sometimes significantly different from the overall average value for the bulk DNA of the organism, with the result that such DNA is of a different buoyant density and will band at a different position in a CsCl density gradient to that of the large part of the bulk DNA (Section 10.1). Such **satellite bands** in CsCl gradients gave satellite DNA its name. The fruit fly *Drosophila virilis* has three related satellite sequences. These are:

5′ ACAAACT 3′   Satellite I

5′ ATAAACT 3′   Satellite II

5′ ACAAATT 3′   Satellite III.

Together, the millions of copies of these three repeats comprise about one third of the entire *D. virilis* genome. This amounts to at least 100 million nucleotide pairs of satellite sequence. Higher eukaryotes may have longer repeat units. These sequences again encode no RNA or protein are assumed at best to fulfil a structural role in chromosome organization, since they are found mainly near the centromere.

5.3.6 DNA sequence polymorphisms

Homologous recombination can occur more or less randomly along a chromosome. One important consequence of this is that the probability of a recombination event occurring between two points on a chromosome is inversely proportional to the distance between them. Thus, if two genes are very close together, there is little chance of a random recombination event happening between them. It is much more likely to occur somewhere else. This phenomenon is called **linkage** and was used as one of the first ways to map genes on chromosomes. In recent times, this phenomenon has been used to aid the diagnosis of genetic disease. Often the nature of the damaged gene causing a particular disease is unknown. However, if one can find a genetic marker of some kind which maps very close to the gene, then it can be used as a diagnostic tool. There is approximately a 1 per cent chance that any base on a chromosome in a particular individual will be different from the corresponding base on the same chromosome from another individual. Therefore, every chromosome has its own 'fingerprint' and if one of these so-called **sequence polymorphisms** can be identified close to the diseased gene, then it can be used to trace that particular gene through the relatives of the bearer. Of course, if the polymorphism creates or removes a restriction enzyme cleavage site, then it can be readily detected. In this way, markers close to several genetic diseases, such as muscular dystrophy and Huntington's disease, have already been isolated. From there is it an arduous, but feasible, task to clone such genes by chromosome walking and, indeed, the chromosomal regions corresponding to both these genetic diseases have now been cloned and sequenced.

One particular type of polymorphism in human DNA has leapt into the limelight recently. The human genome contains many copies of so-called **minisatellite** sequences. Each minisatellite region is comprised of tandem repeats. Individual repeats are very similar to one another in a particular cluster, but the consensus sequence of the repeat from one cluster is very different from that in another (Fig. 5.34). A radioactive copy of a minisatellite sequence is used to probe a restriction digest of total human DNA which has been subjected to electrophoresis on an agarose gel and then transferred to a filter. This very important technique is called **Southern blot analysis** (Section 10.10.2). A unique fingerprint pattern of bands corresponding to restriction fragments containing minisatellite sequences is generated. Since these repeats are inherited in a normal Mendelian manner, it is possible to test the genetic relationship between people by following this procedure. It is also possible to generate this fingerprint from relatively small amounts of tissue or body fluid, especially when PCR (Section 10.7.3) is used to

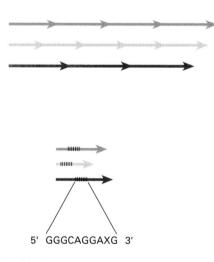

5' GGGCAGGAXG 3'

Fig. 5.34 Hypervariable minisatellite sequences.

amplify the DNA. Alec Jeffreys has developed this procedure to make a powerful impact on forensic medicine.

Summary

Eukaryotic genomes are more complicated than those of prokaryotes. They are in general much bigger than seems necessary for the approximate number of genes they contain. Much of the DNA in eukaryotes is repetitive in sequence, does not encode essential information for the host, and is not expressed as stable protein or RNA.

Eukaryotic genes are transcribed by one of three polymerases. RNA polymerase I transcribes ribosomal RNA from promoters which are poorly conserved in evolution. RNA polymerase II transcribes mRNA and some small RNAs. RNA polymerase II promoters can be very complex and contain one or more upstream activator sequences, enhancers or silencers. RNA polymerase III transcribes small RNAs from promoters located inside the genes.

Many eukaryotic genes contain intervening sequences (introns) which are not present in the mature RNA. The antiquity of introns argues that they are as old as or older than recognized genes. They may be a by-product of gene evolution catalysed by DNA rearrangements and RNA self-splicing. Eukaryotic genes are sometimes duplicated by unequal crossing over. This process has given rise to multi-copy gene families encoding the globins and immunoglobulins. SINE and LINE elements account for a significant part of the dispersed repetitive DNA in higher eukaryotes. Satellite DNA is comprised of millions of tandemly repeated copies of simple sequences and is found at centromeres. DNA sequence polymorphisms are powerful tools in the diagnosis and mapping of genetic disease in humans and in forensic medicine.

5.4 DNA rearrangements in prokaryotes and eukaryotes

One of the first discoveries in genetics was the observation that, before donation to the offspring, the different parental genes are shuffled. How do these 'beads on a string' become assorted? The answer is by **homologous recombination**. If a cell contains more than one copy of a given chromosome, then segments from one copy can recombine with corresponding segments of the other. As we shall see below, this recombination is ultimately dependent upon the DNA sequence homology between the two chromosomes (Fig. 5.35).

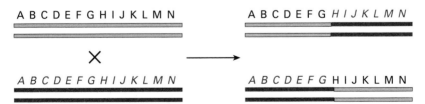

Fig. 5.35 Homologous recombination.

What advantages does homologous recombination provide? It enables the host organism to assort alleles (differing copies of the same gene) into novel groups. If a copy of a particular gene in a fruit fly has randomly acquired a mutation that yields a more efficient enzyme, then Darwinian natural selection can operate upon that gene provided it is not shackled to all the other genes on the chromosome. Thus, favourable and unfavourable alleles can be shuffled randomly and then the many combinations in the population can be tested by natural selection. Another advantage which recombination provides is the ability to repair a damaged gene in an otherwise favourable chromosome. If no ability to assort different alleles existed, then a single unfavourable mutation in a chromosome would consign the whole of it to oblivion. Lastly, homologous recombination is used in many discrete instances to regulate the expression of genes. On a more subjective note, the recombination process is often used as a tool in the laboratory to aid the researcher.

5.4.1 The mechanism of homologous recombination

Homologous recombination is linked to DNA replication, but does not occur at replication forks. Rather, it takes place between intact double helices. Genetic and DNA sequence analysis has shown that recombination is precisely accurate to one base-pair. The inference from this is that base-pairing is involved during the process. Damage to DNA stimulates recombination, strongly suggesting that homologous recombination is initiated from broken DNA strands. Presumably the cell normally creates such breaks enzymatically in order to aid this process.

5.4.2 The Holliday junction

Numerous models have been proposed for the mechanism of homologous recombination. A key intermediate in many of these is the **Holliday junction** (named after the person who first proposed it, Fig. 5.36) (Section 2.3.3). One of the major properties of this structure is its ability to move along the DNA helices. This movement is called **branch migration**. A reversal of the Holliday junction formation results in the exchange of a segment of one strand of the DNA duplex with the corresponding segment of the other duplex. If one DNA strand (colour) has a slightly different sequence from the other strand (black), the mismatch thus created will be repaired. The effect of this repair is either to 'fix' the recombination event or prevent it.

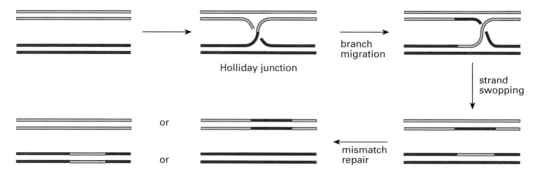

Fig. 5.36 The Holliday junction in recombination.

The Holliday junction is a topologically symmetrical structure. A few simple manipulations of it result in another possible fate (Fig. 5.37). Thus rotation, cleavage at points B, and repair of nicks by ligation result in a recombination event (see Fig. 5.35). This mechanism requires enzymes which recognize and cleave Holliday junctions. Such enzymes are known, for instance bacteriophage T$_4$ endonuclease 7. So if a Holliday junction is formed, it is a plausible substrate for recombination.

The choice between which bonds are cleaved (Fig. 5.37) determines whether a strand swap or a recombination event occurs. This choice may be influenced by the DNA sequence at the junction, firstly because the enzymes may display sequence specificity for cleavage and secondly because the three-dimensional structure of the junction is not a simple tetrahedron but is distorted by the sequence. Choice of cleavage might be influenced by this geometry.

But how could such a structure be created in the first place? One possibility is as a consequence of the nicking of single strands of DNA (Fig. 5.38). We know of enzymes in *E. coli* which can catalyse this single-strand nicking and strand invasion process. The former is achieved by the RecBCD enzyme complex, a large protein complex of about 300 kDa molecular mass. RecBCD can only bind to a free DNA duplex end but, once bound, it moves along the duplex, unwinding the helix as it goes and rewinding the DNA behind it (Fig. 5.39). If RecBCD encounters a specific sequence termed a **chi site** as it moves along the DNA, it cuts close to this site. RecBCD will continue unwinding the DNA, but the rewinding is prevented by the nick. This leaves a single-stranded region (Fig. 5.40).

The second phase of the process, namely **strand invasion**, is catalysed by the RecA protein. RecA binds to the single-stranded region and inserts it into a DNA duplex with which it is homologous (Fig. 5.41). Thus a combination of RecA and RecBCD proteins can catalyse the formation of a Holliday junction.

5.4.3 The implications of recombination

We have already seen that recombination offers advantages to the organism by assorting alleles and repairing genes. What other advantages does it give? One major effect is the potential for expanding or contracting the number of copies of genes. If a chromosome contains a sequence which is repeated several times along its length, then different recombination products are possible. For example, the sequence shown below (Fig. 5.42) contains two copies of the segment CD on the upper chromosome. Recombination between either of these two segments and the single copy on the lower chromosome yields either the normal products or a duplication of the DEFGHC segment on one chromosome plus a corresponding deletion on the other.

The loss of such a sequence would probably be lethal in haploid organisms since they contain only one copy of each gene per cell. In diploid cells, however, the story would be different since one of the two chromosomes would retain the sequence. Each combination (addition and subtraction of DEFGHC) could then be tested by natural selection. This process, which is called **unequal crossing-over**, is extremely important in the evolution of genomes and has led to the multiplication of single genes into gene families and superfamilies (see Section 5.3).

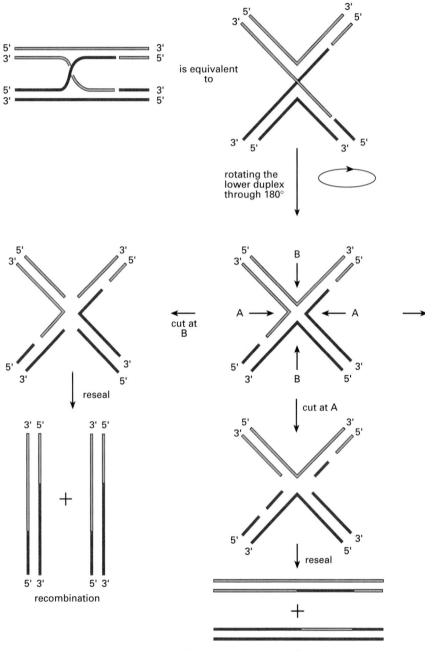

Fig. 5.37 Possible products of cleavage of the Holliday junction.

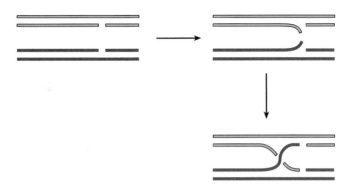

Fig. 5.38 Creation of the Holliday junction.

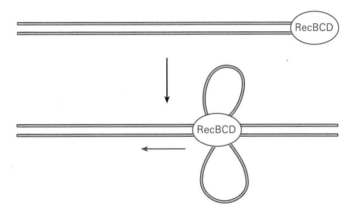

Fig. 5.39 The effect of the RecBCD complex on a DNA duplex.

5.4.4 Gene conversion

The normal product of homologous recombination is the swapping of DNA between two duplexes (see Fig. 5.35). Sometimes, however, two copies of one duplex are produced while the other is destroyed. This process is termed **gene conversion** (Fig. 5.43). It provides a mechanism for genes to edit and correct one another while remaining unchanged themselves. It is also used in some specific instances to control gene expression. We will see in the next section how the yeast *Saccharomyces cerevisiae* uses this process to control its mating type.

5.4.5 Examples of recombination controlling gene expression

Recombination is used in many ways to control gene expression. Several very well characterized examples are described below.

Bacteriophage λ lysogeny

When bacteriophage λ virus infects *E. coli*, two outcomes are possible. Either a lytic growth of the virus results in the destruction of the host cell and release of many virus particles or the

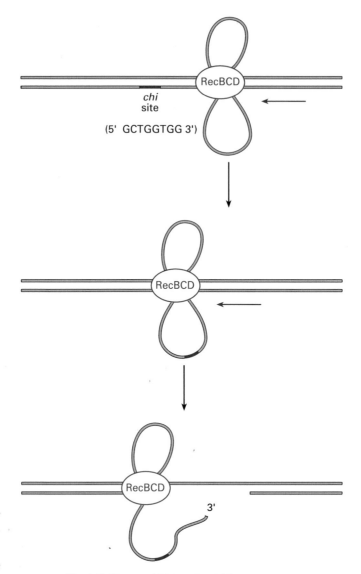

Fig. 5.40 Nicking of DNA by RecBCD at a *chi* site.

bacteriophage remains dormant in the cell (lysogeny). In the latter case, the viral DNA becomes integrated into the *E. coli* chromosome by a specific recombination event. The recombination sites on λ and *E. coli* are unique and the process is catalysed by a λ-encoded enzyme
called an **integrase** (Fig. 5.44). The exact sequences (which are called *att* sites) are:

5′ CTGGT TCAG<u>CT T T T T T TATACTAA</u>GT TGGCAT 3′   λ

5′ TGAAGCCT<u>GCT T T T T T TATACTAA</u>CT TGAGCG 3′   *E. coli*

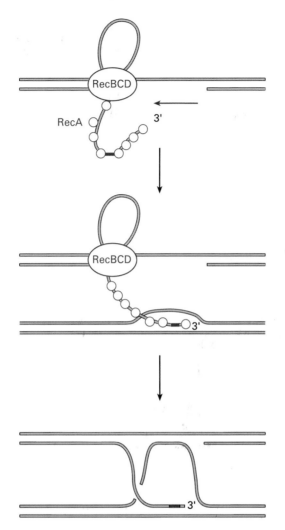

Fig. 5.41 Strand invasion catalysed by RecA protein.

This process can be made to occur *in vitro* using only the two DNAs, Mg^{2+}, integrase, and an *E. coli* protein called IHF (integration host factor). In this way it has been shown that the integration event, which is virtually a one-step process, involves breakage and re-ligation of eight phosphodiester bonds in a synchronous manner. Thus there is no net enthalpy loss. All that happens is that eight phosphodiester bonds are replaced by eight new bonds. In this way it resembles the type II topoisomerase-catalysed reactions (Section 2.3.5) and even self-splicing RNA introns (see Section 6.5). Lastly, this process can be reversed with the aid of another λ-encoded protein called an **excisionase** in addition to integrase.

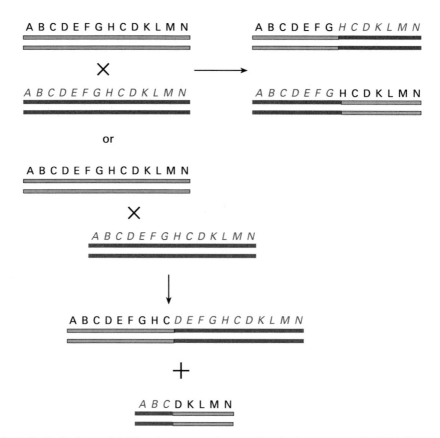

Fig. 5.42 Duplication and deletion of a sequence by recombination between repetitive DNA elements.

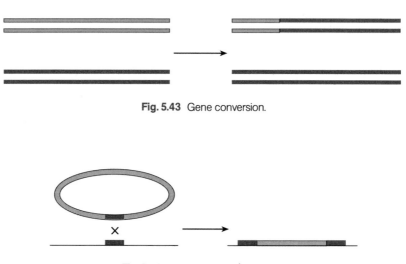

Fig. 5.43 Gene conversion.

Fig. 5.44 Bacteriophage λ lysogeny.

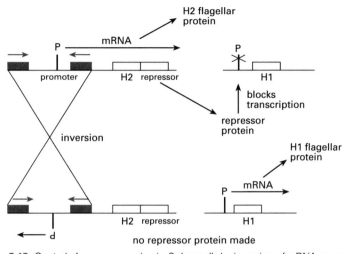

Fig. 5.45 Control of gene expression in *Salmonella* by inversion of a DNA segment.

Control of gene expression in *Salmonella* by inversion of a DNA segment

Salmonella typhimurium is closely related to *E. coli*. It controls the expression of its two flagellar genes (which are involved in motility) by a recombination event. Recombination between the two inverted repeat segments (Fig. 5.45, colour) causes inversion of the region between them. Note that recombination between direct repeats leads to excision of the intervening sequence. This is exactly what happens when the λ prophage DNA excises from the *E. coli* genome (an exact reversal of Fig. 5.44).

Yeast mating type

The yeast *Saccharomyces cerevisiae* is a eukaryote with a primitive sex life. It can exist either as single haploid cells (containing one copy of each chromosome) or as a diploid cell. The switch controlling this interconversion resides at the *MAT* locus which can contain one of two 'cassettes' termed **a** and **α** (Fig. 5.46). These are short DNA stretches of about 700 bases. If an **a**

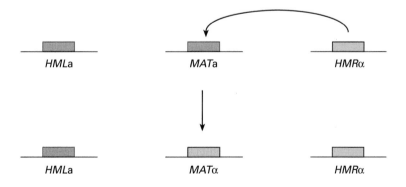

Fig. 5.46 Control of gene expression at the *MAT* locus in yeast by gene conversion.

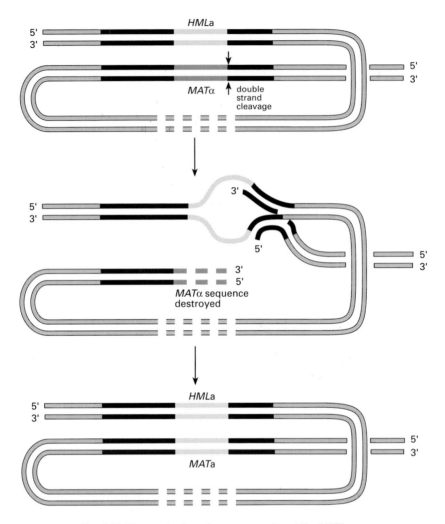

Fig. 5.47 The mechanism of gene conversion at the *MAT* locus.

cassette is present at *MAT*, the cell is of **a** mating type and can mate only with an **α** cell. If an **α** cassette is present the converse is true. The donor **a** and **α** cassettes lie at distant locations on the same chromosome as *MAT* (Fig. 5.46). These distant loci, termed *HML***a** and *HMR***α**, are unaffected by the switching process. There is therefore always a silent copy of **a** and **α** in the yeast cell in addition to the expressed copy at *MAT*. If we look at the switch we find this process to be a specialized form of gene conversion since one copy at *HML* or *HMR* alters the sequence at *MAT* without being affected itself.

The mechanism of this process is initiated by a double-stranded cut at a specific site at *MAT* (Fig. 5.47, upper). The enzyme that catalyses this cut is called the **HO endonuclease** and this is the only site recognized by this enzyme in the whole yeast genome. The ends thus generated probably engage in strand invasion into the *HML* or *HMR* donor cassettes (Fig. 5.47, middle and lower).

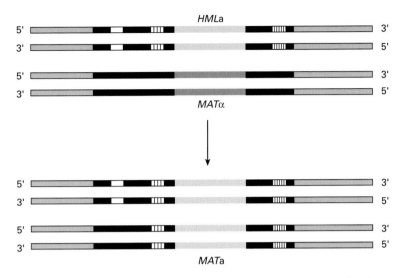

Fig. 5.48 Co-conversion of flanking markers during the mating type switch in *Saccharomyces cerevisiae.*

Another consequence of this process (and other gene conversion events) is the conversion of sequences near to the primary conversion site. The probability of any nucleotide being converted is inversely proportional to the distance from the primary site. In the example shown (Fig. 5.48), two out of three small, polymorphic nucleotide substitutions are co-converted during the mating-type switch.

The omega intron of *Saccharomyces cerevisiae*

The large rRNA gene of mitochrondria in the yeast *Saccharomyces cerevisiae* is found in two configurations. The rRNA gene of some strains (termed Ω^+) contains an intron, but other strains lack this insertion. If Ω^+ and Ω^- strains are mated, the resultant yeast strains contain far more mitochondrial DNAs with introns than would be expected from simple mixing. Such biased transmission is another example of gene conversion (Fig. 5.49) and flanking polymorphisms may also be co-converted in an analogous manner to the mating type conversion. This event is also initiated by a double-stranded cut. In this case, the cleavage occurs at the intron-insertion site in the target rRNA gene and the endonuclease that carries out this cut is encoded by the omega intron itself. Thus an 'infectious' intron potentiates its own proliferation by use of a gene contained in it.

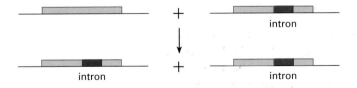

Fig. 5.49 The omega intron of *Saccharomyces cerevisiae.*

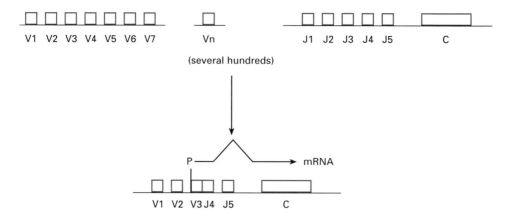

Fig. 5.50 The generation of antibody diversity by DNA rearrangement.

Rearrangements of antibody genes

All vertebrates possess immune systems. The major problem for these is how to recognize and respond to an almost infinite variety of invading entities. Antibody genes in most, if not all, vertebrates are multiple and varied. This solves part but by no means all of the problem. How is the additional diversity created?

Each immunoglobulin gene is a composite of a number of segments. These segments are brought together during the process of development from a germ cell into one which expresses an immunoglobulin gene. There are at least several (and sometimes many more) segments to choose from in the construction of an expressing immunoglobulin gene. For example, in the mouse κ light chain immunoglobulin locus, each developing B lymphocyte chooses one V and one J DNA segment, say V3 and J4, and joins them together (Fig. 5.50). Transcription from a promoter a short distance upstream of V3 produces the V3/J4 hybrid spliced to the C region. There are thus 5 times n possible combinations of V and J, generating $5n$ different antibody molecules. It should also be noted that random mutations in the rearranged genes arise at an abnormally high rate during this process and this increases the variety of possible combinations still further to the order of 10^8. The importance of these discoveries was recognized by the award of the 1987 Nobel Prize for Medicine to Susumu Tonegawa.

The mechanism of this DNA rearrangement is still under active study but it is known to involve partially homologous sequences (H and H') which bracket regions to be joined (Fig. 5.51).

Chromosomal rearrangements and cancer

In several types of human leukaemia, the cancerous cells contain specific chromosomal rearrangements. For instance, most cases of Burkitt's lymphoma involve a reciprocal translocation between chromosomes 8 and 14 such that a part of chromosome 8 is moved to chromosome 14. It turns out that one of the break points for this translocation lies at an immunoglobulin locus and presumably the translocation is due to an aberrant antibody-

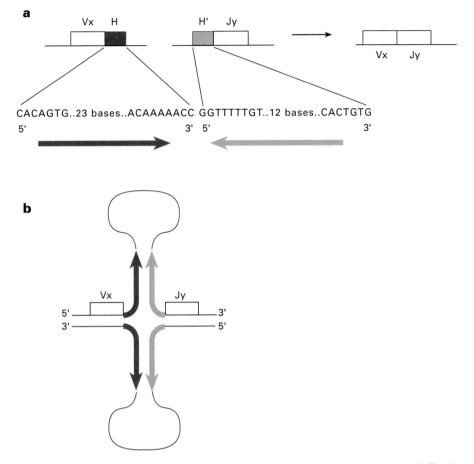

Fig. 5.51 DNA sequences involved in the recombination events at the immunoglobulin loci. (**a**) The H and H′ sequences contain inverted repeats. The hypothetical structure (**b**) places the V and J regions close to each other.

generating rearrangement. Other translocations also have break points lying at other immunoglobulin loci. The reason such rearrangements cause malignant transformation is not fully understood, but probably stems from aberrant expression of the genes at the other break point. In the case of Burkitt's lymphoma, this gene is the oncogene *myc* (see 5.4.7).

5.4.6 Illegitimate recombination

So far we have only considered homologous recombination. However, there is another major class of DNA rearrangement which is not dependent upon sequence homology. This is the movement of a piece of DNA called an **insertion sequence** (or IS element) into, or out of, the bacterial chromosome (Fig. 5.52).

A small piece of DNA (typically 4–12 nucleotide pairs, but constant for a given IS element, indicated by a coloured box in Fig. 5.52) is duplicated as a result of IS insertion. Other features

Fig. 5.52 The IS element.

important to the transposition process are small inverted repeats at each end of the IS element (23 bases for IS1, indicated by arrows in Fig. 5.52) and a **transposase** gene inside the IS element which encodes the enzyme that catalyses the process. Transposition depends upon endonucleolytic cleavage by the transposase enzyme at both the ends of the IS element and at the target site (Fig. 5.53). It is postulated that the small target site duplication arises at a result of staggered nicking of the target DNA. For a given IS element, the target sequence may be effectively random. More frequently, however, there are preferred 'hotspots' for insertion.

Transposon is the general term for a genetic element which moves by non-homologous recombination (such as an IS element). More complicated transposons exist in bacteria. These comprise a central region flanked by long repeats which are either direct or inverted with respect to each other. These long repeats are IS elements or derivatives thereof. An example of the former is Tn9 and of the latter, Tn10 (Fig. 5.54).

The central region of transposons contain antibiotic resistance genes, such as chloramphenicol (cam^r) or tetracycline (tet^r) and it is the transposition of these mobile DNAs that cause the rapid dissemination of antibiotic resistance between different strains and species of bacteria. They are therefore of considerable medical importance. It is presumed that complex transposons evolved from the fortuitous juxtaposition of two IS elements. In the laboratory, the IS elements which comprise a complex transposon can sometimes be mobilized independently. Such events, however, are much less common than transposition of the entire genetic element. Transposition in bacteria can result in two fundamentally different outcomes. In the first case, a transposon simply leaves one site on the one chromosome and enters another. In the second case, the original transposon is not lost, but a new mobile element appears at a

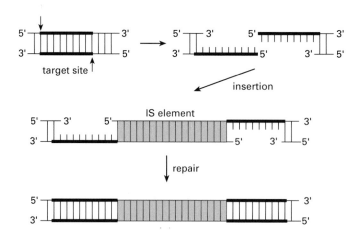

Fig. 5.53 Generation of flanking duplications during transposition of an IS element.

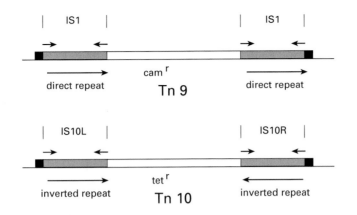

Fig. 5.54 Complex transposons in *E. coli*.

distant site. Transposition in this instance is therefore replicative. Some transposons, including IS1, can use both mechanisms, suggesting that these two processes are mechanistically linked. One way in which this could be achieved is shown in (Fig. 5.55).

The **Shapiro transposition intermediate** is as important to models of transposon movement as the Holliday junction is central to homologous recombination mechanisms. It was postulated in 1978 by Shapiro and its existence was recently proven during studies on transposition intermediates. It is important to note that replicative transposition results in the donor and recipient strands being joined to form a structure called a **cointegrate**. This can be seen during transposition from one plasmid to another and necessitates a homologous recombination step to separate the two duplexes (Fig. 5.56). This latter step is catalysed by a transposon-encoded **resolvase** enzyme. The resolvase stimulates recombination at a specific site (called *res*) inside the transposon by endonucleolytic cleavage in a similar manner to that used by λ integrase (Section 5.4.5). Furthermore, the sequence of *res* is similar to that of the core sequence used by λ.

res   GATAATTTATAATAT

att   GCTTTTTTATACTAA

A final point to note from Fig. 5.55 is that during non-replicative transposition the donor chromosome is left broken and is usually lost.

In addition to simple transposition, transposons can catalyse rearrangements of the DNA surrounding them. Deletion or inversion of flanking DNA is often seen. It seems probable that such rearrangements are all abortive by-products of the normal transposition process.

5.4.7 Eukaryotic transposons

There exists a wide variety of different types of transposon in eukaryotes. Probably all eukaryotes contain at least one type and the variety of size and structure is bewildering. Virtually the only property shared by these elements is their ability to transpose, though most possess small inverted terminal repeats and generate small direct repeats of the target site during

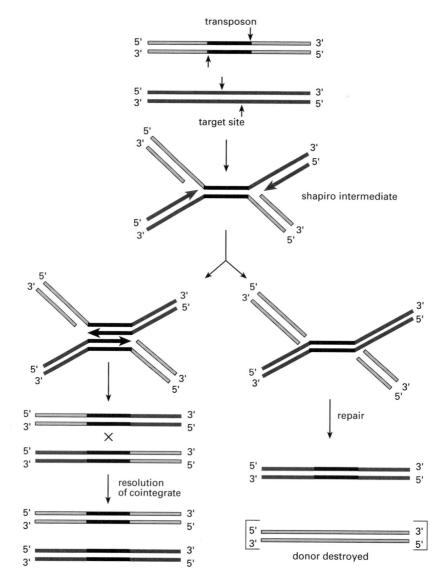

Fig. 5.55 Involvement of the Shapiro intermediate in prokaryotic transposition.

integration. We will only consider two types here. Others have been mentioned earlier (Section 5.3) because they form a significant part of the eukaryotic genome.

The P element transposon

The fruit fly *Drosophila melanogaster* is host to a wide variety of transposons which together comprise at least 10 per cent of its total genome size. One of the most interesting transposons is the **P element**. This is responsible for a complicated series of genetic phenomena which include lowered fertility, aberrant recombination, and increased mutation frequencies. These aberrations only occur in the progeny of a mating between a male whose genome contain P elements and a female lacking these transposons. No other combination of matings works. Furthermore, the damage seems to be confined only to the germline cells (those which give rise to the sperm or egg). All of the diverse genetic effects are believed to be the direct or indirect result of the transposition of the P element in the germ cells.

Two fundamental questions are raised by these observations. First, why does the P element only transpose when introduced via a sperm into an egg which lacks such sequences? Secondly, why is transposition restricted to germ cells? We now have a good idea of the answers to both questions, thanks to the studies of Gerald Rubin. We must first consider the structure of the P element and the nature of the proteins encoded by it. Functional P elements are 2907 base-pairs long and contain the usual features of inverted repeats and duplications of flanking sequence at their ends (Fig. 5.57). One strand of the P element contains four open reading frames. Mutations in any of these frames abolish the ability to transpose, suggesting strongly that they are all spliced together during RNA maturation to form a single mRNA which encodes a transposase enzyme. In support of this notion, the only RNAs encoded by P which are seen in fruit flies are those which contain all four regions. P RNA is synthesized in all flies containing these transposons and is made in all their tissues. If this RNA encodes the transposase, what is stopping transposition in most tissues of progeny flies and the whole body of the donor organism? We know that it is not due to the inability of the message to be translated, so the answer lies elsewhere.

It turns out that the protein made in somatic tissue (the entire body minus the germ cells) of the fly is not a transposase and it only contains sequence information from the first three introns of the P element (see Fig. 5.57). This truncated protein is ubiquitous in those cells of the fly where transposition is not occurring, which suggests that it may be a repressor of the transposase. What is the nature of the transposase? This was answered elegantly by the crea-

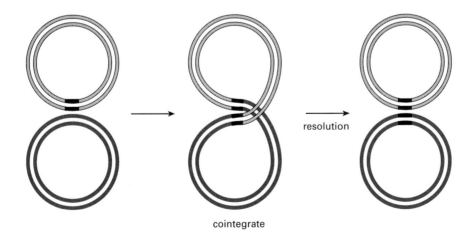

cointegrate

Fig. 5.56 Cointegrate formation and resolution.

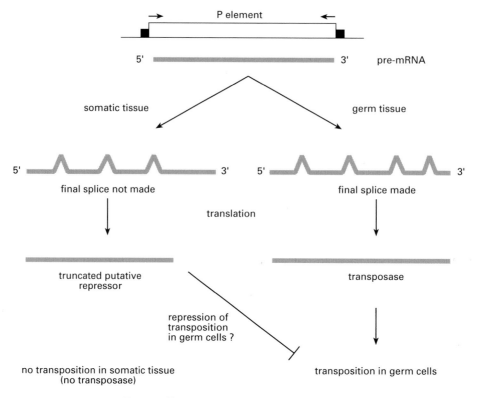

Fig. 5.57 The P element transposon of *Drosophila*.

tion of a genetically engineered P element which lacked the intron separating the last two exons of the P element (Fig. 5.57). This mutated P transposon gained the ability to transpose in all tissues of the fly, thus proving that the four open reading frames together encode transposase and that the block to production of this protein is due to the removal of the third intron during mRNA splicing.

Now we have a clear picture of events. If an egg laid by a fly which lacks P elements is fertilized by a sperm whose DNA contains these sequences, transcription of P proceeds in all tissues, but complete RNA maturation is blocked at the level of splicing in all tissues except germ cells. In germ cells, active transposase is made and the result is massive transposition of the P elements and associated genetic lesions. In the rest of the fly, no active transposase is made and the P elements are immobile. When the embryo reached adulthood, if it is female it lays eggs with a certain amount of the truncated putative repressor protein in them. This may have the effect of inhibiting transposition of P elements even in the germ cells. This resistance takes several generations to build up, for reasons which are unclear.

The P element transposes by a 'cut and paste' mechanism identical to that used for IS elements (Fig. 5.53). In fact, with its single transposase gene, inverted repeats, and flanking host duplications, the P element is best considered as a eukaryotic IS element. Other simple transposable elements of this type exist in plants. The best-studied of these is the Ac element of maize.

Retroviruses as transposable elements

Retroviruses were first identified as agents involved in the onset of cancer about 80 years ago. More recently the AIDS epidemic has been shown to be due to the HIV retrovirus (see Section 4.7.2). In the early 1970s, it was discovered that retroviruses had the astonishing ability to replicate their RNA genomes via conversion into DNA which became stably integrated in the DNA of the host cell. It is only comparatively recently that retroviruses have been recognized as particularly specialized forms of eukaryotic transposons. In effect they are transposons which move via RNA intermediates that usually can leave the host cells and infect other cells. The integrated DNA form (or **provirus**) of the retrovirus bears a marked similarity to a transposon (Fig. 5.58).

Comparison of the general structure of the provirus with that of Tn9 (Fig. 5.54) shows few differences. The long direct repeats of retroviruses rarely encode proteins and, if they do, these are not transposases. The central region does not contain antibiotic resistance, but it may contain cellular genes. Retroviruses which contain cellular genes are usually acutely cancer-inducing (oncogenic) because of these genes (called **oncogenes**). The normal counterparts of viral oncogenes (which are called **proto-oncogenes**) are genes vital for the correct functioning of the organisms they inhabit. It is the perturbation of their functioning as a result of being captured inside the retrovirus that causes cell transformation leading to cancer. One such oncogene, *myc*, is found at the other end of the translocation which causes Burkitt's lymphoma in humans (see Section 5.4.5). Aberrant *myc* expression can therefore be caused by chromosomal rearrangement and retrovirus capture with both processes leading to tumorigenesis (Fig. 5.59). Another way in which retroviruses can disturb *myc* function and cause cancer is found in the infection of chickens by avian leukosis retrovirus (ALV). Infected chickens often develop a type of leukaemia after several months. An analysis of the tumour DNA shows that ALV DNA has become inserted near to the chicken *myc* gene. Again, it is the disturbance in *myc* gene expression resulting from the insertion of the retroviral genome that is presumably involved in the generation of this cancer.

Other retrovirus-related mobile elements inhabit organisms as diverse as yeast, flies, and plants. Many of these do not possess the ability to leave their host cell but still use an intracellular virus-like particle as a transposition intermediate.

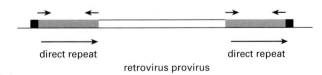

Fig. 5.58 Structure of the retrovirus provirus.

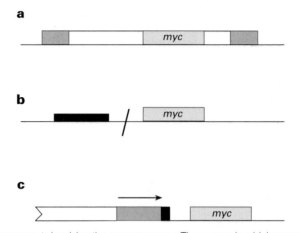

Fig. 5.59 Rearrangements involving the *myc* oncogene. Three ways in which sequence rearrangements connected with the *myc* oncogene can cause cancer: (**a**) capture by a retrovirus; (**b**) translocation close to an immunoglobulin gene; and (**c**) insertion of a retrovirus.

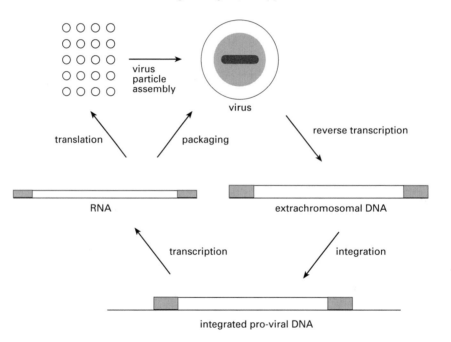

Fig. 5.60 Retrovirus transposition intermediates.

The transposition cycle of retroviruses has other similarities to prokaryotic transposons, which suggest a distant familial relationship between these two types of transposon. Crucial intermediates in retrovirus transposition are extrachromosomal DNA molecules (Fig. 5.60). These are generated by copying (or reverse transcribing) the RNA of the virus particle into DNA by a retrovirus-encoded polymerase called **reverse transcriptase**. The extrachromosomal linear DNA is the direct precursor of the integrated element and the insertion mechanism bears a strong similarity to 'cut and paste' transposition.

Summary

Two major types of recombination occur in all living organisms. Homologous recombination depends upon sequence complementarity and enables DNAs to exchange homologous sequences. The Holliday junction is a crucial intermediate in this process. Gene conversion can be associated with homologous recombination. In prokaryotes, homologous recombination is used as one way to control gene expression.

The second major class of recombination found in all living organisms is illegitimate recombination. It is practised by a wide variety of genetic elements which are collectively termed transposons. Transposons move into or out of stretches of DNA with which they share no direct sequence homology. They usually encode specific enzymes which catalyse this movement. A key intermediate in the mechanism of transposition of many of these sequences is the Shapiro intermediate. Two well studied classes of eukaryotic transposon are the P element of *Drosophila* and the retroviruses.

Further reading

5.1

Bachmann, B. J. (1983). Linkage map of *E. coli*- K12, Edition 7. *Microbiol. Rev.*, **47**, 180–230.

Cairns, J. (1983). The bacterial chromosome and its manner of replication as seen by autoradiography. *J. Mol. Biol.*, **6**, 208–13.

Lehman, I. R. and Kaguni, L. S. (1989). DNA polymerase α. *J. Biol. Chem.*, **264**, 4265–8.

McClure, W. R. (1985). Mechanism and control of transcription initiation in prokayotes. *Ann. Rev. Biochem.*, **54**, 171–204.

Miller, J. H. and Reznikoff, W. S. (ed.) (1978). *The operon*. Cold Spring Harbor Laboratory, New York.

Platt, T. (1986). Transcription termination and the regulation of gene expression. *Ann. Rev. Biochem.*, **55**, 339–72.

Ptashne, M. (1986). *A genetic switch: gene control and phage λ*. Cell Press, Massachusetts and Blackwell Scientific Publications, Oxford.

Ptashne, M. (1989). How gene activators work. *Sci. Amer.*, **260** (1), 24–31.

Rosenberg, M. and Court, D. (1979). Regulatory sequences involved in the promotion and termination of RNA transcription. *Ann. Rev. Biochem.*, **13**, 319–53.

Smith, G. R. (1981). DNA supercoiling: another level for regulating gene expression. *Cell*, **24**, 599–600.

5.2

Coverley, D. and Laskey, R. A. (1994). Regulation of eukaryotic DNA replication. *Ann. Rev. Biochem.*, **63**, 745–76.

DePamphilis, M. L. (1993). Eukaryotic DNA replication: anatomy of an origin. *Ann. Rev. Biochem.*, **62**, 29–64.

Fersht, A. R. (1980). Enzymatic editing mechanisms in protein synthesis and replication. *Trends Biochem. Sci.*, **5**, 262–5.

Joyce, C. M. and Steitz, T. A. (1994). Function and structure relationships in DNA polymerases. *Ann. Rev. Biochem.*, **63**, 777–822.

Kornberg, A. (1980). *DNA replication*. Freeman, San Francisco.

Marians, K. J. (1992). Prokaryotic DNA replication. *Ann. Rev. Biochem.*, **61**, 673–720.

McHenry, C. S. (1988). DNA polymerase III holoenzyme of *Escherichia coli*. *Ann. Rev. Biochem.*, **57**, 519–50.

Modrich, P. (1989). Methyl-directed DNA mismatch correction. *J. Biol. Chem.*, **264**, 6597– 600.

Ogawa, T. and Okazaki, T. (1980). Discontinuous DNA replication. *Ann. Rev. Biochem.*, **49**, 421– 57.

Radman, M. and Wagner, R. (1988). The high fidelity of DNA replication. *Sci. Amer.*, **259** (2), 24–31.

Tomizawa, J. and Selzer, G. (1979). Initiation of DNA synthesis in *E. coli. Ann. Rev. Biochem.*, **48**, 999–1034.

5.3

Brutlag, D. (1980). Molecular arrangement and evolution of heterochromatic DNA. *Ann. Rev. Genet.*, **14**, 121–44.

Lewin, B. (1994). *Genes V.* Wiley, New York, pp. 677–909.

Paranjape, S. M., Kamakaka, R. T., and Kadonaga, J. T. (1994). Role of chromatin structure in the regulation of transcription by RNA polymerase II. *Ann. Rev. Biochem.*, **63**, 265–97.

Rogers, J. (1984). The origin and evolution of retroposons. *Int. Rev. Cytol.*, **93**, 188– 280.

Watson, J., Hopkins, N., Roberts, J., Steitz, J., and Weiner, A. (ed.) (1987). *Molecular biology of the gene.* Benjamin Cummings, Menlo Park, California, pp. 621–73.

Zawel, L. and Reinberg, D. (1995). Common themes in assembly and function of eukaryotic transcription complexes *Ann. Rev. Biochem.*, **64**, 533–61.

5.4

Berg, D. E. and Howe, M. M. (ed.) (1989). *Mobile DNA.* American Society for Microbiology, Washington, DC.

Bradshaw, R. A. and Prentis, S. (ed.) (1987). *Oncogenes and growth factors.* Elsevier, Amsterdam.

Derbyshire, K. M. and Grindley, N. D. F. (1986). Replicative and conservative transpositions in bacteria. *Cell*, **47**, 325–27.

Finnegan, D. J. (1985). Transposable elements in eukaryotes. *Int. Rev. Cytol.*, **93**, 281–325.

Lewin, B. (1994). *Genes V.* Wiley, New York, pp. 967–1110.

Old, R. W. and Primrose, S. B. (1985). *Principles of gene manipulation.* Blackwell, Oxford.

Pai, E. F., Kabasch, W., Krengel, U., Holmes, K. C., John, J., and Wittinghofer, A. (1989). Structure of the guanine-nucleotide-binding domain of the Ha-*ras* oncogene product p21 in the triphosphate conformation. *Nature*, **341**, 209–14.

Watson, J. D., Hopkins, N. H., Roberts, J.W., Steitz, J., and Weiner, A. M. (ed.) (1987). *The molecular biology of the gene*, Vol. 1. Benjamin Cummings, Menlo Park, California, pp. 313–38.

West, S. C. (1992). Enzymes and molecular mechanisms of genetic recombination. *Ann. Rev. Biochem.*, **61**, 603–40.

RNA SEQUENCE INFORMATION AND TRANSMISSION

6.1 Occurrence and isolation of RNA

We know that genetic information is stored in the form of nucleotide sequences in DNA and that this information is copied into RNA before it can be used. RNAs thus play a central role in the transmission of genetic information and the process of copying the sequence information from DNA into RNA is called **transcription**.

Some viruses have RNA as their genetic material. In some cases, for example in retroviruses, the RNA genome is converted into double-stranded DNA during replication. This process is called **reverse transcription**.

6.1.1 Abundance and size of RNA species

At least three major classes of RNAs have been identified in cells. These are **messenger RNAs** (mRNAs), **transfer RNAs** (tRNAs), and **ribosomal RNAs** (rRNAs). All three classes are essential for protein biosynthesis (Section 6.6). Messenger RNAs contain the information necessary to specify the amino acid sequence of proteins. This information is present in the form of nucleotide sequences in mRNA. The process by which the information present in mRNA is read into protein is called **translation**.

Approximately, 80–85 per cent of the total RNA in cells is present as rRNA, 10–15 per cent is tRNA, and only 2–5 per cent consists of mRNAs. Ribosomes from bacteria such as *E. coli* contain three different rRNAs. These are the 5S rRNA, 16S rRNA, and the 23S rRNA (the S stands for the **sedimentation rate** often determined by sucrose gradient centrifugation of the RNAs (Section 10.3.1)). The 5S rRNA of *E. coli* is 120 nucleotides long whereas the 16S and the 23S rRNAs are 1542 and 2904 nucleotides long respectively. Eukaryotic cells also contain the corresponding three types of rRNAs, in this case the 5S, 18S, and 28S rRNAs, and in addition a fourth, 5.8S rRNA. The human ribosomal RNAs are 145 (5S), 1869 (18S), 5025 (28S), and 157 (5.8S) nucleotides long. Although eukaryotic cells contain this 'extra' rRNA in the ribosomes, in bacteria this rRNA species is included as a part of the 23S rRNA. Consequently, the 5.8S rRNA does not endow the eukaryotic ribosome with any particular properties which are lacking in the bacterial ribosome.

Most cells contain 60–70 different tRNAs, the sizes of which range from 74–95 nucleotides except for some mitochondrial tRNAs which are slightly smaller.

The number of different mRNAs in a cell can range from several hundred to a few thousand, depending on the number of proteins made by that cell. Most cells, however, have several hundred different mRNAs. Those few eukaryotic cells which specialize in the production of specific proteins are the only exceptions to this rule. For example, some adult reticulocytes are specialized for production of haemoglobin and contain predominantly two types of mRNAs encoding the α- and β-chains of haemoglobin. Since the size of proteins made in a cell varies over a wide range, the size of mRNAs in a cell also spans a wide range, from a few hundred to several thousand nucleotides.

In bacteria such as *E. coli*, mRNAs are being synthesized, utilized, and degraded continuously which makes mRNAs subject to continual **turnover**. The half-life of a typical bacterial mRNA is around two minutes and so the concentration of any particular mRNA in a bacterial cell is low. By contrast, tRNAs and rRNAs are much more stable and, since there are only a few species of tRNAs and rRNAs, the concentrations of these RNAs in a cell are much higher than those of mRNAs.

Eukaryotic cells contain specialized organelles such as mitochondria (for respiration) and chloroplasts in plant cells (for photosynthesis). Such organelles contain DNAs which are different both from each other and from DNA in the nucleus. Their protein synthesizing systems are also distinct from each other and from the one which operates in the eukaryotic cytoplasm. Consequently, eukaryotic cells such as plants can have as many as three different sets of mRNAs, tRNAs, and rRNAs which are distinct from each other and which operate in different compartments within the same eukaryotic cell.

Besides mRNAs, tRNAs, and rRNAs, there are many other RNAs in a cell with important cellular functions. Several RNAs, collectively called **small nuclear RNAs** (snRNAs), function inside the nucleus of eukaryotic cells and are involved in mRNA biosynthesis, in removal of intervening sequences, and in polyadenylation, etc. (Section 6.4.3). A second class of RNAs called **small nucleolar RNAs** (snoRNAs) function in the processing of rRNA precursors in eukaryotic nuclei. Another RNA, designated 7S RNA, is an essential component of a cytoplasmic ribonucleoprotein (RNP) particle which is involved in the transport of proteins through phospholipid membranes. In addition, there are examples of specific RNA molecules which are essential components of enzymes. In many cases, the RNA alone has the catalytic activity (Section 6.5). Thus, RNAs are a highly versatile class of molecules possessing a wide range of biological activities.

6.1.2 Isolation of RNA

Although RNA is not as long as DNA, care needs to be taken in isolating this class of macromolecule. While the isolation procedures for the different classes vary according to the properties of the RNAs, some generalizations can be made. As in the case of DNA (Section 10.1), the process can be divided into four steps: (1) isolation of cells or organelle (e.g. chloroplasts), (2) lysis, (3) removal of protein, and (4) isolation of the individual class of RNA.

Since RNAs are much more susceptible than DNAs to degradation by nucleases, a major consideration in RNA isolation is the need to eliminate, or at least reduce, the nuclease activity normally present in any cell extract. This is achieved by rapid phenol extraction of the cell lysate, a treatment which denatures most nucleases and other proteins. The aqueous phase contains the nucleic acid derivatives, including nucleotide co-enzymes, while the phenol phase (and the interface) contains most of the protein. In addition, the presence of divalent metal ions during the purification is beneficial, as they are important for preserving the structure and hence the biological properties of RNA.

It is often difficult to isolate biologically active RNA from organs such as the pancreas, because it is difficult to homogenize the tissue. In such cases, guanidinium isothiocyanate is used to dissolve the tissue and this process also inactivates any ribonucleases that are present.

The separation of tRNA and ribosomal RNA relies on the different solubilities of the molecules. The aqueous supernatant of the phenol extraction is adjusted to 1 M in salt, which precipitates the rRNA. The tRNA remains in solution and is purified by anion exchange chromatography. Usually the 5S ribosomal RNA is also in the tRNA fraction and this can be separated by gel filtration or gel electrophoresis.

Most eukaryotic messenger RNAs possess a stretch of adenylate residues (from 20 to 200 long) at their 3'-ends (Section 6.4.4). The interaction of this **poly A** tail with **oligo-dT cellulose** forms the basis of an affinity method for the purification of mRNA, the development of which was crucial for some of the advances in molecular biology. Oligo-dT cellulose chromatography is based on the principle that oligo-dT (chain length 9–20) attached to cellulose provides a solid matrix on to which poly A-containing RNA can be absorbed (as a result of base-pairing with the oligo-dT stretches) and thereby purified in one step from other RNAs.

Summary

The three major classes of RNA in cells are messenger RNA, transfer RNA, and ribosomal RNA of which rRNAs are the most abundant. Prokaryotic cells contain three species of rRNA whereas eukaryotic cells contain four, but the sets of rRNAs confer similar properties to both prokaryotic and eukaryotic ribosomes. Most cells contain 60–70 different tRNAs, whereas the number of mRNAs can vary from several hundred to a few thousand and these can be of widely varying length. Mitochondria and chloroplasts have their own sets of mRNA, tRNA, and rRNA. Small nuclear RNAs play an important role in processing of mRNA precursors. Some RNA molecules possess catalytic activity.

Isolation of RNA involves isolation of cells or organelles from tissue, lysis, removal of protein (including ribonucleases), and separation of the individual class of RNA. Purification of eukaryotic mRNA is facilitated by use of oligo-dT cellulose chromatograhy.

6.2 Determination of RNA sequence

Because methods for determining DNA sequences are so rapid, one approach to RNA sequence determination is to clone the gene for an RNA (Section 10.7) and then sequence the cloned DNA (Section 10.6). However, during biosynthesis most RNA molecules are made as longer transcripts (Section 6.4) and such precursor RNAs are then cut by specific nucleases at their 5'- and 3'-ends to produce the mature functional RNA molecules. In addition, there are enzymes which attach special nucleotide sequences to the 5'- and/or 3'-ends of some RNAs. As a result, the DNA sequence alone cannot tell us about the nature of the ends of a mature biologically functional RNA molecule. Also, RNAs often contain modified nucleotides in which the structures of the four common nucleotides, A,C,U, and G have been changed in various ways by enzymes. Since these modified nucleotides are often important in the biological activity of the RNAs, it is necessary to identify and locate them.

6.2.1 Methods involving reverse transcriptase

RNA can be used as a template for the synthesis of DNA using the enzyme reverse transcriptase. This enzyme copies the RNA sequence into a DNA sequence in the presence of an appropriate DNA primer. The resulting DNA (**cDNA**) can then be cloned and sequenced. This method has been used widely for sequence determination of eukaryotic mRNAs.

If the required RNA is abundant within an RNA preparation, it can be sequenced directly by application of the Sanger method of DNA sequencing using chain terminators (Section 10.6) except that RNA is used as the template and reverse transcriptase is used to copy it. In this case, some part of the RNA sequence must already be known so that a complementary oligodeoxyribonucleotide can be designed for use as a primer. This method is particularly useful in sequence determination of viral RNAs where some part of the sequence is highly conserved between species. Neither of these methods identifies modified nucleotides.

6.2.2 Direct methods for sequence determination

The sequence analysis of an RNA species by direct methods predates reverse transcriptase methods and was first achieved for a tRNA by Holley in 1965. Soon after, Sanger and Brownlee developed a more general sequencing method which required the RNA to be radiolabelled with ^{32}P at each phosphate. The RNA was then subjected to complete cleavage by ribonuclease T1, which cuts after every G residue, and separately by pancreatic ribonuclease, which cuts after every pyrimidine residue (Table 6.1). The products of each digestion were separated by a two-dimensional electrophoresis/chromatography procedure to generate characteristic **fingerprints**. Individual radioactive RNA fragments were isolated from each fingerprint and further degraded using phosphodiesterases (Section 10.7.2) to establish their base sequences.

Nowadways, there are three methods that are commonly used for direct sequencing of an RNA. They all require the RNA to be available in pure form.

Use of base-specific ribonucleases

In contrast to the Sanger and Brownlee method, a single ^{32}P-radioactive phosphate is first introduced at either the 5′-end of the RNA using [γ-^{32}P]-ATP and T4 polynucleotide kinase (Section 10.7.2) or at the 3′-end of the RNA using [5′-^{32}P]-pCp and T4 RNA ligase (Section

Table 6.1. Enzymes commonly used in RNA sequencing

Ribonuclease	Base specificity
T1	G
A	U or C
U2	A > G
B. cereus (B.c.)	U,C
Chicken liver (CL3)	C ≫ A or U
Physarum M (PhyM)	U,A or G≫C
T2	U,C,A, or G

3.5.4). This allows the identification of all the oligonucleotide degradation products of the RNA which contain the labelled end.

The second step is to treat the ^{32}P-labelled RNA in separate reactions with different base-specific ribonucleases, which cut the RNA phosphodiester bonds only on the 3′-side of specific nucleotides. Enzymes commonly used for this purpose cut RNAs after G residues, after A residues, after C residues, or after both U and C residues (Table 6.1). In a parallel reaction, the RNA is treated with alkali which cuts randomly after any residue. All of the RNA cleavages, whether by enzymes or by alkali, are carried out to a limited extent, so that each susceptible phosphodiester bond is cut in only a small fraction of the ^{32}P-labelled RNA molecules and not more than once per molecule.

For example, consider an RNA with the sequence *pAGUCGAAGUU (*p denotes ^{32}P-phosphate). If this RNA is completely digested with ribonuclease T1, which cuts after every G residue, the only radioactive oligonucleotide produced will be *pAG. However, if the same RNA is digested only partially, a series of radioactive oligonucleotides *pAG, *pAGUCG, *pAGUCGAAG is obtained. Each of these oligonucleotides contains a G residue at the 3′-end. Similarly, the use of the enzyme ribonuclease U2, which cuts an RNA after A residues, gives a different series of radioactive oligonucleotides, each of which has an A residue at its 3′-end. Use of the other two nucleases gives rise to two other series of ^{32}P-labelled oligonucleotides which either end with C or end with U or C. In the case of alkaline treatment, cleavage occurs at all possible ester sites to produce a complete series of ^{32}P-labelled oligonucleotides. Such a pattern of RNA is called a **ladder**.

Each of the different series of ^{32}P-labelled oligonucleotides can be separated according to their size by use of polyacrylamide gel electrophoresis. Based on the pattern of radioactive bands obtained (Fig. 6.1), one can locate the positions of G, A, U, and C residues relative to the ^{32}P-labelled 5′-end. Thus we can read the sequence of the RNA simply by following the pattern of oligonucleotides from the shortest towards the longest.

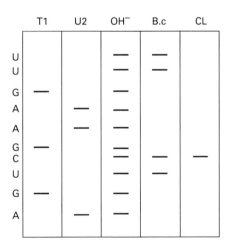

Fig. 6.1 Autoradiographic pattern obtained after polyacrylamide gel electrophoresis of partial digests of 5′-^{32}P-labelled RNA with RNase T1 (T1), RNase U2 (U2), alkali (OH⁻), *Bacillus cereus* RNase (B.c.) and chicken liver RNase (CL). The sequence deduced from the pattern is shown on the left.

Use of base-specific chemicals

This method is very similar to the Maxam–Gilbert method for sequencing DNA (Section 10.6). Four different base-specific chemical reactions are used to modify either G, A, U, or C residues in an RNA. Subsequent treatment of each of the incompletely modified RNAs with aniline results in cleavage of the RNA at the site of modification. Once again, the ^{32}P-labelled oligonucleotides are separated according to their size by polyacrylamide gel electrophoresis (Section 10.3.3) and the resulting pattern of radioactive bands determines the positions of G, A, U, and C residues with respect to the labelled end. Reagents that are used for the base-specific reactions are somewhat different from those used in sequence determination of DNA (Table 6.2). Also, aniline cleavage of the RNA at the site of modification has to be carried out under rather mild conditions to prevent cleavage at unmodified sites (cf. DNA which is much more stable to alkali, Section 3.2.2). These conditions result in formation of a 5'-phosphate group at the nucleoside on the 3'-side of the cleavage site, but the nucleoside on the 5'-side is converted into a mixture of species. Thus, the chemical method is more suitable for sequence analysis of 3'-^{32}P-labelled RNA rather than 5'-^{32}P-labelled RNA.

Use of random chemical cleavage

Neither of the above methods permits the direct identification of a modified base naturally present in the RNA. To identify modified bases, one can use a method first described by Stanley and Vassilenko (Fig. 6.2).

In the first step, the RNA is partially hydrolysed under alkaline conditions to generate a ladder. Under such conditions, one obtains a homologous series of oligonucleotides each of which contains the 3'-terminal end of the original RNA and another series each of which contains the 5'-terminal end of the original RNA. Each of the oligonucleotides that contain the 3'-terminal end of the original RNA has a free 5'-hydroxyl group at its 5'-end. By treatment of the entire mixture of oligonucleotides with the enzyme T4 polynucleotide kinase in the presence of [γ-^{32}P]-ATP, a radioactive phosphate group is transferred to each of these 5'-hydroxyl groups. All the ^{32}P-labelled oligonucleotides are now separated according to their size by polyacrylamide gel electrophoresis.

Table 6.2. Chemical reagents used in RNA sequencing

Reagent	Base specificity
Dimethyl sulfate	G
Diethyl pyrocarbonate	A ≫ G
Hydrazine (aqueous)	U ≫ C
Hydrazine + NaCl	C ≫ U

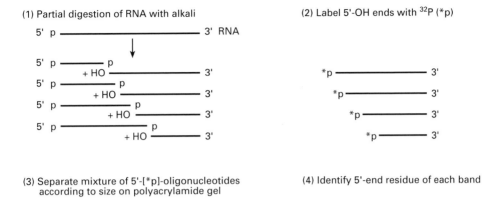

(1) Partial digestion of RNA with alkali

(2) Label 5'-OH ends with ^{32}P (*p)

(3) Separate mixture of 5'-[*p]-oligonucleotides according to size on polyacrylamide gel

(4) Identify 5'-end residue of each band

Fig. 6.2 A scheme for sequencing RNA by the random chemical cleavage method.

In the final step, the 5'-terminal nucleotide of each radioactive band is identified by complete degradation of the oligonucleotide to mononucleotides either by use of alkali or by use of a non-specific ribonuclease. The 5'-terminal nucleotide, which in each case contains all of the ^{32}P-label, is identified by further separation using thin layer chromatography. Because the 5'-end of each oligonucleotide is identified directly, it is now possible to determine whether the 5'-residue is one of the four normal nucleotides (A, C, U, or G) or whether it is a modified nucleotide by comparison of its chromatographic and electrophoretic behaviour using marker nucleotides of known structure.

Summary

The sequence of an RNA can be determined indirectly by cloning and sequence analysis of the corresponding DNA or by application of the Sanger method of DNA sequencing to the purified RNA using chain terminators and the enzyme reverse transcriptase.

The sequence of a purified RNA can be directly determined by methods that involve the use of either base-specific ribonucleases or base-specific chemicals to generate fragments of RNA which can be separated by electrophoresis in polyacrylamide gels. The identification of modified bases naturally present in an RNA is possible using the random chemical cleavage method.

6.3 Secondary and tertiary structure of RNA

The many biological functions of RNA are based on very specific three-dimensional RNA structures. What are these structures and how are they formed and maintained?

In contrast to DNA, RNA occurs mostly as a single polynucleotide chain. However, by virtue of the inherent ability of RNA to make up different conformations (Section 2.4), many RNAs exist in elaborate, defined structures possibly as distinct as those of proteins.

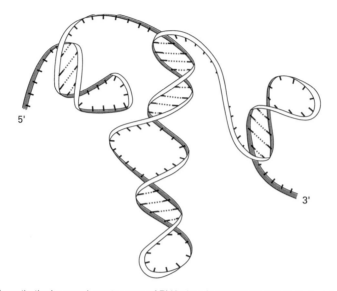

Fig. 6.3 A hypothetical secondary structure of RNA showing short double-helical regions separated by single-stranded regions. Dotted lines indicate base-pairs formed between the strands.

RNA molecules usually contain short double-helical regions connected by single-stranded stretches (Fig. 6.3). The helical hairpin regions can form because of the antiparallel orientation of some complementary sequences found in different parts of the RNA chain. For instance, in the secondary structure of tRNA (Fig. 6.5, Section 6.3.2) the 5'-end of the tRNA forms a helix with nucleotides located near the 3'-terminus. Such double-helical structures contain not only the standard A·U and G·C base-pairs but may contain also a number of energetically less stable G·U base-pairs. The single-stranded regions (loops and bulges) are stabilized in some cases by additional non-Watson–Crick hydrogen bonding interactions (see Section 6.3.1).

A large RNA can have many possible secondary structures based on the frequent occurrence in different parts of the RNA of short sections of complementary bases. Thus to determine the most probable secondary structure of RNase P RNA (Fig. 6.4), it was necessary for supporting evidence to be obtained for the various stem–loop regions in the RNA (see below). Additional pairing between bases distant in the secondary structure generates, through formation of tertiary base-pairs, the structure of RNase P RNA.

The maintenance of RNA structure is dependent on factors such as salt concentration, pH, temperature, and the presence of specific ions (e.g. Mg^{2+} or polyamines). If the temperature is raised sufficiently, then all the base-pairs in the helical regions will disociate to yield the fully 'denatured' state of the RNA.

6.3.1 The determination of RNA structure

Ultimately, X-ray crystallography will provide us with high-resolution structures of RNA molecules, either alone or complexed with proteins. At the present time, however, only a

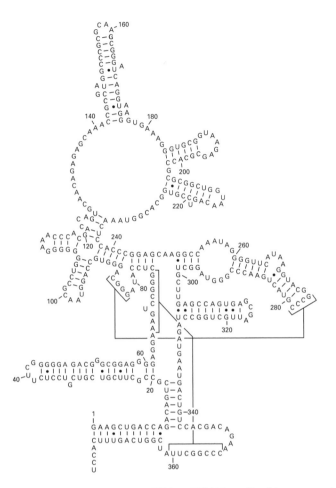

Fig. 6.4 Proposed secondary structure of RNase P RNA from *E. coli*. Long-range base-pairs are shown by extended lines (from Haas, E. S., Brown, J. W., Pitulle, C. and Pace, N. R. (1994). *Proc. Natl Acad. Sci. USA*, **91**, 2527–31).

few RNA species, predominantly tRNA, have given crystals which diffract at high resolution. Fortunately, a combination of other techniques may be used which, together with some of the principles of RNA structure learnt from the crystal structure of tRNA, allow meaningful structures to be derived for many RNA molecules.

NMR spectroscopy is also providing a powerful approach for the determination of the structure of small RNA fragments about 12–25 nucleotides long. For example, the NMR structure of an unusually stable oligonucleotide, which consists of a short RNA helix connected by a loop of four nucleotides (Fig. 6.5), shows that it is stabilized by base-stacking as well as by other unusual hydrogen bonding interactions. In this so-called **tetra-loop**, only U6 in the loop is not hydrogen bonded. U5 and G8 form a non-Watson–Crick base-pair while C7 forms a hydrogen bond to a phosphate oxygen between U5 and U6. Two of the ribose residues adopt a 2′-*endo* sugar pucker (the type normally found in B-DNA) and the base at G8 adopts a *syn* conformation. These features help in the formation of an extremely sharp, two nucleotide turn. Interestingly, the same stem–loop sequence as the corresponding DNA shows none of these unusual features and is very flexible. Thus the presence of a 2'-hydroxyl group in ribose allows RNAs to adopt intricate structures stabilized by hydrogen bonding interactions.

Biophysical methods, such as NMR spectroscopy, X-ray crystallography, or neutron scattering, can provide structural information when a relatively large amount of an RNA is available. In most other cases where only small amounts of RNA are available, two other methods are especially useful and particularly so when used in combination.

RNA modification by chemical or enzymatic reagents

RNA (or an RNA : protein complex) can be subjected to limited modification by structure-selective probes. These are chemicals or enzymes having selectivity for a particular base in either a single-stranded (i.e. unpaired) environment or in a base-pair (secondary or tertiary) (Table 6.3). If such modifications are introduced randomly at a low level and give rise directly or indirectly to chain scission, then gel electrophoretic methods can be used to separate the fragments (Section 6.2.2). If the RNA is end-labelled before modification, then the size of the labelled oligonucleotide fragment identifies the nucleotides in the chain which was modified. This method is powerful and requires little material. For example, it was established by this technique that the bases in the D-loop of tRNA (Section 6.3.2) are in exposed single-stranded regions. Under certain conditions, tertiary base-pair interactions (Section 6.3.2) can also be distinguished. These interactions are less stable than the standard Watson–Crick pairs and are broken under mild denaturing conditions.

Table 6.3[a]. Some probes for RNA secondary structure

Probe	Specificity
RNase T1	unpaired G
RNase U2	unpaired A>G
RNase CL3	unpaired C≫A>U
RNase T2	any unpaired base
Nuclease S1	any unpaired base
RNase VI	any paired or stacked base
Dimethyl sulfate	N^3C, N^1A, N^7G
Diethyl pyrocarbonate	N^7A
CMCT[b]	N^3U, N^1G
Kethoxal	N^1G, N^2G
Bisulfite	unpaired C ($\rightarrow$U)
Ethylnitrosourea	phosphates

[a] Adapted from C. Ehresmann *et al.* (1987). *Nucleic Acids Research,* **15**, 9109.
[b] 1-Cyclohexyl-3-(2-morpholinoethyl)carbodiimide metho-*p*-

In an extension of this technique, called **reverse transcriptase mapping**, chemically modi-fied RNA can be used as a template for DNA synthesis catalysed by reverse transcriptase. Base residues which are accessible to reaction with a modifying reagent can be identified by a block to DNA synthesis at each site of modification, since base-pairing is disrupted between the modified base and the base's normal complement. Only those residues in single-stranded re-gions of the RNA are modified and hence identified. The method has been particularly useful for the analysis of rRNA and also for the identification of sites of interaction of RNA with antibiotics and proteins.

Phylogenetic comparisons

This approach can be used to refine a structure determined by RNA modification experi-ments. The method is based on a comparison of the nucleotide sequences of homologous RNA species from closely and more distantly related organisms. Such a comparison, which was crucial in establishing the accepted structure of many RNA molecules (e.g. RNase P RNA, Fig. 6.4), shows that helical regions are usually conserved. Although changes in sequence are often found in double-stranded regions, base-pairing (including G·U pairs) is preserved as a result of compensating mutations in the opposite strand of the duplex. Such phylogenetic comparisons have shown that the secondary structures of some RNA molecules are more conserved than their primary sequences or their lengths (e.g. 16S and 23S rRNA, RNase P RNA).

6.3.2 Transfer RNA: an example of the intricate features of RNA structure

The sequences of several hundred tRNAs from many organisms have now been established. Transfer RNA is also a class of RNA molecule for which X-ray crystallographic structural in-formation is currently available. In addition, since tRNA is small and easily prepared, and

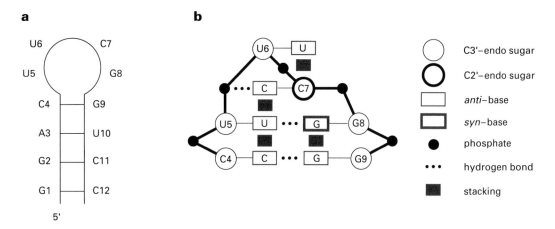

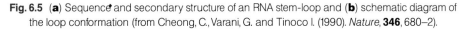

Fig. 6.5 (**a**) Sequence and secondary structure of an RNA stem-loop and (**b**) schematic diagram of the loop conformation (from Cheong, C., Varani, G. and Tinoco I. (1990). *Nature*, **346**, 680–2).

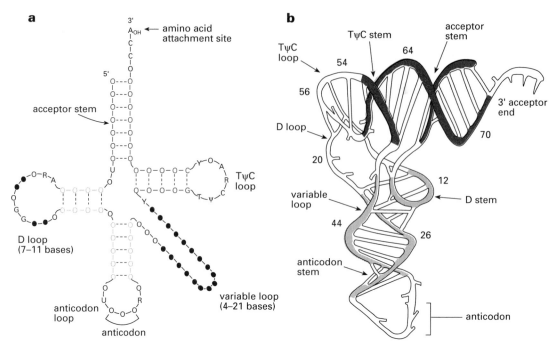

Fig. 6.6 (**a**) Clover leaf secondary structure of tRNA and (**b**) three-dimensional structure of tRNAs. The horizontal part of the L-shaped structure in (**b**) is formed by colinear stacking of base-paired regions in the acceptor stem and the TψC stem whereas the vertical part has base-pairs in the D-stem stacked over the anticodon stem.

plays a central role in protein synthesis, it has been the object of many biochemical and biophysical studies.

All tRNAs contain a common sequence (. . . CCA) at their 3′-termini, the **acceptor end**. During protein synthesis, the ribose of the 3′-terminal A residue becomes attached to an amino acid through an ester linkage. In addition, there are certain nucleotides which are highly conserved between different tRNAs (Fig. 6.6a). Most of these nucleotide residues are involved in tertiary base-pair interactions which stabilize the three-dimensional structure of the tRNA.

Although the sizes of tRNAs vary (74–95 bases), they can all be folded into a common secondary structure (**cloverleaf**), in which four or five base-paired regions (**stems**) are separated by single-stranded regions (**loops**) (Fig. 6.6a). The sizes of the **acceptor stem** (7 base-pairs), the **anticodon loop** and stem (7 in the loop and 5 base-pairs in the stem), and the TψC loop and stem (7 in the loop and 5 base-pairs in the stem) are constant in all tRNAs. The variation in size between tRNAs is due to differences in the number of nucleotides in the D loop and stem and/or in the variable loop, leaving a common secondary structure intact. Additional hydrogen bonds between nucleotides quite distant in the primary sequence cause the cloverleaf to fold into a stable tertiary structure in which two helical segments stack on top of each other to give the appearance of two continuous (yet not covalently joined) double-helices. The D-loop

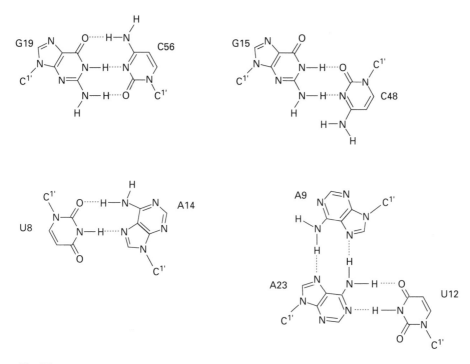

Fig. 6.7 A sample of the tertiary hydrogen bond interactions found in tRNA. Except for the G19 : C56 pair, all of the others involve non-Watson–Crick interactions.

and the TψC-loop provide the hinge in the characteristic L-shaped structure (Fig. 6.6b) in which the anticodon loop and the acceptor end are the furthest apart. The X-ray crystallographic analyses of two tRNA aminoacyl–tRNA synthetase complexes show that the L-shaped structure is retained in the complex (Fig. 9.42).

Many of the teriary hydrogen-bonding interactions show structures very different from the standard A·U and G·C base-pairs (Fig. 6.7) and involve Hoogsteen and other types of pairs (Section 2.1.2). In addition, the phosphates and 2′-hydroxyl groups of the RNA backbone are also involved in tertiary interactions. Further stability of the tRNA molecule comes from stacking interactions between the heterocyclic bases. As a result, tRNAs are very compact, stable, yet flexible molecules with structures that accommodate all the different tRNA sequences while allowing for slight conformational changes which might be crucial for specific interactions with proteins.

6.3.3 The RNA pseudoknot

One example of a tertiary structural motif in RNA is the **pseudoknot**. This is formed when nucleotides in a short single-stranded region in the RNA engage in base-pair interactions with looped nucleotides of a hairpin structure (Fig. 6.8). Pseudoknots are important in many biological processes, such as retroviral replication and translational regulation, and have been proposed to occur also in RNase P RNA (Fig. 6.4), rRNA, and plant viral RNAs.

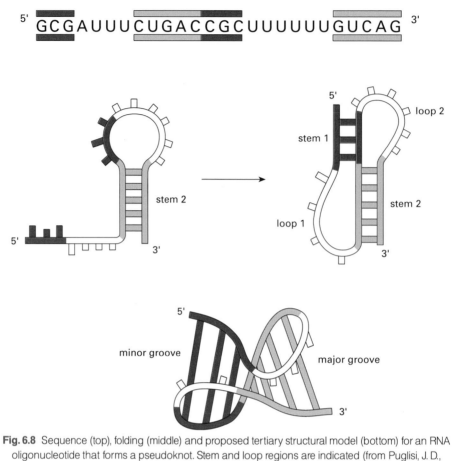

5' G C G A U U U C U G A C C G C U U U U U U G U C A G 3'

Fig. 6.8 Sequence (top), folding (middle) and proposed tertiary structural model (bottom) for an RNA oligonucleotide that forms a pseudoknot. Stem and loop regions are indicated (from Puglisi, J. D., Wyatt, J. R. and Tinoco, I. (1990). *J. Mol. Biol.*, **214**, 437–53).

Summary

RNAs adopt specific three-dimensional structures which are important for their interactions with proteins and for their biological functions. Secondary structures of RNAs can be determined by a combination of techniques including the use of chemical and enzymatic reagents selective for bases in either single-stranded environments or in base-pairs and the use of phylogenetic comparisons. Reverse transcriptase mapping depends on the identification of blocks to reverse transcriptase-catalysed DNA synthesis that arise as a result of chemical modifications carried out in bases in RNA.

NMR spectroscopy and X-ray crystallography are helping to define the tertiary structure of small RNAs. Transfer RNAs are folded into a cloverleaf secondary structure containing stems and loops that are mostly conserved in size. Additional tertiary interactions occur between nucleotides distant in the primary sequence and involve unusual types of base-pairs. These interactions are crucial for the folding of tRNA into an L-shaped tertiary structure. One example of a tertiary structural motif is the pseudoknot which is formed by intramolecular base-pairing between looped nucleotides and a single-stranded region.

6.4 Biosynthesis of RNA

RNA is produced in cells by transcription of DNA. The initial products of transcription in most cases are longer than the mature products. The additional nucleotides in these longer RNAs, called **precursor RNAs**, may occur in regions **upstream** (5′, leader) or **downstream** (3′, trailer) of the sequence found in the mature RNA or even within the precursor RNA (called **introns** or **intervening sequences, IVS**). During the processing of RNA (**maturation**) these extra sequences are removed (**splicing** in the case of introns) by a variety of different nucleases, each being specific for a certain reaction on a specific class of RNA. Some of these enzymes are proteins, others are ribonucleoprotein particles, and yet others are ribonucleoproteins in which the RNA component contains the catalytic activity. In some cases the precursor RNA contains its own processing activity (Section 6.5.2)! The maturation process may also involve post-transcriptional addition of nucleotides not encoded in the gene. For example, the 5′-end of a eukaryotic mRNA undergoes a **capping reaction** (Section 3.3.2 and Fig. 3.52) and the 3′-end of eukaryotic mRNA is extended by a stretch of poly A (Section 6.4.3).

Another processing reaction that occurs sometimes in RNA is called **RNA editing**. Unlike most other RNA processing reactions, RNA editing results in a change in the nucleotide sequence of the RNA. The most common form of editing is deamination of C to U or additions and/or deletions of U or C residues within an RNA. RNA editing is found to occur primarily, although not exclusively, in mitochondria of plants and protozoa and in chloroplasts of plants. In some cases it can be quite extensive. An extreme example is editing of some pre-mRNAs in the mitochondria of protozoa, such as trypanosomes, where up to 50 per cent of nucleotides in some mRNAs may be different from the sequences of the corresponding pre-mRNAs.

6.4.1 Transfer RNA

The biosynthesis of transfer RNA is a process that is relatively well understood. In bacteria, tRNA transcripts contain either single tRNA sequences (monomeric) or many tRNA sequences (multimeric). In eukaryotes the transcripts are mostly monomeric, since each tRNA gene possesses its own internal promoter for transcription by RNA polymerase III.

The precursor tRNA sequences fold into tRNA-like structures. This folding appears to be necessary for recognition of the precursors by processing enzymes which then act to remove the leader and trailer sequences. In the majority of cases, the maturation of the 5′-end precedes that of the 3′-end, but only after both ends have been matured are any intervening sequences excised. However, the number of tRNA genes that contain intervening sequences is only a small fraction of all tRNA genes.

The processing of a monomeric or a multimeric tRNA transcript involves the enzyme RNase P which acts to release the mature 5′-end of the tRNA. This enzyme is a non-covalent complex of RNA and protein in which the RNA component is essential for activity (Section 6.5.2). At the 3′-end, either exo- or endonucleases trim the trailer sequence. In the cases where the 3′-terminal universal CCA sequence is not encoded by the tRNA gene (as in all eukaryotic tRNA genes), the CCA sequence is added by a tRNA nucleotidyl transferase.

Most tRNAs contain a number of modified bases. These are introduced post-transcriptionally, usually by direct enzyme modification of a pre-existing nucleoside. However, in some cases, an unmodified base is excised by the action of glycosylases and replaced by a modified one. Some modified bases are present in virtually all tRNAs, for example, ribothymidine (T, 5-methyluridine) and pseudouridine (ψ, an isomer of uridine with a C5-to-C1$'$ glycosylic bond) (Section 3.1.1).

6.4.2 Ribosomal RNA

Ribosomal RNA is the most abundant RNA in the cell. Its synthesis is highly regulated because in periods of rapid cell growth the number of ribosomes increases. Biosynthesis is regulated in such a way that the three rRNAs required for assembly of ribosomes are produced in equal amounts. Thus in a bacterial cell, one large RNA transcript contains one copy each of the 5S, 16S, and 23S rRNAs. This transcript also contains some specific tRNAs (Fig. 6.9).

RNase III is an enzyme specific for cleaving double-stranded regions and also appears to be involved in rRNA maturation. It cuts the transcript in the short double-helical regions which are found on each side of the 16S and of the 23S rRNA sequence. It is interesting to note that the ribosomal proteins start to assemble with the RNAs soon after the beginning of transcription and well before RNase III cleavages occur.

In eukaryotic cells, the genes for 5.8S, 18S, and 28S rRNAs are linked. These three genes are transcribed as a long precursor by RNA polymerase I, whereas the gene for 5S RNA is transcribed separately by RNA polymerase III. In a few cases, rRNA genes contain intervening sequences (Section 6.5.2).

In both bacterial and in eukaryotic cells, assembly of ribosomal proteins with RNA takes place before the precursors are processed. Since transcription in a eukaryotic cell takes place in the nucleus and translation takes place in the cytoplasm, ribosomal proteins must enter the nucleus and assemble there on the rRNA precursor. The fully assembled ribosomes then leave the nucleus and participate in protein synthesis in the cytoplasm.

6.4.3 Messenger RNA

The formation of mRNA is a process that requires initiation and termination at precise sites on a DNA template (Sections 5.1 and 5.3). Most mRNA transcripts begin with either guanosine

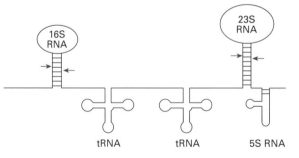

Fig. 6.9 Schematic depiction of an *E. coli* rRNA precursor. Arrows indicate sites of processing by RNase III.

5′-triphosphate or adenosine 5′-triphosphate and at the 3′-end there is a free hydroxyl group.

In bacteria and other prokaryotes, no processing or modification reactions are necessary to produce mature mRNA. In eukaryotic cells, transcription and translation take place in different cellular compartments, and thus mRNA is synthesized in the nucleus and transported to the cytoplasm for use in protein synthesis. Organelles such as mitochondria and chloroplasts in plant cells have independent machinery for mRNA synthesis.

Eukaryotic mRNA transcripts are extensively modified before transport. At the 5′-end, a highly unusual structure called a **cap** is introduced. To achieve this, the 5′-triphosphate of the mRNA transcript is hydrolysed to a diphosphate and a guanosine 5′-phosphate is transferred from GTP to the 5′-end to give a 5′–5′ tripolyphosphate linkage. The N-7 position of the terminal guanine is then methylated using *S*-adenosylmethionine. This resultant cap structure (7MeG(5′–5′)-pppXpY . . .) (Fig. 3.52) is further methylated in some cases at the 2′-hydroxyl position of nucleotide X or both nucleotides X and Y. At the 3′-end of the mRNA, a **poly A** tail of up to 200 adenylate residues is added by the enzyme poly A polymerase. The addition of poly A occurs after specific endonucleolytic processing close to the 3′-end of the mRNA transcript. Both the cap and the poly A appear to be important to translation and to the stability of the mRNA, but the poly A may also be necessary for transport.

Another feature of eukaryotic mRNA transcripts is the frequent occurrence of **intervening sequences** (**introns**) (Section 5.3) which are located between **coding sequences** (**exons**) and which are removed by a process known as **splicing** (Fig. 6.10). All intervening sequences begin with the sequence G·U and end with A·G.

Splicing can be considered to be a two step transesterification reaction. In the first step, the 2′-hydroxyl group of an A-residue, found within the consensus sequence PyPyPuAPy in the intron, attacks the phosphodiester bond located at the 5′-exon–intron junction. This results in the cleavage of the phosphodiester bond and the formation of a **branch structure** (**lariat**) in the intron. In the second step, the newly generated 3′-hydroxyl group in the 5′-exon (exon I of Fig. 6.10) attacks the phosphodiester bond located at the 3′-intron–exon junction. This results in an RNA in which the two exons have been spliced together and the intron originally located between them removed as a lariat structure.

Some pre-mRNAs contain more than one intron which interrupt the coding sequence. In special cases, these introns are spliced out in more than one way (**alternative splicing**) to produce different mRNAs. For example, such mRNAs may code for proteins which differ only slightly in their activities, but which are located in different compartments of a cell. However, when alternative splicing results in a change in the reading frame of the mRNA, the proteins produced will be quite different. Alternative splicing often plays a very important role in cell metabolism.

The chemical principle of the splicing reaction is simple: it is the exchange of one substituent of a phosphodiester linkage for another. However, the machinery of splicing is very complex, currently under investigation, and not clearly understood. Many different proteins, RNAs, and complexes thereof are thought to be involved in splicing. Of these, small nuclear ribonuclearproteins (**snRNPs**) are most crucial. Each snRNP ('snurp') is characterized by a unique RNA (e.g. U1 RNA, Fig. 6.11). The splicing reaction occurs in a large macromolecular complex (**spliceosome**) where many snRNPs interact with the nuclear pre-mRNA by comple-

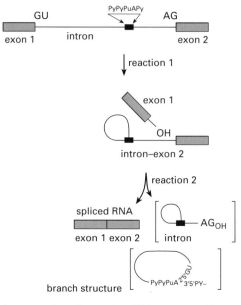

Fig. 6.10 A scheme for processing of introns in mRNA precursors. The two exons are shown as coloured bars and the intron as a thin line. The solid box within the intron contains a consensus sequence and the 2′-hydroxyl group of the A residue within it is involved in the formation of a branch point during splicing (adapted from Brody, E. and Abelson, J. (1985). *Science*, **228**, 963–7. Copyright (1985) AAAS).

mentary base-pairing of the pre-mRNA with the specific RNA component. The spliceosome provides a macromolecular surface where the snRNPs recognize the splice sites, position the exons to be joined, and hold together the resultant pieces.

The catalytic site for the splicing reaction is masked by base-pairing between two of the snRNP RNAs (U4 and U6) and at the appropriate time becomes unmasked so that U6 RNA can act as a catalyst. Whether the phosphodiester transfer activity is an intrinsic property of U6 RNA is not yet established. The splicing of exons from two different RNAs into a single mRNA (**trans splicing**) is also known to occur (e.g. in trypanosomes). This is reminiscent of the trans-cleavage activity of ribozymes (Section 6.5). It is interesting to note that both mRNA generation and protein biosynthesis (Section 6.6) involve intricate macromolecular assemblies consisting of both RNA and protein.

Summary

The initial products of transcription in most cases are RNA precursors which are longer than the mature products.

Transfer RNA precursors fold into tRNA-like structures and both 5′-leader and 3′-trailer sequences are removed enzymatically. In those cases where the 3′-terminal CCA sequence is not encoded by the tRNA gene, the CCA sequence is added by a tRNA nucleotidyl transferase. Modified bases are introduced post-transcriptionally.

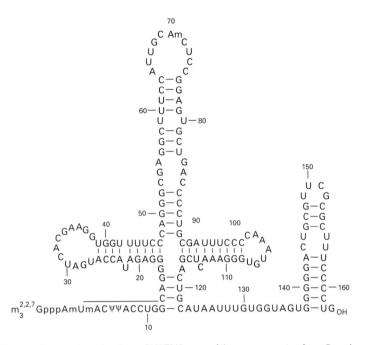

Fig. 6.11 Proposed secondary structure of U1 RNA, one of the components of small nuclear ribonucleoprotein particles (snRNPs) (adapted from Mount, S. M. and Steitz, J. A. (1981). *Nucleic Acids Research*, **9**, 6351–68).

Ribosomal RNA transcripts in bacterial cells contain one copy each of 5S, 16S, and 23S rRNA which are released from the transcript after cleavage by RNase III. Ribosomal proteins begin to assemble with the transcripts before RNA processing takes place. In a eukaryotic cell, transcription takes place in the nucleus and ribosomal proteins are imported from the cytoplasm for assembly. Fully assembled ribosomes are exported to the cytoplasm for use in protein synthesis.

The 5′-end of an mRNA transcript contains either GTP or ATP. Bacterial mRNA does not require further processing. In eukaryotic cells, a cap is introduced at the 5′-end and a poly A tail at the 3′-end. Intervening sequences are removed by splicing, a process involving a number of different complexes of proteins and RNAs, the most crucial being small ribonucleoproteins (snRNPs).

6.5 Catalysis by RNA

The catalysis of reactions in biology has long been considered exclusively to be the property of proteins. However, much excitement has been generated by the discovery of several cases of catalysis by RNA and crowned by the award of the 1990 Nobel Prize for Chemistry to Sidney Altman and Tom Cech. Classically, an enzyme is defined as a macromolecule which (1) enhances the rate of a reaction by lowering its activation energy, (2) is specific with regard to substrates and products, and (3) is a true catalyst (i.e. is not consumed in the reaction). Although proteins have a wider range of functional groups than does RNA, there is clearly enough chemical variety in an RNA to provide specific substrate binding in well defined sites as well as functionalities for general acid–base catalysis. RNA enzymes are sometimes called **ribozymes**.

RNA can act as an enzyme either by itself or when complexed to a protein. To date, all well-characterized reactions catalysed by RNA involve other RNAs as substrates although an artificial DNA substrate containing a single ribonucleotide residue can sometimes be used. There have been suggestions that RNA catalysis may be also involved in the degradation of glycogen.

6.5.1 Ribonuclease P

RNase P is an endonuclease which is responsible for producing the 5'-ends of mature tRNA (Section 6.4.1). This ribonucleoprotein contains an RNA which is essential for catalytic activity, since the activity is lost if the RNA component is destroyed by nuclease digestion. There is much evidence to suggest that the enzyme is present in every cell and organism, but the best understood enzyme is that from *E. coli* which consists of an RNA of 377 nucleotides and a protein subunit of 14 kDa. At high magnesium ion concentration (non-physiological) *in vitro*, the RNA by itself can carry out accurate cleavage of the precursor tRNAs, albeit at a somewhat slower rate than when the protein is present. It is believed that the high ionic strength overcomes the repulsion between the anions of the RNA enzyme and of the substrates. The protein by itself has no enzyme activity.

The enzymatic reaction does not need energy, but it is dependent on the presence of native RNA structure in both the enzyme and substrate. There is also a specific requirement for divalent metal ions (magnesium or manganese). By use of mutagenesis techniques and by phylogenetic comparisons, a model has been derived of the secondary structure of the RNA component of RNase P from *E. coli* (Fig. 6.14). Although the primary sequence of this RNA varies widely between different organisms, its secondary structure is well conserved. While the RNA component of eukaryotic RNase Ps is also essential for enzyme function, catalytic activity of the RNA by itself has not yet been demonstrated.

6.5.2 The ribosomal RNA of *Tetrahymena*

An exciting story has emerged from the study of the biosynthesis of the nuclear 26S rRNA from the protozoa *Tetrahymena thermophila*. Gene sequencing studies showed that the single rRNA gene had a 413 bp intron. Attempts to purify the enzyme responsible for removing the intervening sequence in the rRNA showed that the splicing activity resides totally within the rRNA precursor molecule. This self-splicing is brought about by two consecutive transesterification reactions (Fig. 6.12). The reaction sequence is initiated when the 3'-hydroxyl group of an external (free) guanosine or guanosine 5'-phosphate displaces the nucleotide at the 3'-end of the 5'-exon and adds itself via a normal 3'-to-5' phosphodiester bond to the 5'-end of the

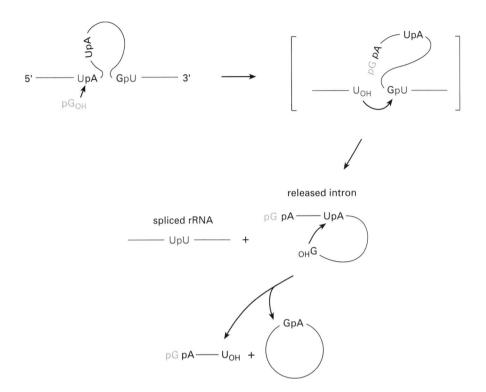

Fig. 6.12   Steps in the self-splicing of *Tetrahymena thermophila* rRNA precursor. Exon sequences are shown in red and the intron sequence in black. The pG$_{OH}$ which initiates the splicing reaction is shown in partial colour (adapted from Zaug, A. J, Grabowski, P. J., and Cech, T. R. (1983). *Nature*, **301**, 578–83. Copyright (1983) Macmillan Magazines Ltd.).

intervening sequence. The newly created 3′-hydroxyl group of the 5′-exon then attacks the phosphodiester bond preceding the 5′-terminal nucleotide of the 3′-exon, thus generating the spliced RNA. The released intervening sequence (L-IVS) then undergoes a cyclization re-action which is initiated by nucleophilic attack of the newly-generated 3′-hydroxyl group on a phosphodiester bond near the 5′-end of the IVS. This creates a short oligonucleotide (15-mer) which at its 5′-end carries the original G residue which initiated the chain of reactions. A circularized and shortened intervening sequence (C-IVS) is also produced.

What are the requirements of these reactions and where does the energy come from? It turns out that no free energy change is needed for this concomitant cleavage and rejoining of RNA. The reaction seems to be driven by a high concentration of guanosine, or a derivative of guanosine, relative to the RNA. However, it is clear that the RNA must form a particular structure in order to render the specific phosphodiester bonds susceptible to cleavage and to provide the binding site for the 'substrate'. The binding site for the initial RNA substrate con-tains a **guide sequence** (CUCUC) which positions the scissile bond in the correct orientation by base-pairing to a complementary sequence in the RNA.

Even more remarkably, it has been shown that a shortened form of the intervening se-quence (L-191VS) can act as a true enzyme and can carry out a range of interesting reactions involving nucleotidyl transferase, phosphoryl transferase, phosphomonoesterase, and phos-phodiesterase activities. In this respect the molecule 'mimics' the generic activity of a number of different protein enzymes.

6.5.3 Group I introns

The *Tetrahymena* rRNA intron (Section 6.5.2) is one example of a group I intron. Other examples are found in rRNA, tRNA, protein coding genes, and mitochondrial mRNAs of plants and fungi. Group I introns undergo self-cleavage *in vitro* and frequently are able to self-splice. Although there is little overall sequence conservation, each intron contains four conserved sequence elements which can form complementary pairs (P, Q, R, and S) and thus give rise to a similar secondary structure (Fig. 6.13a). In the case of the *Tetrahymena* rRNA intron, a tertiary structural model has been proposed (Fig. 6.13b).

6.5.4 Group II introns

The mitochondrial mRNA precursors and chloroplasts contain a second group of self-splicing intervening sequences. These RNAs lack the characteristic group I consensus elements and they splice by a different mechanism which does not require an external nucleotide substrate. Like the splicing of introns in nuclear pre-mRNA (Section 6.4.3), the mechanism proceeds through a branched 'lariat' intermediate structure (Fig. 6.10) which is produced by the attack of a 2'-hydroxyl group of an internal adenylate on the phosphate of the 5'-splice site. However, unlike nuclear pre-mRNA splicing, the splicing activity of Group II introns is an inherent ac-tivity of the RNA and can proceed *in vitro* without added proteins or ribonucleoprotein com-plexes.

Group II introns also share a conserved secondary structure required for catalytic activity that consists of six complex helical domains.

6.5.5 RNA self-cleavage

Another type of catalysis reaction involves intramolecular cleavage of RNA. The mechanism of cleavage involves a transesterification reaction where the 2'-hydroxyl group attacks the adjacent 3'-phosphodiester leading to formation of 5'-hydroxyl and 2',3'-cyclic phosphate ends. The reaction proceeds with inversion of configuration at phosphate.

Yeast phenylalanine tRNA cleavage by lead ions

The first example of sequence-specific cleavage by Pb^{2+} was found to occur in the D-loop of yeast phenylalanine tRNA. Here, it was demonstrated that the three-dimensional structure of

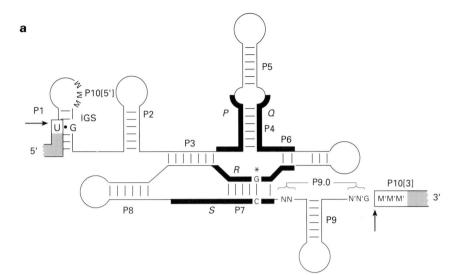

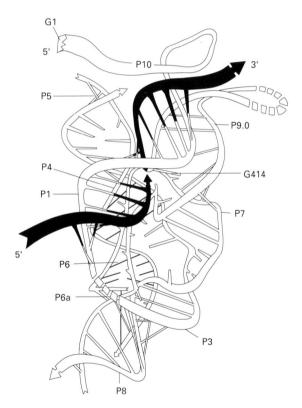

Fig. 6.13 (**a**) Secondary structure of group I introns. Arrows indicate splice sites and conserved sequence elements P, Q, R, and S are indicated by bold lines. The guanosine binding site is shown by an asterisk. Note also the wobble base-pair close to the 5′-cleavage site and the internal guide sequence (IGS). Complementary pairing also takes place between NN and N′N′ (P9.0) and between MMM and M′M′M′ (P10) (from Saldhana, R., Mohr, G., Belfort, M. and Lambowitz, A. M. (1993). *FASEB J.*, **7**, 15–24). (**b**) Proposed tertiary structural model of the *Tetrahymena* rRNA intron. The exons are in black (from Michel, F. and Westhof, E. (1990) *J. Mol. Biol.*, **216**, 585–610).

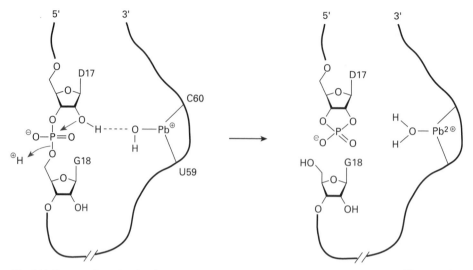

Fig. 6.14 Proposed mechanism for the cleavage of a phosphodiester bond in yeast tRNA^{Phe} by Pb^{2+} bound to a specific site on the tRNA (adapted from Cech, T. R. and Bass, B. L. (1986), *Annu. Rev. Biochem.*, **55**, 599–629. Copyright (1986) Annual Reviews Inc.).

the tRNA influences the reactivity of a particular phosphodiester bond. Analysis of the crystal structure of the complex revealed how Pb^{2+} is bound to a specific pocket in the RNA comprising bases U59 and C60 in such a way that the metal ion is oriented properly towards the polynucleotide chain (Fig. 6.14). The active species is probably (PbOH)$^{+}$ which has a pK_a of about 7. Removal of the proton from the 2′-hydroxyl group of dihydrouridine at position-17 leads to nucleophilic attack on the phosphodiester bond between D17 and G18 resulting in the specific cleavage.

The hammerhead ribozyme

A number of plant satellite RNAs and viroids undergo a self-cleavage reaction that is a crucial step in their replication. The RNAs contain sequences which can be folded into a particular secondary structure called a **hammerhead**. One example is the plus (genomic) strand of the satellite RNA of tobacco ringspot virus (sTobRV, Fig. 6.15a), which is only 359 nucleotides long. Self-cleavage is rapid and the only external requirement is for magnesium ions. All hammerheads contain three base paired sections. Whereas stems I and II do not show sequence conservation between different hammerheads, other nucleotides are strictly conserved (shown boxed in Fig. 6.15a). It is believed that the hammerhead forms a specific three-dimensional structure in which magnesium ion binding plays an important structural and catalytic role. A recent X-ray crystal structure of a non-cleavable hammerhead RNA–DNA inhibitor complex has revealed a Y shape where stems I and II are in close proximity, the conserved CUGA sequence undergoes a sharp turn, and helix II is extended by non-Watson-Crick G·A and A·G base-pairs (see picture on the back cover of this book). However, the precise details of

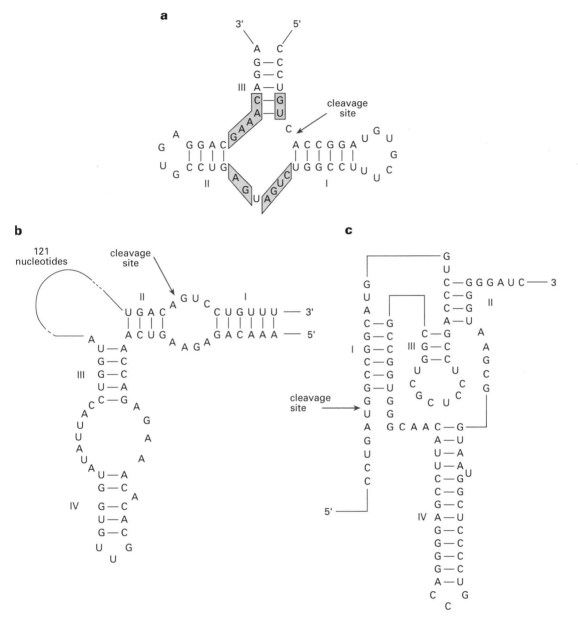

Fig. 6.15 Three self-cleaving ribozymes showing the sites of cleavage and the major elements of secondary structure. (**a**) The hammerhead ribozyme showing the conserved residues (boxed), (**b**) the hairpin ribozyme, and (**c**) hepatitis delta virus RNA ribozyme.

catalytic cleavage remain unclear. Some RNAs, such as the plus strand of avocado sunblotch viroid and the RNA transcript of newt satellite DNA II, have only the two conserved base-pairs in stem III of their hammerhead structures and probably dimerize to form double hammerheads as their self-cleaving forms.

The hairpin ribozyme

The minus (antigenomic) strand of sTobRV self-cleaves via a different mechanism where the RNA folds into a **hairpin** structure (Fig. 6.15b). The exact sequences in stems I–IV appear not to be essential provided that base-pairing is maintained whereas the identities of many of the other residues in the loop regions are critical. Like the hammerhead, a specific three-dimensional structure is likely for the hairpin and there is also a requirement for magnesium ions.

Hepatitis delta virus RNA

The human hepatitis delta virus, which is a satellite virus of hepatitis B virus, contains a circular RNA of about 1700 nucleotides where both genomic and antigenomic strands self-cleave. Of several recently proposed secondary structures, a pseudoknot-like structure (Section 6.3.3) seems the most likely for the cleavage domain (Fig. 6.15c). The magnesium ion-dependent cleavage is thought to take place in a similar way to that of hammerhead cleavage, but the three-dimensional structure of the RNA is likely to be very different.

6.5.6 *Trans* cleavage and ribozyme targeting

It has become possible to harness the cleavage activity of ribozymes so that they can be targeted towards the destruction of other RNAs. For example, Haseloff and Gerlach showed how the hammerhead ribozyme can be modified so that it will cleave a substrate RNA in *trans* (Fig. 6.16). Stems I and III are created by hybridization of the ribozyme with the target RNA whereas stem-loop II is provided by the ribozyme. Hammerhead ribozymes can be designed to cleave at any site in an RNA containing the sequence GUX (where X is C, U, or A). *Trans* cleavages by ribozymes are now being utilized to suppress the function of particular cellular mRNAs in an effort to understand their roles in metabolic pathways and also to destroy viral RNA within virally infected cells as potential therapeutic agents.

6.5.7 RNA catalysis and evolution

The increasing number of examples of reactions catalysed by RNA suggests that RNA catalysis is a widespread phenomenon. A number of enzymes with essential RNA components have now been identified. There is also evidence that RNA components of ribonucleoprotein complexes may play a catalytic role in reactions involved in splicing (Section 6.4.3) and in peptide bond formation (Section 6.6).

The discovery of enzymatic activity in RNA has also provided a solution to a long-standing problem of how specificity in protein synthesis and replication evolved. The existence of an 'RNA world' composed of RNA molecules acting as enzymes and carrying out vital life functions (e.g. self-replication, protein synthesis) is an attractive idea. These early enzymes might not have been particularly specific or efficient and therefore might be expected to have given way to proteins during evolution.

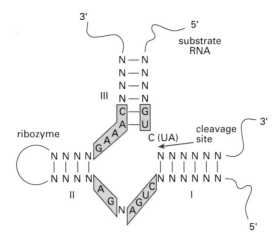

Fig. 6.16 General secondary structure of a hammerhead ribozyme arranged for *trans* cleavage of a substrate RNA. Boxed residues are those whose identity must be preserved for maximal cleavage activity. N represents any of the four ribonucleotides provided that Stems I, II and III remain intact.

Further credence for this view has come recently from a powerful technique called *in vitro* **selection**. Large pools of synthetic DNAs containing sections of randomly mixed nucleotide sequence (10^6–10^{15} different sequences) are prepared (Section 3.4) and transcribed into RNA by use of T7 RNA polymerase (Section 3.5.4). The rare RNA sequences capable of binding a particular ligand (e.g. an organic dye) or capable of carrying out a particular enzymatic reaction are selected from the randomized mixture by use of affinity chromatography techniques (Fig. 6.17). These RNA molecules are then reverse transcribed into DNA, amplified by PCR (Section 10.7.3) and transcribed back into RNA by T7 RNA polymerase. After a number of such cycles, a unique sequence often emerges that is capable of binding the ligand with high affinity. Using such techniques Bartel and Szostak have recently selected an RNA ribozyme capable of carrying out template-directed phosphodiester bond formation at a rate 7 million times faster than the uncatalysed reaction.

Summary

RNA can act as an enzyme either by itself or in combination with protein. A catalytic RNA is sometimes called a ribozyme and usually involves RNA as a substrate.

Ribonuclease P contains an RNA subunit essential for catalytic activity. RNase P from *E. coli* can cleave precursor tRNA by itself *in vitro*, but RNase Ps obtained from eukaryotes require the protein component also for catalytic activity.

The ribosomal RNA from *Tetrahymena* carries out self-splicing. The excised intervening sequence has a variety of enzymatic activities.

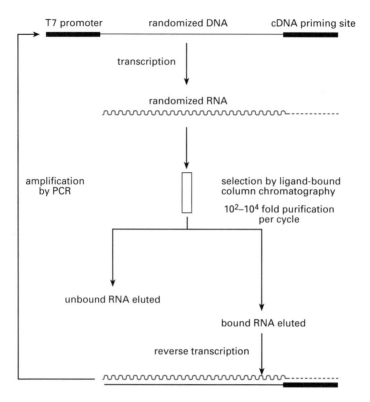

Fig. 6.17 Scheme for *in vitro* selection of RNA sequences that bind to a particular ligand.

Some mitochondrial mRNA and rRNA precursors contain intervening sequences that self-splice and are classed as group I introns. These are analogous to the *Tetrahymena* rRNA. A second class of introns occurs in mitochondrial mRNA precursors. These group II introns self-splice by a mechanism which is analogous to the splicing of nuclear mRNA precursors. Some plant satellite and viroid RNAs and the satellite RNA from hepatatis delta virus undergo self-cleavage reactions. The RNAs can be folded into unique secondary structures. Ribozymes can be arranged to cleave a target RNA in *trans*. By *in vitro* selection techniques, artificial RNAs can be found that will bind a ligand or carry out unusual enzymatic reactions.

6.6 Protein synthesis

During protein synthesis, a linear sequence of nucleotides in mRNA is translated by the **ribosome** into a specific sequence of amino acids in protein. Transfer RNAs play a central role by acting as adapter molecules between the nucleotide sequences (**codons**) on the mRNA and the amino acids specified by them. Amino acids destined for incorporation into protein are first attached to tRNA and brought to the ribosome as **aminoacyl-tRNAs**. On the ribosome (Fig. 6.18), codon 1 on the mRNA binds to a specific tRNA, which in turn is attached to the first amino acid. Similarly, codon 2 on the mRNA binds to another tRNA which is attached to the second amino acid. The amino group of the second amino acid then attacks the ester of the first aminoacyl-tRNA to form a peptide bond. The order of amino acids in the protein is thus specified by the sequence of nucleotides in the mRNA. The role of tRNAs in translation depends on two important properties:

- specific base-pairing between three nucleotides (**anticodon**) on the tRNA and the mRNA codon
- Attachment of specific amino acids to specific tRNAs.

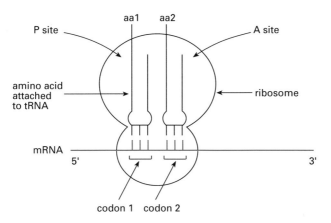

Fig. 6.18 Schematic depiction of a ribosome translating an mRNA into protein. The three vertical lines on the mRNA (red) indicate the nucleotides of the codon and the three vertical lines on the tRNA indicate the complimentary anticodon sequence. The A site and P site are the two well-characterized tRNA binding sites on ribosome.

6.6.1 Aminoacylation

Since there are 20 amino acids in most proteins, at least 20 different tRNA species are needed for protein synthesis. In fact, most cells contain from 60 to 70 different tRNAs. Thus, the same amino acid can be attached to more than one species of tRNA. Such different tRNA species which can be aminoacylated by the same amino acid are called **isoacceptor tRNAs**. Individual tRNA species are usually identified by their amino acid specificity. Thus, tRNA^{Ala} specifies a tRNA which can be aminoacylated with the amino acid alanine. The enzyme which aminoacylates this tRNA is called alanyl-tRNA synthetase (abbreviated as AlaRS).

The accuracy of protein synthesis depends crucially on the covalent attachment of a specific amino acid to a specific tRNA. This reaction, called the **aminoacylation reaction**, is catalysed by a group of enzymes called **aminoacyl-tRNA synthetases**. There are 20 aminoacyl-tRNA synthetases in most cells, each specific for one particular amino acid. These fall into two classes with ten enzymes in each class. The two classes differ in the structural domain responsible for binding ATP. Each enzyme is also highly specific for tRNA and can discriminate among tRNAs which have different sequences, but similar three-dimensional structures. Only the correct amino acid becomes attached to the corresponding tRNA. For example, the enzyme alanyl-tRNA synthetase will attach the amino acid alanine to tRNA^{Ala} but not to any

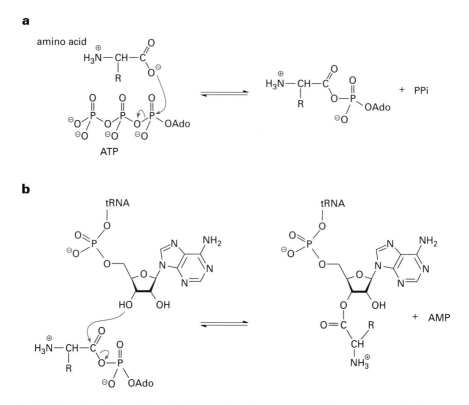

Fig. 6.19 Steps in aminoacylation of tRNAs catalysed by aminoacyl-tRNA synthetases. (**a**) Activation of the amino acid; (**b**) transfer of the activated amino acid to the correct tRNA. Note that the reaction of aminoacyl-adenylate with tRNA can occur either on the 2′ or 3′-hydroxyl group of the terminal adenosine. This depends on the aminoacyl-tRNA synthetase.

other tRNA. The discrimination of tRNAs by aminoacyl-tRNA synthetases involves a few critical nucleotides in each tRNA. Depending on the individual tRNA species, these positions are localized near the acceptor end, or in the anticodon, or are distributed throughout the molecule (Section 9.3.7).

Aminoacylation of tRNAs occurs in two steps. First, the aminoacyl-tRNA synthetase catalyses the reaction of the amino acid and ATP to form an activated aminoacyl-adenylate (Fig. 6.19a). Secondly, the enzyme catalyses the transfer of the amino acid to the terminal adenosine of the correct tRNA to form an aminoacyl-tRNA, AMP is released, and the enzyme is free to recycle (Fig. 6.19b).

Following attachment of the amino acid to a tRNA, the amino acid can, in some cases, be further modified by enzymes which are highly specific with respect to the tRNA. For example, methionyl-tRNA is converted into formyl methionyl-tRNA by the enzyme methionyl-tRNA transformylase. This enzyme uses N^{10}-formyltetrahydrofolate to transfer a formyl group onto the methionine tRNA species that is to be used for initiation of protein synthesis. All cells have at least two methionine tRNA species, one of which is specially used to initiate protein synthesis.

6.6.2 Ribosomes

The ribosome is a large ribonucleoprotein which is an essential component of the protein synthesizing system. The *E. coli* 70S ribosome contains 51 different proteins and the three rRNAs. The ribosome can be reversibly dissociated into a 30S (small) subunit, containing 21 proteins and the 16S rRNA, and a 50S (large) subunit containing the remaining 30 proteins and the 5S and 23S rRNAs. The ribosome subunits dissociate and re-associate during their function in protein synthesis. The binding domain for mRNA and the ancticodon region of tRNA lies in the 30S subunit whereas the rest of the tRNA-binding domain and the catalytic domain for peptide bond formation is in the 50S subunit. There are two well-characterized tRNA binding sites on the ribosome. These are adjacent to one another and are called the **A site** (aminoacyl-tRNA site) and the **P site** (peptidyl-tRNA site) (Fig. 6.18). There is also good evidence for a third site, called the **E site** (exit site), thought to be located mostly on the 50S subunit.

The 80S ribosome in eukaryotic cytoplasm is larger than the *E. coli* ribosome. It can reversibly dissociate into 40S and 60S subunits. The 40S subunit contains the 18S RNA, whereas the 60S subunit contains the 5S, 5.8S, and 28S rRNAs. Sequence relationships between the eukaryotic rRNAs and proteins and those of prokaryotic origin suggest that the eukaryotic ribosome is simply an elaboration of the prokaryotic ribosome.

6.6.3 Messenger RNAs

In bacteria such as *E. coli*, mRNA synthesis (**transcription**) and utilization (**translation**) are coupled. Even before synthesis of the mRNA is completed, part of it is translated by the ribosome. This is possible because bacteria do not contain different cellular compartments for transcription and translation.

Most bacterial mRNAs are **polycistronic**, which means that a single long transcript codes for several proteins. Here, ribosomes can usually bind independently to the various translational start sites on the mRNA and initiate protein synthesis. A polycistronic mRNA may code for a number of proteins which are all part of a related biochemical pathway, as in the case of enzymes involved in tryptophan biosynthesis (Section 5.3.3). On the other hand, it may code for proteins which have no obvious relationship to one another. For example, in *E. coli* a single mRNA codes for three proteins: the *rpsU* protein involved in protein biosynthesis the *rpoD* protein involved in RNA biosynthesis, and the *dnaG* protein involved in DNA biosynthesis. The relative amounts of proteins which are made from a polycistronic mRNA can also be quite different. In contrast to most bacterial mRNAs, mRNA from eukaryotic cells are mostly mono-cistronic and each mRNA has a single **ribosome binding site** and codes for a single protein.

Several factors determine the relative activities of mRNAs in protein synthesis. These include the secondary and tertiary structure of mRNA and sequences upstream (on 5′-side) of the translational start site in mRNAs. In a polycistronic mRNA with independent translational initiation sites, the affinity of ribosomes for one site may be quite different to that for another site, probably because of differences in accessibility of the mRNA to the ribosome. In such a case, some proteins encoded by the mRNA would be made in significantly different amounts to others.

In the case of the mRNAs coding for most of the ribosomal proteins in *E. coli*, apart from the ribosome binding site nearest to the 5′-end, all the others are inaccessible because of particular mRNA secondary and tertiary structure. As the ribosomes translate this segment and approach the next coding region, the mRNA structure downstream is opened up and the downstream translational start site now becomes accessible. This process is called **sequential translation** and is one of the strategies used by *E. coli* to ensure that proteins which are part of a multisubunit complex, such as the ribosome, are made in equal amounts.

Sequences upstream of the translational start site can also have a major effect on relative efficiencies of mRNA translation. In *E. coli*, base-pairing between a string of pyrimidine nucleotides near the 3′-end of 16S RNA and a purine-rich sequence about ten nucleotides upstream of the translational start site on the mRNA helps to align the initiating AUG codon on the ribosome for translation. The extent of base-pairing here is one of the factors which determines the relative affinity of the ribosome for a translational start site on the mRNA.

In eukaryotic cells, sequences which immediately flank the AUG start site are also important in translation. Since eukaryotic mRNAs are monocistronic, the first AUG near the 5′-end of mRNA is normally used as the start site for translation. Here, the optimal sequence for initiation has been identified as ACC**AUG**G, where AUG is the initiator codon (Section 6.6.8). It is not yet established whether a sequence in 18S rRNA of eukaryotic ribosomes, complementary to the ACC sequence, plays a role analogous to that of *E. coli* 16S rRNA in aligning the mRNA on the eukaryotic ribosome.

6.6.4 Protein factors

Several other proteins, called **initiation factors**, **elongation factors**, and **termination factors**, function at specific stages of protein synthesis. Each of these factors associates with the ribosome, carries out a particular function, and then dissociates from the ribosome to allow the binding of the next protein factor. GTP acts as an important co-factor for many of these proteins. At each step, GTP is hydrolysed to GDP and inorganic phosphate is released. The hydrolysis of GTP plays a crucial role in recycling of the protein factors from the ribosome.

6.6.5 Steps in protein synthesis

There are three distinct stages in protein synthesis: **initiation**, **elongation**, and **termination**. In initiation, ribosomes bind to the initiation site on the mRNA and associate with fMet-tRNA to form an **initiation complex** (step 1 of Fig. 6.20). In *E. coli*, this reaction requires the participation of three initiation factors IF-1, IF-2, and IF-3. Formation of an initiation complex consists of several steps. First, 70S ribosomes dissociate to form 30S and 50S subunits. IF-3 promotes this dissociation by binding to the 30S subunit to form a 30S·IF-3 complex and by preventing reassociation of the 30S and 50S subunits. The 30S·IF-3 complex then binds both to mRNA at a particular AUG codon and also to formylmethionyl-tRNA (fMet-tRNA). This reaction requires

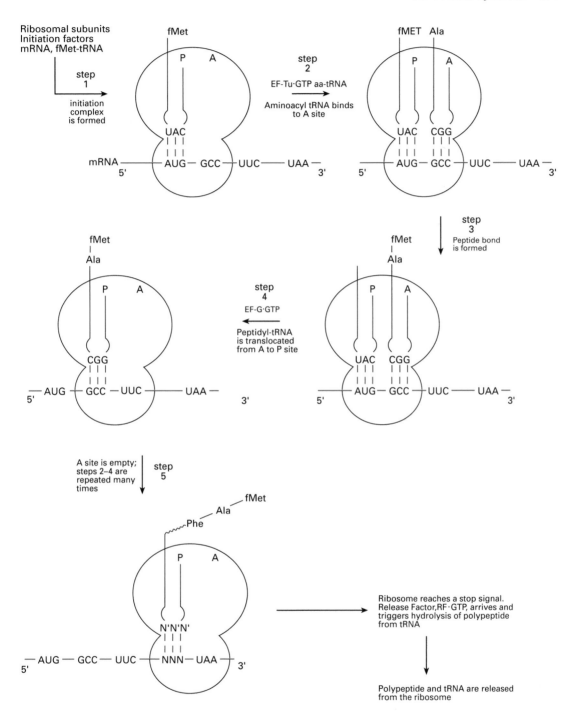

Fig. 6.20 Schematic depiction of the steps in protein synthesis.

the other two initiation factors, IF-1 and IF-2, as well as GTP. At this point, IF-3 is released allowing the 50S ribosomal subunit to join the 30S·mRNA·fMet-tRNA complex to form the 70S initiation complex. Subsequently, or concomitantly, GTP is hydrolysed to GDP, and IF-1 and IF-2 are released from the ribosome. At this stage, the fMet-tRNA is bound to the P site on the ribosome.

The elongation cycle of protein synthesis consists of three distinct processes which are repeated many times (Fig. 6.20, step 2 to step 4). At each elongation cycle, a peptide bond is formed and an amino acid is added to a continuously growing polypeptide chain.

In step 2, an appropriate aminoacyl-tRNA (in this case alanyl-tRNA) binds to the A site (as directed by the particular codon–anticodon interaction) in the form of a ternary complex with elongation factor Tu (EF-Tu) and GTP. Once the aminoacyl-tRNA is bound, GTP is hydrolysed to GDP and EF-Tu is released from the ribosome as EF-Tu·GDP complex. In order to generate EF-Tu·GTP, the GDP bound to EF-Tu must be exchanged with GTP. This reaction is facilitated by elongation factor EF-Ts which binds to EF-Tu to form EF-Tu·EF-Ts and helps dislodge GDP from EF-Tu. The EF-Tu·EF-Ts complex then combines with GTP to form EF-Tu·GTP and EF-Ts.

In step 3, peptidyl transferase on the ribosome catalyses the formation of a peptide bond. This involves attack by the free amino group of aminoacyl-tRNA bound to the A site on the carbonyl group of fMet-tRNA bound to the P site, resulting in formation of a peptide bond at the expense of an ester bond (Fig. 6.21). The peptide chain which was linked to the tRNA on the P site is thus extended by one amino acid and the incoming amino acid is added to the C-terminus of the growing peptide chain.

In step 4, the peptidyl-tRNA on the A site is **translocated** to the P site, the displaced, deacylated tRNA on the P site moves to the E site (not shown in Fig. 6.20) and eventually falls off the ribosome, and the A site is empty once again. The translocation step thus moves the ribosome three nucleotides along with respect to the mRNA and brings the next codon into the reading frame. The translocation step is catalysed by another elongation factor, EF-G, which also uses GTP as a co-factor. In the process of translocation, GTP is hydrolysed to GDP and EF-G is released from the ribosome.

After many cycles of elongation, involving steps 2, 3, and 4, the ribosome eventually reaches a codon on the mRNA, in this case UAA, which specifies termination of protein synthesis. This leads to binding of an appropriate termination factor to the ribosome, more commonly called a release factor (RF) as an RF·GTP complex. Binding of the RF results in hydrolysis of the ester linkage between the completed polypeptide chain and the last tRNA. The polypeptide and tRNA are released from the ribosome and once again GTP is hydrolysed to GDP. RF is then released from the ribosome. In the absence of any tRNA bound to the ribosome, the 70S ribosome falls off the mRNA, associates with IF-3 to form a 30S·IF-3 complex, and is now free to participate in another round of protein synthesis.

Initiation of protein synthesis in eukaryotic cytoplasm involves many more initiation factors than in *E. coli*. Another major difference lies in the way in which the 40S ribosomal subunit locates the AUG initiator codon on the mRNA. In the first step, the eukaryotic initiation factor eIF-2 binds to GTP and the initator Met-tRNA to form a 'terniary' complex (Fig. 6.22). This complex associates with the 40S ribosomal subunit which in turn binds to the 5′-end of mRNA at the cap site. The ribosome : initiator Met-tRNA complex then moves down the

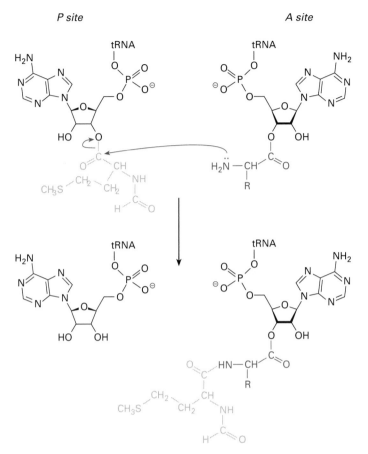

Fig. 6.21 Mechanism of peptide bond formation. The aminoacyl portion is shown in full colour and the fMet portion in partial colour.

mRNA until it reaches the first AUG codon. This process is called **scanning**. In almost all cases the first AUG codon near the 5′-end is used as the initiator codon. The 60S ribosomal subunit then joins this complex to form an 80S complex and the remaining steps in protein synthesis are essentially the same as in *E. coli*.

6.6.6 Polyribosomes

After a ribosome becomes attached and moves along the mRNA, another ribosome can follow and bind at the same initiation site. Thus, a single mRNA is translated by many ribosomes at the same time. An mRNA which is being translated is often found as part of a complex called a polyribosome. In a polyribosome, the individual ribosomes and the nascent polypeptide chains are held together by mRNA (Fig. 6.23). Gentle treatment of such a complex with ribonuclease breaks the mRNA chain at points where it is not protected by bound ribosomes. This results in conversion of a polyribosome into individual ribosomes which contain fragments of mRNA.

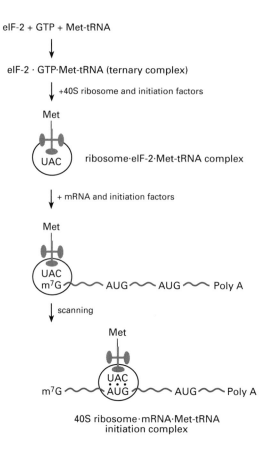

Fig. 6.22 Steps in formation of the eukaryotic initiation complex.

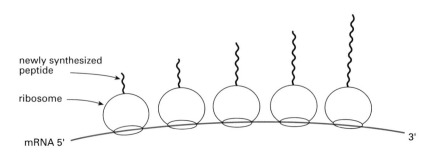

Fig. 6.23 An mRNA (red) being translated simultaneously by many ribosomes to give a polyribosome.

6.6.7 Processing and transport of proteins

Virtually every protein is altered in one way or another following its synthesis on the ribosome. Protein synthesis is always initiated with either formylmethionine in the case of bacterial, mitochondrial, and chloroplast proteins or with methionine in the case of eukaryotic, cytoplasmic proteins. However, few proteins contain fMet or Met at their N-termini. The reason for this is that cells contain enzymes that remove the fMet or Met residue from the N-terminus, often as the nascent polypeptide chain is being elongated on the ribosome. Besides these changes, there are several other examples of covalent modifications that are introduced into a protein. These include hydroxylation, methylation, acetylation, phosphorylation, addition of sugar residues, and cross-linking of one or more polypeptide chains. A common and important modification involves cleavage of proteins by proteases.

6.6.8 Genetic code

The genetic code was established in 1965 largely through the work of Nirenberg, of Khorana, and of Ochoa (Section 1.5). They showed that the codon for an amino acid consisted of three consecutive nucleotides (**codons**) (Fig. 6.24). Of the 64 possible codons, 61 specify the 20 amino acids and 3, UAA, UAG, or UGA, specify termination of protein synthesis. AUG is the codon used to specify initiation, but occasionally GUG and UUG are also used.

first nucleotide (5'-end)	middle nucleotide				third nucleotide (3'-end)
	U	C	A	G	
U	Phe	Ser	Tyr	Cys	U
	Phe	Ser	Tyr	Cys	C
	Leu	Ser	STOP	STOP	A
	Leu	Ser	STOP	Trp	G
C	Leu	Pro	His	Arg	U
	Leu	Pro	His	Arg	C
	Leu	Pro	Gln	Arg	A
	Leu	Pro	Gln	Arg	G
A	Ile	Thr	Asn	Ser	U
	Ile	Thr	Asn	Ser	C
	Ile	Thr	Lys	Arg	A
	Met[a]	Thr	Lys	Arg	G
G	Val	Ala	Asp	Gly	U
	Val	Ala	Asp	Gly	C
	Val	Ala	Glu	Gly	A
	Val	Ala	Glu	Gly	G

[a]AUG is also used as a start signal

Fig. 6.24 The genetic code. The 64 codons are divided into 16 four-codon boxes. The four codons of a codon box differ only in their 3'-terminal nucleotide. Full colour shows where an amino acid is specified by all four codons of a codon box, partial colour shows where an amino acid is specified by two (or in one case three) of the four codons, and grey shows where an amino acid is specified by a single codon.

In a few mRNAs UGA can also specify selenocysteine, which differs from cysteine in having CH_2SeH as the amino acid side chain instead of CH_2SH. Selenocysteine is found in both prokaryotes and eukaryotes but only in a few proteins, such as redox enzymes, that require this unusual amino acid for their function. Unlike amino acid analogues, such as hydroxyproline and methyl lysine, which are produced by covalent modification of proline and lysine respectively (post-translational modification), selenocysteine is incorporated directly into the nascent polypeptide chain. Thus this amino acid represents an addition to the 20 amino acids commonly found in proteins and is sometimes called amino acid number 21.

A remarkable feature of the genetic code is that, with very few exceptions, it is totally universal. The genetic code in bacteria, fungi, plants, fish, and animals is the same. Except for methionine (AUG) and tryptophan (UGG), all the other amino acids are specified by more than one codon, most often by two, four, or six codons. The codons that specify the same amino acid are often closely related and differ only in third position. The two codons for histidine, CAU and CAC, differ only in that the third nucleotide can be either of the pyrimidines. Similarly, the two codons for glutamine, CAA and CAG, differ only in that the third nucleotide can be either of the purines. Likewise, the four codons for proline, CCU, CCC, CCA, and CCG, differ only in that the third nucleotide can be either U, C, A, or G.

It might be expected that at least 61 different tRNAs would be needed to read the genetic code. This, however, is not the case. Although selection of correct tRNAs on the ribosome depends on base-pairing between mRNA codon and tRNA anticodon, this base-pairing need not always be strictly of the Watson–Crick type for all three base-pairs. A slightly relaxed form of base-pairing in one of the base-pairs reduces the minimum number of tRNAs needed to read the genetic code to 32. The rules for base-pairing between the codon in mRNA and the anticodon in tRNA were formulated by Crick as the so-called **wobble hypothesis**. According to this hypothesis, the first two base-pairs, between the first and second nucleotides of the codon and the third and second nucleotides of the anticodon (**a** to **a′** and **b** to **b′** in Fig. 6.25) are of the Watson–Crick type, whereas the third base-pair (**c** to **c′**) can be more flexible (Section 2.1.2). The kinds of base-pairs allowed are listed (Table 6.4). G in a tRNA can pair with either U or C, a modified derivative of U in tRNA can pair with either A or G, and I in tRNA can

Table 6.4. Possible base-pairs between the third nucleotide of the mRNA codon and the first nucleotide of the tRNA anticodon

First nucleotide (5′-end) of anticodon	Third nucleotide (3′-end) of codon
C	G
G	U or C
U*	A or G
I	U, C or A

U* a derivative of U (Section 7.8.1).

codon   5' — a — b — c
                · · ·   · · ·   · · ·
anticodon 3' — a'— b'— c' -

Fig. 6.25 Base-pairing between the nucleotides of an mRNA codon and a tRNA anticodon. Base-pairing of **a** to **a'** and of **b** to **b'** is strictly of the Watson–Crick type whereas that between **c** and **c'** can be somewhat more flexible.

pair with either U, C, or A in an mRNA. The wobble pairs that are proposed to form between tRNA and mRNA are G·U, I·U, I·C, and I·A (Figs. 6.26 and 2.9). Thus, given the organization of the genetic code in which the four codons of a codon box specify either the same amino acid or two different amino acids, two tRNAs can read all four codons of a codon box. For example, a histidine tRNA with the anticodon sequence GUG can read both CAU and CAC codons for histidine. Similarly, a glutamine tRNA with U*UG can read both CAA and CAG codons for glutamine. Also in cases where a codon box specified a single amino acid, two tRNAs can read all four codons. The two tRNAs could either have G to read U and C, with U* to read A and G, or I to read U, C and A, with C to read G. This choice depends on the organism.

In mitochondria, a simpler mechanism to read the genetic code further reduces the number of tRNAs from 32 to 24. A single tRNA with U in the first anticodon position can pair with either U, C, A, or G and read all four codons of a codon box. Since there are eight four-codon boxes each of which specifies a single amino acid (Fig. 6.23), this reduces the number of tRNAs needed by eight.

Mitochondrial protein synthesis is also unusual in providing an exception to the universality of the genetic code. First, there are some differences between the genetic code in mitochondria and the 'universal' genetic code and secondly there are variations in the genetic code among mitochondria from different organisms.

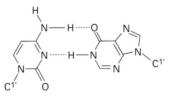

Fig. 6.26 Base-pairing between C and I. Other wobble pairs are shown in Fig. 2.9.

Summary

In protein synthesis, a linear sequence of nucleotides in mRNA is translated by the ribosome into a specific sequence of amino acids in protein. Amino acids are carried to the ribosome as aminoacyl-tRNA. Each tRNA is aminoacylated by a specific amino acid in a reaction catalysed by a specific aminoacyl-tRNA synthetase. One of the two methionyl-tRNAs is formylated and is involved in initiation of protein synthesis.

Ribosomes consist of two subunits. The mRNA and fMet-tRNA are bound by the 30S subunit. The aminoacyl-tRNA binding domain and the catalytic domain are mostly in the 50S subunit. Most bacterial mRNAs are polycistronic whereas most eukaryotic mRNAs are monocistronic. Each translational start site in polycistronic mRNA may be translated independently and may have a different accessibility to ribosomes. mRNAs for ribosomal proteins are usually translated sequentially. Sequences upstream of the AUG start site can substantially affect the efficiency of translation of both prokaryotic and eukaryotic mRNAs.

The three distinct stages of protein synthesis are initiation, elongation, and termination. Each stage involves a series of reactions that require a number of protein factors. Initiation results in fMet-tRNA becoming bound to the P site on the ribosome. In elongation, an aminoacyl-tRNA becomes bound to the A site on the ribosome, the fMet (or peptide chain) at the P site is transferred to the amino acid of the aminoacyl-tRNA at the A site, and the elongated peptidyl-tRNA is translocated to the P site. After many cycles of elongation, a termination codon is reached which signals the release of the polypeptide and the last tRNA from the ribosome. A single mRNA is translated simultaneously by many ribosomes as a complex called a polyribosome. Many proteins are post-translationally modified before use.

The genetic code is the key that relates the sequence of nucleotide triplets (codons) on the mRNA to the sequence of amino acids in the corresponding protein. Because of flexibility in base-pairing between codons in the mRNA and anticodons in tRNA (the wobble hypothesis), only 32 tRNAs are required to read the 61 codons that specify the 20 amino acids found in proteins. In mitochondria, the number of tRNAs required is reduced to 24.

Further reading

Gesteland, R. F. and Atkins, J. F. (eds.) (1993). *The RNA world.* Cold Spring Harbor Press, New York.

Gray, M. W. and Cedergren, R. J. (eds.) (1993). The new age of RNA. *FASEB. J.*, **7**, 1–239.

6.1

Poulson, R. (1977). In Stewart, P. R. and Letham, D. S. (eds.) *The ribonucleic acids*, pp. 333–67. Springer-Verlag, Weinheim.

6.2

RajBhandary, U. L. (1980). Recent developments in methods for RNA sequencing using *in vitro* ^{32}P-labelling. *Fed. Proc.*, **39**, 2815–21.

Stanley, J. and Vassilenko, S. (1978). A different approach to RNA sequencing. *Nature*, **274**, 87–9.

6.3

Cheong, C., Varani, G. and Tinoco, I., Jr (1990). Solution structure of an unusually stable hairpin. *Nature*, **346**, 680–2.

Ehresman, C., Baudin, F., Mougel, M., Romby, P., Ebel, J.-P., and Ehresman, B. (1987). Probing the structure of RNAs in solution. *Nucleic Acids Res.*, **15**, 9109–28.

James, B. D., Ohlsen, G. J., Liu, J., and Pace, N. R. (1988). The secondary structure of RNase P RNA, the catalytic element of a ribonucleoprotein enzyme. *Cell*, **52**, 19–26.

Moazed, D., Stern, S., and Noller, H. F. (1986). Rapid chemical probing of conformation in 16S RNA and 30S ribosomal subunits using primer extension. *J. Mol. Biol.*, **187**, 399–416.

Pleij, C. W. A. (1990). Pseudoknots: a new motif in the RNA game. *Trends Biochem. Sci.*, **15**, 143–7.

Pugilisi, J. D., Wyatt, J. R. and Tinoco, I. Jr. (1991). RNA pseudoknots. *Acc. Chem. Res.*, **24**, 152–8.

Rich, A. and Kim, S-H. (1978). The three-dimensional structure of transfer RNA. *Sci. Amer.*, **238**, 52–62.

Woese, C. R., Gutell, R. R., Gupta, R., and Noller, H. F. (1986). Detailed analysis of higher order structure of 16S-like ribosomal RNAs. *Microbiol. Rev.*, **47**, 621–69.

6.4

Gray, M. W. and Covello, P. S. (1993). RNA editing in plant mitochondria and chloroplasts. *FASEB J.*, **7**, 64–71.

Hadjuk, S. L., Harris, M. E., and Pollard, V. W. (1993). RNA editing in kinetoplastid mitochondria. *FASEB J.*, **7**, 54–63.

McClain, W. H. (1977). Seven terminal steps in a biosynthetic pathway leading from DNA to transfer RNA. *Acc. Chem. Res.*, **10**, 418–25.

Maden, B. E. H. (1990). The numerous modified nucleotides in eukaryotic ribosomal RNA. *Prog. Nucleic Acids Res. Mol. Biol.*, **39**, 241–303.

Moore, M. J., Query, C. C., and Sharp, P. A. (1993). Splicing of precursors to mRNA by the spliceosome. In *The RNA world* (ed. R. F. Gesteland and J. F. Atkins), pp. 303–57. Cold Spring Harbor Press, New York.

Sharp, P. A. (1994). Split genes and RNA splicing. Nobel Lecture.

Stryer, L. (1988). *Biochemistry*, 3rd edn, pp. 703–6. Freeman, San Francisco.

6.5

Bartel, D. P. and Szostak, J. W. (1993). Isolation of new ribozymes from a large pool of random sequences. *Science*, **261**, 1411–18.

Burke, J. M. and Bernal-Herranz, A. (1993). In vitro selection and evolution of RNA: applications for catalytic RNA, molecular recognition and drug discovery. *FASEB J.*, **7**, 106–12.

Cech, T. R. (1990). Self-splicing of group I introns. *Annu. Rev. Biochem.*, **59**, 543–68.

Chastain, M. and Tinoco, I. (1991). Structural elements in RNA. *Prog. Nucleic Acids Res. Mol. Biol.*, **41**, 131–77.

Haas, E. S., Brown, J. W., Pitulle, C., and Pace, N. R. (1994). Further perspective on the catalytic core and secondary structure of ribonuclease P RNA. *Proc. Natl. Acad. Sci. USA*, **91**, 2527–31.

Haseloff, J. and Gerlach, W. L. (1988). Simple RNA enzymes in new and highly specific endoribonuclease activities. *Nature*, **334**, 585–91.

Pley, H. W., Flaherty, K. M., and McKay, D. B. (1994). Three-dimensional structure of a hammerhead ribozyme. *Nature*, **372**, 68–74.

Symons, R. H. (1994). Ribozymes. *Curr. Opin. Struct. Biol.*, **4**, 322–30.

6.6

Böck, A., Forchhammer, K., Heider, J. and Baron, C. (1991). Selenoprotein synthesis: and expansion of the genetic code. *Trends Biochem. Sci.*, **16**, 463–7.

Breitenberger, C. A. and RajBhandary, U. L. (1985). Some highlights of mitochondrial research based on analysis of *Neurospora crassa* mitochondrial DNA. *Trends Biochem. Sci.*, **10**, 478–83.

Cavarelli, J. and Moras, D. (1993). Recognition of tRNAs by aminoacyl-tRNA synthetases. *FASEB J.*, **7**, 79–86.

Crick, F. H. C. (1966). The genetic code: III. *Sci. Amer.*, **215**, Oct. 55–62.

Hou, Y-M., Francklyn, C., and Schimmel, P. (1989). Molecular dissection of a transfer RNA and basis for its identity. *Trends Biochem. Sci.*, **14**, 233–7.

Khorana, H. G. (1973). Nucleic acid synthesis in the study of the genetic code. *Nobel Lectures: Physiology or Medicine 1963–1970*, pp. 341–69. American Elsevier, New York.

Nomura, M. (1984). The control of ribosome synthesis. *Sci. Amer.*, **250**, 102–14.

Stryer, L. (1988). *Biochemistry*, 3rd edn, pp. 91–116. Freeman, San Francisco.

Watson, J. D., Hopkins, N. H., Roberts, J. W., Steitz, J. A., and Weiner, A. M. (1988). *Molecular biology of the gene*, pp. 360–462. Benjamin-Cummings, Menlo Park, CA.

Zubay, G. L. (1988). *Biochemistry*, 2nd edn, pp. 928–73. Macmillan, New York.

COVALENT INTERACTIONS OF NUCLEIC ACIDS WITH SMALL MOLECULES

The simple purpose of this chapter is to provide an outline of the more important examples of covalent interactions of small molecules with nucleic acids. Topics relevant to modifications of intact nucleic acids have been chosen, especially as they relate to mutagenic and carcinogenic effects. While much of the early information has come from studies on nucleosides, in many cases the net effect of a reagent on an intact nucleic acid may be quite different from either the sum or the average of its interactions with separate components.

Above all, we have to recognize that studies on the subtle effects of DNA and RNA secondary and tertiary structures on covalent interactions are only in their infancy.

7.1 Hydrolysis of nucleosides, nucleotides, and nucleic acids

Nucleic acids are readily denatured in aqueous solution at extremes of pH or on heating. The phosphate ester bonds are only slowly hydrolyzed (Section 3.2.2), but the *N*-glycosylic bonds are more labile. Purine nucleosides are cleaved faster than pyrimidines while ribonucleosides are more stable than deoxyribonucleosides. Thus, dA and dG are hydrolysed in boiling 0.1 M hydrochloric acid in 30 min, rA and rG require 1 hour with 1 M hydrochloric acid at 100°C, while rC and rU have to be heated similarly with 12 M perchloric acid (Fig. 7.1). It follows that the glycosylic bonds of carbocyclic nucleoside analogues (Section 3.1.2), which cannot donate electrons from a 4′-oxygen, are much more stable to acidic hydrolysis and this property has been used to advantage in many applications.

Formic acid has been used to prepare apurinic acid, which has regions of polypentose phosphate diesters linking pyrimidine oligonucleotides. Such phosphate diesters are relatively labile since the pentose undergoes a *β*-elimination in the presence of secondary amines such as diphenylamine. This gives tracts of pyrimidine oligomers with phosphate monoesters at both 3′- and 5′-ends. Total acidic hydrolysis with minimum degradation of the four bases is best achieved with formic acid at 170°C.

DNA is resistant to alkaline hydrolysis but RNA is easily cleaved because of the involvement of its 2′-hydroxyl groups (Section 3.2.2).

7.2 Reduction of nucleosides

Purine and pyrimidine bases are sufficiently aromatic to resist reduction under the mild conditions used, for example, in the hydrogenolysis of benzyl or phenyl phosphate esters.

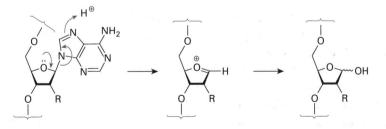

Fig. 7.1 Mild acidic hydrolysis of purine glycosides in DNA (R = H) and RNA (R = OH).

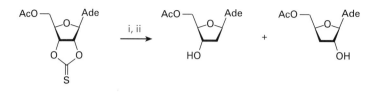

Fig. 7.2 Barton reduction of adenosine to 2'- and 3'-deoxyadenosines.
Reagents: (i) Bu_3SnH, DMA, AIBN; (ii) NH_3, MeOH.

However, hydrogenation with a rhodium catalyst converts uridine or thymidine into 5,6-dihydropyrimidines. Alternatively, sodium borohydride in conjunction with ultraviolet irradiation gives the same products which can lead on by further reduction in the dark and cleavage of the heterocyclic ring.

Reduction of ribonucleosides directly to 2'-deoxynucleosides can be accomplished by one of the several Barton procedures involving tributyltin hydride. A nice example is the synthesis of a mixture of 2'- and 3'-deoxyadenosines which are easily separable (Fig. 7.2). This type of reduction has been widely employed to transform various extensively modified ribonucleosides and their nucleotide analogues into the corresponding deoxyribonucleosides.

2',3'-Dideoxynucleosides are valuable for use in DNA sequence analysis and also showed some promise for AIDS therapy, both features being related to their chain-terminator activity (Section 5.2). One synthesis involves hydrogenation of 2',3'-unsaturated nucleosides or an appropriate precursor (Fig. 7.3).

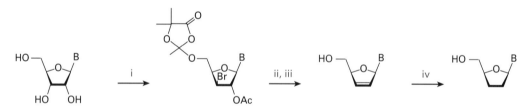

Fig. 7.3 Synthesis of 2',3'-dideoxynucleosides by reduction.
Reagents: (i) $Me_2C(OAc)COBr$; (ii) Cr^{2+}, $(CH_2NH_2)_2$, -75°; (iii) KOH; (iv) H_2/PdC.

7.3 Oxidation of nucleosides, nucleotides, and nucleic acids

In general, nucleoside bases are destroyed by strong oxidizing agents such as potassium permanganate. Hydrogen peroxide and organic peracids can be used to convert adenosine into its *N*-1-oxide and cytidine into its *N*-3-oxide while the 5,6-double bond of thymidine is a target for oxidation by osmium tetroxide, forming a cyclic osmate ester of the *cis*-5,6-dihydro-5,6-glycol. This reaction is sensitive to steric hindrance and has been employed to study the details of cruciform structure in DNA (Section 2.3.3).

Recent studies on the oxidation of DNA with hypochlorate and similar oxidants has identified the formation of 8-hydroxyguanine residues as an important mutagenic event.

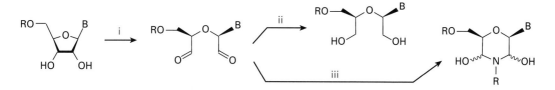

Fig. 7.4 Periodate cleavage of 3′-terminal nucleotide and subsequent modification. Reagents: (i) IO_4^-, pH 4.5; (ii) $NaBH_4$; (iii) RNH_2.

The pentoses are sensitive to free radicals produced by the interaction of hydrogen peroxide with Fe(III) or by photochemical means, which causes strand scission (Section 7.9.1). Dervan has made this process sequence specific *in vitro* by linking radical-generating catalysts to a groove-binding agent (Section 8.8.2) and he has also employed it as a 'footprinting' device by linking an Fe–EDTA complex to an intercalating agent such as methidium. Other useful oxidative reactions of the pentose moieties are typical of the chemical reactions of primary alcohols and *cis*-glycols. In particular, periodate cleavage of the ribose 2′,3′-diol gives dialdehydes that have been either stabilized by reduction to give a ring-opened diol or condensed with an amine or with nitromethane to give ring-expanded products (Fig. 7.4). Such procedures have been adopted frequently to make the 3′-terminus of an oligonucleotide inert to 3′-exonuclease degradation.

Summary

The glycosylic bond in the nucleosides is hydrolysed by acids and shows increasing stability: dA, dG < rA, rG < dC, dT < rC, rU ≪ carbocyclic nucleosides. RNA is readily hydrolysed in alkali *via* rapid formation of nucleoside 2′,3′-cyclic phosphates.

The 5,6-double bond of thymine is the most sensitive site for reduction of the bases. The pentoses are rather more difficult targets for reduction though some selective procedures provide access to unnatural deoxy-pentose nucleosides.

With appropriate oxidizing agents, nucleic acids can be oxidized at the thymine 5,6-double bond, at C-8 of guanine, or in the pentose. The latter is especially useful for effecting strand scission by free radical oxidizing agents.

7.4 Reactions with nucleophiles

In general, nucleophiles can attack the pyrimidine residues of nucleic acids at C-6 or C-4 while reactions at C-6 of adenine or C-2 of guanine are more difficult. α-Effect nucleophiles, such as hydrazine, hydroxylamine, and bisulfite, are especially effective reagents for nucleophilic attack on pyrimidines.

Hydrazine adds to uracil and cytosine bases first at C-6 and then reacts at C-4. The bases are converted into pyrazol-2-one and 3-aminopyrazole respectively leaving an *N*-ribosylurea

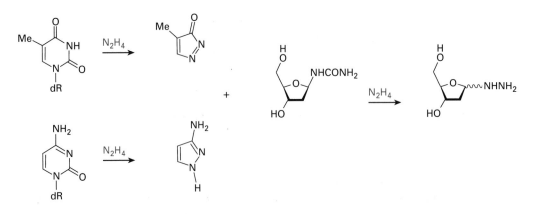

Fig. 7.5 Hydrazinolysis of pyrimidine nucleosides.

that can react further to form a sugar hydrazone. These reactions are used in the Maxam–Gilbert method of DNA sequence determination (Section 10.6) where subsequent treatment of the modified ribose residues with piperidine causes β-elimination of both 3′- and 5′-phosphates at the site of depyrimidination (Fig. 7.5).

Under mild, neutral conditions, cytosine and its nucleosides react with hydroxylamine, semicarbazide, and methoxylamine to give N^4-substituted products. The mechanism of this process involves reaction with the cytosine cation, as illustrated for hydroxylamine (Fig. 7.6). The formation of N^4-hydroxydeoxycytidine is an important mutagenic event in DNA since this base exists to a significant extent in the unusual imino-tautomeric form (Section 2.1.2) and is capable of mispairing with adenine.

A third addition reaction at C-6 of cytosine and uracil residues involves the bisulfite anion, which adds reversibly. The resulting non-aromatic heterocycles can undergo a variety of chemical changes of which the most important are: transamination of cytosine at C-4 by various primary or secondary amines, hydrogen isotope exchange at C-5, and deamination of cytosine to uracil.

This third process explains the basis for the mutagenicity and cytotoxicity of bisulfite (which is equivalent to aqueous sulfur dioxide). Such mutations are best carried out at pH

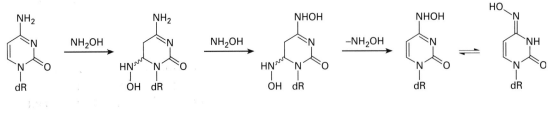

amino-   tautomer   imino-

Fig. 7.6 Reaction of hydroxylamine with deoxycytidine leading to tautomerization of N^4-hydroxy-2′-deoxycytidine.

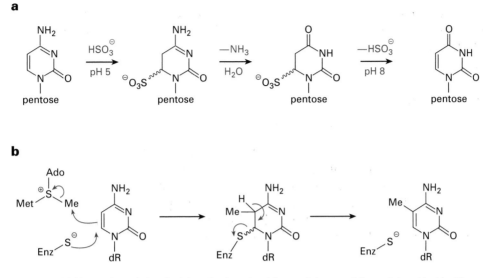

Fig. 7.7 (**a**) Mechanism of chemical deamination of cytidine and deoxycytidine catalysed by bisulfite. (**b**) Schematic mechanism for the restriction methylation of deoxycytidine by *S*-adenosyl methionine catalysed by M. *Hha*I.

5–6 to bring about deamination and then at pH 8–9 to eliminate bisulfite (Fig. 7.7a). Another substitution reaction which requires the addition of a thiol to C-6 is the cysteine-catalysed *in vitro* incorporation of deuterium at C-5 into cytosine.

This addition of sulfur to C-6 of the pyrimidine ring is employed in the biological methylation of pyrimidines. Santi has established that the mechanism of action of thymidylate synthase involves addition of a catalytic cysteine to C-6 of the deoxyuridylate in conjunction with electrophilic addition of the methylene group of tetrahydrofolate to C-5. It is this process which results in the suicide substrate activity of the anticancer drug, 5-fluorouracil (Section 4.7.1). In a similar fashion, Roberts has shown that cytosine-specific DNA restriction methylases, such as M.*Hha*I, add a catalytic thiol to C-6 of a deoxycytidine residue in conjunction with transfer of the methyl group of *S*-adenosylmethionine to C-5 (Fig. 7.7b, and cf. Section 9.3.7).

Summary

Cytosine is the most sensitive of the bases to nucleophilic attack, especially by α-effect nitrogen nucleophiles such as hydrazine, hydroxylamine, etc. Under mild conditions, transamination is easily accomplished. Under more vigorous conditions, transformation of the pyrimidine ring for C, T, and U gives a pyrazole with extrusion of a pentosyl-urea moiety.

Bisulfite addition to C-6 of cytosine is reversible, but the adduct can easily be deaminated with provides a mutagenic route from C to U.

The nucleophilic addition of a cysteine residue to C-6 of a pyrimidine deoxynucleotide is the initial step in C-5 alkylation catalysed by thymidylate synthase and by cytosine-specific restriction DNA methylases.

7.5 Reactions with electrophiles

7.5.1 Halogenation of nucleic acid residues

Chlorine and bromine react directly with uracil, adenine, and guanine and so offer easy routes to 5-chloro- (or bromo-) uridines and 8-chloro- (or bromo-) purines (the latter are readily converted into 8-azidopurine nucleosides for use as photoaffinity labels). It is much more difficult to control the use of elemental fluorine, though fluorine gas can be used in anhydrous acetic acid **with care** to prepare 5-fluorouracil and 5-fluorouridine. The 5-iodo-uridines are best made by the action of iodine on 5-mercuri- or 5-palladium-uridines.

7.5.2 Reactions with nitrogen electrophiles

The normal reaction of nitrous acid in the deamination of primary amino groups converts deoxyadenosine into deoxyinosine, deoxycytidine into deoxyuridine, and deoxyguanosine into deoxyxanthosine. In each case, the reaction provides an alternative tautomeric form of the base which can lead to altered base-pairing. The **transitions** $dA \cdot dT \rightarrow dI \cdot dC$ and $dC \cdot dG \rightarrow dU \cdot dA$ are characteristic of the mutagenic action of nitrous acid (Fig. 7.8).

Aromatic nitrogen cations are a second important class of nitrogen electrophiles. These species are derived either from aromatic nitro compounds by metabolic reduction or from aromatic amines by metabolic oxidation. In both cases, an intermediate hydroxylamine species interacts with purine residues in DNA or RNA either on C-8 or at N-7 (Section 7.6.1).

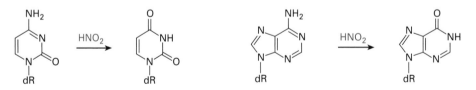

Fig. 7.8 Mutagenic deamination of dC → dU and of dA → dI by nitrous acid.

7.5.3 Reactions with carbon electrophiles

There is a huge variety of reagents which form bonds from carbon to nucleic acids. The simplest are species like formaldehyde and dimethyl sulfate. The most complex are carcinogens such as benzo[a]pyrene which require transformation by three consecutive metabolic processes before they can become bound to purine bases in DNA or RNA. Not surprisingly, there is a wide range in selectivity for the sites of attack of these reactive species, some of which have been rationalized in terms of Pearson's **HSAB theory** (HSAB = hard and soft acids and bases). Frontier orbital analysis can provide a more rigorous picture of the problem, but requires a deeper insight into theoretical chemistry. Other relevant factors may relate to the degree of steric access of the electrophile to exposed bases or to intercalation of reagents prior to bonding to nucleotide residues.

Formaldehyde

Covalent interactions of formaldehyde with RNA and its constituent nucleosides take place in a specific reaction of the amino bases. Formaldehyde first adds to the N^6-amino group of adenylate residues to give a 6-(hydroxymethylamino)purine and with guanylate residues to give a 2-(hydroxymethylamino)-6-hydroxypurine. These labile intermediates can react slowly with a second amino group to give cross-linked products which have a stable methylene bridge joining their two amino groups. All three possible species, pAdo-CH_2-pAdo, pAdo-CH_2-pGuo, and pGuo-CH_2-pGuo, have been isolated from RNA which has been treated with formaldehyde and then hydrolysed with alkali (Fig. 7.9). The detailed mechanism of formaldehyde mutagenicity is not yet clear.

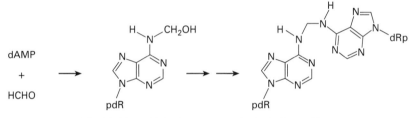

Fig. 7.9 Formaldehyde mutagenesis of adenine residues.

Alkylating agents

All of the nitrogen and oxygen residues of the nucleic acids bases, with the exception of purine-N^9 and pyrimidine-N^1, can be alkylated in aqueous solution at neutral pH. 'Soft' electrophiles, such as dimethyl sulfate (DMS), methyl methanesulfonate (MMS), and alkyl halides (such as MeI) react in an S_N2-like fashion. Alkylation takes place mainly at nitrogen sites with a general selectivity $G-N^7 > A-N^1 > C-N^3 > T-N^3$. A key parameter for 'softness' is the very low ratio of $G-O^6 : G-N^7$ methylation of 0.004:1. In double-stranded DNA, the major alkylation site for DMS with adenines is at N-3 with lesser substitution at N-7. 'Hard' electrophiles, such as N-methyl-N-nitrosourea (MNU) and its ethyl homologue ENU, are S_N1-like alkylating agents. MNU methylation of phosphate esters in nucleic acids can account for up to 50 per cent of total alkylation and also gives higher ratios for $G-O^6 : G-N^7$ products, ranging from 0.08 in liver to 0.15 in brain DNA. Other sites for O-alkylation include $T-O^2$, $T-O^4$, and $C-O^2$.

In contrast to the C-methylation of nucleic acids by various enzymes, such as thymidylate synthetase (Section 4.4), products arising from C-alkylation using chemical agents have not been detected.

Many alkylating agents are known to be primary carcinogens (agents that act directly on nucleic acids and that do not require metabolic activation). An extensive list includes DMS, MMS, and their ethyl homologues, β-propiolactone, 2-methylaziridine, 1,3-propanesultone, and ethylene oxide. The list of bifunctional carcinogenic agents includes bischloromethyl ether, bischloroethyl sulfide, and epichlorohydrin along with such 'first generation' anti-cancer agents as myleran, chlorambucil, and cyclophosphamide (Section 4.7.1). In general, 'hard' alkylating agents have been found to be a greater carcinogenic hazard than 'soft' ones.

Bis-(2-chloroethyl)sulfide

This is the mustard gas of World War I as well as of more recent conflicts. It is a typical bifunctional alkylating agent in addition to being a proven carcinogen of the respiratory tract. In the early 1960s, Brookes and Lawley showed that it can cross-link two bases either in the same or in opposite strands of DNA. The typical products isolated (Fig. 7.10) have a five-atom bridge between N-7 of a guanine joined either to a second guanine or to adenine-N^1 or to cytosine-N^3. Similar products are formed on alkylation of DNA with 2-methylaziridine. These reagents show little sequence selectivity though nitrogen mustard, $MeN(CH_2CH_2Cl)_2$, shows some preference for alkylation of internal residues in a run of guanosines.

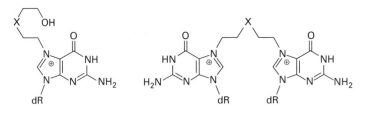

Fig. 7.10 Mono- and bi-functional products of dG with sulfur (X = S) and nitrogen (X = NH, NMe) mustard reagents.

Chloroacetaldehyde

This reagent combines the reactivity of formaldehyde and the alkyl halides. It reacts with adenine and cytosine residues, converting them into etheno-derivatives which have an additional five-membered ring fused on to the pyrimidine ring (Fig. 7.11). These modified bases are strongly fluorescent and have been used to probe the biochemical and physiological modes of action of a range of adenine and cytosine species.

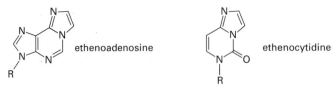

R = ribofuranosyl

Fig. 7.11 Fluorescent ethenoderivatives of adenosine and cytidine from chloroacetaldehyde.

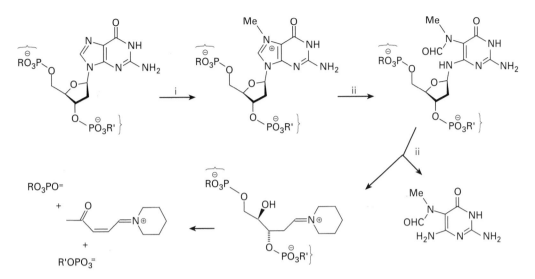

Fig. 7.12 Mechanism for the cleavage of a deoxyoligonucleotide strand at G residues by dimethyl sulfate and hot aqueous piperidine in the Maxam–Gilbert sequencing of DNA. Reagents: (i) dimethyl sulfate; (ii) piperidine, 90°C, 30 min.

Dimethyl sulfate

The methylation of deoxyguanosine is a major feature of the Maxam–Gilbert method for DNA sequence determination (Section 10.6). Following the formation of the 7-methyldeoxyguanosine residue, treatment of the oligonucleotide with 1 M piperidine at 90°C for 30 min leads to opening of the imidazole ring followed by glycosylic bond cleavage and β-elimination of the phosphate residue, effecting strand cleavage either side of the dG residue (Fig. 7.12).

7.5.4 Metallation reactions

Mercury(II) acetate and chloride readily substitute C-5 of uridine and cytidine nucleosides. The products can be converted into organo-palladium species that are useful intermediates in the synthesis of a range of 5-substituted pyrimidine nucleosides. These include C-5 allyl-, vinyl-, halovinyl-, and ethynyluridines, all of which have been much studied for possible antiviral activity.

One of the most important recent applications of such chemistry is for the synthesis of fluorescent chain-terminating dideoxynucleotides, used in a rapid DNA sequencing method developed by Du Pont chemists (Fig. 7.13). 5-Iodo-2′,3′-dideoxyuridine is coupled to N-trifluoroacetylpropargylamine using palladium(0) catalysis and the resulting amine is then condensed with a protected succinylfluorescein dye. Deprotection provides the dideoxythymidine terminator species T-526 (which has a fluorescent emission maximum at 526 nm). Related fluorescent derivatives of dideoxycytidine, C-519, dideoxyadenosine, A-512, and dideoxyguanosine, G505, provide a family of chain-terminators which are incorporated with

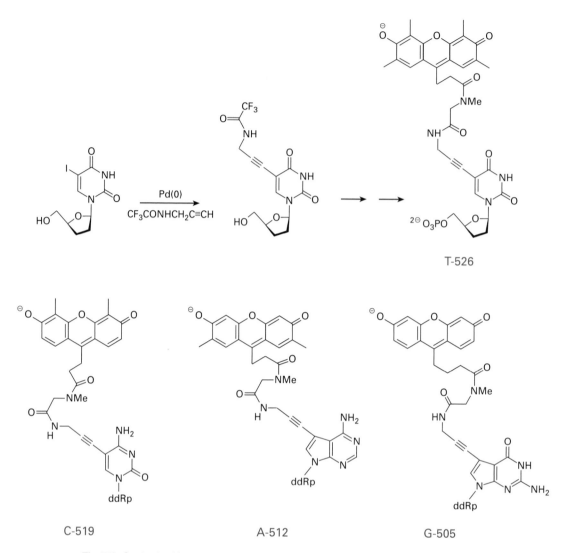

Fig. 7.13 Synthesis of fluorescent base T–526 using Pd(0) catalysis; structures of fluorescent dideoxynucleotides related to A, C, and G for use in rapid, single-lane DNA sequence analysis.

efficiencies comparable to those of unsubstituted ddNTPs and can be used for rapid, single-lane DNA sequencing (Section 10.6).

Rosenberg's discovery of the cytotoxicity of *cis*-diaminedichloroplatinum(II) has been carefully developed to make cisplatin the reagent of choice for the successful chemotherapy of testicular (and other forms of) cancer. The reagent bonds to N-7 in one guanine residue and then joins it to a second purine. It binds selectively to d(pGpG) and d(pApG) sequences, but not to d(pGpA) sites, and so forms intra-strand cross-links. Cisplatin is also capable of binding to two guanines separated by another base, as in d(pGpNpGpG). X-Ray structures have been solved for its complexes with d(pGpG) and d(CpGpGp) (Fig. 7.14) and show some agreement with solution structures determined by NMR. The platinum is *cis*-linked to N-7 in both guanines which lie in planes almost at right angles and this breaks up the base-stacking and the base-pairing patterns of the DNA helix. It is noteworthy that *trans*-diaminedichloroplatinum binds to DNA as well as does the *cis*-isomer and while the resulting lesions may be more readily excised by repair enzymes than in the case of cisplatin, the *trans*-isomer has useful biological activity.

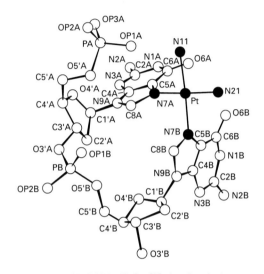

Fig. 7.14 Molecular structure of *cis*-[Pt(NH₃)₂d(pGpG)] showing the two guanines in perpendicular planes (adapted from Sherman, S. E., Gibson, D., Wang, A. H.-J., and Lippard, S. J. (1985). *Science*, **230**, 412–7. Copyright (1985) AAAS).

Summary

The pyrimidine bases are less aromatic than the purines and their chemistry is dominated by addition reactions to the 5,6 double bond, some of which have a mutational effect.

Electrophiles are capable of attacking most nitrogens and oxygens of the bases: the position of attack of soft electrophiles is N^7-G > N^1-A > N^3-C > N^3-T. For hard electrophiles there is an increasing proportion of alkylation on oxygen, especially at O^6-G. Aromatic nitrogen electrophiles bond to C-8 of guanine.

Many carcinogens directly act on DNA as alkylating agents, resulting in base modification. Formaldehyde adds to the adenine N^6-amino group while bifunctional alkylating agents, such as cisplatin and nitrogen mustards, mainly link two guanines together through N-7.

Methylation of deoxyguanosine residues at N-7 is a key feature of the Maxam–Gilbert method for sequence determination of DNA.

7.6 Reactions with metabolically activated carcinogens

Many synthetic chemicals and natural products are known to be carcinogens or mutagens. While these do not react directly with nucleic acids *in vitro*, they are transformed *in vivo* by metabolic processes to give electrophiles that bond covalently to DNA, RNA, and also to proteins. Most of these transformations are carried out by the mixed-function cytochrome P-450 oxidases, enzymes whose 'proper' function seems paradoxically to be directed towards the detoxification of alien compounds! The following four classes of metabolically activated compounds are representative of an intensive study of a problem of very grave significance.

7.6.1 Aromatic nitrogen compounds

Investigations of the binding of *N*-aryl carcinogens to nucleic acids began in the 1890s with an investigation of the epidemiology of bladder cancer among workers in a Basel dye factory. The list of chemicals banned today is extensive, but by no means complete, and some examples of proscribed aromatic amines, nitro compounds, and azo dyes are illustrated (Fig. 7.15).

Aromatic amines of this sort are substrates for oxidation by cytochrome P-450 isozymes which give either phenols, which are inactive and safely excreted, or hydroxylamines. Conjugation of the latter by sulfotransferase or acetate transferase enzymes converts these proximate carcinogens into ultimate carcinogens that bind covalently to nucleic acid bases, especially guanine.

The competition between such alternative 'safe' and 'hazardous' metabolic processes is illustrated for 2-acetylaminofluorene (Fig. 7.16). The guanine nucleoside adducts have been isolated and identified and are formally derived from a hypothetical nitrenium ion. This, as an ambient cation, bonds either from nitrogen to guanine-C^8 or from carbon to guanine-N^2.

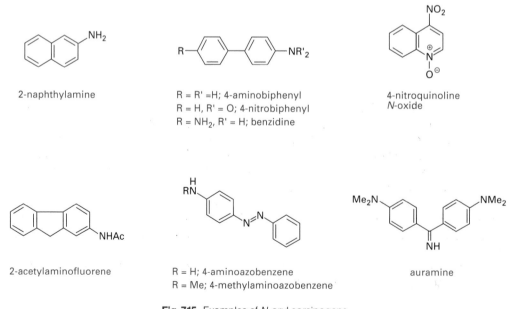

2-naphthylamine

R = R' =H; 4-aminobiphenyl
R = H, R' = O; 4-nitrobiphenyl
R = NH₂, R' = H; benzidine

4-nitroquinoline
N-oxide

2-acetylaminofluorene

R = H; 4-aminoazobenzene
R = Me; 4-methylaminoazobenzene

auramine

Fig. 7.15 Examples of *N*-aryl carcinogens.

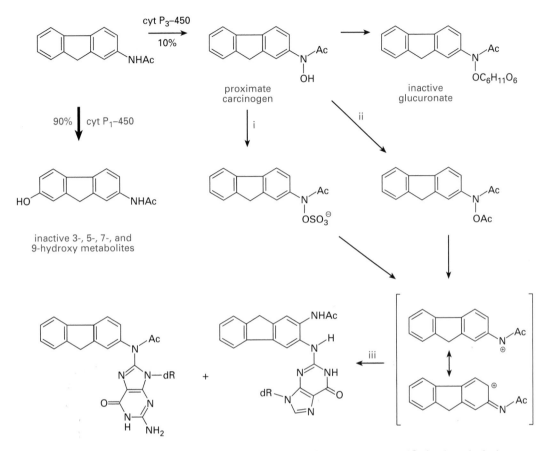

Fig. 7.16 Metabolic activation of 2-acetylaminofluorene, AAF, and its binding to dG *via* a hypothetical nitrenium intermediate.
Processes: (i) sulfotransferase; (ii) acetyl transferase; (iii) DNA binding *in vitro* or *in vivo*.

Similar adducts have been identified for many other amines. Azo dyes are first cleaved *in vivo* by an azoreductase to aromatic amines and then activated as described above.

One example of considerable significance is the mutagenicity of two types of heterocyclic amine which are found in cooked meats, where they are formed by the pyrolysis of tryptophan and glutamine. Sugimura has identified guanine adducts which are generated by metabolic activation and binding to nucleic acids (Fig. 7.17). Thus, grilled beef has been estimated to contain nearly 1 ppm of Trp-P-1, while up to 80 ng of this carcinogen has been isolated from the smoke of a single cigarette!

Aromatic nitro compounds are present in diesel engine emission, urban air particles, and photocopier black toners, and some have been identified as mutagens in the Ames test. They can be reduced to aryl hydroxylamines by anaerobic bacteria in the gut, by xanthine oxidase, or by cytochrome P-450 reductase to give substrates for the processes described above. The best-studied example is 4-nitroquinoline N-oxide, which is first reduced to a hydroxylamine and then binds to DNA *in vivo*, forming characteristic guanine adducts (Fig. 7.18).

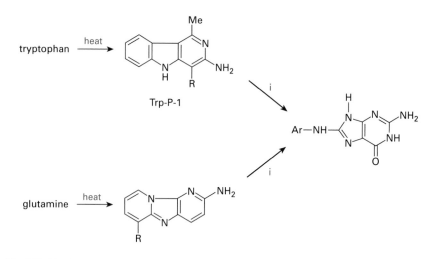

Fig. 7.17   Metabolic activation of heterocyclic amines from cooked food and their binding to DNA;
R = H or Me.
Processes: (i) cytochrome P-450, DNA binding, hydrolysis.

7.6.2 *N*-Nitroso compounds

Nitrosoureas, nitrosoguanidines, and nitrosourethanes hydrolyse to give methyldiazonium hydroxide, Me–N=N–OH, which is a 'hard' methylating agent. (In the case of methyl *N*-nitrosoguanidine (MNNG), thiol groups may catalyse the *in vivo* methylation of DNA by this carcinogen). The same methylating species is produced as a result of the cytochrome P-450 oxidation of a wide range of *N*-methyl-nitrosamines, of which dimethylnitrosamine was the first to be identified as a carcinogen in 1956. The common metabolic pattern is hydroxylation of one alkyl residue to form a carbinolamine which breaks down to give methyldiazohydroxide (Fig. 7.19). Very many *N*-nitroso compounds of this type have proved to be

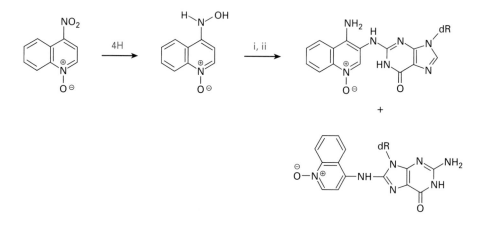

Fig. 7.18   Reductive activation of 4-nitroquinoline *N*-oxide and products resulting from its binding to DNA *in vivo*.
Reagents: (i) DNA *in vivo*; (ii) hydrolysis.

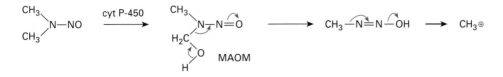

Fig. 7.19 Cytochrome P-450 oxidation of dimethylnitrosamine and its conversion into methyl azo-oxymethanol (MAOM) *en route* to DNA methylation.

carcinogenic in animals and lead to methylation, ethylation, or propylation of DNA, as described above (Section 7.5.3).

7.6.3 Polycyclic aromatic hydrocarbons

The polycyclic aromatic hydrocarbons (PAHs) provided the first example of an industrial carcinogen, benzo[*a*]pyrene (BaP). Its identification marked the first stage in the molecular analysis of hydrocarbon carcinogenesis, which had begun with Percival Pott's study of scrotal cancer in chimney sweeps in 1775. BaP becomes covalently bound to DNA *in vivo* following a series of three metabolic changes. In the first, cytochrome P_1-450 adds oxygen to BaP to give the two enantiomers of BaP-7,8-epoxide. Next, these are used as substrates for an epoxide hydrolase that converts them into the two enantiomeric *trans*-dihydrodiols. Finally, both diols are substrates for cytochrome P_1-450 and are converted into three of the four possible stereoisomers of the 9,10-epoxide, BPDE (Fig. 7.20).

The carcinogenicity and DNA-binding capability of such dihydrodiol epoxides is closely linked to 'Bay Region' architecture, so called because of the concave nature of this edge of the PAH, which appears to be strongly recognized by the metabolizing enzyme. Among the

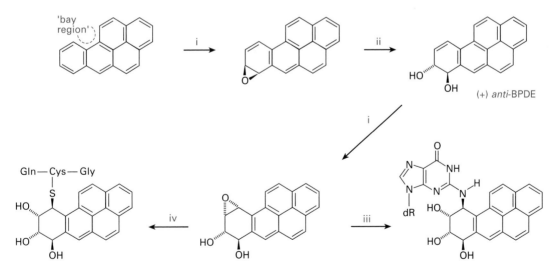

Fig. 7.20 Metabolic activation of B[a]P to BPDE (major sterioisomer illustrated) and its binding to DNA *in vivo* to give guanine adducts (major product shown).
Processes: (i) Cyt P_1-450; (ii) epoxide hydrolase; (iii) DNA binding *in vivo*; (iv) gluthathione-*S*-transferase.

products that have been characterized are adducts with guanine-N^2 (the major product), guanine-N^7, guanine-O^6, and adenine-N^6 (minor products). These are formed as a result of a rapid intercalation of the BPDE into $d(A \cdot T)_n$-rich parts of the DNA helix, which is manifest as a red-shift in UV absorption of the hydrocarbon and a negative CD spectrum for the complex. A rate-determining protonation of the C-10 hydroxyl group then leads to the formation of a carbocation that reacts predominantely (90 per cent) with water to give the harmless 7,8,9,10-tetra-ol or less frequently (10 per cent) binds to a proximate base, most often a dG residue.

The resulting covalent adducts appear to be of two distinct types. The minor 'site I' adducts have the hydrocarbon still intercalated in an intact DNA helix. The major 'site II' adducts appear to have the hydrocarbon lying at an angle to the helix axis, either in the minor groove of a DNA helix or forming a wedge-shaped intercalation complex. Similar results have been found for chrysene, while the Bay Region dihydrodiol epoxide of 3-methylcholanthrene appears to be too bulky to intercalate into DNA.

This type of epoxidation is not restricted to synthetic chemicals. One of the most potent groups of carcinogens are the aflatoxins, which are fungal products from *Aspergillus flavus*. A dose of less than 1 ppm of aflatoxin B_1 can cause lung, kidney, and colon tumours in rats and is directly attributable to its oxidation to an epoxide that binds covalently to guanine residues in DNA (Fig. 7.21).

While both *endo-* and *exo*-epoxides are formed metabolically, only the *exo*-isomer (Fig. 7.21) is mutagenic. It seems likely that this metabolite intercalates into the DNA helix with optimal orientation for an S_N2 reaction with N-7 of a proximate guanine residue. By contrast, intercalation of the non-mutagenic *endo*-isomer places its epoxide in an orientation that precludes reaction.

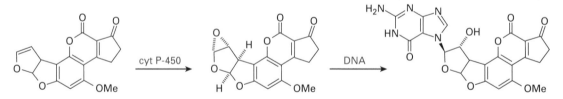

Fig. 7.21 Metabolic oxidation of aflatoxin B_1 and binding of the *exo*-epoxide to DNA.

Summary

Many carcinogenic compounds are converted by metabolic processes into alkylating agents which form covalent bonds to nucleic acid bases. For aromatic nitrogen compounds the hazardous species are aryl hydroxylamine derivatives, which can be formed either from amines by biological oxidation or from nitro compounds by metabolic reduction. Activation of azo dyes involves first reduction and then oxidation.

Dialkylnitrosamines are converted into alkyldiazonium hydroxides as a consequence of α-hydroxylation by cytochrome P-450, and these behave as 'hard' alkylating agents.

Many polycyclic aromatic hydrocarbons and some natural products, for instance the aflatoxins, are also metabolized by cytochrome P-450 oxygenation and lead directly or indirectly to epoxides as ultimate carcinogens.

7.7 Reactions with anti-cancer drugs

A large number of 'first generation' anti-cancer drugs were designed to combine a simple alkylating function such as a nitrogen mustard, an aziridine, or an alkanesulphonate ester with another function designed to direct the agent towards the target tissue. Most of these compounds turned out to be less tumour-selective than one might have hoped and, what is worse, many of them have proved to be carcinogens which have given rise to new tumours some time after the termination of chemotherapy for the original cancer. As a result, their general use is now viewed with some suspicion.

A group of 'second generation' compounds has emerged, many of them natural products, but now augmented by a growing number of rationally designed, synthetic drugs. Their common feature is that they appear to form an initial physical complex with DNA before covalently bonding to it. This heterogeneous group of compounds includes aziridines such as mitomycin C, several pyrrolo[1,4]benzodiazepines, and spirocyclopropanes such as CC-1065. Their vital purpose is to kill bacteria by disrupting the synthesis of DNA and RNA, but many of them have also shown useful anti-tumour activity, which must arise from selective toxicity. This can be attributed to DNA-binding specificity or to preferential metabolic activation by tumour cells.

7.7.1 Aziridine antibiotics

An assortment of naturally occurring antibiotics, each having an aziridine ring, has been isolated from *Streptomyces caespitonis*. The most interesting of them in clinical terms is **mitomycin C**. This compound requires enzymatic reduction of its quinone function to initiate the processes that cause it to alkylate DNA. It seems likely that the second step is elimination of methanol which potentiates either monofunctional or bifunctional alkylation (Fig. 7.22). The antibiotic has been shown to interact with DNA at O^6-G $> N^6$-A $> N^2$-G and forms one cross-link for about every ten monocovalent links. The primary process is bonding of the 2-amino group of a guanine residue to C-1 of the reductively activated mitomycin (Fig. 7.22). This reaction shows a selectivity for 5'-CG sequences. Cross-linking is completed by alkylation of the 2-amino group of the second guanine to C-10 of the mitomycin. This has been accurately analysed by Patel in NMR studies on the adduct of mitomycin C to the hexamer d(TACGTA) · d(TACGTA) in which the two guanines are crosslinked with the mitomycin molecule positioned in the minor groove of the duplex.

Many drugs which act on DNA exhibit a requirement for reductive activation, including adriamycin, daunomycin, actinomycin, streptonigrin, saframycin, bleomycin (Section 8.7.1), and tallysomycin in addition to mitomycin C. While there is no common factor uniting the chemistry of DNA modification by these agents, the fact that tumour tissues seem to have a

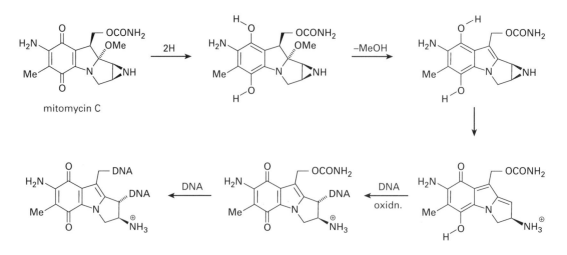

Fig. 7.22 Activation of mitomycin C by metabolic reduction and bifunctional alkylation of DNA at the 2-amino group of adjacent inter-strand deoxyguanines.

higher reducing potential than normal tissue has led to the concept of bioreductive drug activation.

Carzinophilin A is also a DNA-alkylating aziridine antibiotic, though it does not appear to need reductive activation. It has been identified as the antibiotic **azinomycin B**, isolated from *Streptomyces griseofuscus*. Carzinophilin operates in the major groove of DNA, causing cross-linking between a guanine residue and a purine residue that is two bases removed in the duplex, as in the sequences d(**G**NT)·d(CN**A**) or d(**G**NC)·d(CN**G**). Good evidence suggests that alkylation is initially at N-7 of a G residue, probably involving the aziridine ring as shown (Fig. 7.23), and is followed by a slower alkylation of the nearby purine by the second alkylating function.

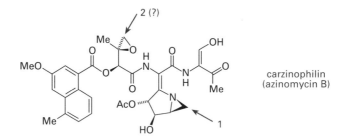

Fig. 7.23 Structure of carzinophilin (=azinomycin B) with probable sites for nucleophilic attack by N-7 of guanine (purine) residues in opposite strands of DNA.

7.7.2 Pyrrolo[1,4]benzodiazepines, P[1,4]Bs

Anthramycin and tomaymycin, along with sibiromycin and neothramycins A and B, are members of the potent P[1,4]B anti-tumour antibiotic group produced by various actinomycetes (Fig. 7.24). The first three of these compounds bind physically in the minor groove of

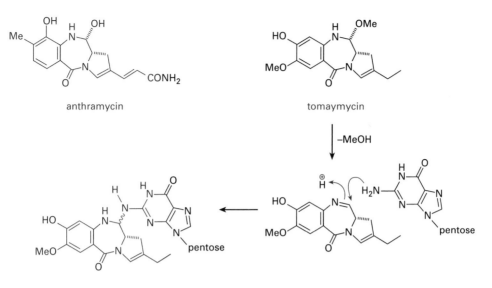

Fig. 7.24 Binding of P[1,4]B antibiotics to the N^2-amino group of guanine.

DNA where they then form covalent bonds to N^2-G, and showing a DNA sequence specificity for 5′-PuGPu sequences.

These P[1,4]Bs seem to interact with DNA in a biphasic process. Initially there is a rapid, non-covalent association which results from a close interaction of the antibiotic with the 'floor' of the minor groove of DNA (Section 8.3). Subsequent loss of water or methanol and covalent addition of N^2-G to C-11 then forms an aminal linkage that is well stabilized by favourable steric and electrostatic interactions. The structure of a condensation product between anthramycin and d(ATGCAT)$_2$ has been partially characterized by NMR analyses and much studied by molecular mechanics. Altogether, the picture that emerges is that the bonding from the guanine N-2 to C-11 is in the (S)-configuration and this makes the aromatic ring of the antibiotic lie in the DNA minor groove on the 3′-side of the modified guanine. In the case of tomaymycin, NMR studies have established the existence of two distinct tomaymycin-d(ATGCAT)$_2$ species in solution. These have the antibiotic oriented in opposite directions in the minor groove according to its (R)- or (S)-configuration at C-11. The resulting lesions appear neither to impede Watson–Crick base-pairing nor to distort the B-DNA helix structure (Fig. 7.25) so that they probably pose difficult recognition problems for DNA repair systems (Section 7.11).

Tomaymycin has been shown to induce greater conformational changes (helix bending and associated narrowing of the minor groove) than anthramycin. It thus appears that sequence-dependent conformational flexibility may be an important factor in determining the selectivity for DNA sequence binding of P[1,4]Bs.

7.7.3 Spirocyclopropane antibiotics

The structurally novel antibiotic **CC-1065** is an extremely potent cytotoxin whose biological activity has been attributed to sequence-selective binding in the minor groove of DNA fol-

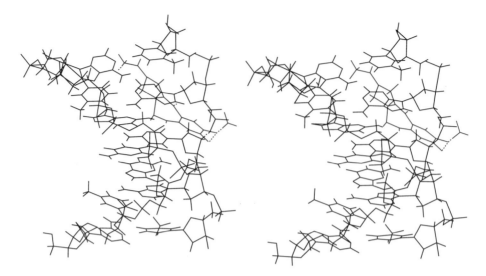

Fig. 7.25 Stereopair diagram of the 11-(*R*) condensation product of tomaymycin with d(ATGCAT)$_2$ (adapted from Cheatham, S., Kook, A., Hurley, L. H. Banklay, M. D. and Remers, W. (1988). *Journal of Medicinal Chemistry*, **31**, 583–90. Copyright (1988) American Chemical Society). (Stereopair for parallel viewing.)

lowed by covalent bonding. This involves attack by adenine N-3 on C-4 of the antibiotic to open the cyclopropane ring and aromatize the indole system (Fig. 7.26). CC-1065 is highly selective for AT rich sequences in DNA, with a strong affinity for 5′-PuNTTA and 5′-AAAAA while molecular modelling studies have suggested that the concave edge of the antibiotic interacts through close van der Waals contacts with the floor of the DNA minor groove. One consequence of alkylation of adenine at N-3 is that its glycosylic bond becomes very labile and thermal treatment of CC-1065 : DNA adducts leads to single-stranded breaks on the 3′-side of modified adenine residues (Fig. 7.26).

While CC-1065 cannot be used as an anti-tumour agent because of its unacceptably toxic side effects, a synthetic analogue, **U-71184**, has enhanced anti-tumour activity and diminished side-effects. The desired biological activity is found only in the enantiomer that corresponds to the stereochemistry of CC-1065 (Fig. 7.26).

7.7.4 Enediyne antibiotics

A range of clinically significant anti-cancer drugs can mediate oxygen-dependent cleavage of the ribose phosphate backbone of DNA. They can be broadly assigned into three classes:

- generators of reactive carbon radicals
- photogenerators of hydroxyl radicals
- metal-mediated activators of O$_2$.

The first class contains the enediyne antibiotics, whose interaction with DNA is more specific than that of many alkylating agents and is irreversible. The second class includes anti-

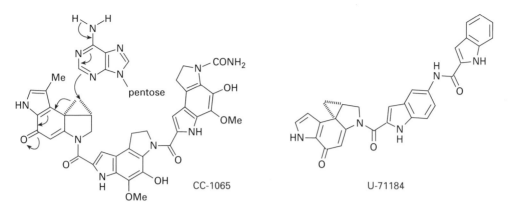

Fig. 7.26 Structures of antibiotics CC-1065 and U-71184 and mode of their covalent binding to adenine-N^3.

biotics such as tetrazomine and quinocarcin, where redox chemistry ultimately results in the reduction of oxygen to superoxide and leads to 'nicking' of DNA. The third class is well represented by the bleomycins, which are discussed in the next chapter as compounds that interact reversibly with DNA (Section 8.7).

The structure of the chromophore of the antibiotic neocarzinostatin, NCS, was established in 1985 and soon followed by those of calicheamicin γ_1, esperamicin C, dynemicin A, kedarcidin and C-1027 (Fig. 7.27a). A common structural feature of these compounds is a highly unsaturated medium-sized ring, which contains a $-C\equiv C-CH=CH-C\equiv C-$ arrangement of multiple bonds. They have thus become identified as the **enediyne antibiotics** which have taken a place at the forefront of research in biology, chemistry, and medicine because this group of compounds contains some of the most potent anti-tumour antibiotics known. They are about a thousand times more active than the clinically used adriamycin and anthracycline antibiotics.

Their mode of action is clearly linked to strand scission of DNA. This involves the formation of radicals rather than of covalent bonds with individual nucleotide residues. The basic reaction is an electrocyclisation, as reported by Bergman in 1972, which generates a 1,4-benzenoid diradical (Fig. 7.27b). In the case of the very unstable antibiotic C-1027, this process takes place merely on warming the antibiotic to 50°C in solution in ethanol. For the other compounds, the rearrangement is 'triggered' by a chemical reaction that releases strain in the enediyne ring of the antibiotics, thereby permitting the electrocyclization to take place spontaneously. In the cases of NCS, calicheamicin, and esperamicin this trigger is the attack of a thiol, possible glutathione, at the site indicated (arrows in Fig. 7.27a). In the case of dynemicin, an initial biological reduction to give a quinol is followed by opening of the epoxide whereupon nucleophilic attack at the position indicated triggers a Bergman rearrangement.

The potent anti-tumour and antibacterial activity of **neocarzinostatin** is exerted primarily through single-strand cleavage of DNA which also requires oxygen. NCS first intercalates its naphthoate rings into the DNA duplex and this positions the remainder of the molecule in the minor groove (Section 8.7.1). Following activation of the molecule by thiol addition at C-12, a Bergman cyclization generates a benzenoid diradical which removes a 5'-hydrogen atom

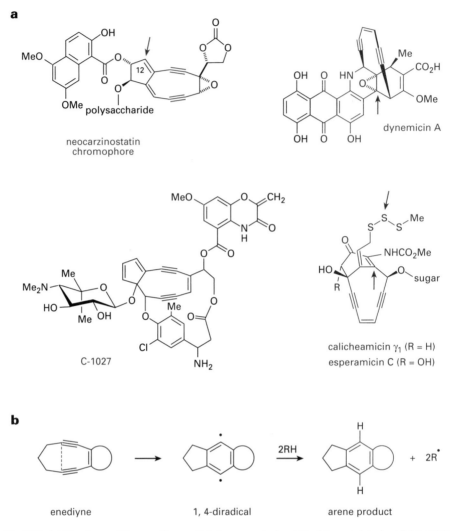

Fig. 7.27 (**a**) Structures of the enediyne antibiotics: neocarzinostatin (chromophore), calicheamicin γ_1 (chromophore), esperamicin A (chromophore), dynemicin A, and antibiotic C-1027. The site of thiol attack is indicated ($\rightarrow$) for the first three antibiotics. (**b**) The Bergman enediyne cyclization gives a 1,4-benzenoid diradical (not quite the same in the case of neocarzinostatin): this then abstracts two hydrogen atoms to give the stable arene product.

from a residue in the DNA recognition sequence. Such action takes place preferentially at adenine and thymine sites with at least 80 per cent of the DNA cleavage resulting in the formation of 5′-aldehydes of A and T residues. Less than 20 per cent of strand breaks result from pathways initiated by hydrogen abstraction from C-4′ or C-1′.

Calicheamicin γ_1 binds in the minor groove of a DNA duplex with specificity for TCCT sites. The structure in solution of the complex formed between calicheamicin and the DNA octamer d(GTGACCTG)·d(CAGGTCAC) has been determined by high resolution NMR by Dan

Kahne (Fig. 7.29). The drug binds in the narrow groove at the recognition sequence ACCT. While the drug appears not to be distorted on binding, the oligopyrimidine sequence appears to change conformation to accommodate the drug, notably by a widening of the minor groove at the CpC step. The mechanism of strand cleavage is triggered through activation by a thiol such as glutathione, which splits the trisulfide group, releasing a thiolate which adds in Michael fashion to the proximate α,β-unsaturated ketone. There is a consequent change in the geometry of the enediyne ring which favours a Bergman cyclization and generates a benzenoid diradical. This is suitably placed to abstract one 5'-hydrogen atom from the first deoxycytidine residue in the recognition sequence d(T**C**CT) and a second hydrogen atom from a deoxyribose in the opposite strand. This leads to a double strand cleavage process. **Esperamicin C** works in a similar fashion but has a low sequence selectivity and favours cleavage at T > C > A > G, giving rather more double than single strand cuts.

It seems likely that **dynemicin A** also binds to DNA by a combination of intercalation and groove binding. It is activated both by thiols (bioreduction) and by light and it also causes both single and double strand DNA cleavage.

The mechanism of strand cleavage by removal of a 5'-hydrogen is common to all of these antibiotics, as shown in Figure 7.28. Processes involving hydrogen abstraction from C-4' or C-1' are illustrated later (Fig. 7.33).

By using a combination of NMR analysis (Section 10.2.5) and molecular modelling (Section 10.5), Kahne has generated a structure for the complex formed between calicheamicin and the non-complementary octamer d[GTGACCTG]·d[CAGGTCAC], in which ACCT is the recognition sequence. The DNA becomes distorted on binding to accommodate the drug through a widening of the minor grove, most strongly at the CpC step of the recognition sequence. He suggests that the selectivity of binding may be a consequence of sequence-dependent DNA

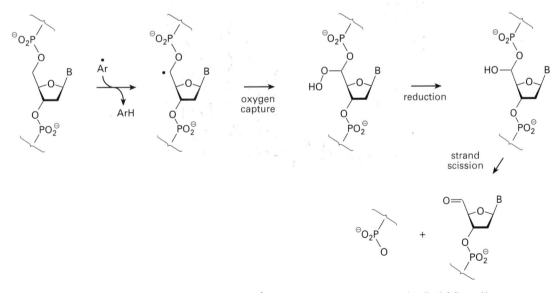

Fig. 7.28 DNA single-strand cleavage by 5'-hydrogen abstraction by an aryl radical, followed by oxygenation and biological reduction.

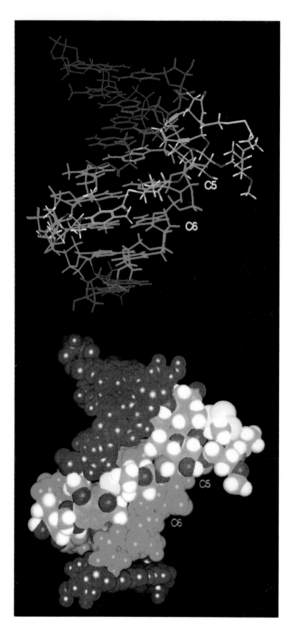

Fig. 7.29 Model for the complex formed between calicheamicin and the octadeoxynucleotide d[GTGACCTG] · d[CAGGTCAC] using a combination of molecular graphics and NMR NOE analysis. The recognition sequence ACCT is shown in blue, other nucleotides in red, and the antibiotic in atom-coloured format. (Line diagram upper, space-filling model lower). Courtesy of Dr D. Kahne with permission from the American Chemical Society (*J. Am. Chem. Soc.*, 1993, **115**, 7954–61).

flexibility, and follows the ability of pyrimidine/purine sequences to become distorted in order to accommodate a relatively inflexible drug (see Fig. 7.29).

7.7.5 Antibiotics generating superoxide

Tetrazomine is a secondary metabolite that is a member of the quinocarcin/saframycin class of antitumour agents. It has antibacterial activity as well as promising *in vivo* activity against leukaemia in mice while **quinocarcin** has been used in clinical trials for a range of solid tumours (Fig. 7.30). These compounds undergo a spontaneous reaction that involves a stereo-specific self-disproportionation of the oxazolidine ring which generates superoxide, $HO_2^{\cdot}$. That leads to a radical-initiated cleavage of DNA whose details are still under examination.

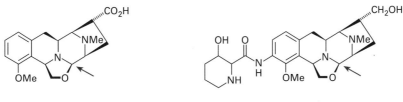

quinocarcin   tetrazomine

Fig. 7.30 Antibiotics that interact with DNA by generation of superoxide leading to strand cleavage, probably through ring opening and formation of a peroxide radical at the positions indicated ($\rightarrow$).

Summary

Considerable progress has been made towards the identification of the mode of action of a range of natural and synthetic anti-tumour antibiotics. Physical binding in either the major or minor groove of DNA gives target selectivity. Covalent bonding to DNA may be a consequence (a) of inherent, weak electrophilic character (CC-1065), (b) of elimination of water or methanol (anthramycin and tomaymycin), or (c) of metabolic reduction (mitomycin C). In some cases the covalently attached antibiotic resides in the DNA groove with little distortion of its conformation (tomaymycin and carzinophilin). In others, strand breaks may result (CC-1065).

The enediyne antibiotics generally form a physical complex with DNA at a recognition sequence that is determined by intercalation or groove binding or both. A 'molecular trigger' causes the formation of a diradical that abstracts a 5'-hydrogen from a proximate nucleoside in the minor groove which leads to single strand breaks and also, in the cases of calicheamicin, dynemicin, and esperamicin, to double strand breaks.

7.8 Photochemical modification of nucleic acids

Our very serious concern about the depletion of the global ozone barrier is directly related to the action of UV light on nucleic acids. Its effect is mutagenic at low doses, cytotoxic at high doses, and is linked to skin cancer where there is chronic, excessive exposure to sunlight among whites, albino blacks, or for people with deficiencies in their repair genes.

7.8.1 Pyrimidine photoproducts

Light of 240–280 nm excites the pyrimidine bases, C, T, and U, to give a higher singlet state ($^1S_1$) which has a lifetime of only a few picoseconds before it gives photohydrates (in which water has added to either face of the 5,6-double bond), decays, or passes into the triplet state. **Uridine photohydrate** (U*) dehydrates slowly to uridine in acidic or alkaline solution and is moderately stable at neutral pH ($t_{1/2}$ 9 h at 50°C). The **cytidine photohydrate** is some tenfold less stable ($t_{1/2}$ 6 h at 20°C) and either reverts to cytidine (90 per cent) or is deaminated to give U* (10 per cent) (Fig. 7.31). This process effects a net conversion of C into U (Section 7.11.2). The formation of photohydrates of thymine has a very low quantum yield and its biological consequences are not significant.

All of the major pyrimidines form **cyclobutane photodimers** on direct irradiation at 260–300 nm. The reaction is a [2 + 2] cycloaddition, mainly involving the triplet state. Of the four possible isomers for thymine dimer, T<>T, the *cis-syn* isomer is formed by irradiation of thymine in an ice matrix and is known to be the major product (>95 per cent) formed by UV irradiation of native DNA. The *trans-syn* isomer, which is one of the four isomeric products produced by the photosensitized irradiation of thymidine in solution and accounts for some 2 per cent of the native DNA T<>T. A larger proportion of this thymine dimer is formed in denatured DNA (Fig. 7.32), where the *trans-syn*, *cis-syn* T<>T and T<>U dimers account for 1.6, 11.0 and 4.2 per cent of total thymine.

The stereochemistry of these [2 + 2] dimers has established that Py<>Py formation in native DNA is predominantly an intrastrand process with photoaddition involving adjacent pyrimidines. Oligonucleotides containing a single T<>T dimer have been analysed by molecular modelling and by NMR methods, which suggest that there is little distortion of the DNA helix apart from the local disruption of hydrogen bonding to adenines. This unexpected result will no doubt be clarified by X-ray structural analysis. The *trans-syn* isomer may well result from regions of Z-DNA, and NMR analysis of an oligomer containing this lesion has indicated that in this structure it causes a larger distortion of helical structure.

These cyclobutane photodimers revert to monomers on irradiation at wavelengths shorter than 254 nm. Dimers containing cytosine residues are easily deaminated and subsequent

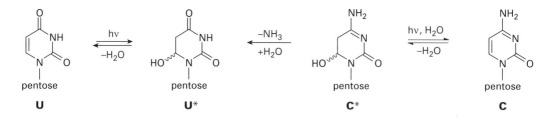

Fig. 7.31 Photohydration of uracil and cytosine nucleosides and deamination of cytosine photohydrate leading to uracil.

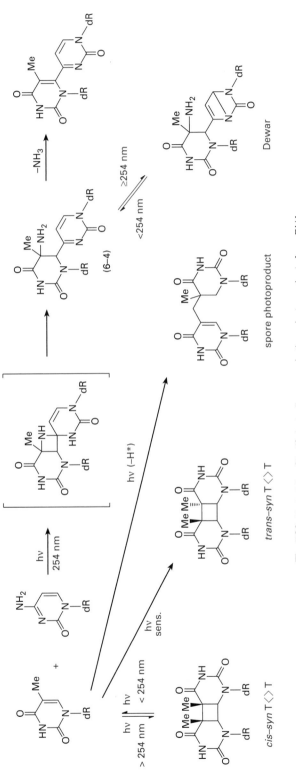

Fig. 7.32 Pyrimidine [2+2]photodimers and other photoproducts from DNA.

photoreversion provides yet another source of the C→T base transition as a result of the following reaction sequence:

$$C + T \xrightarrow{h\nu} C<>T \xrightarrow{H_2O} U<>T \xrightarrow{h\nu} U + T$$

Many other types of pyrimidine photoproduct have been isolated from irradiated DNA. The most noteworthy are the pyrimidine(6-4)pyrimidone photoadducts and the 'spore photoproduct'. The latter is formed from a radical generated by loss of a hydrogen atom from the methyl group of thymine (Fig. 7.32) and is the major UV-induced photoproduct in the dehydrated state of DNA that is found in bacterial spores (Fig. 7.32).

These pyrimidine(6-4)pyrimidone photoproducts are formed by UVC with an incidence about 25 per cent that of the cyclobutane photodimers, with T(6-4)C being twice as abundant as T(6-4)T. With UVB these (6-4) products isomerize to give a 'Dewar-benzene' structure that reverts to its precursor on UVC irradiation (Fig. 7.32).

Both the cyclobutane photodimers and the (6-4) products are causes of cell death, as based on four criteria:

- at low doses, dimers (80 per cent) and (6-4) products (20 per cent) are the major photoproducts
- the action spectra for formation of both types of photoproduct correlate well with that for cell death
- enzymatic excision repair of both lesions enhances cell survival
- cells deficient in excision repair are hypersensitive to the lethal effects of UV radiation.

The mutagenic nature of cyclobutane dimers, (6-4) products, and photohydrates is a complex combination of the relative yields for their photochemical formation (approx. 100 : 25 : 1), the capability of DNA polymerase to read through the lesion, the errors that may result, and the ability of enzymes to repair that lesion. It appears that bacterial replication past cyclobutane dimers is rather accurate whereas bypass of T(6-4)T products is highly mutagenic and leads to T → C transitions. In yeast, it is Py<>Py that is the most mutagenic lesion. In mammalian systems, such as hamster cells, GC → AT transitions predominate (50 per cent) with large amounts of transversions (23 per cent) and tandem and non-tandem double mutations (20 per cent). C → T transitions generally arise through deamination of C<>T or T<>C followed by monomerization to U rather than by the deamination of a cytosine photohydrate.

In general, it now appears that:

- the (6-4) photoproduct is more mutagenic than Py<>Py while T<>T is poorly mutagenic
- in bacteria the (6-4) product is likely to be the major premutagenic lesion
- in mammalian cells, the (6-4) product is repaired more quickly than the Py<>Py making the cyclobutane dimers the major premutagenic lesion (Section 7.11)
- in repair-deficient strains, the (6-4) product may be the dominant mutational lesion even though its photochemical yield is lower than for Py<>Py.

7.8.2 Psoralen–DNA photoproducts

Psoralens are furocoumarins that have been widely used in the phototherapy of psoriasis and other skin disorders. Their photochemical cross-linking of DNA involves intercalation followed by two successive photochemical [2+2] cycloadditions which result in the formation of two cyclobutane linkages. Thymines are the preferred target so that cross-linking occurs mainly at d(TpA) sites. The products are predominantly the *cis-syn* stereoisomers which have an overall S-shape as a result of one thymine being above and the other below the plane of the psoralen (Fig. 7.33).

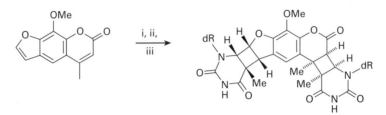

Fig. 7.33 Photochemical binding of 4-methyl-8-methoxypsoralen to DNA and isolation of dithymidine photoproducts.
Procedures: (i) DNA; (ii) hν 320–400 nm; (iii) H$^+$.

NMR analysis of the conformation of a psoralen adduct of the octanucleotide, d(GGGTACCC), has shown a degree of distortion of the B-DNA helix both for the psoralen-bonded thymine residues and the adenines. This distortion is also shared by their nearest neighbours. However, the outer G-1, G-2, C-7, and C-8 residues still form a regular B-helix stack. The overall effect of psoralen cross-linking is to unwind the helix by 63° and impose a kink of 45° on its axis (Fig. 7.34).

Psoralen cross-linking of a duplex has been developed for the examination of triple-strand helix formation in homopurine-homopyrimidine DNA sequences (Section 2.4.5). Psoralen is attached by its C-5 position to a 5′-thiophosphate on a homopyrimidine undecamer nucleotide. On incubation with DNA which has the complementary sequence followed by UV irradiation, the two parent strands of the DNA become cross-linked at a TpA step present at the junction between the duplex and the triplex. Such psoralen-oligonucleotide conjugates are probes for sequence-specific helix formation and may have application in site directed mutagenesis or control of gene expression.

7.8.3 Purine photoproducts

Adenine and guanine are intrinsically more photostable than the pyrimidines and tend to transfer photochemical excitation energy to neighbouring pyrimidines in DNA duplexes. General purine photoreactivity is most marked at C-8 as illustrated by three examples:

- adenosine and guanosine form C-8 substitution products with secondary alcohols such as isopropanol by UV or γ-radiation induced processes
- UV irradiation of 8-bromo-adenine or -guanine generates purinyl radicals that either couple to form dimers or add to an adjacent purine to form 8,8′-dipurinyl products

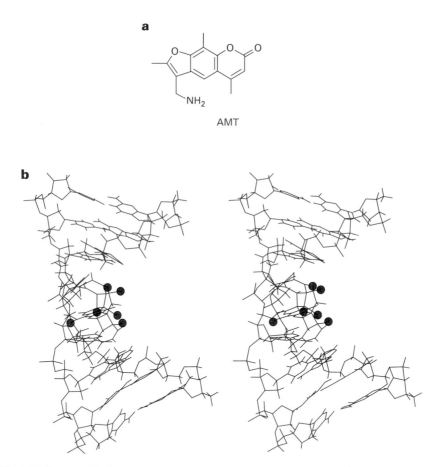

Fig. 7.34 Structure (**b**) of the octanucleotide, d(GGGTACCC), cross-linked to 4′-aminomethyl-4,5,8-trimethylpsoralen, (**a**) (AMT), as determined by NMR. All five methyl groups and the amino group of AMT are indicated by filled circles (adapted from Tomic, M. T., Wemmer, D. E., and Kim, S.-H. (1987). *Science*, **238**, 1722–5. Copyright (1987) AAAS). Stereopair diagram for parallel viewing.

- UV irradiation of coenzyme B_{12} causes scission of the bond from cobalt to adenosine C-5′ and results in the formation of 8,5′-cyclo-5′-deoxyadenosine.

7.8.4 DNA and the ozone barrier

Life is shielded from DNA-damaging solar radiation by the **ozone layer**. The amount of ozone involved is very small indeed. It is formed photochemically in the stratosphere by UVC radiation and is distributed throughout the atmosphere. Its maximum density is at some 20–25 km above the earth's surface, where its pressure is about 130 nbar. Were all the ozone to be concentrated at sea level it would form a layer only 3 mm thick!

The absorption spectrum of ozone is very similar to that of DNA (Fig. 7.35), so it generally serves to prevent the short-wavelength UVC radiation that does the most damage to DNA,

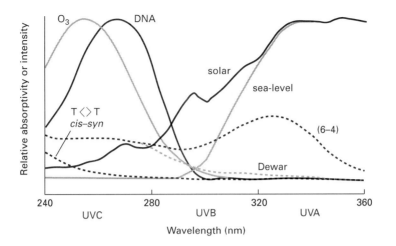

Fig. 7.35 Absorption spectra for DNA and ozone (colour) and the various photoproducts (broken lines) compared to the incidence of sunlight at sea level and above the atmosphere.

from reaching the earth's surface. To put that in perspective, it has been estimated that one hour's exposure of DNA to sunlight at sea level would generate about seven *cis-syn* thymine dimers per 1000 reactive sites. Were the ozone level to be halved, that time would be reduced to 10 minutes. In the complete absence of ozone the same DNA damage would occur in 10 seconds. Other calculations have estimated that for every 1 per cent decrease in the ozone column there could be a 4 per cent increase in the incidence of skin cancer.

One of the major causes for concern is the depletion of ozone resulting from the photochemical behaviour of chlorofluorocarbons in the upper atmosphere. This interferes with the formation of ozone through the capture of oxygen atoms by chlorine atoms, as shown by the following equations.

Ozone formation $\quad O_2 + h\nu \rightarrow O^{\cdot} + O^{\cdot}$

$\quad\quad\quad\quad\quad\quad\quad O^{\cdot} + O_2 + N_2 \rightarrow O_3 + N_2$

Ozone depletion $\quad Cl^{\cdot} + O^{\cdot} \rightarrow ClO$

$\quad\quad\quad\quad\quad\quad\quad ClO + O_3 \rightarrow Cl^{\cdot} + O_2 + O_2$

A more cautious view is that the dominant variables that determine the UV incidence at ground level integrated over 24 hours are the elevation of the sun above the horizon and the duration of daylight. These two factors, which ultimately relate to the earth's solar orbit, lead to variations that far exceed the changes predicted to occur at any middle latitude location as a consequence of any decline in the column of ozone since 1970. The situation over the Antarctic is, however, a special case. The large depletion in ozone over that area in springtime has resulted in UVB irradiances that are substantially larger than existed in that part of the world before the 1980s.

7.9 Effects of ionizing radiation on nucleic acids

X-Rays, γ-radiation, and high-energy electrons all interact indirectly with nucleic acids in solution as a result of the formation of hydroxyl radicals, of solvated electrons, or of hydrogen atoms from water. In aerobic conditions, the most important processes result from HO˙ radicals which abstract a hydrogen atom to give a radical which then captures O_2. Measurements of the efficiency of these processes indicate that for every 1000 eV of energy absorbed, about 27 HO˙ radicals are formed of which 6 react with the pentoses and 21 react almost randomly with the 4 bases.

7.9.1 Deoxyribose products in aerobic solution

The hydroxyl radical can abstract a hydrogen atom from C-4′ or any of the other four carbons in the sugar. The resultant radicals capture oxygen to give a hydroperoxide radical at C-4′ or C-5′ which leads directly to cleavage of a phosphate ester. Similar reactions at C-1′ or C-2′ lead to alkali-labile phosphate esters that are cleaved on incubation with 0.1 M sodium hydroxide at room temperature in 10 minutes. In both cases, the bases are released intact (Fig. 7.36).

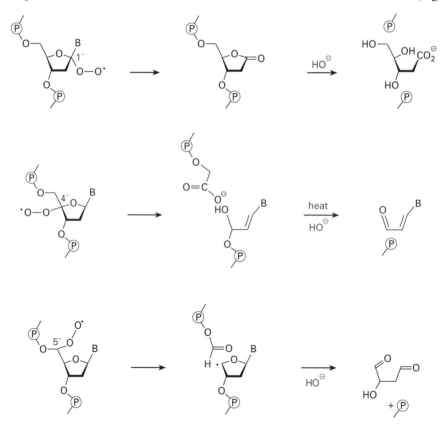

Fig. 7.36 Breaks in DNA resulting from hydrogen atom abstraction and peroxide radical formation at C-1′ (upper), C-4′ (centre), and C-5′ (lower) in deoxynucleotides followed by mild alkaline treatment (0.1 M NaOH, 10 min. 20°C).

7.9.2 Pyrimidine base products in solution

At least 24 different products have been isolated from irradiation of thymidine in dilute, aerated solution and many more are formed in anoxic conditions or in the solid state. Cytosine and the purines show a similar diversity. The situation is simplified by limiting the study to oxygenated solutions when the principal site for reaction with the pyrimidines is the 5,6-double bond. Under these conditions, hydroxyhydroperoxides are formed that are semi-stable for thymidine, but break down rapidly for deoxycytidine to give a range of products, of which the major ones are illustrated (Fig. 7.37). About 10 per cent of thymine modification occurs at the methyl group with the formation of 5-hydroxymethyldeoxyuridine.

In anaerobic solution, γ-radiolysis gives 5,6-dihydrothymidine as the major product with a preference for formation of the 5(R)-stereoisomer.

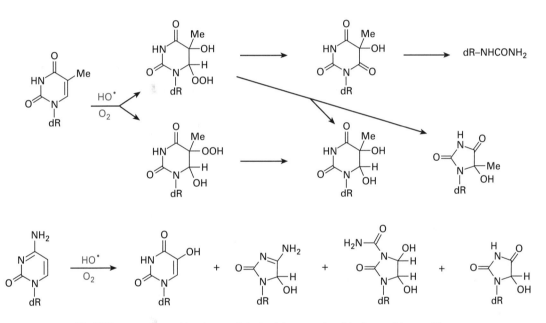

Fig. 7.37 Products resulting from radiolysis of deoxynucleosides in aerobic conditions.

7.9.3 Purine base products

The radiation chemistry of the purines is less well understood than that of the pyrimidines. Deoxyadenosine can add the HO˙ radical at C-8 either to give 8-hydroxydeoxyadenosine or, via cleavage of the imidazole ring, to give a 5-formamido-4-aminopyrimidine derivative of deoxyribose (Fig. 7.38). Guanine has been even less well studied but reports of the formation of 8-hydroxyguanine from anaerobic irradiation coupled to much interest in this modified base as a strongly mutagenic lesion may change this situation.

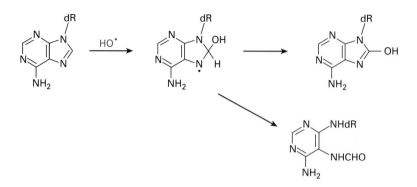

Fig. 7.38 Major γ-radiolysis products from deoxyadenosine.

Summary

The photochemistry of nucleic acids is dominated by the formation of thymine and cytosine dimers, Py<>Py, of which the principal isomer obtained from DNA has the *cis-syn* cyclobutane structure. Dimers containing cytosine can be deaminated to give uracil-containing dimers. While T<>T appears to be not very mutagenic, C<>T and T<>C are important sources of C → T transition mutations. Pyrimidine(6-4)pyrimidone photoproducts are formed to a lesser extent than Py<>Py but their mutagenic significance can be at least as high as that of Py<>Py dimers. U and C form photohydrates, U* and C*, through addition of OH to position-6 and of H to position-5. Cytosine hydrate can be deaminated to give U* which through dehydration to U causes a C → T base transition.

The purines are relatively photostable.

Psoralens can act as photochemical cross-linking agents and form cyclobutane adducts involving two thymine residues in opposite strands of DNA.

Ionizing radiation in aerated solution generates hydroxyl radicals which react with all four bases giving a wide range of products. The pentoses are also attacked by hydrogen atom abstraction, which leads to single-stranded breaks.

7.10 Biological consequences of DNA alkylation

7.10.1 *N*-Alkylated bases

The major site of DNA alkylation is at the N-7 position of guanine. However, methylation of this site appears not to change the base-pairing of G with C and so is an apparently harmless lesion. Moreover, the glycosylic bond of a 7-alkylguanine residue appears to be slowly and spontaneously hydrolysed, and so creates an apurinic site which is a target for repair (Section 7.11.2). By contrast, the cross-linking of two neighbouring guanines at N-7 by a

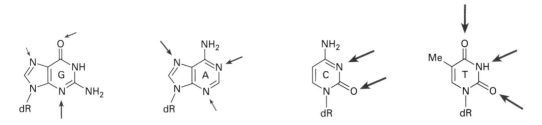

Fig. 7.39 Major sites of methylation of bases in double-stranded DNA by hard alkylating agents such as MNU and NMNG. Major sites shown by bolder arrows.

nitrogen mustard or equivalent bifunctional alkylating agent is an important cell-killing event, both for interstrand and for intrastrand cross-links.

3-Alkyladenines are the major toxic lesion resulting from monofunctional alkylation of DNA. The alkyl group lies in the minor groove of the DNA double helix where it can block the progress of DNA polymerases (Section 7.11.3). 3-Alkylguanines have a similar physiological effect but are much less prevalent, being formed ten times less frequently than 3-alkyladenines (Fig. 7.39). Some 1-methyladenine is generated by the methylation of DNA, but while this lesion as a 'non-instructional base' must necessarily interfere with A·T pairing, it appears to be only slowly excised *in vivo*.

7.10.2 *O*-Alkylated lesions

O^6-Alkylguanines are locked in the *enol*-tautomeric form while guanine in DNA is normally in the *keto*-form (Chapter 2.1.2). It appears that O^6-methylguanine forms an O^6-MeG·C base-pair that is more stable in a DNA duplex than the O^6-MeG·T base-mispair (Fig. 2.3.2). In practice, O^6-MeG residues do not block DNA polymerase and have been shown to direct preferential incorporation of thymidine on replication. This unexpected result has led to a suggestion that the less stable O^6-MeG·T base-mispair may have a more Watson–Crick-like geometry and so better satisfy the demands of the DNA polymerase. The result is a G → A transition. This type of mutation is common in cells exposed to hard alkylating agents. In particular, it has been identified with a single base transition for activation of the Ha-*ras*-1 proto-oncogene in the process of initiation of mammary tumours in rats with methylnitrosourea (MNU), as a result of the specific conversion of G^{35} into O^6-MeG^{35}.

O^4-Alkylthymines exist in the *enol*-tautomeric form and therefore they can base-pair with guanine. In model studies, both O^4-methylthymine and O^4-ethylthymine form base-mispairs with guanine that do not block the replication of a defined DNA sequence *in vitro*. However, alkyl pyrimidines are very minor products of DNA alkylation and their biological effects appear to be of low significance.

Lastly, the *O*-alkylation of the phosphate diesters in DNA gives phosphate triesters but these are repairable (Section 7.11.1) and do not seem to be important either as cell-killing or as mutagenic lesions.

Summary

The alkylation of a single DNA base can have three consequences: either no change in its coding property, a change to pairing with a different base, mispairing, or a change to a non-coding base, a non-instructional site. These are illustrated by the formation of 7-methylguanine, 4-O-methylthymine, and 6-O-methylguanine respectively. Covalent bonding of a base with a larger molecule, such as acetylaminofluorene or BaP-diolepoxide, can have more complex effects.

The overall effect on DNA of chemical modification by other external agents is one of five types of damage: (1) introduction of strand breaks; (2) loss of a base leaving an unpaired partner; (3) covalent modification of a base, especially by the addition of bulky groups; (4) conversion of one base into another with changed Watson–Crick pairing; (5) covalent linking of bases in the same or opposite strands.

7.11 DNA repair

All living cells possess a range of DNA repair enzymes in order to correct damage resulting from radiation and external chemical agents. Humans appear to be more effective than rodents in repairing DNA and are also better able to resist mutagenic agents. A striking similarity has emerged between repair systems found in many species from bacteria to humans, though much of our knowledge comes from studies on bacteria or yeasts. Two general types of repair process have been identified (Sections 7.11.1 and 7.11.2).

7.11.1 Direct reversal of damage

Photoreactivation is catalysed by an enzyme which uses tryptophan residues to sensitize the photochemical (300–400 nm) cleavage of *cis-syn* cyclobutane photoproducts to their constituent pyrimidines. This photolyase does not cleave other types of pyrimidine photoproducts (Fig. 7.34) and there is no evidence for any DNA phosphorylase activity in humans.

O-Demethylation is carried out by an O^6-methylguanine-DNA methyltransferase. This sacrifical enzyme uses the SH group of a Pro–Cys–His sequence near its C-terminus as a methyl acceptor (Fig. 7.40). The same enzyme can also dealkylate ethyl, 2-hydroxyethyl, and 2-chloroethyl O^6-derivatives of guanine. This repair reaction changes the enzyme into an

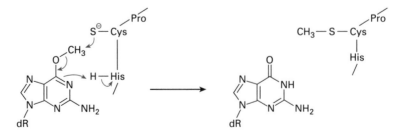

Fig. 7.40 O^6-Methylguanine demethylation by the Pro-Cys-His component of methyltransferase.

S-alkylcysteine protein which is both inactive and also not reactivated. It follows that one molecule of the enzyme can only repair one *O*-methylated base. Human lymphoid cells which are resistant to alkyl-nitrosoureas have 10 000–25 000 copies of this enzyme per cell while mutants deficient in the repair of O^6-MeG have no enzyme. It is easy to see that the threshold of tolerance for alkylating agents may vary greatly between different species of cells.

In *E. coli*, the methyltransferase from the ada^+ gene differs in three respects from the mammalian enzyme. Its C-terminal cysteine demethylates both O^6-MeG and O^4-MeT; there is a second active cysteine near the N-terminus which demethylates DNA phosphate triesters (but only of the (*S*)-configuration); and the system shows an adaptive response to alkylating agents which may be triggered by the demethylation of phosphate triesters. The structure of the 19 kDa C-terminal domain of this protein has been determined by X-ray diffraction and reveals that the active site cysteine is buried. It has therefore been suggested that the protein must undergo a significant conformational change in order to bind DNA and effect methyl transfer (Section 9.3.7).

7.11.2 Excision repair of altered residues

In mammalian cells, excision repair is the most important mechanism and involves enzyme recognition of the modified base. There are two distinct types of excision repair process and both appear to be error-free: nucleotide excision and base excision.

In repair by **nucleotide excision**, an endonuclease initiates the process by making a single-stranded incision close to the damaged nucleotide. The best understood repair system of this sort is that from *E. coli*, whose *uvrABC* genes have been cloned and their protein products purified. The uvrA protein (114 kDa) is an ATP-dependent DNA binding protein that recognizes and binds to the DNA photolesion and to many other types of bulky base modification. The uvrB (84 kDa) and uvrC (70 kDa) proteins can now bind on and initiate the repair process by nicks in the damaged strand which are on either side of the lesion and some 12 bases apart. The excised nucleotide containing the lesion is now released by uvrD (DNA helicase II) while it is still bound to the protein complex. That leaves a single-stranded gap which is filled by DNA polymerase I which binds and fills the gap from the 3′-end until it approaches the 5′-terminus. Ligation completes this **short-patch repair** (Fig. 7.41a).

Repair by **base excision** uses a DNA-glycosylase. These enzymes hydrolyse the glycosylic bonds of modified purine or pyrimidine nucleosides. 3-Methyladenine is the principal target for a pair of 3-methyladenine-DNA glycosylases from *E. coli*, and one of them also acts on 3-MeG, O^2-MeC, and O^2-MeT residues. Among several other glycosylases is a uracil-DNA glycosylase which efficiently cuts out deaminated cytosines in DNA. A pyrimidine dimer–DNA glycosylase works on thymine dimers, cleaving only the glycosylic bond of the 5′-thymine base.

The next step is incision of the phosphate backbone by an AP endonuclease, which operates close to apurinic and apyrimidinic sites. Subsequent exonuclease action then excises nucleotides to create a gap at least 30 residues long. In some cases, such as endonuclease III from *E. coli*, the same enzyme acts both as a glycosylase towards oxidized pyrimidine residues and as an endonuclease.

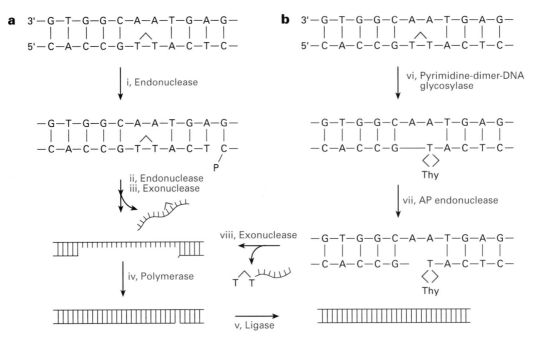

Fig. 7.41 Excision repair—illustrating two alternative modes of excising *cis-syn* thymine photodimers. (**a**) (left) (i) endonuclease makes an incision on 5′-side of dimer; (ii) incision of 3′-side and removal of a 12-base oligomer containing the lesion; (iii) gap enlarged by exonuclease; (iv) polymerase resynthesis; (v) ligation to complete patch (colour). (**b**) (right) (vi) a glycosylase cleaves the 5′-thymine from its deoxyribose; (vii) AP endonuclease hydrolyses the exposed phosphate ester; (viii) and exonuclease makes a gap of *ca.* 30 nucleotides, to allow (iv) and (v) as in (**a**).

In the repair step, this single-stranded region acts as a template for DNA polymerase to fill in the excised sequence, and repair is completed by DNA resynthesis and ligation as before (Fig. 7.41b). This repair system is also short-patch and, like the direct base-repair systems, it is error-free. Excision repair appears to be the dominant type of repair process in bacteria. In humans, defects in this multi-gene system can lead to abnormal sensitivity to sunlight and to the high incidence of skin cancer associated with *Xeroderma pigmentosum*.

The DNA repair processes of eukaryotes are much less understood at a molecular level. The fact that their DNA is organised into higher-order structures (Section 2.6.2) presumably has implications for their DNA repair processes, as has already been established for their transcription (Section 5.1.4) and replication (Section 5.2). What is clear is that a large number of loci is involved in the excision repair process for eukaryotes as compared to prokaryotes, but the mammalian system is remarkably similar to the bacterial system.

7.11.3 Preferential repair of transcriptionally active DNA

It has been known for over 20 years that DNA repair is a heterogeneous process. In 1985, work from the laboratory of Hanawalt established that a transcriptionally active gene is

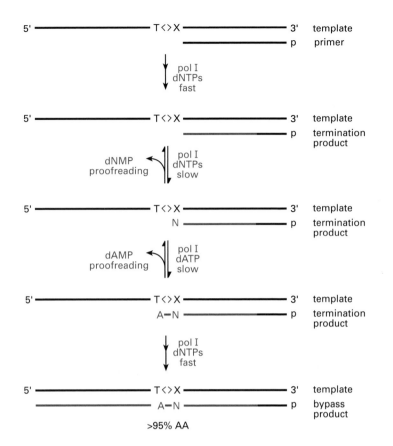

Fig. 7.42 Bypass mutagenesis illustrated by the scheme employed to investigate the kinetics and mutagenic consequences of the bypass of DNA cyclobutane photoproducts by polymerase I *in vitro*. (where X = T or U).

repaired preferentially to the genome overall. The term **preferential repair** was introduced to describe the phenomenon that is caused by interactions between a stalled pol II transcription complex and the excision repair enzymes. For example, 70 per cent of dimers in an active *DHFR* gene in Chinese hamster ovary cells are excised in a 24 hour period while only 10–20 per cent of dimers are removed from the genome overall. Later studies established that this phenomenon is confined to the transcribed DNA strand while virtually no repair occurs in the non-transcribed strand.

One suggestion advanced to explain this phenomenon is that RNA polymerase complexes that have become blocked at a UV lesion act as a signal for repair. Additional support for this hypothesis has come from studies on bacteria where such **strand-specific repair** has been identified with the mfd gene product in *E. coli* under conditions of induction by isopropyl thio-galactoside.

Selective repair is not confined to cyclobutane dimers and (6–4) photoproducts are also removed preferentially from active genes. On the other hand, it appears that *N*-methylpurine repair is not coupled to transcription while the situation regarding covalent modification by carcinogens, such as *N*-AAF (Section 7.6.1), remains to be clarified.

7.11.4 Post-replication repair

The repair systems described in the preceding sections maintain the integrity of the genetic message in the cell as long as repair is rapid and is accurately completed before DNA polymerase attempts to copy the damaged DNA. What happens if repair is slow or deficient? It appears that when replication by DNA polymerase is frustrated by a region of damage, a major response is the activation of 'error-prone repair'.

SOS repair is the name given to such enhanced repair (ER), which is induced by a wide variety of types of DNA damage. This is how it seems to work in bacteria. When the advancing DNA polymerase reaches a lesion which blocks its further progress, such as T<>T or Me^{3}A, a very sizeable gap is left opposite the lesion before replication starts again. This gap can either be made good by recombination processes, known as recombination-repair (Section 5.4), or by the induction of SOS repair.

The SOS process has been most thoroughly studied in *E. coli* and is one of the functions of the *recA* gene protein. Several recA proteins identify and bind to any long, single-stranded stretch of DNA which has been formed by the recBCD proteins. RecBCD produces ssDNA using its helicase/nuclease activities. Through interaction with ATP, recA now becomes activated as a protease towards the lexA protein, which is known to be a multifunctional repressor in control of several DNA repair enzymes. Among the many consequences of hydrolysis of lexA are the derepression of *ruvAB* and *recA* genes and a reduction in the $3' \rightarrow 5'$ exonuclease proof-reading activity of DNA polymerase III. The net result is that pol III operates with decreased fidelity, not only in its ability to read through regions of damage such as T<>T but also elsewhere in the cell, and this situation persits until the activation of recA as a protease ends. At this point SOS repair is rapidly switched off.

The mechanism of repairing these long gaps, which can be over 1000 bases long, is highly variable. In particular, the specificity of elongation varies with several factors: the type of polymerase, the activity of the $3'-5'$ editing nuclease, and the particular type of DNA lesion. For example, in cases where the base modification behaves as a 'non-instructional' site there is a strong tendency for the incorporation of an adenine residue. This explains why transversion mutations are seen when guanines are modified by aflatoxin or by acetylaminofluorene derivatives. Even then, further elongation in the gap is dependent on the base-sequence in the template strand on the 5′-side of the lesion. The overall result is that 'long-patch' repair seems likely to be idiosyncratic for each type of lesion in each mutable site and, above all, is error-prone.

7.11.5 Bypass mutagenesis

A single mutation that arises from damage to DNA can be described by a two-step mechanism. **Misincorporation** is the inclusion in the daughter strand of a nucleotide different from the Watson–Crick complement to the original residue(s) in the primer strand. **Bypass** is the process of continued chain elongation beyond the misinserted base at the site of the lesion. This process has been carefully examined in *E. coli* through the insertion of synthetic cyclobutane photodimers into oligomers and their use as templates for pol I in primer extension reactions *in vitro* and in phage and bacterial replication.

When pol I arrives at a lesion in the template, chain extension of the primer strand can either be terminated or retarded. Three types of result have now been observed for bypass replication beyond a cyclobutane dimer lesion. Firstly, in the case of the lesion T<>T, pol I is able to bypass the dimer and incorporates adenines opposite the lesion with >95 per cent efficiency (Fig. 7.40). This explains why T<>T appears to be inherently non-mutagenic. Secondly, adenines are also incorporated when the dimer is T<>U, and this strongly supports the idea that T<>C and C<>T photodimers cause $G \to A$ transition mutations either through deamination of the cytidine residue to give a deoxyuridine photoproduct prior to replication, **deamination bypass**, or through mispairing of the T<>C dimer with the 5,6-saturated cytosine moiety in its imino tautomeric form, **tautomer bypass** (Section 2.1.2). Thirdly, when T<>T dimers are incorporated into oligo(dT) tracts, specifically at positions -2 and -3 from the 5'-end of the T_n tract, both −1 and −2 deletions are observed.

Studies of this nature can be expected to uncover the role of bipyrimidine (6-4) photoproducts and other DNA lesions in the future.

Summary

Cells have a large repertoire of repair kits to deal with damage by various external agents and to safeguard DNA against their consequences. Direct reversal of the damage is the simplest response, as in the case of photolyase cleavage of Py<>Py dimers or demethylation of O^6-MeG. Excision of a base lesion can involve either a glycosylase, that removes the damaged base, or nucleotide excision, that removes an oligonucleotide of about 12 residues containing the lesion.

DNA replication prior to such repair can either result in misincorporation or be stalled at the lesion. Some photochemical lesions may be bypassed, with slow misincorporation that can lead to $G \to A$ transitions or to deletions. Alternatively, an SOS repair system can be activated to fill in long, single-strand gaps that follow termination at lesions. This activates the recA protein to switch on the *uvrABC* repair genes. At the same time there is a reduction in the fidelity of repair. These changes all enhance the chances of survival of the cell. Quite long patches are made in the daughter strand opposite base lesions and these are error-prone. The error frequency remains generally high for all DNA synthesis and persists until levels of recA protein return to normal.

Genetic deficiencies in some repair systems are associated with the human disease *Xeroderma pigmentosum* that is associated with an exceptionally high sensitivity to sunlight and early mortality from skin cancer.

Further reading

7.1–7.5

Brown, D. M. (1974). Chemical reactions of polynucleotides and nucleic acids. In *Basic principles of nucleic acids chemistry* (ed. P.O.P. Ts'o), Vol. II, pp. 2–90. Academic Press, London.

Schwartz, A., Marrot, L., and Leng, M. (1989). Conformation of DNA modified at a (dGG) or a (dAG) site by the antitumour drug *cis*-diamminedichloroplatinum(II). *Biochemistry*, **28**, 7979–84.

7.6

Blackburn, G. M. and Kellard, B. (1986). Chemical carcinogens. *Chem. Ind. (Lond.)*, 607–13, 687–95, 770–9.

Harvey, R. G. and Geacintor, N E. (1988). Intercalation and binding of carcinogenic hydrocarbon metabolities to nucleic acids. *Accts. Chem. Res.*, **21**, 66–73.

Iyer, R. S., Coles, B. F., Raney, K. D., Thier, R., Guengerich, F. P., and Harris, T. M. (1994). DNA adduction by the potent carcinogen aflatoxin B_1 : mechanistic studies. *J. Am. Chem. Soc.*, **116**, 1603–9.

Singer, B. and Grunberger, D. (1983). *Nucleic acid alkylation and molecular biology of mutagens and carcinogens.* Plenum Press, New York.

Walker, G. C. (1984). Mutagenesis and inducible repair responses to DNA damage in *E. coli. Microbiol. Revs.*, **48**, 60–93.

7.7

Frederick, J. E. (1993). Ultraviolet sunlight reaching the earth's surface. *Photochem. Photobiol.*, **57**, 175–8.

Armstrong, R. W., Salvati, M. E., and Nguyen, M. (1992). Novel interstrand cross-links induced by the antitumour antibiotic carzinophilin/azinomycin B. *J. Am. Chem. Soc.*, **114**, 3144–5.

Hurley, L. H. and Needham-Van Devanter, D. R. (1986). Covalent binding of anti-tumour antibiotics in the minor groove of DNA, mechanism of action of CC-1065 and the pyrrolo[1,4]benzodiazepines. *Accts. Chem. Res.*, **19**, 230–7.

Kizu, R., Draves, P. H., and Hurley, L. H. (1993). Correlation of DNA sequence specificity of anthramycin and tomaymycin with reaction kinetics and bending of DNA. *Biochemistry*, **32**, 8712–22.

McFarland, M. and Kaye, J. (1992). Chlorofluorocarbons and ozone. *Photochem. Photobiol.*, **55**, 911–29.

Neidle, S. and Waring, M. J. (ed.) (1983). *Molecular aspects of anti-cancer drug action*, Vol. 3. Macmillan, London.

Norman, D., Live, D., Saastry, M., Lipman, R., Hingerty, B. E., Tomasz, M., Broyde, S., and Patel, D. J. (1990). NMR and computational characterisation of mitomycin cross-linked to adjacent deoxyguanosines in the minor groove of the d(TACGTA)·d(TACGTA) duplex. *Biochemistry*, **29**, 2861–75.

Stubbe, J. (1987). Mechanisms of bleomycin-induced DNA degradation. *Chem. Revs.*, **87**, 1107–36.

Walker, S., Murnick, J., and Kahne, D. (1993). Structural characterisation of a calicheamicin-DNA complex by NMR. *J. Am. Chem. Soc.*, **115**, 7954–61.

Tomasz, M. (1995). Mitomycin C: small, fast and deadly (but very selective). *Chem. Biol.*, **2**, 575–80.

Williams, R. M., Flanagan, M. E., and Tipple, T. N. (1994). O_2-Dependent cleavage of DNA by tetrazomin. *Biochemistry*, **33**, 4068–92.

7.8

Cadet, J. and Vigny, P. (1990) The photochemistry of nucleic acids. In *Bio-organic Photochemistry* (ed. H. Morrison), Vol. 1, pp. 1–272. Wiley, New York.

Mitchell, D. L. (1988). The relative cytotoxicity of (6-4) photoproducts and cyclobutane dimers in mammalian cells. *Photochem. Photobiol.*, **48**, 51–7.

Sage, E. (1993). Distribution and repair of photolesions in DNA: genetic consequences and the role of the sequence context. *Photochem. Photobiol.*, **57**, 163–74.

Takasugi, M., Guendouz, A., Chassignol, M., Decout, J. L., L'homme, J., Thuong, N. T., and Hélène, C., (1991). Sequence-specific photo-induced cross-linking of the two strands of double-helical DNA by a psoralen covalently linked to a triple-helix-forming oligonucleotide. *Proc. Natl. Acad. Sci. USA.*, **88**, 5602–6.

Tomic, M. T., Wemmer, D. E., and Kim, S.-H. (1987). Structures of psoralen cross-linked DNA in solution by NMR. *Science*, **238**, 1722–4.

Wang, S. Y. (ed.) (1976). *The photochemistry and photobiology of nucleic acids*, Vol. 1. Academic Press, New York.

7.9

Hutchinson, F. (1985). Chemical changes induced in DNA by ionising radiation. *Progr. Nucleic Acid Res.*, **32**, 115–54.

von Sonntag, C. (1986). *The chemical basis of radiation biology*. Taylor and Francis, London.

7.10

Leonard, G. A., Thomson, J., Watson, W. P., and Brown, T. (1990). High resolution structure of a mutagenic lesion in DNA. *Proc. Natl. Acad. Sci. USA*, **87**, 9573–6.

Lipscomb, L. A., Peek, M. E., Morningstar, M. L., Verghis, S. M., Miller, E. M., Rich, A., Essigmann, J. M., and Williams, L. D. (1995). X-Ray structure of a DNA decamer containing 7,8-dihydro-8-oxoguanine, *Proc. Natl Acad. Sci. USA*, **92**, 719–23.

7.11

Bohr, V. A., and Wassermann, K. (1988). DNA repair at the level of the gene. *Trends Biochem. Sci.*, **13**, 629–33.

Friedberg, E. C., Walker, G. C., and Siede, W. (1995). *DNA repair and mutagens*. A.S.M. Press, Washington.

van Houten, B. (1990). Nucleotide excision repair in *E. coli. Microbiol. Rev.*, **54**, 18–51.

Lindahl, T. (ed.) (1985). Carcinogens and DNA. *Cancer Surveys*, **4(3)**, 491–624.

Moore, M. H., Gulbis, J. M., Dodson, E. J., Demple B., and Moody, P. C. E. (1994). Crystal structure of a suicidal repair protein: the Ada O^6-methylguanine–DNA methyltransferase from *E. coli. EMBO Journal*, **13**, 1495–501.

Sancar, A. (1994). Mechanisms of DNA excision repair. *Science*, **266**, 1954–6.

Savva, R., McAuley-Hecht, K., Brown, T., and Pearl, L. (1995). The structural basis of specific base-excision repair by uracil–DNA glycosylase, *Nature*, **373**, 487–9.

Taylor, J-S. (1994). Unravelling the molecular pathway from sunlight to skin cancer. *Accts. Chem. Res.*, **27**, 76–82.

Terleth, C., van de Putte, P., and Brouwer, J. (1991). New insights into DNA repair: preferential repair of transcriptionally active DNA. *Mutagenesis*, **6**, 103–11.

Walker, G. C. (1985). Inducible DNA repair systems. *Ann. Rev. Biochem.*, **54**, 425–58.

REVERSIBLE INTERACTIONS OF NUCLEIC ACIDS WITH SMALL MOLECULES

8.1 Introduction

Nucleic acids interact reversibly with a broad range of chemical species that include water, metal ions and their complexes, small organic molecules, and proteins. In this chapter we focus on the non-covalent interactions of DNA with small molecules of molecular mass less than approximately 1000 Da. Molecules and ions in this group represent a wide variety of chemical types from simple to complex metal species, a variety of drugs, carcinogens, and complex antibiotics. Of the wide variety of examples which exist, only those will be chosen which give a fairly clear and detailed illustration of as many different features as possible.

At the outset, we should consider the importance of reversible interactions on nucleic acid structure and function. First, all of the intricate nucleic acid conformations which exist are stabilized by and only possible because of reversible interactions with water, metal ions, and/or organic cations. Dramatic structural transitions in nucleic acids are brought about by changes in water activity, salt concentration (ionic strength), or by interaction with organic molecules. The duplex–triplex equilibrium, B–Z transition, hairpin–cruciform formation (Section 2.5), and packaging of nucleic acids into virus particles and chromatin (Section 9.2.1) come particularly to mind as being quite sensitive to reversible interactions.

Second, one of the most important lines of drug development and of current chemotherapy against some cancers, viral, and parasitic diseases involves drugs which interact reversibly with nucleic acids. Natural antibiotics such as adriamycin and synthetic drugs such as amsacrine which interact with DNA are widely used in clinical treatment of a variety of neoplastic diseases. Much of the drive to understand nucleic acid interactions has come from the interest in understanding the mode of action of existing medicinal agents and from the desire to develop a new generation of superior drugs. Synthetic oligopeptides and oligonucleotides offer exciting new possibilities as nucleic acid recognizing drugs and as potential 'nucleases' of high sequence specificity.

Third, because of their relative simplicity, the interactions of small molecules with nucleic acids have provided much of our most accurate information about nucleic acid binding specificity, ligand-induced conformational transitions, the molecular basis of co-operativity in binding, the interaction of aromatic amino acid side chains with nucleic acid bases, and other similar critical features of nucleic acid interactions and chemistry.

More physical studies have been conducted on interactions of small molecules with DNA than with RNA. Part of this emphasis on DNA rather than RNA arises from the potential of DNA as a target for anti-cancer drugs. Because of the ease of synthesis of DNA relative to RNA oligomers and polymers, it is also easier to obtain DNA than RNA model systems for high-resolution physical studies. This chapter will, thus, necessarily include more studies with DNA, although advances in RNA synthesis are rapidly expanding the RNA interaction database.

8.1.1 Types of reversible interaction

Molecules and ions interact with duplex nucleic acids in three primary ways which are significantly different:

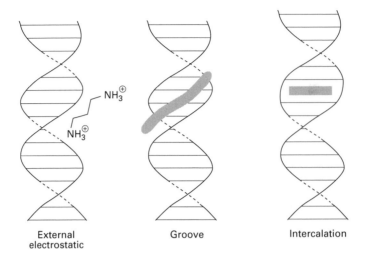

Fig. 8.1 The three primary binding modes are illustrated with a double-helical nucleic acid.

- binding along the exterior of the helix through interactions which are generally non-specific and are primarily **electrostatic** in origin;
- **groove-binding** interactions which involve direct interactions of the bound molecule with the edges of base-pairs in either of the (major or minor) grooves of nucleic acids; and
- **intercalation** of planar or approximately planar aromatic ring system between base-pairs (Fig. 8.1).

The first two binding modes do not require a nucleic acid conformational change, but may induce structural transitions on complex formation. Intercalation requires changes in sugar-phosphate chain torsional angles for separation of adjacent base-pairs by a distance (typically near 3.4 Å) sufficient to allow insertion of the intercalating ring system. This can be accompanied by other changes in the helical parameters such as unwinding, bending, etc. Both DNA and RNA can engage in external and intercalation interactions in a similar manner. However, the grooves of DNA and RNA are quite different and interact with very different agents (Sections 2.2.4 and 2.4.2). Examples of the types of molecules and ions that bind to nucleic acids by the three different modes are shown (Fig. 8.2). These three types of interactions will be discussed separately.

Duplex structures of nucleic acids are not alone in their ability to engage in reversible interactions. Single-stranded RNA can form extensive intramolecular duplex regions that contain base bulges, base-pair mismatches, hairpin loops, and similar structures that arise from folding the RNA strands from biological systems such as ribosomes, tRNA, or the genomic RNA of some viruses. These perturbed duplex conformations can undergo very specific interactions and in pathogenic RNA viruses, such as HIV-1, offer an exciting potential target in drug design. Several antibiotics that act on ribosomes have been discovered, and in many cases the primary target appears to be ribosomal RNA.

a   External, electrostatic interactions

$$Na^{\oplus} \qquad Mg^{2\oplus} \qquad H_3\overset{\oplus}{N}CH_2CH_2CH_2\overset{\oplus}{N}H_3$$

b Groove-binding

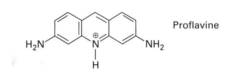

Netropsin

c Intercalation

Proflavine

Fig. 8.2 Examples of cations which bind by the three primary modes.

8.2 External electrostatic interactions

8.2.1 Condensation type interactions

Nucleic acids are highly charged polyelectrolytes whose phosphate groups strongly affect their structure and interactions. Manning and co-workers have shown that simple ions, such as those of the alkali metals, associate with nucleic acids largely as a function of the polymer charge density. In B-DNA, for example, if the molecule is modelled as a line of equally spaced charges there is one anionic charge per 1.7 Å distance (two charges per base-pair and approximately ten base-pairs per turn of the duplex). If the line charge spacing becomes less than approximately 7 Å, the molecule becomes unstable. Polymer conformations with charges that are more closely spaced must thus associate with counter-ions from solution to achieve stability. The association of ions with the polyelectrolyte is called **counter-ion condensation** and causes an unfavourable entropy term in the overall polymer conformational free energy summation. This unfavourable term is more than outweighed by the numerous favourable interactions in the folded polymer (such as the DNA double helix). Counter-ions 'condense' until the charge density is reduced to the stable level of approximately one charge per 7 Å. Additional counter-ions are associated with the remaining charges on the polyanion through Debye–Hückel interactions. An important prediction of the condensation theory, which has received experimental support, is that the counter-ions condensed per phosphate charge remain relatively constant for a particular conformation as the solution salt concentration is varied over a fairly wide range.

The initial association of counter-ions is referred to as condensation, since the ions are associated with the general charge density of the polyelectrolyte and are not bound at specific sites. The ions retain their inner sphere water of hydration and move rapidly along the sugar-phosphate backbone of the nucleic acid. Secondary hydration layers of both the polyelectrolyte and counter-ion are affected by this interaction. The Poisson–Boltzmann equation, applied to nucleic acids as rod-like polyanions, predicts a similar number of strongly associated counter-ions. If the phosphate charges of DNA are 'deleted', for instance by esterification of the phosphate groups or by replacement with methylphosphonate linkages (Section 3.4.6), the molecule still can form a B-like double helix, but, as expected, it does not exhibit the salt-dependent properties of normal B-DNA. With the B-form duplex structure of DNA, the condensation theory predicts an average of 0.76 monovalent counter-ions (such as Na^+) condensed per phosphate group and a total of 0.88 counter-ions associated per phosphate group in condensation plus Debye–Hückel type interactions (see equations 8.2.5 and 8.2.6). Predictions by other theoretical methods arrive at similar numbers of associated counter-ions.

The associated counter-ions reduce the effective charge on nucleic acids and strongly affect the solution properties and binding interactions of the polymer. A significant portion of the binding free energy of species ranging from small cations to large proteins can result from the neutralization of nucleic acid charges (ion pair formation) in the complex and the resulting favourable entropic effect of release of counter-ions. At a constant temperature, this entropic effect resulting from counter-ion release can lead to nucleic acid denaturation as well as to increased binding of cationic ligands as the bulk solution salt concentration is decreased. Thus, both the T_m of nucleic acids and the observed equilibrium constant for binding of cationic ligands depend strongly on salt concentration. Thomas Record and co-workers have developed a particularly useful formulation for the effects of salt concentration on nucleic acid equilibria. In particular they have shown that for a cationic ligand, L, binding to a nucleic acid site, D, the thermodynamic equilibrium in the presence of sodium counter-ions is

$$D + L \rightleftharpoons C + m'\psi_c[Na^+], \qquad (8.2.1)$$

where L makes m' ion pairs with DNA phosphate groups, C is the complex, and ψ_c is the average fraction of sodium ions condensed per phosphate group. The observed equilibrium is simply

$$D + L \rightleftharpoons C, \qquad (8.2.2)$$

and is described by K_{obs}, an observed equilibrium constant. K_{obs} is experimentally evaluated at constant salt concentration and temperature by determining the concentrations of the species D, L, and C specified in equation (8.2.2). The variation of K_{obs} with sodium ion concentration is

$$\frac{\delta \log K_{obs}}{\delta \log [Na^+]} = -m'\psi, \qquad (8.2.3)$$

where ψ represents the average fraction of a sodium associated with each phosphate group through condensation and Debye–Hückel type screening processes. ψ is related to the average linear phosphate group spacing, b, in nucleic acids (Sections 2.2.3 and 2.2.4) through the dimensionless parameter ξ:

$$\xi = e^2/\varepsilon kTb,\qquad(8.2.4)$$

where e is the electronic charge magnitude, ε is the bulk solution dielectric constant, k is Boltzmann's constant, and T is the temperature in K. The relations of ψ and ψ_c to ξ are

$$\psi_c = \left(1 - \xi^{-1}\right)\qquad(8.2.5)$$

$$\psi = \left(1 - (2\xi)^{-1}\right)\qquad(8.2.6)$$

For B-DNA, b is 1.7 Å and ψ and ψ_c are 0.88 and 0.76 respectively. As b increases on binding of specific cationic ligands or on DNA denaturation, ψ and ψ_c decrease.

The negative sign in equation (8.2.3) indicates that the observed equilibrium constant decreases with increasing sodium ion concentration. The magnitude of the decrease will depend on the charge on the ligand through m', the number of charged groups which can interact with phosphate groups on the nucleic acid. It is important to note that most experimental equilibrium constants, K_{obs}, have no meaning unless the solution conditions are carefully specified. It is also useless to compare K_{obs} (or T_m changes) for different ligands unless they are measured under the same conditions. A better comparison method is to determine K_{obs} (or T_m) at several salt concentrations and plot log K_{obs} versus log $[Na^+]$ according to equation (8.2.3). Comparison of several compounds can then be made under more defined standard state conditions, physiological salt concentration, etc. A plot of log K_{obs} versus log $[Na^+]$ also allows a determination of m' from the slope.

As stated above, conformational changes of nucleic acids, such as denaturation, are also dependent on the salt concentration of the solution. Any ligand-binding process which causes a conformational change will thus have an additional salt dependence above that given by equation (8.2.3). For example, all intercalators lengthen the double helix of DNA and this causes release of counter-ions since the longer helix has a larger spacing between phosphate groups and a resulting lower intrinsic charge density. For a neutral intercalator, this will be the only source of counter-ion release since no ion pairs can form between the DNA phosphate groups and the uncharged ligand. Cationic intercalators release counter-ions from DNA both as a result of the conformational change induced in the double helix and because of phosphate neutralization through ion pairing.

Multiply charged simple cations such as magnesium and cations of simple organic amines such as 1,3-diaminopropane interact with DNA more strongly than monovalent cations (sodium, potassium, etc.) and displace the monocations from DNA. Much, and in some cases all, of the binding free energy of these simple multi-cations is electrostatic in nature and their interaction with nucleic acids can be modelled as a condensation process. Cations of other metals, such as zinc, copper, mercury, etc., interact with DNA partly through a non-specific condensation type binding, but these more complex metal species can also bind directly to base-pairs to form a site-bound complex. Although increasing concentrations of simple ions cause increases in the T_m of DNA, the more direct base-binding interactions of metals frequently decrease the T_m of DNA since the bases in the denatured state can interact either more strongly or at additional sites with metals (Section 10.10.1).

Any cation can associate with nucleic acids through a condensation interaction. Since condensation creates a non-specific, mobile-type complex along the exterior of the double helix, the kinetics of association and dissociation of such a complex are quite rapid. Kinetic studies indicate that complex cations, which at equilibrium bind to DNA through groove or intercalation complexes with very specific interactions, initially associate with the duplex through condensation. The association pathway may involve initial Debye–Hückel type interactions followed by condensation and fast diffusion along the duplex backbone to the specific binding site. This process is somewhat analogous to 'sliding' of protein molecules along the DNA helix as they search for their most favoured binding site (Section 9.1.5). Dissociation could also involve condensation as an intermediate step between the site-bound ligand and release of the bound molecule into the bulk solution. The binding mechanism in specific cases may have additional steps.

Water is also bound along the exterior of nucleic acid structures (Section 2.2.4). Its specific interactions with polar groups on bases and sugars as well as with the charged phosphate groups are essential for the stability of nucleic acid conformations. At a stage weaker than these specific interactions, the water of solvation begins more closely to approach the properties of bulk solvent. Release of strongly bound water by association of ligands at specific nucleic acids sites can provide both favourable (increase in entropy) and unfavourable (increase in enthalpy through loss of specific interactions) contributions to the free energy of binding. The relative magnitude of these contributions will depend on the number of water molecules released and the types of interactions which are broken during complex formation.

8.2.2 Non-specific outside stacking

Planar aromatic molecules can stack on each other to form dimers and higher aggregates. When the compounds are charged, as with proflavine (Fig. 8.2), they repel each other electrostatically. If, however, the cations stack along the anionic DNA sugar-phosphate chain, the charge repulsion is decreased and this type of binding leads to non-specific **outside stacking** of planar cations along the double helix. Such stacking has many charge interactions and releases a large fraction of condensed counter-ions. It is therefore very dependent on salt concentration and is generally quite weak at salt concentrations of 0.1 M and above. Because this binding mode is a type of extended self-association, it can be highly co-operative and will generally be more favourable at high ratios of the aromatic cation to DNA phosphate groups.

Summary

Nucleic acids are highly charged polymers which must 'condense' a significant number of cations from solution to exist in stable conformations. Partial release of these cations on denaturation or on binding of cationic ligands to specific nucleic acid sites is a major factor in the strong dependence of these processes on salt concentration.

8.3 Groove binding molecules

8.3.1 Characteristics of groove binding

The general concept of 'groove binding' has been illustrated above (Fig. 8.1), and netropsin (Fig. 8.2) is a typical molecule which interacts with DNA by this mechanism. The major and minor grooves differ significantly in electrostatic potential, hydrogen bonding characteristics, steric effects, and hydration. Many proteins exhibit binding specificity primarily through major groove interactions while small groove binding molecules in general prefer the minor groove of DNA. No similar reversible interactions in the grooves of RNA have been identified as yet.

Typically, minor groove binding molecules have several simple aromatic rings such as pyrrole, furan, or benzene connected by bonds with torsional freedom. This creates compounds which, with the appropriate twist, can fit into the helical curve of the minor groove with displacement of water from the groove. The DNA minor groove is generally not as wide in A·T-rich relative to G·C-rich regions (Section 2.3.3) and may 'fit' aromatic molecules better at A·T than at G·C sequences. Given the correct twist of its linked aromatic rings, a molecule can fit snugly into the minor groove and form van der Waals contacts with the helical chains which define the 'walls' of the groove. Additional specificity in the binding comes from contacts between the bound molecule and the edges of the base-pairs on the 'floor' of the groove. Hydrogen bonds can be accepted by A·T base-pairs from the bound molecule to the C-2 carbonyl oxygen of T or the N-3 nitrogen of A. Although similar groups are present on G·C base-pairs, the amino group of G presents a steric block to hydrogen bond formation at N-3 of G and at the C-2 carbonyl of C. The hydrogen bond between the amino group of G and the carbonyl oxygen of C in G·C base-pairs lies in the minor groove and sterically inhibits penetration of molecules into this groove in G·C-rich regions. Thus, the aromatic rings of many groove binding molecules form close contacts with A-H[2] protons in the minor groove of DNA and there is no room for the added steric bulk of the G-NH$_2$ group in G·C base-pairs. Pullman and co-workers have shown that the negative electrostatic potential is greater in the A·T minor groove than in G·C-rich regions of DNA, and this provides an additional important source for A·T-specific minor groove binding of cations. It is possible to enhance G·C binding specificity by designing molecules which can accept hydrogen bonds from the G-NH$_2$ group. Synthesis and analysis of such molecules is an active area of research.

A possibility with groove binding molecules, which does not exist with intercalators, is that they can be extended to fit over many base-pairs along the groove and have very high sequence-specific recognition of nucleic acids. Additional pyrrole groups could be added to netropsin, or an oligopyrimidine could be designed to form a triple helix with a specific homo-purine–homopyrimidine duplex sequence (Section 2.4.5). By linking DNA cleavage reagents to the oligopyrimidine, a highly specific 'nuclease' can be created (Section 8.7.2). Oligonucleotides are also being used as 'antisense' anti-viral and anti-cancer drugs which specifically recognize single-stranded cellular nucleic acids.

8.3.2 Netropsin and distamycin

Dickerson and co-workers obtained a crystal structure of netropsin (Fig. 8.2) bound to the DNA duplex d(CGCGAATTCGCG) (Fig. 8.3) which has provided considerable molecular detail about complex formation in the minor groove. Netropsin binds at the AATT centre of the duplex and displaces the spine of hydration seen in that region of the free oligomer (Fig. 8.3). The three amide NH groups point inward and form bifurcated hydrogen bonds with N-3 of A and O-2 of T. The molecule is held in the centre of the groove by van der Waals contacts with the atoms of DNA which form the walls of the groove. These contacts hold the pyrrole rings approximately parallel to the walls of the groove and, as a consequence of the helical twist of the groove, the two pyrrole rings are twisted by approximately $33°$ with respect to one another (Fig. 8.3).

The two cationic ends of netropsin are also centred in the minor groove and are associated with N-3 on the outer A bases of the central four A·T base-pairs. Steric interactions between the pyrrole-CHs and the DNA bases prevent netropsin from moving more deeply into the groove. As a consequence, some of the hydrogen bond lengths between netropsin and the A·T base-pairs are quite long (3.3–3.8 Å) compared to standard values (less than 3 Å). Binding of netropsin causes a slight widening of the minor groove in the AATT region and a bending of the helix axis away from the site of binding. No other characteristic helical parameters are significantly changed in the complex. As indicated above, the amino group of G prevents molecules of this type from sliding deeply into the minor groove and forming hydrogen bonds with the bases. Netropsin binding to G·C-rich regions of DNA is thus weaker than to A·T sequences. Other factors such as water and counter-ion release contribute to the overall free energy of binding of netropsin, but probably have little effect on binding specificity.

Lown and Dickerson have devised an interesting new series of sequence-specific binding reagents, the **lexitropsins**, which are designed to recognize both A·T and G·C base-pairs. They noted that the pyrrole ring CH-groups of netropsin point into the minor groove and make close contact with A·T base-pairs. In the lexitropsins one or more of the pyrrole rings is replaced by hydrogen-bonding acceptor heterocycles such as imidazole. The exchange of a pyrrole-CH for an imidazole-N function alleviates the steric clash with the G-NH$_2$ group which prevents netropsin from binding in G·C regions. In its place, a specific hydrogen bond can form between the G-NH$_2$ and an imidazole nitrogen which should increase G·C interactions with appropriately substituted lexitropsins. Nuclear magnetic resonance and footprinting experiments have confirmed increased G·C base-pair recognition specificity for synthetic lexitropsins although A·T base-pairs are are still permitted at the imidazole substituted sites.

Distamycin (Fig. 8.4) has a structure and binding specificity similar to netropsin, and its DNA complex has many characteristics quite similar to the netropsin complex. Rich and

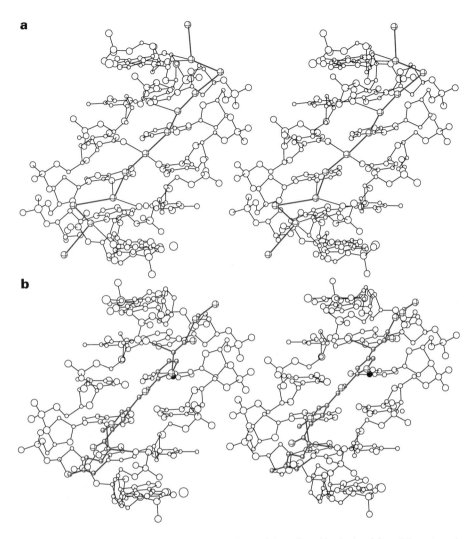

Fig. 8.3 Stereoviews of B-DNA minor groove bindings of the spine of hydration (**a**) and the netropsin molecule (**b**). In each case, only the central six base-pairs are drawn for the dodecamer of sequence: CGCGAATT-BrC-GCG, as established by single-crystal X-ray structural analysis. Base-pair G4 · C21 is at the top and C9 · G16 is at the bottom (both chains numbered from the 5'-end). Open circles are DNA atoms and red circles are water or netropsin atoms. The guanidinium end of the netropsin molecule is at the top, and the amidinium end is at the bottom. Distances from adenine N-3 or thymine O-2 atoms to water oxygens, or to netropsin amide nitrogens, are drawn as thin lines only if they at 3.5 Å or less (adapted from Kopka, M. L., Yoon, C., Goodsell, D., Pjura, P., and Dickerson, R. E. (1985). *Proc. Natl Acad. Sci. USA*, **82**, 1376–80). Stereopair figure for parallel viewing.

Wang have solved crystal structures for the oligomer d(CGCA$_3$T$_3$GCG) and its distamycin complex. Distamycin has a crescent shape which closely matches the curvature of the minor groove which is the binding site of the drug. In the crystal structure, the molecule is twisted in a complementary manner to the natural helical twist of the minor groove. As with netrop-

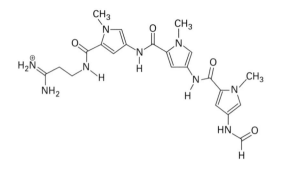

Fig. 8.4 Structure of distamycin. The crescent shape is similar to that observed in the crystal structure with the oligomer d(CGCA₃T₃GCG).

sin, the molecule makes close van der Waals contacts with the 'walls' of the minor groove and all solvent is displaced from the groove at the binding site.

Five NH groups are on the inside of the 'crescent' shape of distamycin and these can form hydrogen bonds with the N-3 of A and O-2 of T. At least three bifurcated hydrogen bonds are formed (to N-3 and O-2) in the crystal structure. The geometric arrangement of NH groups on distamycin does not exactly match the helical displacement of hydrogen bond accepting groups on A·T base-pairs in DNA, and consequently not all distamycin NH groups can be in optimum hydrogen bonding positions simultaneously. The exact nature of the distamycin–DNA hydrogen bonding will, no doubt, depend on the local sequence and helix geometry. The distamycin molecule has three pyrrole rings, compared to two in netropsin, and covers five A·T base-pairs in the complex rather than four in the case of netropsin.

The oligomer was found to have an unusual conformation in the central d(A₃T₃) sequence of the crystalline complex. The base-pairs in this region have a high positive propeller twist which places the amino group of A between the O^4 group of its Watson–Crick complementary T and the O^4 group of its 5′-neighbour T. Thus, this amino group of A can form bifurcated hydrogen bonds with the two O^4 residues. A similar unusual conformation was found in crystals of this oligomer lacking distamycin and also by Klug and co-workers in the crystal structure of d(CGCA₆GCG)d(CGCT₆GCG).

Groove binding molecules and intercalators both exhibit unusual thermodynamic parameters: large positive enthalpies and entropies are found for binding to poly(dA)·poly(dT). Enthalpies are generally negative for binding to other sequences. These unusual thermodynamic results can be explained if the polymer has bifurcated hydrogen bonds for non-alternating A·T base-pair sequences, as observed for the oligomers. Any weakening or breaking of the precise geometry of these bifurcated hydrogen bonds would result in a positive enthalpy contribution, but the enhanced flexibility of the structure could result in a positive entropy. The release of water molecules which are bound to the poly(dA)·poly(dT) helix and which are displaced on binding other molecules could also contribute to the unusual thermodynamic properties of this sequence.

In NMR studies of DNA complexes with minor groove drugs, David Wemmer and co-workers have discovered a surprising and very interesting additional binding mode for distamycin and related compounds. In sequences that contain more than four consecutive A·T base-pairs, such as d(CGCAAATTTGCG), two distamycin molecules can bind into the central minor groove A·T-rich sequence. In sequences with four A·T base-pairs (as in Fig. 8.3) the complex with a single bound distamycin is favoured. In the sequence shown above with six A·T base-pairs, however, a second distamycin binds with higher affinity than the first to favour the dimer complex with two bound distamycin molecules. The NMR results suggest that the distamycin molecules are stacked on each other with their charged groups oriented in opposite directions. Both distamycin molecules in the dimer lie deep in the minor groove and can form hydrogen bonds with A-N^3 and T-O^2 acceptor groups as in the complex with a single bound distamycin.

The most surprising feature of this complex is that it requires the width of the DNA minor groove to be extended to approximately twice its width in the free oligomer or in the single distamycin complex. The width of the groove in the A·T-rich region of the oligomer is less than the width in the classical B-form helix. With the second bound distamycin, the width increases to a value larger than the B-form width. Although this represents a positive free energy contribution for distortion of the DNA, it is clearly more than compensated for by the excess negative free energy of binding two distamycin molecules instead of one distamycin in the favoured narrow minor groove complex. The flexibility of the minor groove of DNA and ability of the molecule to adapt to two quite different distamycin complexes reinforces the view of the conformational mobility of the B-form DNA duplex. Such sequence-dependent conformational adaptability may also play an important role in the specific recognition of DNA by proteins.

8.3.3 Aromatic diamidines

A number of aromatic diamidines (Fig. 8.5) bind in the DNA minor groove in A·T sequences in a manner similar to netropsin. As a class these compounds have excellent biological activity as drugs against a variety of microorganisms. Berenil and compounds of related structure have anti-trypanosomal activity while pentamidine is in clinical use against *P. carinii* pneumonia, a common opportunistic infection of AIDS patients. Neidle and co-workers have obtained X-ray and NMR structures for berenil and for pentamidine bound to A·T regions of DNA oligomers. Pentamidine, with its polymethylene linker, is somewhat different from the other minor groove binding agents, but in the X-ray structure the amidine groups form hydrogen bonds to A-N^3 acceptors at the floor of the minor groove as observed with other A·T-specific agents. In the pentamidine complex each phenyl-amidinium group is an approximately planar unit which is inserted deep into the minor groove and is aligned parallel to the walls of the groove. The two planar units are twisted by 35° with respect to each other as they follow the curvature of the groove. The phenyl rings are in close contact with A-H^2 protons, and substitution of G for A at these positions would disrupt the complex due to the steric bulk of the G-NH_2 groups. The A·T-specificity of pentamidine, as with other minor groove binding agents, is thus a combination of positive (hydrogen bonding, van der Waals, electrostatic) and negative (steric repulsion) effects. The methylene chain of pentamidine fits snugly into

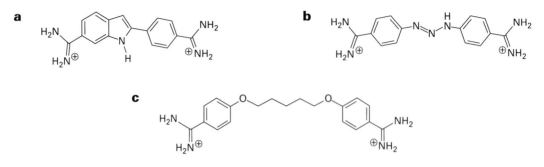

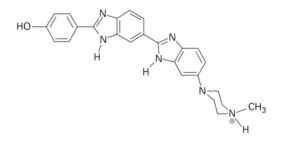

Fig. 8.5 Structures of the aromatic diamidines (**a**) DAPI (4′,6-diamidino-2-phenylindole), (**b**) berenil, and (**c**) pentamidine.

the minor groove and assumes a conformation to allow hydrogen bonding of the amidines of pentamidine with A-N^3 groups at the floor of the groove.

The amidine group can be replaced by similar planar structures such as imidazolines, but replacement by other cationic groups (such as $-CH_2NH_3^+$) leads to a dramatic decrease in both binding strength and A·T specificity. The planar aromatic amidine type unit is thus a key A·T recognition component of these drugs.

8.3.4 Hoechst 33258

Hoechst 33258 (Fig. 8.6) is an antibiotic and chromosome stain, quite different in structure from netropsin and distamycin, that also binds to A·T minor groove sequences of DNA. As with netropsin the molecule has a crescent shape with hydrogen bond donating groups on the inner face. The structure of DNA oligomer complexes with Hoechst 33258 has been investigated by X-ray and NMR methods.

In an X-ray structure with the sequence d(CGCGAATTCGCG) by Andrew Wang and co-workers, the Hoechst 33258 molecule is located in the AA·TT centre of the oligomer as with netropsin. The binding of Hoechst 33258 to DNA involves hydrogen bonds from the benzimidazole-NH groups to O-2 of T and N-3 of A and electrostatic interaction of the cationic dye with the anionic oligomer. The Hoechst 33258 molecule is also highly curved to match the curvature of the minor groove. The dye forms numerous favourable contacts with the walls of the minor groove and these interactions provide significant free energy of stabilization for

Fig. 8.6 Structure of Hoechst 33258.

the complex. The phenol ring of Hoechst 33258 makes an angle of 8° with the benzimidazole ring to which it is attached, but the two benzimidazole ring planes are twisted 32° with respect to each other. The piperazine is only slightly puckered and lies almost in the plane of the benzimidazole to which it is attached (dihedral angle 14°). The $O^{4'}$ atoms of deoxyribose in the minor groove are in a favourable position to interact with the π-electron system of the dye as it sits in van der Waals contact with the walls of the minor groove. This type of interaction may represent a general principle for binding of aromatic rings in the DNA minor groove.

Some variations are observed in the conformation of the Hoechst 33258 molecule in the X-ray and NMR structures, and in its orientation in the minor groove. These differences probably reflect different low-energy conformations that Hoechst 33258 can assume when bound to natural DNA sequences. One interesting feature that arises from an analysis of the complex dynamics by NMR is that although the phenol ring of Hoechst 33258 is tightly sandwiched between the walls of the DNA minor groove, it is undergoing ring flips (180° rotations about the phenol-benzimidazole bond) faster than the dissociation rate of the complex. This means that dynamic breathing of the DNA duplex occurs that transiently widens the minor groove such that the phenol ring can flip 180°. Similar fast rotation is observed for the Hoechst 33258 N-methylpiperazine group in the bound state. This is another indication of the flexibility of DNA, and indicates that extensive dynamic motions of the minor groove occur at a rapid rate (Section 2.5.4).

8.3.5 SN 6999

A range of biophysical studies with the drug SN 6999 (Fig. 8.7) indicate that it also binds in a minor groove complex with high A·T specificity. It is a very interesting compound with a fused bicyclic quinoline ring which, in the antimalarial drug chloroquine and other similar

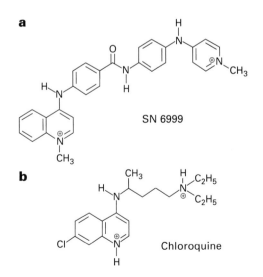

Fig. 8.7 Structure for (**a**) the groove-binding compound SN 6999 and (**b**) the intercalator chloroquine.

compounds, is capable of intercalation. The structures of the two quinoline compounds are shown (Fig. 8.7) for comparison.

Two-dimensional NMR methods can provide considerable structural detail about nucleic acids and their complexes. Leupin, Wüthrich, and co-workers have conducted a detailed investigation of the solution structure of the 1:1 complex of SN 6999 with the oligomer sequence d(GCATTAATGC) by such methods. Intermolecular nuclear Overhauser effect (NOE) results indicate that the compound binds in the minor groove and confirm that it does not intercalate despite the presence of the quinoline ring. As with other groove binding molecules, SN 6999 interacts with the central A·T base-pairs of the oligomer sequence. Bound and free SN 6999 are in slow exchange on the NMR timescale near 0°C, but move into fast exchange near 25°C. While the self-complementary oligomer is symmetrical when free, it becomes unsymmetrical in the complex with SN 6999 because of the lack of symmetry in the drug. The two exchange orientations of the drug in the oligomer A·T region and the intermolecular contact points between the base-pairs and the bound molecule have been deduced from NOE studies as shown (Fig. 8.8). The drug is curved in a crescent shape to fit into the helical minor groove and to form hydrogen bonds with A·T base-pairs as with other minor groove binding molecules.

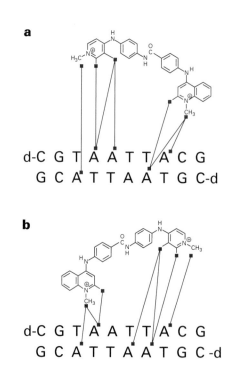

Fig. 8.8 Binding orientations and contacts for SN 6999 with the oligomer d(CGATTAATGC) as determined by NOE distance measurements (adapted from Leupin, W., Chazin, W. J., Hyberts, S., Denny, W. A., and Wüthrich, K. (1986). *Biochemistry,* **25**, 5902–10. Copyright (1986) American Chemical Society).

8.3.6 Groove binding versus intercalation

The similarity in structures but difference in the binding modes of chloroquine and SN 6999 raise an important question: 'What determines the selection between intercalation and groove binding interactions?' Based on model building studies, the quinoline ring of SN 6999 could form favourable contacts with the DNA base-pairs and perhaps electrostatic interactions with the DNA phosphate groups in an intercalation complex. Relative to groove binding, it would lose significant free energy of hydrogen bonding, of non-bonded contacts with the walls of the groove, and of solvent release from the groove. The same is true of netropsin and other similar groove binding molecules. At least for structures of this type, binding in the minor groove is much more favourable than intercalation. Chloroquine, on the other hand, does not have an optimum structure for groove interactions and it binds to DNA in an intercalation complex. Even so, the intercalation binding constant of chloroquine is low and it is classified overall as a weak DNA-binding molecule.

8.3.7 G·C specific minor groove agents

Chromomycin (Figure 8.9), mithramycin, and related members of the aureolic group of antitumour antibiotics exert their biological effects through complex formation with DNA. Chromomycin complexes have been studied in detail by two-dimensional NMR methods and the compound binds to the DNA minor groove as a magnesium ion linked dimer. It has a quite

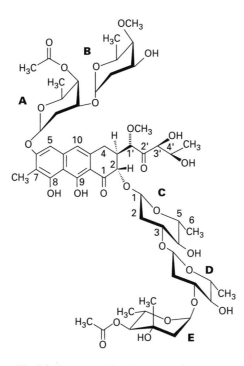

Fig. 8.9 Structure of the chromomycin monomer.

different structure from the A·T-specific groove binding agents discussed above. Strong binding is obtained with the DNA sequence 5'XGCY3', where X and Y can be any base. It is generally impossible to isolate a complex with a single bound chromomycin, and magnesium ion is an essential component of the dimer complex. The dimer is large compared to the standard width of the B-form minor groove, and Patel and co-workers have suggested that the DNA shifts to a more A-like helical conformation at the chromomycin binding site with significant widening of the minor groove. In the oligomer complex the hydrophilic edge of the chromomycin ring and the C—D—E sugar chain contact the edges of the G·C base-pairs at the floor of the minor groove. The G·C specificity arises through specific chromomycin-G hydrogen bonds, and the DNA recognition motif is completely different from that observed with minor groove binding agents such as netropsin which are specific for A·T base-pairs.

Summary

Molecules that bind in A·T sequences of the DNA minor groove are typically crescent shaped with hydrogen bonding NH groups on the interior of the crescent. These NH groups form hydrogen bonds with A·T base-pairs in the minor groove but are excluded from similar interactions with G·C base-pairs by the amino group of G.

Such molecules typically contain several small aromatic ring systems linked with torsional freedom to allow a twist complementary to that of the DNA minor groove. Electrostatic interaction of cationic groups with the negative electrostatic potential in the minor groove plus close van der Waals contacts with the 'walls' of the minor groove provide additional favourable components to the free energy of binding. Compounds related to chromomycin bind at G·C sequences of the DNA minor groove as magnesium ion linked dimers.

8.4 Intercalation

8.4.1 The classical model

In the early 1960s, Lerman conducted a number of physical studies on the interactions of DNA with planar aromatic cations, and concluded that planar aromatic molecules could bind to DNA by a process which he termed **intercalation** (Fig. 8.1). This mode of binding has now been established for a large number of polycyclic aromatic cations (example structures are shown in Figs 8.2 and 8.10). At the same time, the classical intercalation model has been extended in molecular detail. Just as the classical B-DNA is now seen as only one of the many possible conformational states of double helical nucleic acids, the classical intercalation model is only one view of the ways in which aromatic compounds can be accommodated between nucleic acid base-pairs.

As a result of rotation about torsional bonds in the DNA backbone, the creation of an intercalation site causes separation of base-pairs with a lengthening of the double helix and a decrease in the helical twist at the intercalation site (unwinding). The increase in length can be detected by hydrodynamic methods (Section 10.3), such as viscosity and sedimentation measurements, using short linear sections of DNA prepared by sonication or enzymatic methods.

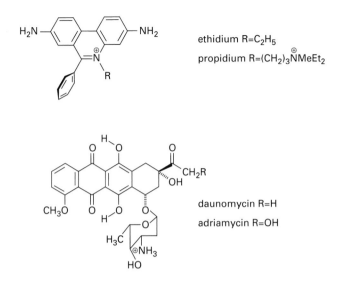

Fig. 8.10 Structures of the phenanthridinium intercalators, ethidium and propidium, and the anthracycline intercalators, daunomycin and adriamycin.

Compounds which bind in the DNA grooves such as netropsin, Hoechst 33258, and SN 6999 do not significantly increase the viscosity of DNA or unwind DNA base-pairs. In the classical model, the helix is lengthened by 3.4 Å, the thickness of typical aromatic ring systems. In practice, the observed length increase of the DNA complex is generally less than the 3.4 Å maximum because of effects such as bending of the helix at the intercalation site. The helix is unwound at the site of an intercalation complex and the normal approximately 36° rotation of one base-pair with respect to the next is decreased as a result of intercalation. The amount of unwinding varies considerably with the intercalator structure and probably with the DNA sequence in a manner that is not well understood. Of the compounds illustrated (Figs 8.2 and 8.10), ethidium and propidium unwind DNA by 26°, proflavine and related acridines by 17°, and the anthracycline drugs daunomycin and adriamycin by 11° per bound molecule.

The amount of unwinding in DNA produced by intercalation is conveniently measured with closed circular supercoiled DNA (Section 2.3.4, Fig. 2.25) and the unwinding angle calculated from such experiments is averaged over all of the intercalation sites in the random sequence DNA. However, actual unwinding angles at each binding site may vary with the local sequences.

An additional test for intercalation is provided by linear dichroism methods. The planar ring systems of intercalators stack with the DNA base-pairs and thus have similar dichroism values to the base-pairs. The dichroism of groove binding molecules is frequently opposite of that of the base-pairs since the groove complexes are bound along the edges of the DNA base-pairs rather than stacked between base-pairs, as is the case with intercalators (Fig. 8.1).

Crothers and his group have measured the linear dichroism both of DNA bases and of bound intercalators in double helical nucleic acids when the complexes are oriented by means of a strong electric field (electric dichroism). Based on such electric dichroism results, they proposed a model in which the propeller twist of base-pairs adjacent to the planar intercalator is flattened in order to stack better with the intercalated ring system. The intercalator and adjacent base-pairs are then tilted by 20–25° so as to optimize stacking interactions with propeller-twisted base-pairs adjacent to the complex. In general the overall DNA duplex seems not to be bent significantly by the bound intercalators, and there must be some other compensating changes which occur to counteract the tilting induced at the intercalation site.

Crystallographic studies have provided additional molecular detail concerning intercalator-induced tilt and long-range compensating effects in the DNA duplex. However, many of the available mono-intercalator crystal structures involve dinucleotides which cannot provide long-range information. With the dinucleotide crystals, only molecules with 5′-pyrimidine-purine-3′ sequences form Watson–Crick base-paired duplex structures with intercalators. Such duplex dinucleotide structures provide support for and extend the detailed pictures of the classical intercalation model. In general, drugs such as proflavine and ethidium (Figs 8.2 and 8.10) are stacked with their long axes parallel to the long axes of the adjacent base-pairs in these crystal structures. The exocyclic amino groups of the intercalators point towards diester oxygens of the DNA phosphate groups at the intercalation site and provide additional electrostatic and hydrogen bonding stabilization of the complex. The base-pair centres are separated by 6.9 Å in the complex (3.4 Å space for the intercalator) and stack well with the intercalated ring system. The base-pairs may be slightly kinked, especially for ethidium complexes where the out-of-plane phenyl group limits full intercalation of the cationic phenanthridinium ring. The phenyl and ethyl substituents of ethidium lie in the minor groove of the complex. Base-pair unwinding angles in these dinucleotide models have ranged from 0° for proflavine to 26° for ethidium. It is not necessary for them to relate closely to observed DNA-unwinding values because of the local sequence effects on unwinding angles in the high molecular weight nucleic acid and end-effects in the dinucleotides.

8.4.2 Anthracycline intercalators

The first crystal structure with a monointercalator and oligonucleotide was obtained by Wang, Rich, and co-workers for a complex of the antibiotic daunomycin (Fig. 8.10) and the oligomer d(CGTACG) (Fig. 8.11). Unlike intercalators studied at the dinucleotide level, daunomycin binds to the duplex with its long axis almost perpendicular to the long axes of adjacent base-pairs at the intercalation site. The daunomuycin amino-sugar, which is attached to ring A of the anthracycline ring system, lies in the minor groove while ring D, which bears a methoxy group, protrudes into the major groove. The daunomycin core, rings B and C, lies between base-pairs. The hexamer has two daunomycin molecules intercalated at each of the two C·G sites at either end of the duplex. These intercalation sites are presented (Fig. 8.11a) with the daunomycin molecules deleted so that the geometry of the intercalation site can be seen more easily. The central A·T base-pairs retain a general B-DNA-like geometry in the complex, but with some backbone distortions. The same view of the duplex is shown containing the intercalated drugs (Fig. 8.11b). Views of the complex into the major and minor grooves (Fig. 8.11c and 8.11d respectively) are also shown.

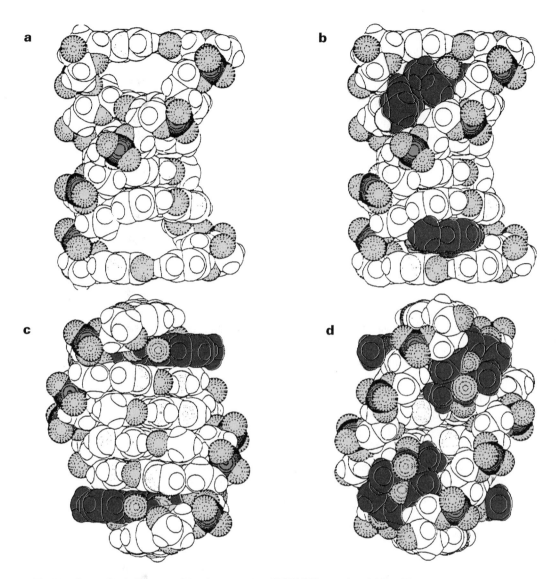

Fig. 8.11 Space-filled diagrams of the daunomycin : d(CGTACG) complex. (**a**) The DNA hexamer intercalation cavity with the daunomycin molecule deleted. This view is perpendicular to the molecular twofold axis, which lies horizontally in the plane of the paper. The minor groove of the distorted right-handed B-DNA is at the upper left of the figure, while the major groove is at the lower right. (**b**) The drug–DNA complex viewed from the same direction as in (**a**). The amino sugar of the daunomycin (colour) fills the minor groove of the double helix. Ring D of the intercalated aglycon chromophore skewers through the base-pairs oand protrudes into the major groove. (**c**) A view looking down the twofold axis from the major groove. The degree of penetration of the aglycon ring through the base-pairs is evident. (**d**) A view looking down the twofold axis from the minor groove. The amino sugars of the daunomycins largely fill the minor groove of the double helix. Note that the disposition of the sugar relative to the aglycon keys the daunomycin for a right-handed helix (adapted from Wang, A. H., Ughetto, G., Quigley, G. J., and Rich, A. (1987). *Biochemistry*, **26**, 1152–63. Copyright (1987) American Chemical Society).

As expected, the C·G intercalation site is opened by 3.4 Å to create the intercalation space. The cationic amino-sugar substituent and ring A largely fill the minor groove and displace water molecules and ions from it. The hydroxyl group on ring A donates a hydrogen bond to N-3 of G and is a hydrogen bond acceptor from the NH_2 group of the same G. This gives increased binding free energy and both base-pair and orientational specificity for the intercalated drug. The conformations of ring A and of the amino-sugar of daunomycin change relative to the unbound molecule and these changes facilitate the snug fit of the molecule into the right-handed minor groove.

The conformation of the duplex is also significantly changed relative to the B-DNA helical structure to accommodate the antibiotic. In addition to the increased separation of base-pairs at the intercalation site, the G·C base-pairs are also shifted laterally towards the major groove so that the helix axis changes position. There is no unwinding of base-pairs at the intercalation site, i.e. they maintain the usual 36° helical twist, but base-pairs at adjacent sites are unwound by 8°. While this agrees reasonably well with the net 11° unwinding angle determined from studies in solution with superhelical DNA, the fact that the unwinding occurs at the base-pairs adjacent to the intercalation complex is unexpected. Such long-range induced conformational changes provide another illustration of the flexibility of DNA, and demonstrate the significant variations that can occur in duplex conformations to accommodate a large intercalator which has structurally complex groove binding substituents. In other intercalation complexes unwinding of DNA does occur at the intercalation site. The manner in which DNA is unwound in intercalation complexes depends on the bound drug structure and probably also on the local DNA sequences.

Footprinting, equilibrium binding, and theoretical studies on daunomycin–DNA complexes have demonstrated that the binding site of the drug consists of three consecutive base-pairs rather than just the two at the intercalation site. The optimum daunomycin binding sequence consists of an A·T base-pair followed by two G·C base-pairs with daunomycin intercalated at the G·C base-pairs. The X-ray model predicts that the sugar residue in the minor groove largely accounts for the base-pair selectivity of this intercalator, and this model is strongly supported by detailed thermodynamic studies by Breslauer and co-workers. Daunomycin has similar thermodynamic binding profiles with poly(dA-dT)$_2$, poly(dA-dU)$_2$ and poly(dI-dC)$_2$ duplexes which have the same minor groove groups but different base substituents in the major groove. Poly(dG-dC)$_2$ and poly(dI-dC)$_2$, which have the same groups in the major grooves but different groups in the minor groove, have quite different thermodynamic binding characteristics. These results are consistent with a model in which minor groove interactions control the specificity of daunomycin interactions.

8.4.3 Binding specificity

There are 10 possible dinucleotide combinations which form different, right-handed, antiparallel intercalation sites for simple intercalators. The ten combinations are (all written $5' \rightarrow 3'$): A-T·A-T; A-A·T-T; TA·TA; A-C·G-T; C-A·T-G; G-C·G-C; G-G·C-C; C-G·C-G; G-A·T-C; and A-G·C-T. For intercalators with substituents which contact base-pairs beyond the intercalation site or which cause distortion in sequences neighbouring the binding site, the number of specific possibilities is even larger. Most simple intercalators display either no binding preference or a slight G·C base-pair preference and this contrasts with the A·T preference of many outside-binding compounds. It hs been suggested that the general G·C preference of simple intercalators is due to the larger intrinsic dipole moment of G·C relative to A·T base-pairs and the resulting ability of G·C base-pairs to induce polarization in the ring system of intercalators. Intercalators with an A·T preference have been synthesized, and in their case overall binding specificity must depend not only on dipole interactions but also on hydrogen-bonding interactions, on the size, hydration, and electrostatic potential of the grooves that interact with substituents on the intercalator.

In comparing groove binding molecules and intercalators, it is clear that groove binders, as a class, display significantly greater binding selectivity than intercalators. Intercalation cavities created at A·T or at G·C base-pairs are quite similar in their potential for interaction with planar aromatic ring systems. Electrostatic, van der Waals, hydrophobic, etc. contributions to binding are similar for the two sites. On the other hand, groove binding molecules can contact more base-pairs as they lie along the groove in a DNA complex, and this gives them an inherently greater recognition potential. As discussed above, the grooves are quite distinct in A·T and G·C regions and this adds to the potential for specific interactions by groove-binding agents.

Hélène and co-workers have dramatically enhanced the binding specificity of intercalators by linking them covalently to oligonucleotides of specific sequence. These covalent adducts can locate complementary single-stranded regions on other nucleic acid molecules with high precision. The intercalator provides increased free energy of binding as well as additional possible specificity in complex formation.

8.4.4 Neighbour exclusion

Daunomucin molecules in the d(CGTAGG) crystal structure bind at both C·G sites with the amino-sugar groups pointing inwards and essentially filling the minor groove (Fig. 8.7). In agreement with this structure, solution studies indicate that each bound daunomycin molecule covers three DNA base-pairs. By contrast, it appears that simpler intercalators with reduced steric constraints such as proflavine and ethidium (Figs 8.2 and 8.8) could bind at every potential intercalation site between base-pairs. At saturation of the double helix with the intercalator this would yield a binding stoichiometry of one base-pair per bound intercalator.

While models of duplex DNA with reasonable backbone torsional angles can indeed be constructed with intercalators at *all* possible sites between base-pairs, solution studies indicate that even the simplest intercalators appear to reach saturation at a maximum of one intercalator per two base-pairs. This empirical observation has led to the **neighbour exclusion principle** which states that intercalators can at most bind at alternate possible base-pair sites on DNA, giving a maximum of one intercalator between every second site. Initially, all spaces between base-pairs are potential binding sites for a non-specific intercalator. When an intercalator binds at one particular site, the exclusion principle states that binding another intercalator at adjacent sites is inhibited.

Various ideas have been developed to explore the molecular basis of neighbour exclusion of binding. One suggestion for the molecular basis of neighbour exclusion is that intercalator binding induces conformational changes at adjacent sites in DNA and the new conformation makes it impossible to intercalate another monointercalator at the adjacent site. An alternative, suggested by Friedman and Manning, is that since intercalators bind to DNA and neutralize some of its charge, the favourable release of condensed ions is reduced as intercalators bind to the helix and the observed equilibrium constant is also reduced. This leads to a curvature in binding isotherms which is similar to that predicted by neighbour exclusion. Since the local release of ions due to intercalation would be greater than ion release at a distance along the DNA molecule, the local release of counter-ions could also lead to a lower free energy of binding at neighbouring sites. It may well be that both conformational and electrostatic forces contribute to the observed neighbour exclusion.

It would seem that these various theories could be tested with simple uncharged intercalators which could not ion-pair with DNA phosphate groups. The difficulty, however, has been to design a simple, planar, uncharged aromatic molecule, e.g. similar to anthracene, which is soluble enough in water for its binding site size with DNA to be investigated quantitatively. Obviously, the vailidity and molecular origin of the neighbour exclusion principle will remain a topic of experimentation and discussion until its molecular basis is better understood.

8.4.5 Synthetic bisintercalators

Bisintercalators have two covalently linked intercalating ring systems with connecting chains of variable length (Fig. 8.12). It is also possible to link three or more ring systems together in the same way. The synthesis of these multiple ring compounds was partially stimulated by the idea that the medicinal activity of intercalating drugs could be enhanced by the significantly higher DNA binding constants and slower dissociation rates from DNA expected for bisintercalators relative to monintercalators.

Another feature anticipated for these molecules is that with short linkers the two intercalating rings might be induced to bind at adjacent base-pair sites and to test and more extensively evaluate the neighbour exclusion principle. For example, with a long linker, both planar rings are capable of intercalation without violation of the neighbour exclusion principle (Fig. 8.13a). With a short linker, both the rings may intercalate at adjacent sites in violation of neighbour exclusion (Fig. 8.13b).

One obvious way to change the linker length in bisintercalators is to vary the number of methylene units as shown by varying n in the bisacridine in Fig. 8.12a. Compounds of this type have been synthesized and it has been found with viscometric methods that for $n < 4$ the molecule binds with only one ring intercalated (as in Fig. 8.13c). When $n = 6$, the molecule apparently binds by bisintercalation with violation of neighbour exclusion (as in Fig. 8.13b). When $n > 8$, the molecule binds by bisintercalation but without violation of neighbour exclusion (as in Fig. 8.13a).

The results for the $n = 6$ compound are not in accord with NMR studies on complexes with oligonucleotides. Analysis of imino proton chemical shifts on adding the bisacridines to the

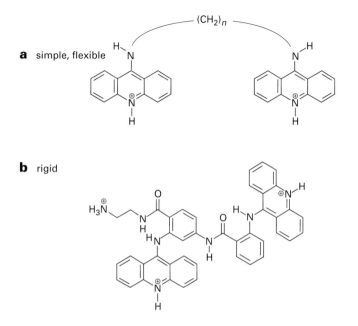

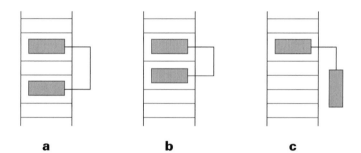

Fig. 8.12 (**a**) Simple, flexible acridine bisintercalators, (**b**) A more rigid acridine bisintercalator.

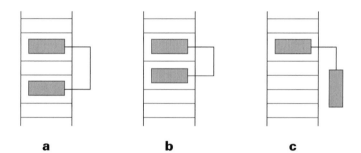

Fig. 8.13 The effect of linker length on the possible binding modes of bisintercalators.

self-complementary oligomer d(AT)$_5$ indicated that while bisintercalation occurred for $7 \leq n \leq 10$, only monointercalation occurred for $4 \leq n \leq 6$.

It is possible that the highly flexible linker of the bisacridines (Fig. 8.12) could allow the $n = 6$ molecule, with some local distortion of DNA, to bind by bisintercalation without violation of neighbour exclusion. If such a distortion of the oligomer was not possible with the sequence used or under the conditions of the NMR experiments, this could explain the disagreement between viscometric and NMR methods.

To eliminate the problem of linker flexibility, Denny and co-workers used a more rigid compound (Fig. 8.12b) which is also a better model for natural antibiotics than are bisintercalators with rigid linkers (Fig. 8.15). Because of its rigid structure, it cannot bind by mono-intercalation of the type shown in Fig. 8.13. Since the two rings of this species are insufficiently

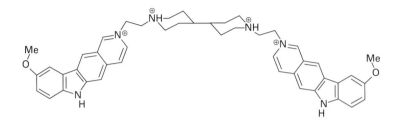

Fig. 8.14 Structure of the anti-cancer bisintercalator ditercalinium.

separated to allow bisintercalation with an intervening site (as in Fig. 8.13a), the compound must either bind as in Fig. 8.13b or by a non-intercalative mode. Viscosity results indicate that the rigid compound does intercalate and, in view of the geometric considerations discussed above, suggest that there is bisintercalation with violation of the neighbour exclusion principle.

Ditercalinium (Fig. 8.14) is a very important bisintercalator, with partial rigidity in the linking chain, that has been studied by X-ray and NMR methods. Ditercalinium is an anti-cancer drug which appears to exert its biological effects through an unusual mechanism in which DNA repair is induced by the bound drug. Since the drug forms a non-covalent complex, it can dissociate and induce a repair response in numerous positions on DNA such that the cell eventually dies due to extensive repair induced DNA degradation. The activity of derivatives of ditercalinium depends strongly on the position and type of substituents, and the monomer is biologically inactive. Compounds with more flexible linkers also have reduced activity.

In NMR and X-ray studies on the ditercalinium complex with the DNA tetramer duplex, $d(CGCG)_2$, the aromatic rings of ditercalinium are intercalated at the C·G sites with the linker in the DNA major groove. This is one of the few examples of intercalators with major parts of the molecule lying in the major groove, and it is interesting that ditercalinium has a different mechanism for its anti-cancer activity than other DNA interactive anti-cancer drugs such as the anthracyclines (Fig. 8.10).

The aromatic systems of ditercalinium stack well with the base-pairs at the intercalation site with the long axes of the aromatic groups roughly parallel to the base-pair long axes. Very little of the drug protrudes into the minor groove and when viewed from the minor groove, the drug molecule resembles a DNA base-pair. The rigidity of the linker causes an approximately 15° bend in the DNA helix axis, and the DNA is unwound as usual for intercalation complexes.

An interesting possibility for the biological activity of the drug is that the kink in DNA induced by the drug and its resemblance to a base-pair from the minor groove side may mimic the structural distortion produced by many DNA lesions, and could account for the ability of ditercalinium to induce DNA repair. This model explains why flexible linkers, which would not cause the kink, and large alkyl substituents at the indole nitrogen position, which would protrude into the minor groove, are biologically inactive.

Crystallographic and high resolution NMR studies of complexes of these molecules with DNA oligomers are difficult because of the low solubility of the complexes. It will be quite interesting to see what perturbations in normal DNA structure are induced by bisintercalation of this type when such high resolution studies become available.

8.4.6 Natural bisintercalators

The quinoxaline antitumour antibiotics, echinomycin and triostin A (Fig. 8.13a), have been studied in considerable detail. Structural analysis, both X-ray and NMR, indicates that the cross-linked cyclic peptides form a relatively rigid plate with the two quinoxaline rings perpendicular to the plate, parallel to each other on the same side and at opposite ends of the plate (Fig. 8.15b). The two aromatic rings are oriented in an optimum configuration for bisintercalation, separated by 10–11 Å, and can accommodate two base-pairs between the rings as required for binding by the neighbour exclusion principle.

In studies with supercoiled DNA, these molecules unwind the duplex by 40–50° and also produce a length increase approximately twice as large as for monointercalators. Both observations are in agreement with a bisintercalation binding mode. Binding and footprinting experiments suggest that echinomycin and triostin A intercalate preferentially around CpG sites in DNA and cover a 6 bp binding region of the type NNCGNN.

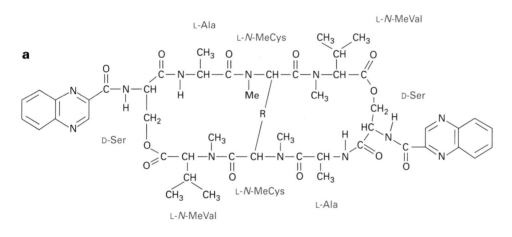

triostin A, R=—CH₂—S—S—CH₂—

echinomycin, R=—CH—S—CH₂—
                |
               SCH₃

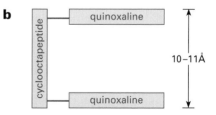

Fig. 8.15 (a) Structures for the natural bisintercalators triostin A and echinomycin. (b) Schematic diagram of the orientation and spacing between the quinoxaline rings.

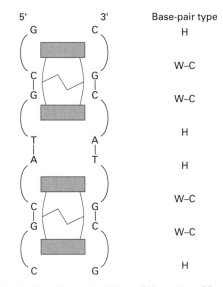

Fig. 8.16 Illustration of the binding sites and positions of Hoogsteen (H) and Watson–Crick (W–C) base-pairs in the crystal structure of triostin A (red) with the oligomer d(GCGTACGC).

Crystal structures of echinomycin and triostin A complexes with d(CGTACG) and of triostin A bound to d(GCGTACGC) have been determined with two antibiotic molecules bound per oligomer in all three cases. In all of the structures the quinoxaline rings are bisintercalated with the C-G sequence sandwiched between the intercalated rings (Fig. 8.16) in agreement with solution studies. The cyclic peptides lie in the minor groove and displace water molecules and cations from the groove at the binding site. The alanine residues of the cyclic peptides are key components in the binding specificity because the NH groups of both L-alanine residues of the cyclic peptide point in to the DNA base-pairs where they form hydrogen bonds with N-3 of the two G residues at the binding sites. The carbonyl group of the L-alanine on the outside edge of the complex also forms a hydrogen bond to the amino group of G.

The two A·T base-pairs in the centre of the oligomer have an unusual conformation in the crystal structure. The two adenines are in *syn* rather than the standard *anti* conformation and, as a result, they form Hoogsteen-type hydrogen bonded base-pairs with their complementary thymines (Fig. 8.16). The width across the helix is less for a Hoogsteen than for a Watson–Crick base-pair and as a result the chains are approximately 2 Å closer in the centre of the complex than they would be in an equivalent Watson–Crick helix. Because the quinoxaline aromatic system consists of only two fused intercalating rings, it also gives a better stacking arrangement with the reduced width Hoogsteen than with normal Watson–Crick base-pairs. By contrast, fused tricyclic intercalators fill the space between Watson–Crick base-pairs better than bicyclic ring systems. The quinoxaline ring system is small (Fig. 8.15) and much of the free energy of binding of complex formation with these antibiotics must come from peptide-base interactions in the minor groove rather than from base–intercalator interactions.

In the triostin A : d(GCGTACGC) crystal structure, the central A·T and the terminal G·C base-pairs adopt the Hoogsteen conformation (Section 2.1.2). This was particularly unexpected with the G·C base-pair because formation of the Hoogsteen base-pair requires protonation of C and, even then, only two hydrogen bonds are formed instead of the three obtained with Watson–Crick G·C base-pairing.

NMR studies by Juli Feigon and co-workers on a triostin A analogue, which does not have the methyl group on the valine nitrogens, indicate that this derivative specifically intercalates at T_pA sequences and does not intercalate in sequences that lack a T_pA step. The free valine NH in this derivative apparently forms a hydrogen bond with the peptide alanine carbonyl group. The resulting peptide conformation positions the alanine NH in the minor groove of the DNA complex to hydrogen bond with the A-N^3 group, and this accounts for the T_pA sequence specificity of the modified bisintercalator. No Hoogsteen base-pairs were observed in the complex of this derivative with DNA, but the sugar of T in the T_pA binding site is converted from the standard C2'-endo to a C3'-endo type conformation.

These observations emphasize two important points. First, DNA is a highly flexible molecule and can assume a wide variety of conformations. Secondly the binding free energy for complex formation with small molecules is sufficient for conversion of DNA into any of a variety of structures which, under normal conditions, are only slightly less stable than the normal B-family of conformations of DNA. It should be noted, however, that solution studies with specific chemical probes have not provided evidence for Hoogsteen base-pairs in longer DNA sequences. Hoogsteen base-pairs have been identified in NMR studies of echinomycin complexes with some short DNA sequences but not in others. The importance of Hoogsteen base-pairs in DNA–bisintercalator complexes may thus depend very strongly on the length and sequence of the DNA under study.

Summary

Typical intercalators are planar, aromatic cations which bind by insertion of the aromatic ring system between DNA base-pairs. They may have non-planar substituents, either cationic or neutral, which protrude into one of the DNA grooves. Generation of the intercalation site causes extension of the DNA duplex, local unwinding of the base-pairs, and other possible distortions in the DNA backbone (such as bends) which are characteristic of the intercalator species.

Bisintercalators have two potential intercalating ring systems connected with linkers which can vary in length and rigidity. The interaction of the ring systems with DNA base-pairs is controlled to a large extent by the characteristics of the linker. It has been proposed that some synthetic bisintercalators, with short linkers, can bind to DNA in violation of neighbour exclusion.

The larger molecular systems of bisintercalators gives them generally larger binding free energies (favourable), and they are able to induce significant distortions into the local DNA structure on complex formation.

8.5 Non-classical intercalators

8.5.1 Threading intercalators

Intercalators which have substituents on opposite sides of the intercalating aromatic system must thread one of the substituents between the base-pairs at the intercalation site in the binding mechanism. If the substituent is bulky and/or polar or charged, this could be an un-favourable step which might profoundly affect the kinetics of association and dissociation of the intercalator. Examples of intercalators of this type are naphthalene-bisimides, cationic porphyrins, and the anti-cancer antibiotic nogalamycin (Fig. 8.17).

The binding mechanism for *simple* intercalators such as proflavine (Fig. 8.2) involves two steps. First, the cationic intercalator interacts with DNA through an external electrostatic complex with the anionic backbone of the double helix. The intercalator can then diffuse in the anionic potential along the surface of the helix until it encounters gaps between base-pairs which have separated as a result of thermal motion to create a cavity for intercalation. The molecule can then bind in an intercalation complex. Threading intercalators with large side chains can require wider openings with significant distortions or breaking of base-pair

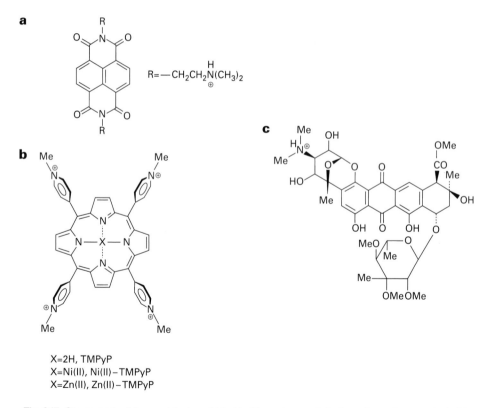

Fig. 8.17 Structures for (**a**) a naphthalene-bisimide (R can be an aliphatic amine or large cation). (**b**) cationic porphyrins and (**c**) nogalamycin. The non-metallo or Ni(II) derivatives in (**b**) are intercalators and the Zn(II) derivative is a groove-binding compound.

hydrogen bonds before they can form an intercalation complex. These larger, dynamic motions of the base-pairs are less likely and the kinetics of intercalation are thus much slower for threading molecules.

It should be emphasized that the unfavourable step for binding of threading intercalating refers to the kinetics and not to the thermodynamics of complex formation. These types of molecules (Fig. 8.17) can have high DNA binding constants indicating that once the side chain slides between the base-pairs, the DNA molecule can assume a conformation that gives a very favourable free energy of complex formation. The side chains present a kinetic barrier to binding, but have favourable interactions in the final complex after they have passed through the double helix. The kinetic barrier to binding depends on the size of the side chain, its orientation, and its polarity.

Porphyrins (Fig. 8.17b) are particularly interesting molecules since they appear to intercalate in G·C-rich regions of DNA, but bind in the minor groove in A·T-rich regions. Their binding properties vary substantially with the cationic substituent and type of metal bound centrally in the porphyrin. Oligonucleotide NMR studies by Marzilli, Wilson, and co-workers have suggested that the intercalating porphyrins have very high selectivity for binding at CpG sequences. This is unusual since the porphyrins do not have the hydrogen bonding capability observed for other molecules which show high selectivity for binding at specific DNA sequences.

NMR and X-ray studies of DNA complexes of the antibiotic and anti-cancer drug nogalamycin (Fig. 8.17) indicate that the amino sugar is in the major groove while the uncharged nogalose sugar is in the minor groove. As with daunomycin, the amino sugar forms a number of hydrogen bonds to G·C base-pairs and the drug has a binding preference for N·G or C·N sequences where N can be any base (it prefers to intercalate at the 5′-side of a G or at the 3′-side of a C). The two sugars point in the same direction in the intercalation complex and overlap the G·C base-pair at the binding site.

In the daunomycin complex the contacts that lead to base-pair specificity occur in the minor groove while with nogalamycin they are largely in the major groove, and the uncharged sugar in the minor groove appears to interact weakly with DNA. As with daunomycin, the long axis of nogalamycin is approximately perpendicular to the long axes of the base-pairs at the intercalation site. This orientation leads to a slight buckling of the base-pairs at the intercalation site such that they wrap around the drug molecule and form better van der Waals contacts. A conformation of this type should be most favourable for intercalators that bind with their long axis perpendicular to the long axes of base-pairs at the intercalation site.

8.5.2 Propeller-twisted intercalators

Several examples of aromatic cations with unfused bicyclic (netropsin) and tricyclic (distamycin and Hoechst 33258) aromatic rings have already been described. These compounds, which are all groove-binding molecules, have torsional freedom to optimize their twist to fit the minor groove of DNA. Strekowski, Wilson, and co-workers have designed, synthesized, and evaluated the DNA interactions of several new classes of unfused aromatic cations. The

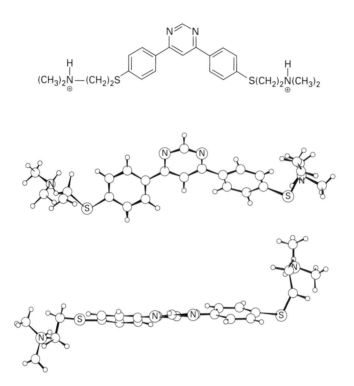

Fig. 8.18 Crystal structure for a propeller-twisted intercalator. The compound was synthesized by Dr. L. Strekowski and the structure was solved by Drs S. Neidle and G. Webster.

structure of one such derivative is shown (Fig. 8.18). Both molecular mechanics calculations and crystal structure determination indicate that this molecule has a significant intrinsic twist of 20° to 25° between the phenyl and pyrimidine planes. The twisted rings and terminal basic functions match the paradigm for a groove-binding molecule. Solution studies, however, indicate that the 4,6-diphenylpyrimidine (Fig. 8.18) and related molecules unwind supercoiled DNA, lengthen linear DNA, and have NMR and linear dichroism characteristics which establish them unequivocally as intercalators. It is interesting that all of these twisted tricyclic intercalators increase the rate of the bleomycin-catalysed cleavage of DNA (Section 8.9).

8.5.3 Intercalation versus groove binding

A comparison of the factors which control groove binding versus intercalation was made earlier, and it is worthwhile to consider these factors again after the above discussion of intercalators. It is now well known that the base-pairs of DNA can have significant intrinsic propeller twist and it seems quite reasonable that intercalation of unfused aromatic ring compounds might complement or perhaps enhance the interactions of base-pairs responsible for propeller twist. What is not yet completely clear is why the 4,6-diphenyl-pyrimidine (Fig. 8.18) is an intercalator while other related compounds such as Hoechst

33258 bind in the DNA minor groove. There are several possible reasons for this difference. First, the major groove is too wide to form many favourable contacts with all of these molecules and it is rejected as a significant binding site. Secondly, in the minor groove the amino group of G provides a steric block to prevent significant binding of many molecules. This leaves either A·T regions in the minor groove or intercalation as preferred binding sites. Thirdly, the choice between minor groove binding at A·T sites and intercalation probably strongly depends on the availability of hydrogen bond donating groups, such as NH, in the small molecule. *Appropriately placed* NH groups can form hydrogen bonds with N-3 of A and/or O-2 of T to form a very strong minor groove complex. With the 4,6-diphenylpyrimidine molecule (Fig. 8.18), however, there are no good hydrogen bond donor groups in or connecting the unfused rings and so there is no strong minor groove binding mode. However, the 4,6-diphenylpyrimidine obviously can form a good stacked complex with DNA base-pairs and it forms a strong intercalation complex with DNA.

Results with DAPI (Fig. 8.5a) have provided insight into the method by which an unfused aromatic cation selects between an intercalation or a groove binding mode. Numerous biophysical studies have shown that DAPI binds in the minor groove at A·T sequences much as netropsin does. Footprinting experiments indicate that DAPI spans approximately 3 bp in the minor groove which is reasonable based on the molecular length of DAPI. This molecule is structurally similar to many unfused intercalators discussed above and a variety of biophysical experiments have provided the surprising result that DAPI binds to G·C base-pairs by intercalation, not by groove interactions.

On review, this finding agrees quite well with what we have learned about DNA structure and interactions. Both A·T and G·C base-pairs form good intercalation sites. However, A·T but not G·C sequences also have a very favourable minor groove binding sites for unfused aromatic cations like DAPI and netropsin. DAPI thus selects intercalation as the binding mode at G·C sites but forms a minor groove complex in A·T regions. The binding constant for DAPI at G·C intercalation sites is similar to the binding constant for other intercalators such as proflavine (Fig. 8.2c). In the A·T minor groove, however, DAPI covers more base-pairs with more specific contacts than at the intercalation site. DAPI thus binds particularly strongly at A·T sites. This type of site selection by DAPI illustrates that intercalation and groove binding should be viewed as a two potential wells on a continuous energy surface. The binding mode with the lower energy will depend on the DNA sequence and structure as well as on the molecular features of the bound compound.

Pentamidine (Fig. 8.5c) binds similarly to DAPI in A·T sequences, but cannot stack well with DNA base-pairs in an intercalation complex. Pentamidine thus binds weakly to G·C-rich DNA sequences by external electrostatic interactions.

Summary

Classical intercalators are planar aromatic cations with no bulky substituents. Non-classical intercalators fall into two general classes. First, there are molecules with bulky substituents, such as porphyrins, which perturb the kinetics of interaction with DNA, but enjoy good stacking of the intercalated ring system with DNA base-pairs. Secondly, some molecules with non-fused, twisted, aromatic ring systems and terminal cationic functions, which more closely resemble the paradigm for a groove binding molecule, bind to DNA by intercalation. The distinction between intercalation and groove binding for unfused-aromatic cations probably depends on (1) placement of hydrogen bonding groups, (2) stacking interactions with DNA base-pairs relative to interactions with the walls of the minor groove, and (3) the degree of twist in the unfused ring system.

8.6 RNA Interactions

8.6.1 Intercalation with RNA duplexes

Simple A-form RNA duplexes, formed from synthetic polymers such as poly(A)·poly(U), bind intercalators with association constants in the same range as B-form DNA duplexes. X-ray structures of intercalator complexes with dinucleotide monophosphate miniduplexes are similar for both DNA and RNA and it is clear that intercalators can form similar stacked complexes with base-pairs in both DNA and RNA duplexes. Classical, threading, and unfused-aromatic intercalators can all form strong complexes with RNA duplexes in much the same way they do with DNA. The different local geometries and charge densities of DNA and RNA modify the intercalation sites such that specific intercalators can, however, have very different binding affinities for RNA and DNA in spite of the similar stacking potential in the binding sites.

The grooves of DNA and RNA have very different structures and charge densities. Compounds such as netropsin, distamycin, and pentamidine that bind strongly to the minor groove in A·T sequences of DNA, bind weakly to *all* sequences in RNA. Minor groove binding agents, such as DAPI, that can bind by intercalation in G·C sequences of DNA, can also bind to RNA by intercalation. Such mixed binding mode compounds can form strong intercalation complexes with RNA and with G·C sequences in DNA, but the binding in A·T sequences of DNA is superior due to the larger number of base-pair interactions in the groove.

8.6.2 Higher-order RNA structures

Natural RNAs are synthesized as single strands, but they typically form extensive base-paired duplex regions that are essential to their biological functions. The base-pairings in tRNA, rRNA, viral genomes, and other similar RNA strands have been described. These RNA duplexes (Section 6.3) can have distortions due to base bulges and base-pair mismatches as well as tertiary interactions that can give each sequence a unique conformation. The tertiary interactions of tRNA are, to a large extent, responsible for the characteristic conformation of that molecule, and specific tertiary structures, such as pseudoknots, are now known to be of wide-spread importance (Section 6.3.3).

Small organic cation interactions with RNA structures can affect biological functions. The best-documented interactions are with antibiotics that bind to ribosomes and inhibit translation. Such compounds are important drugs in the treatment of microbial infections. Two such compounds that have been studied in detail are streptomycin and kanamycin (Fig. 8.19).

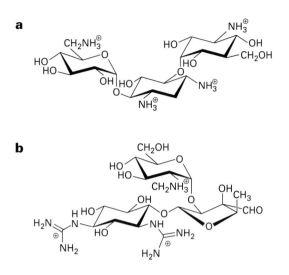

Fig. 8.19 Structures for the aminoglycoside antibiotics (**a**) kanamycin A and (**b**) streptomycin A.

Although these drugs could potentially inhibit translation by binding to either ribosomal proteins or RNA, considerable evidence suggests that specific interactions with 16S RNA (Section 6.4.2) are important for their function.

Ribosomes are large and complex structures and it is not yet clear how the antibiotics bind to ribosomal RNA to inhibit translation. The situation for interactions with complex RNA structures is more analogous to enzyme–inhibitor complexes, and simple classifications such as intercalation and groove binding modes, which is possible with DNA and RNA duplexes, is not possible with complex RNAs.

It is interesting that splicing of the sun Y intron RNA is inhibited by streptomycin but not by kanamycin. Thus, there are significant similarities and differences in the effects of antibiotics on splicing and on translation (Section 6.4.3). It is not clear whether the similarities are due to structural analogies between 16S ribosomal RNA and the intron RNA or to some other factor. The result does suggest that drugs can be designed to bind to RNA and selectively inhibit biological processes in microorganisms. Such agents could also be developed as antiviral agents against pathogenic RNA viruses.

Summary

Intercalators can bind to RNA duplexes through complexes that are similar to those formed with B-form DNA. The grooves of RNA are quite different from those of DNA, however, and agents that bind strongly in the minor groove of DNA bind only very weakly to either of the RNA grooves. No strong and specific RNA groove binding agents have been identified.

Antibiotics can interact with the tertiary structure of natural RNAs, and the interactions cause inhibition of the biological functions of the RNA. Such interactions are more analogous to enzyme–inhibitor complexes than to DNA complexes and reinforce our concept of the enzyme-like properties of RNA.

8.7 Multistrand structures

8.7.1 Types of structures and interactions

Shortly after the discovery of the B-form helix by Watson and Crick, triple-helical structures of DNA and RNA were described. The initial structures were composed of duplexes with a purine and a pyrimidine strand, and the triple helix was formed on binding of an additional pyrimidine strand in the major groove of the double helix (Sections 2.3.6 and 2.4.5). Hoogsteen base-pairing occurs between the duplex purines and the pyrimidines of the third strand, and the third strand binds in a parallel orientation with respect to the purine strand. Formation of the Hoogsteen pairs requires protonation of C to form a $C^+ \cdot G \cdot C$ base-triple along with $T \cdot A \cdot T$ base interactions (Fig. 8.20). Initial fibre diffraction studies suggested an A-form (C3'-endo sugar conformation) type helical structure for the triplexes, but NMR studies indicate that the sugar-phosphate backbone of the triplex can contain both C3'-endo and C2'-endo sugar conformations.

Of the bases found in nucleic acids, G can pair with itself to form dimers and tetramers. Four repeating sequences of G can fold to form four-stranded helices whose stability depends on the type and concentration of counter-ions. Such structures are models for the telomere conformations in eukaryotic chromosomes. Oligomer models have been synthesized for these structures, and are being studied in molecular detail (Section 2.5). Although four-stranded structures could form attractive targets in drug design, very little is known about their interaction with small organic compounds.

8.7.2 Triple-helix interactions

Interest in triple-helical conformations has been stimulated by several recent discoveries. First, Frank-Kamenetskii and co-workers demonstrated that mirror-repeat homopurine–homopyrimidine sequences in supercoiled plasmids could form an intramolecular triplex by dissociation of half of the mirror repeat duplex followed by triplex formation between the free pyrimidine strand and the remaining half of the mirror repeat. Such conformations could occur in cells in similar regions and may represent a gene control mechanism.

The potential for triplexes to affect gene expression lead to the antigene therapy concept described in detail by Hélène and co-workers. In this approach an externally added nucleic acid strand can bind to genomic DNA to inhibit selected processes in a diseased cell (e.g. a neoplastic cell or a cell with an integrated viral genome) to terminate the results of the disease process or simply kill the cell. A sufficiently long DNA strand can bind with very high selectivity to a desired duplex gene, and additional agents to cross-link or cleave (see below) the genomic DNA could permanently inactive the gene with minimal toxic side effects. The promise of this method has lead to a major effort to discover new triplex recognition modes that can increase the diversity of duplex sequences that can bind a third strand.

One of the difficulties of the antigene approach is that although triplexes can be formed with high specificity, they have generally low stability under physiological conditions. The third strand increases the negative charge on the highly anionic duplex and relatively high cation concentrations are needed to stabilize triplexes. A major search is under way to find

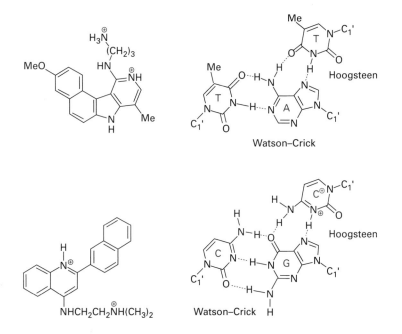

Fig. 8.20 Structures (right) for T·A·T and C$^+$·C·G base-triple interactions. Triplex-specific compounds designed and studied by Hélène and coworkers (top left) and by Strekowski, Wilson and coworkers (bottom left) are shown.

small molecules that can stabilize triplex conformations selectively. As described above, the grooves of triplex structures are very different from the duplex grooves and compounds such as netropsin, which bind strongly in the duplex minor groove, bind very weakly to triplex structures, as with RNA.

The search for triplex-stabilizing compounds has therefore focused on intercalators. A number of studies have been conducted on complexes of the prototype intercalator ethidium (Fig. 8.10) with triplex structures. Ethidium intercalates into triplexes and binds more strongly to triplexes composed of T·A·T base-triples than to the corresponding duplex. When G·C base-pairs are introduced into the duplex, the ethidium cation binds more weakly to the corresponding triplex than to the duplex, presumably because of the positive charge on the C$^+$·G·C base-triple (Fig. 8.20).

Two intercalators that dramatically stabilize triplex relative to duplex structures are shown in Fig. 8.20. The compounds have two common features that are important for selective triplex stabilization: they are both dications and they both have a large aromatic stacking surface to match the surface of the triple-base interactions in triplex intercalation sites. The quinoline compound is an unfused aromatic system and this may help it match the propeller twist in the triplex base interactions as discussed above for duplex interactions.

A triplex therapeutic strategy could be incorporated into antisense methods to recognize mRNA. In standard antisense therapy, a nucleic acid strand is added to form a duplex with mRNA and inhibit translation (Section 2.4.3). A longer added strand could fold back (hairpin) and form a three-stranded structure at appropriate sequences. Such a hairpin triplex could form an especially stable and specific clamp to inhibit translation.

8.8 DNA cleavage reagents

8.8.1 Antibiotics

Nucleases that cleave nucleic acids are well-known (Section 10.7.2). There are also small molecules that carry out similar phosphate diester hydrolyses. A thoroughly studied DNA cleavage reagent is the glycopeptide antibiotic and anti-cancer drug bleomycin A2 (Fig. 8.21). A number of closely related antibiotics has been discovered with slight modifications in different regions of the bleomycin molecule.

The 2,4'-bithiazole rings, a common feature of bleomycins, are linked to a small side chain which usually has a positive charge. The extensive remaining portion of the molecule provides a metal-complexing domain which is responsible for DNA strand cleavage. Removal of the bithiazole rings and cationic side chain leaves a metal-binding domain which neither significantly binds to nor degrades DNA at reasonable concentrations. Despite considerable study, the mechanisms by which the bithiazoles and side chain interact with DNA are not clear. NMR and hydrodynamic studies show that the bithiazole rings upon cleavage from the metal-binding domain can bind to DNA in an intercalation complex. Other such studies however, strongly suggest that the bithiazoles bind to DNA through a groove binding mechanism when they are part of the antibiotic. It may be, as with DAPI, that both binding modes are possible, and the ratio selected depends on the DNA sequence.

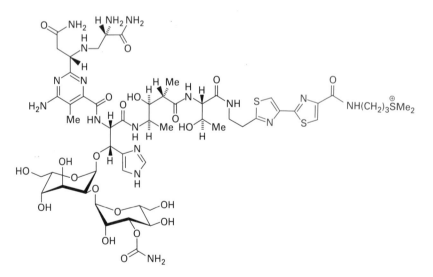

Fig. 8.21 The structure of bleomycin A2. The bithiazole rings and cationic side chain (colour) which direct the binding of bleomycin to DNA at the right. The metal-complexing portion of the molecule is to the left.

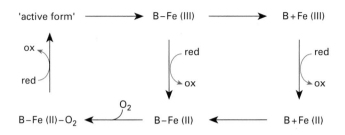

Fig. 8.22 A catalytic cycle for the bleomycin-induced oxidative degradation of DNA (bleomycin is designated as B). This scheme requires the presence of organic reducing reagents (red). Depending on the reducing agent and conditions, B–Fe(III) may be directly reduced or may first dissociate to free bleomycin and Fe(III).

There is a selectivity for bleomycin cleavage of DNA at purine–pyrimidine sequences with a higher than expected occurrence of double-strand nicks (i.e. cleavage of both DNA strands at sites in close proximity). Bleomycin can complex with several different metal ions to cleave DNA, but most work has been done with the Fe(II)–bleomycin species. The initial step in degradation of DNA is very selective abstraction of a deoxyribose 4′-hydrogen (Figs 8.22, 8.23; Section 7.9.1). Since H-4′ lies in the minor groove of DNA, this site-selectivity indicates that bleomycin binds in the minor groove.

The species directly responsible for the removal of the 4′-hydrogen is an 'activated' bleomycin–Fe(II)–O₂ ternary complex whose activation requires a one-electron reduction. The electron can come from another bleomycin–Fe(II)–O₂ complex (producing the activated species and an inactive bleomycin-Fe(III) complex) or from organic reducing agents such as ascorbic acid or thiols (Fig. 8.22, see Section 7.9.1). Hydrogen peroxide can produce the activated com-

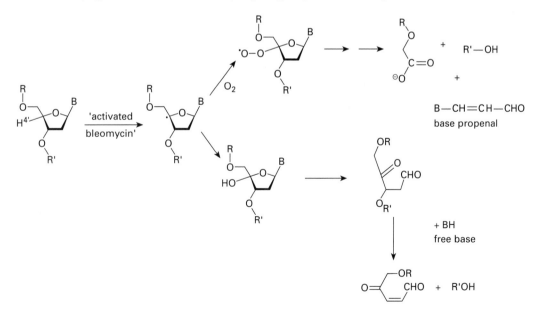

Fig. 8.23 Mechanisms for release of free bases and base-propenals by attack of 'activated' bleomycin on the C-4′ position of DNA.

plex directly from bleomycin–Fe(III). In the absence of DNA, the activated species slowly decomposes with destruction of the bleomycin molecule.

A particularly interesting feature of the bleomycin-induced degradation of DNA is the fact that the rate of strand cleavage by bleomycin can be enhanced significantly by some compounds such as the unfused intercalator shown above (Fig. 8.20 lower left). Most of the unfused intercalators have been shown to bind with their side chains in the DNA major groove so that they do not block bleomycin entry to the minor groove. Formation of the intercalation complex expands the grooves and facilitates bleomycin cleavage of DNA in the minor groove. Since clinical use of many anti-cancer drugs is limited by toxic side effects not necessarily related to DNA interactions, the ability to design amplifiers which increase the effects of the drugs at DNA but do not increase the toxic side effects offers the potential of significantly increasing the therapeutic value of the drugs.

In the basic bleomycin–DNA reaction scheme (Fig. 8.22) catalytic cleavage of DNA in the presence of organic reducing agents is a two-electron process involving two one-electron reductions of bleomycin–iron species. The selective reaction at the $C^{4'}$-H bond of deoxyribose in DNA suggests that a specific orientation of the bleomycin complex in the minor groove of DNA is responsible for the chemistry and that no mobile intermediates, such as hydroxyl radicals (cf. Section 7.9), are involved in the reaction. The oxidative degradation of DNA produces two types of base products: free bases and base-propenals. Formation of base-propenals

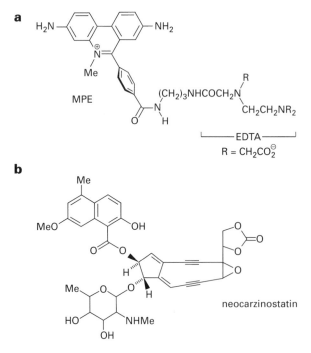

Fig. 8.24 Structures for (**a**) the ethidium analogue, methidiumpropyl-EDTA (MPE) and (**b**) the neocarzinostatin chromophore.

requires additional oxygen, but formation of free bases does not. Mechanisms are shown for formation of both products, through a common reaction intermediate at C-4' (Fig. 8.23). Both processes result in single chain breaks by the destruction of the deoxyribose sugar.

Neocarzinostatin, a peptide antibiotic quite different from bleomycin, also causes strand cleavage of DNA. The active part of the antibiotic is a non-protein chromophore containing a substituted naphthalenecarboxylic acid capable of intercalation with DNA (Fig. 8.24). The strand cleavage of DNA by the neocarzinostatin chromophore is activated by O_2 and thiol reagents. Cleavage is proposed to occur through removal of a 5'-hydrogen from deoxyribose groups in DNA (Section 7.7.4 and Fig. 7.28).

8.8.2 Synthetic cleavage reagents

Synthetic metal-complexing reagents such as porphyrin derivatives, 1,10-phenanthroline metal complexes, and an analogue of ethidium with an Fe(II)–EDTA containing system bonded to the *para*-position of the phenyl substituent (Fig. 8.24a) have also been found to bind to DNA and cause strand cleavage. This MPE derivative, designed by Peter Dervan, is now widely used as a footprinting reagent because of its relatively non-selective cleavage. It acts through an initial non-selective intercalation complex of the phenanthridinium ring system in which the EDTA–Fe(II) side chain lies in the minor groove. It is this group which is entirely responsible for cleavage of nucleic acid duplexes.

Unlike the highly specific proton abstraction observed with the natural antibiotics, MPE–Fe(II)–O_2 produces a diffusible species, presumably a hydroxyl radical, which can react at several nucleic acid sites (Section 7.9.1). The species is short-lived and attacks the deoxyribose groups of DNA in close proximity to the intercalated ring of MPE. As expected, a larger variety of products is produced than with bleomycin cleavage. Free bases are observed in the product mixture in amounts related to their percentage in the DNA sample. The Fe(III) produced in formation of hydroxyl radical can be reduced again to MPE–Fe(II), so that DNA cleavage can occur repeatedly if a reducing agent and oxygen are available. No cleavage occurs if Fe(II) in MPE is replaced by Ni(II) or Zn(II).

Phenanthroline cleavage reagents have been particularly useful in probing unusual structures of DNA and RNA. In the presence of H_2O_2, 1,10-phenanthroline-copper can give selective cleavage at unusual secondary structures, sequence selective cleavage of B-form DNA, and specific cleavage at loop region in RNA. The reagent, as with EDTA–iron, can be linked to sequence specific binding agents to give highly specific cleavage of DNA.

Jacqueline Barton has designed several metal–phenanthroline cleavage reagents for specific recognition of nucleic acids. Complexes such as $Rh(DIP)_3^{3+}$ (Fig. 8.25) are rigid and lack hydrogen bonding or other direct base interaction modes with nucleic acids. These reagents are designed to be *shape-selective* recognition units which bind to nucleic acid sites that are complementary to the shape and symmetry of the metallo species. Irradiation of the reagent induces strand cleavage of the nucleic acid at the binding site. Since detection of cleavage sites in isotope-labelled nucleic acids is very efficient, strong binding of the reagent is not essential, but highly selective binding is required. The metallo reagent gives highly specific cleavage at base-pair mismatches in A-form RNA but not at unperturbed sections of A-form duplex.

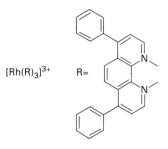

$[Rh(R)_3]^{3+}$ R=

Fig. 8.25 A schematic of the structure of tris(4,7-diphenyl-1,10-phenanthroline)rhodium(III), $[Rh(DIP)_3{}^{3+}]$.

Sequences in DNA that define the boundaries of introns are also hyper-reactive with $Rh(DIP)_3^{3+}$. The structural rationale for these highly specific cleavage events has not been discovered. Recognition of different structures can be obtained by variation of the metal and of the phenanthroline substituents.

One of the most exciting areas of research on synthetic cleavage reagents involves the design of highly specific artificial nucleases. The EDTA group, for example, has also been attached to distamycin and related A·T-specific groove binding molecules and, in the presence of Fe(II) and O_2, these reagents cause DNA cleavage with high specificity in A·T-rich regions. The EDTA group has also been covalently linked to homopyrimidine oligonucleotides. These reagents can cause DNA cleavage through selective binding to sequences of double-helical DNA which can form a triple-helix (Section 8.7) with the homopyrimidine–EDTA strand. The chemistry of DNA cleavage is similar for all of these reagents and depends on the formation of diffusible radicals at the EDTA–Fe(II)–O_2 complex.

Summary

A number of natural and synthetic reagents has been found which have the ability to bind to nucleic acids and cause strand cleavage. A metal ion associated with the bound molecule frequently initiates the chemistry of the cleavage reaction and O_2 is involved in some activated form. Specific cleavage of nucleic acids can be obtained by linking a metal-complexing reagent to a molecule which binds to nucleic acids with high intrinsic specificity. Unusual nucleic acid conformations that selectively bind one of these agents can be detected even in the presence of a large number of other structures.

8.9 Recognition of nucleic acids and drug design

The discussion in this chapter illustrates how research in reversible interactions of small molecules with nucleic acids can be divided into three primary areas:

- analysis of model systems at high resolution for the development of a better understanding of the molecular basis for nucleic acid interactions in general;
- development and analysis of drugs which interact with nucleic acids;

- design of new highly specific agents whose sequence-recognition properties can be selected as desired.

Improvements in DNA and RNA synthetic techniques as well as design and synthesis of additional DNA interactive drugs and model compounds have made available molecules which are providing new information on the structure and energetics of nucleic acid complexes and are advancing dramatically the field of reversible nucleic acid interactions. In addition, it is clear that both DNA and RNA are prime targets as receptors in drug development and a number of clinically useful drugs act at the nucleic acid level. There is particularly exciting progress in the design of oligopeptides and oligonucleotides for use as drugs and as reagents for the recognition of specific nucleic acid sequences. Oligonucleotides can recognize single-stranded DNA and RNA as well as duplex nucleic acids through triple helix formation. This gives oligonucleotides the potential to act as highly specific drugs, as drug carriers, or as site-specific nucleases, when linked to a cleavage reagent such as Fe(II)–EDTA.

The base-pair specificity and nucleic acid binding constants for the types of molecules discussed in this chapter can vary quite widely. Simple monointercalators, such as an anthracene ring system with a cationic side chain, have DNA binding constants of about 10^4 M^{-1} at neutral pH and 0.1 M NaCl, and there is very little specificity in their DNA interactions. Bisintercalators or monointercalators with more complicated side chains, which make specific contacts in the DNA grooves, can have significantly higher binding constants and recognition specificity. Groove binding molecules, such as the oligopeptides and oligonucleotides discussed above, can be designed to interact with the hydrogen bond accepting and donating groups of base-pairs which face into the DNA grooves and can thus have very high DNA binding constants and recognition specificity. The ability to extend such groove binding agents over many base-pairs gives them the ability to recognize a single sequence in the DNA of any organism.

In general, for a drug to be active it must not only bind to a nucleic acid but must also inhibit some protein activity at the nucleic acid (e.g. polymerase, topoisomerase, enhancer, etc.). Fairly simple intercalators have shown good ability to permeate cells and inhibit DNA associated enzymes, but their lack of specificity can cause non-selective toxicity. Groove binding molecules have shown good binding specificity and binding strength which can be extended to a range of sequences, for example, the lexitropsins (Section 8.3.2). A very interesting class of compounds, called combilexins are formed by linking an intercalator to a groove binding unit. The goal is to incorporate the advantages of both type units into a combilexin to obtain a highly selective drug. Because of the different structural requirements for intercalation and groove binding, care must be taken in the design of these agents to make sure that they can interact with the nucleic acid in the desired manner.

The chemistry described here gives just a few examples of the many very exciting ideas being tested in development of nucleic acid interactive drugs. As our knowledge of nucleic acid structure and interactions increases, particularly in areas such as RNA and multi-stranded-nucleic acids, we can look forward to additional ideas for the design of nucleic acid interactive drugs. The time when chemotherapy can target the genomic nucleic acid of selected cells or viruses is at hand.

Further reading

General

The following three books each belong to a series which contain many excellent and detailed articles on nucleic acid structure and interactions. Each series provides a very good source for more in-depth reading on topics covered in this and other chapters.

Hurley, L. H. (ed.) (1992). *Advances in DNA sequence specific agents.* Vol. 1. JAI Press, Greenwich, Connecticut.

Neidle, S. N. and Waring, M. J. (eds.) (1993). *Molecular aspects of anticancer drug–DNA interactions.* Vol. 2. The Macmillan Press, Ltd., London.

Pullman, B. and Jortner, J. (Eds.) (1990). *Molecular basis of specificity in nucleic acid–drug interactions: Proceedings of the 23rd Jerusalem Symposium on Quantum Chemistry and Biochemistry.* Kluwer Academic Publishers, Dordrecht.

8.1

Dervan, P. B. (1986), Design of sequence-specific DNA-binding molecules. *Science,* **232,** 464–71.

Hurley, L. H. and Boyd, F. L. (1987). Approaches toward the design of sequence-specific drugs for DNA. *Ann. Rep. Med. Chem.* **22,** 259–68.

Nielsen, P. E. (1991) Sequence selective DNA recognition by synthetic ligands. *Bioconjugate Chem.,* **2,** 1–12.

Wang, A. H.-J. (1987). Interactions between antitumor drugs and DNA. In *Nucleic acids and molecular biology* (ed F. Eckstein and D. M. J. Lilley), Vol. 1. Springer-Verlag, Berlin. pp. 53–69.

Waring, M. (1981). Inhibitors of nucleic acid synthesis. In *The molecular basis of antibiotic action* (ed. E. F. Gale *et al.*), 2nd edn. Wiley, London, pp. 274–341.

Wilson, W. D. (1987). Cooperative effects in drug–DNA interactions. *Progr. Drug Res.,* **31,** 193–221.

8.2

Bloomfield, V. A., Crothers, D. M., and Tinoco, I. (1974). *Physical chemistry of nucleic acids.* Harper and Row, New York.

Friedman, R. A. G. and Manning, G. S. (1984). Poly-electrolyte effects on site-binding equilibria with application to the intercalation of drugs into DNA. *Biopolymers,* **23,** 2671–714.

Record, M. T., Jr., Lohman, T. M., and deHaseth, P. (1976). Ion effects on ligand–nucleic acid interactions. *J. Mol. Biol.,* **107,** 145–58.

8.3

Coll, M., Frederick, C. A., Wang, A. H.-J., and Rich, A. (1987). A bifurcated hydrogen-bonded conformation in the d(A · T) base-pairs of the DNA dodecamer d(CGCAAATTTGCG) and its complex with distamycin. *Proc. Nat. Acad. Sci. USA,* **84,** 8385–9.

Edwards, K. J., Jenkins, T. C., and Neidle, S. (1992). Crystal structure of a pentamidine–oligonucleotide complex: implications for DNA-binding properties. *Biochemistry,* **31,** 7104–9.

Jenkins, T. C., Lane, A. N., Neidle, S., and Brown, D. G. (1993). NMR and molecular modeling studies of the interaction of berenil and pentamidine with d(CGCAAATTTGCG)$_2$. *Eur. J. Biochem.,* **213,** 1175–84.

Kopka, M. L., Yoon, C., Goodsell, D., Pjura, P., and Dickerson, R. E. (1985). The molecular origin of DNA-drug specificity in netropsin and distamycin. *Proc. Nat. Acad. Sci. USA,* **82,** 1376–80.

Lavery, R., Zakrzrewska, K., and Pullman, B. (1986). Binding of non-intercalating antibiotics to B-DNA: a theoretical study taking into account nucleic acid flexibility. *J. Biomol. Str. Dyn.,* **3,** 1155–70.

Leupin, W., Chazin, W. J., Hyberts, S., Denny, W. A., and Wüthrich, K. (1986). NMR studies of the complex between the decadeoxynucleotide d(GCATTAATGC) and a minor-groove-binding drug. *Biochemistry,* **25,** 5902–10.

Lown, J. W. (1988). Lexitropsins: rationale design of DNA sequence reading agents as novel anti-cancer agents and potential cellular probes. *Anti-Cancer Drug Design,* **3,** 25–40.

Nelson, H. C. M., Finch, J. T., Luisi, B. F., and Klug, A. (1987). The structure of an oligo(dA) · oligo (dT) tract and its biological implications. *Nature,* **330,** 221–6.

Pelton, J. S. and Wemmer, D. E. (1990). Binding modes of distamycin A with d(CGCAAATTTGCG)$_2$ determined by 2D NMR. *J. Am. Chem. Soc.,* **112,** 1393–9.

Searle, M. S. and Embrey, K. J. (1990). Sequence-specific interaction of Hoechst 33258 with the minor groove of an adenine-tract DNA duplex studied in solution by NMR spectroscopy. *Nucl. Acid Res.*, **18**, 3753–62.

Teng, M.-K., Usman, N., Fredrick, C. A., and Wang, A. H.-J. (1988). The molecular structure of the complex of Hoechst 33258 and the DNA dodecamer d(CGCGAATTCGCG). *Nucleic Acids Res.*, **16**, 2671–90.

Zimmer, C. and Wahnert, U. (1986). Nonintercalating DNA-binding ligands: specificity of the interaction and their use as tools in biophysical, biochemical and biological investigations of the genetic material. *Prog. Biophys. Molec. Biol.*, **47**, 31–112.

8.4

Address, K. J., Gilbert, D. E., Olsen, R. K., and Feigon, J. (1992). Proton NMR studies of [N-MeCys³, N-MeCys⁷] TANDEM binding to DNA oligonucleotides: sequence-specific binding at the TpA site. *Biochemistry*, **31**, 339–50.

Assa-Munt, N., Denny, W. A., Leupin, W., and Kearns, D. R. (1985). Proton NMR study of the binding of bis(acridines) to d(AT) · d(AT). 1.Mode of binding. *Biochemistry*, **24**, 1441–9.

Atwell, G. J., Stewart, G. M., Leupin, W., and Denny, W. A. (1985). A diacridine derivative that binds by bisintercalation at two contiguous sites on DNA. *J. Amer. Chem. Soc.*, **107**, 4335–7.

Chaires, J. B., Fox, K. R., Herrera, J. E., Britt, M., and Waring, M. J. (1987). Site and sequence specificity of the daunomycin-DNA interaction. *Biochemistry*, **26**, 8227–36.

Egli, M., Williams, L. D., Frederick, C. A., and Rich, A. (1991). DNA–nogalamycin interactions. *Biochemistry*, **30**, 1364–72.

Gao, Q., Williams, L. D., Egli, M., Rabinovich, D., Chen, S.-L. Quigley, G. J. and Rich, A. (1991). Drug-induced DNA repair: X-ray structure of a DNA–ditercalinium complex. *Proc. Natl. Acad. Sci. USA*, **88**, 2422–6.

Hogan, M., Dattagupta, N., and Crothers, D. M. (1978). Transient electric dichroism studies of the structure of the DNA complex with intercalated drugs. *Biochemistry*, **17**, 280–8.

Remeta, D. P., Mudd, C. P., Berger, R. L., and Breslauer, K. J. (1993). Thermodynamic characterization of daunomycin–DNA interactions: comparison of complete binding profiles for a series of DNA host duplexes. *Biochemistry*, **32**, 5064–73.

Wang, A. H.-J. (1992). Intercalative drug binding to DNA. *Current Opinion in Structural Biology*, **2**, 361–8.

Wakelin, L. P. G. (1986). Polyfunctional DNA intercalating agents. *Med. Res. Rev.*, **6**, 275–340.

Wang, A. H.-J., Ughetto, G., Quigley, G. J., Hakoshima, T., van der Marel, G. A., van Boom, J. H., and Rich, A. (1984). The molecular structure of a DNA-triostin A complex. *Science*, **225**, 1115–21.

Wang, A. H.-J., Ughetto, G., Quigley, G. J., and Rich, A. (1987). Interactions between an anthracycline antibiotic and DNA: molecular structure of daunomycin complexed to d(CGTACG) at 1.2 Å resolution. *Biochemistry*, **26**, 1152–63.

Wilson, W. D. and Jones, R. L. (1981). Intercalating drugs: DNA binding and molecular pharmacology. *Adv. Pharmac. Chemother.*, **18**, 177–222.

Wilson, W. D., Krishnamoorthy, C. R., Wang, Y.-H., and Smith, J. C. (1985). Mechanism of intercalation: ion effects on the equilibrium and kinetic constants for the interaction of propidium and ethidium with DNA. *Biopolymers*, **24**, 1941–61.

8.5

Marzilli, L. G., Banville, D. L., Zon, G., and Wilson, W. D. (1986). Pronounced proton and phosphorus NMR spectral changes on *meso*-tetrakis-(N-methylpyridinium-4-yl) porphyrin binding to poly[d(G·C)] · poly[d(G·C)] and to three tetradeca-oligodeoxyribonucleotides: evidence for symmetric selective binding to 5′CG3′ sequences. *J. Amer. Chem. Soc.*, **108**, 4188–92.

Strekowski, L., Wilson, W. D., Mokrosz, J. L., Strekowska, A., Koziol, A. E., and Palenik, G. J. (1988). A non-classical intercalation model for a bleomycin amplifier. *Anti-Cancer Drug Design*, **2**, 387–98.

Tanious, F. A., Yen, S.-F., and Wilson, W. D. (1991). Kinetic and equilibrium analysis of a threading intercalation mode: DNA sequence and ion effects. *Biochemistry*, **30**, 1813–19.

Wilson, W. D., Tanious, F. A., Barton, H. J., Jones, R. L., Fox, K., Wydra, R. L., and Strekowski, L. (1990). DNA sequence dependent binding modes of 4′,6-diamidino-2-phenylindole (DAPI). *Biochemistry*, **29**, 8452–61.

8.6

Cundliffe, E. (1987). On the nature of antibiotic binding sites in ribosomes. *Biochimie*, **69**, 863–9.

Moazed, D. and Noller, H. F. (1987). Interaction of antibiotics with functional sites in 16S ribosomal RNA. *Nature*, **327**, 389–94.

Tanious, F. A., Veal, J. M., Buczak, H., Ratmeyer, L. S., and Wilson, W. D. (1992). DAPI (4′,6-diamidino-2-phenyindole) binds differently to DNA and RNA: minor-groove binding at AT sites and intercalation at AU sites. *Biochemistry*, **31**, 3103–12.

Wilson, W. D., Ratmeyer, L., Zhao, M., Strekowski, L., and Boykin, D. (1993). The search for structure-specific nucleic acid-interactive drugs: effects of compound structure on RNA versus DNA interaction strength. *Biochemistry*, **32**, 4098–104.

von Ahsen, U. and Noller, H. F. (1993). Footprinting the sites of interaction of antibiotics with catalytic group I intron RNA. *Science*, **260**, 1500–3.

8.7

Hélène, and Toulme, J.-J. (1990). Specific regulation of gene expression by antisense, sense and antigene nucleic acids. *Biochim. Biophys. Acta*, **1049**, 99–125.

Mergny, J.-L., Collier, D., Rougée, M., Montenay-Garestier, T., and Hélène, C. (1991). Intercalation of ethidium bromide into a triple-stranded oligonucleotide. *Nucl. Acids Res.*, **19**, 1521–6.

Mergny, J.-L., Duval-Valentin, G., Nguyen, C. H., Perrouault, L., Faucon, B., Rougée, M., Montenay-Garestier, T., Bisagni, E., and Hélène, C. (1992). Triple helix-specific ligands. *Science*, **256**, 1681–4.

Scaria, P. V. and Shafer, R. H. (1991). Binding of ethidium bromide to a DNA triple helix. *J. Biol. Chem.*, **266**, 5417–23.

Wilson, W. D., Tanious, F. A., Mizan, S., Yao, S., Kiselyov, A. S., Zon, G., and Strekowski, L. (1993). DNA triple-helix specific intercalators as antigene enhancers: unfused aromatic cations. *Biochemistry*, **32**, 10614–21.

8.8

Chow, C. S. and Barton, J. K. (1992). Recognition of G-U mismatches by tris(4,7-diphenyl-1,10-phenanthroline) rhodium(III). *Biochemistry*, **31**, 5423–9.

Hecht, S. M. (1986). The chemistry of activated bleomycin. *Acc. Chem. Res.*, **19**, 383–91.

Lee, I. and Barton, J. K. (1993). A distinct intron-DNA structure in simian virus 40 T-antigen and adenovirus 2 E1A genes. *Biochemistry*, **32**, 6121-7.

Moser, H. E. and Dervan, P. B. (1987). Sequence-specific cleavage of double helical DNA by triple-helix formation. *Science*, **238**, 645–50.

Nielsen, P. E. (1990). Chemical and photochemical probing of DNA. *J. Mol. Recog.*, **3**, 1–25.

Sigman, D. S., Bruice, T. W., Mazumder, A., and Sutton, C. L. (1993). Targeted chemical nucleases. *Acc. Chem. Res.*, **26**, 98-104.

Strekowski, L., Strekowska, A., Watson, R. A., Tanious, F. A., Nguyen, L. T., and Wilson, W. D. (1987). Amplification of bleomycin-mediated degradation of DNA. *J. Med. Chem.*, **30**, 1415–20.

Stubbe, J. and Kozarich, J. W. (1987). Mechanism of bleomycin-induced DNA degradation. *Chem. Revs.*, **87**, 1107–36.

8.9

Atwell, G. J., Baguley, B., and Denny, W. A. (1989). Potential antitumour agents. 57. 2-Phenylquinoline-8-carboxamides as 'minimal' DNA-intercalating antitumour agents with *in vivo* solid tumor activity. *J. Med. Chem.*, **32**, 396–401.

Bailly, C. and Henichart, J.-P. (1991). DNA recognition by intercalator–minor groove binder hybrid molecules. *Bioconjugate Chem.*, **2**, 379–93.

Maher, L. J., Wold, B., and Dervan, P. B. (1989). Inhibition of DNA binding proteins by oligonucleotide-directed triple helix formation. *Science*, **245**, 725–30.

Zon, G. (1988). Oligonucleotide analogues as potential chemotherapeutic agents. *Pharm. Res.*, **5**, 539–49.

INTERACTIONS OF PROTEINS WITH NUCLEIC ACIDS

9.1 Perspective

9.1.1 Why study proteins?

The importance of nucleic acids in the cell has been pointed out many times, but the importance of proteins in the structure and function of nucleic acids is often ignored. From synthesis to degradation, nucleic acids are rarely found alone: proteins package them and mediate their transformations. Thus, although nucleic acids are the message of the cell, proteins are the medium through which that message is expressed. One means little without the other.

Rather than discuss the interaction of proteins and nucleic acids in general, we concentrate on the molecular basis of protein–nucleic acid complexes, focusing on systems where the structure of the protein–nucleic acid complex is known at high resolution. Consequently, we discuss neither site-specific nor site-general recombination systems, even though they are structurally and biochemically fascinating. The incomplete selection of systems we do discuss reflects just how difficult it is to obtain information on macromolecular complexes, especially those where the binding of protein to nucleic acid causes large conformational changes. High-resolution structural studies of protein–nucleic acid interactions used to be the sole purview of X-ray crystallographers: in the last 10 years, macromolecular NMR spectroscopy has been very effective for systems whose molecular mass is under about 25 kDa. Both techniques are equally valid but neither is sufficient without detailed kinetic, thermodynamic, and site-directed mutagenesis studies.

9.1.2 History of structure determination

To understand how a molecule works, one must know its structure. Probably the most famous example of this is the structure of DNA, which provided much more than a list of the positions of the atoms in the double helix. It explained the stability of DNA and the Chargaff rules (Section 1.4), and provided a model for how DNA stores genetic information.

Proteins are much less regular than DNA and RNA, and so understanding them is more difficult. The first structures for nucleic acid binding proteins were of the stable and abundant nucleases. These structures provided much information about how proteins interact with single-stranded oligonucleotides. Work on more complicated protein–nucleic acid complexes, such as repressors, polymerases, and tRNA synthetases, required two advances: molecular biological techniques for overexpressing normally scarce proteins, and chemical techniques for synthesizing large amounts of oligonucleotides of a known sequence. Only in the late 1970s did it become reasonable to try to determine the structure of an oligonucleotide, the protein it interacts with, and the complex between the two.

9.1.3 The forces between proteins and nucleic acids

Intermolecular forces determine how proteins interact with ligands. How proteins and nucleic acids achieve their mutual specificity is much more complicated than how proteins interact with small molecules; generalizations from the structure of the protein or nucleic acid alone have therefore been only partially correct. Furthermore, both proteins and nucleic acids are flexible: one must consider each molecule in isolation, how their conformations change as the complex forms, and how both bound and unbound forms are affected by solvent and dissolved ions. As a result, assessing how mutations change the free energy of association ($\Delta\Delta G$s) is difficult. Many proteins, furthermore, bind to nucleic acids in more than one way thus increasing the number of states to be analysed. Thus, DNA-dependent polymerases have both proof-reading and polymerase modes and transcription factors bind specific and non-specific sequences differently.

The forces between proteins and nucleic acids can be classified into four types.

Electrostatic forces: salt bridges

Salt bridges are electrostatic interactions between groups of opposite charge. They typically provide about 40 kJ mol^{-1} of stabilization per salt bridge. In protein–nucleic acid complexes, they occur between the ionized phosphates of the nucleic acid and either the ϵ-ammonium group of lysine, the guanidinium group of arginine, or the protonated imidazole of histidine. Only histidine can exist in both acidic and basic form at physiological pHs. Salt bridges are influenced by the concentration of salt in the solution: as it increases, the strength of the salt bridge decreases and they are much stronger in the absence of water molecules between the ionized groups because water has a high dielectric constant (cf. Section 8.2.1).

Compared to the other forces between proteins and nucleic acids, salt bridges are relatively long range. Their strength is proportional to the inverse square of the separation of the charges but is rather insensitive to the relative orientation of the charges; they therefore confer little specificity on the interaction between protein and nucleic acid. Especially in B-DNA, changes in sequence perturb the average structure only subtly. Salt bridges alone, therefore, can not distinguish one B-DNA sequence from another. Patterns of salt bridges, however, could clearly be used to distinguish single-stranded from double-stranded nucleic acids, and B-DNA from Z-DNA.

Although individual salt bridges are not very specific, the overall electrostatic field of the protein does orientate the polyanionic nucleic acid. The surface of the protein touching the nucleic acid has a positive electrostatic potential whose three-dimensional shape can help when modelling the interaction between protein and nucleic acid (Fig. 9.1). Such models are, at best, correct only in general outline, not in detail.

Dipolar forces: hydrogen bonds

Hydrogen bonds are a result of dipole–dipole interactions:

$$\overset{\delta-}{X} - \overset{\delta+}{H} \;\text{-----}\; \overset{\delta-}{Y} - \overset{\delta+}{R}$$

The strength of a hydrogen bond falls off with the inverse third power of the H–Y distance and also decreases greatly if the bond is bent (i.e. if X, H and Y are not colinear). X and Y are almost always nitrogen or oxygen in biological macromolecules (Section 2.1.9).

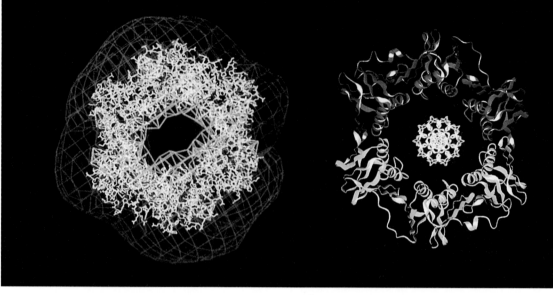

a   **b**

Fig. 9.1 The structure and electrostatic properties of the β subunit of *E. coli* DNA polymerase III. Both views look down the two-fold axis of the protein. (**a**) The electrostatic potential of the dimer, resulting from its charged groups and α- helix dipoles. The regions of positive potential are blue and the regions of negative potential are red. The contour level is ± 2.5 *kT/e*; the positive electrostatic potential creates a path for the DNA through the centre of the doughnut. The atoms of the protein are in yellow. (**b**) A ribbon representation, with one monomer coloured yellow and the other red. The α-helices are shown as spirals and the β-strands as ribbons. A stick model of B-form DNA has been built into the centre of the β-subunit doughnut with phosphorus atoms coloured yellow; oxygen red; nitrogen blue; and carbon green (reprinted with permission from Kong, X.-P., Onrust, R., O'Donnell, M., and Kuriyan, J. (1992). *Cell*, **69**, 425–37. Copyright (1992) Cell Press).

Hydrogen bonds, which occur between the amino acid side chains , the backbone amides and carbonyls of the protein, and the bases and sugar-phosphate oxygens of the nucleic acid, are sometimes mediated by bound water (Section 9.3.2). When the molecules are not complexed, all their exposed hydrogen bond donors (X) and acceptors (Y) form linear hydrogen bonds to water. When the complex forms, there is little change in the free energy due to hydrogen bond formation if the linear hydrogen bonds to water are replaced by similar ones between the macromolecules. By contrast, forming bent hydrogen bonds, or *not* forming some of them, carries a free energy penalty of up to 4 kJ mol^{-1} per hydrogen bond. Thus hydrogen bonds are very important in making sequence-specific protein–nucleic acid interactions.

Entropic forces: the hydrophobic effect

The hydrophobic effect is due to the behaviour of water at an interface. Any molecule in water creates a sharply curved interface and so orders a layer of water molecules around itself. When molecules aggregate, the ordered water molecules at the interface are released and become part of the disordered bulk water, thus stabilizing the aggregate by increasing the entropy of the system. The change in free energy is roughly proportional to the number of water molecules released and, therefore, to the surface area buried.

Polar surfaces, where the enthalpy loss tends to offset the entropy gain on desolvation, are less likely to aggregate than non-polar ones. Molecules of water left at the interface between a protein and a nucleic acid obviously decrease the entropy of the system. Consequently, the surfaces of the protein and nucleic acid tend to be exactly complementary so that no unnecessary water molecules remain when the complex forms. However, specific water molecules with defined functions in sequence recognition are often bound in the interface (see *met* repressor, Section 9.3.5).

Base stacking: dispersion forces

Base stacking is caused by two kinds of interaction: the hydrophobic effect mentioned above and dispersion forces (Section 2.3.1). Molecules with no net dipole moment can attract each other by a transient dipole-induced dipole effect first described by London in the 1920s. Such dispersion forces decrease with the inverse sixth power of the distance separating the two dipoles and so are very sensitive to the thermal motion of the molecules involved.

Despite their extreme distance dependence, dispersion forces are clearly important in maintaining the structure of double-stranded nucleic acids because they help to cause base stacking. Furthermore, they also help single-stranded regions in nucleic acids bind to proteins because aromatic side chains can intercalate between the bases of a single-stranded nucleic acid.

Summary

Electrostatic forces are long range, not very structure-specific, and contribute substantially to the overall free energy of association. Hydrogen bonds are dipolar, short-range interactions that contribute little to the stability of the complex but much to its specificity.

Hydrophobic forces are due to changes in the solvation of macromolecules when they interact: they are short range, sensitive to structure, proportional to the size of the macromolecular interface, and contribute to the free energy of association.

Dispersion forces have the shortest range but are very important in base stacking in double-stranded nucleic acids and in the interaction of protein with single-stranded nucleic acids. All such forces occur when proteins and nucleic acids interact, but it is very difficult to ascribe precise changes in free energy of association to specific interactions between protein and nucleic acid.

9.1.4 Geometric constraints imposed by the nucleic acid

Nucleic acids contain a number of different molecular groups: the phosphates, the bases, and the sugars. There are gross differences in the three-dimensional arrangement of these groups depending on whether the nucleic acid is single- or double-stranded, A-, B-, or Z-form (Section 2.2). In B-form structures, which are quite flexible, there are more subtle sequence-dependent differences which can affect either the average structure (e.g. the bend in poly(dA · dT)) or its flexibility (Section 2.3.1). RNA has many irregular arrangements of the bases and backbone such as Hoogsteen pairs, base-triples, and bulges (Section 2.4.4). Any changes might influence how a protein and nucleic acid interact and, consequently, allow a protein to discriminate between one nucleic acid and another.

Double-stranded B-DNA

Double-stranded right-handed B-DNA has a highly repetitive, negatively charged polyphosphate surface; protein domains that interact with it have a complementary positive electrostatic field. They also have polar or charged sidechains that interact with the phosphate oxygens of the backbone: about half of the identified contacts between transcription factors and B-DNA are to the backbone. Such contacts occur both in sequence-specific protein–DNA complexes, such as the λ-repressor–operator complex (Section 9.3.2) and in complexes that are not sequence-specific, such as the DNA polymerase I 3′-5′ exonuclease interacting with DNA (Section 9.2.4).

There are a number of ways of achieving sequence-specific recognition. The most common involves **direct readout** of the DNA sequence by hydrogen bonds between the protein and bases in either the major or the minor groove of the DNA. Such readout can also be mediated by water molecules, as in *met* repressor (Section 9.3.5). **Indirect readout**, where the protein is sensitive to the average conformation or flexibility of the DNA is often important (see CAP, Section 9.3.2), but it is difficult to know how important without detailed energetic analyses and structures of both protein and nucleic acid in bound and unbound forms.

Simple model-building predicted two of the many ways in which proteins interact with B-DNA: an antiparallel β-ribbon in the minor groove hydrogen-bonding to the phosphate backbone (Fig. 9.2) and an α-helix interacting with bases in the major groove (Fig. 9.20). The latter is the most common method of interaction—possibly because more hydrogen bond donor and acceptor sites are accessible in the major than in the minor groove. Furthermore, although no single site uniquely distinguishes a base-pair in either groove, the patterns of sites are more dissimilar for each base-pair in the major than in the minor groove (Fig. 9.3). Thus, to distinguish the cognate sequence from all others by direct readout alone, proteins must form more than one hydrogen bond to some of the base-pairs in the major groove. This can be achieved, for instance, by an arginine or glutamine forming bridging hydrogen bonds to two successive bases.

Single-stranded nucleic acids

Single-stranded nucleic acids that do not fold up to compact structures are much more flexible than double-stranded nucleic acids and the predominantly hydrophobic bases are much more exposed. A single-stranded nucleic acid binding protein will therefore have a much more hydrophobic nucleic acid-binding surface than one that binds double-stranded nucleic acids: the hydrophobic surface often contains aromatic groups because they interact more effectively with the nucleic acid bases (Section 9.1.3). It must also have an electrostatic field that neutralises the charge of the phosphate backbone.

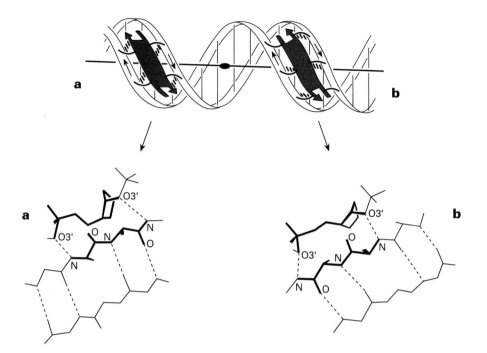

Fig. 9.2 Schematic representation of a β-sheet in the minor groove of DNA. The β-sheets are located in the minor grooves so that their dyad symmetries match that of B-DNA. In (**a**) the polarities of the nucleic acid and the protein are parallel whereas in (**b**) they are antiparallel. In the more detailed drawings (lower), hydrogen bonds are shown as dashed lines and only one strand of the DNA is shown (from Church, G.M., Sussman, J.L., and Kim, S.-H. (1977). *Proc. Natl Acad. Sci. USA*, **74**, 1458–62).

Ribonucleic acids

Single-stranded RNA, in contrast to DNA, forms complex secondary and tertiary structures such as bubbles, bulges, and pseudoknots (Section 2.4.4). In A-form RNA, the major groove is deeper and the bases are more inaccessible than in B-form DNA, so bulges may help recognition by widening the major groove and making the bases more accessible. Alternatively, a protein may recognize the loop of an RNA hairpin. For example, in the crystal structure of the RNA binding domain of V1A, a small ribonucleoprotein, a ten-nucleotide RNA loop binds to the surface of a four-stranded β-sheet as an open structure.

Possibly because the structure of RNA varies more than that of DNA, proteins seem to recognize RNAs in more ways than they recognize DNAs. For example, two high-resolution RNA–protein complex structures (Section 9.3.7) contain completely different structural solutions to the problem of recognizing a specific tRNA: almost the only similarity is that both distort the RNA. RNAs, even more than DNAs, may be recognized by indirect readout: the correct RNA can distort to fit the protein whereas the incorrect one can not.

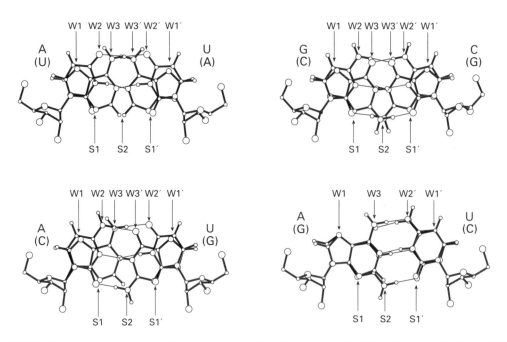

Fig. 9.3 The hydrogen bond donors and acceptors accessible in the grooves of B-DNA. All four combinations of A·U and G·C base-pairs are shown, indicating where hydrogen bonds can be used to discriminate between various base-pairs. The upper letters refer to the base-pair in red lines, the lower ones in parentheses to the base-pair in black lines. **W** refers to potential hydrogen bonds in the major groove; **S** refers to potential ones in the minor groove. **W** and **S** sites are related to **W′** and **S′** by the two-fold symmetry of a DNA double helix (from Seeman, N.C., Rosenberg, J.M., and Rich, A. (1976). *Proc. Natl Acad. Sci. USA*, **73**, 804–8).

9.1.5 The kinetics of forming protein–nucleic acid complexes

Two factors affect the rate of formation of all protein–nucleic acid complexes: random thermal diffusion and long-range, directional electrostatic attraction. A simple model, therefore, of how a nucleic acid binding protein finds its cognate sequence would be random association of the protein with the nucleic acid, followed by dissociation and reassociation elsewhere if the sequence is not correct. Such a model is a three-dimensional random walk through the contents of the entire cell. However, in even the smallest bacterial genome sequence-specific protein–DNA complexes form too rapidly for this model to be correct.

An alternative model, which *can* account for the observed rates, is a one-dimensional walk. The protein first binds non-specifically to the DNA and then diffuses or jumps along the DNA until it finds the appropriate sequence. How non-specific and specific binding are distinguished is not yet clear because, firstly, there are no structures of a *sequence-specific* protein binding to a completely non-specific sequence (but see Section 9.3.3 and Fig. 9.27b) and, secondly, there may be as many different ways to discriminate as there are DNA-binding protein motifs. However, in the glucocorticoid receptor complex structure, (Section 9.3.3) there are effects on both direct readout (fewer hydrogen bonds made to the non-cognate half-site) and indirect readout (a loss of co-operativity in DNA binding). Nonetheless, the difference in binding constant between cognate and non-cognate sequences appears to be small. The specific complex may make more ionic interactions than the non-specific complex, as in *Eco*RV (Section 9.3.6). Finally, many proteins, on binding their cognate sequence, undergo a conformational change which may be different or absent when they bind non-cognate sequences. The two protein conformations would have different properties: the non-specific binding conformation would increase the rate of diffusion along the DNA and the specific binding conformation would lock down more tightly on the DNA once the cognate sequence was reached.

Summary

All nucleic acids have repeating polyanionic backbones and so all proteins that bind them have strategically placed arginines and lysines that create an electrostatic field to neutralize the negative charge. The most common method of interacting with B-DNA is to insert an α-helix into the major groove and form hydrogen bonds to the bases. In specific protein B-DNA complexes, about half of the hydrogen bonds are to the bases and the other half to the phosphate backbone.

Contacts to the bases are called **direct readout** because what contacts form depends directly on the sequence of the nucleic acid; distinguishing sequences by how the sequence affects the distortability or conformation of the nucleic acid is called **indirect readout**. To interact with single-stranded nucleic acids, aromatic side chains are used to stack against the nucleic acid bases. Protein–RNA interactions are more varied than protein–DNA interactions because RNA adopts more conformations than B-DNA. Consequently, indirect readout, where the protein recognizes its cognate RNA by distorting it, may be very important.

Sequence-specific protein–DNA complexes must be formed by the protein first binding loosely to the incorrect sequence and then diffusing along the DNA in a **one-dimensional random walk** until it finds the correct sequence. Thus all sequence-specific DNA binding proteins may bind DNA in two ways: one for tight, sequence-specific binding and the other for looser, non-sequence specific binding.

9.2 Non-specific interactions

9.2.1 The need for packaging

An *E. coli* bacterium contains about 4×10^6 bp of DNA, which corresponds to about 1 mm of double-stranded B-DNA (many times the circumference of the bacterium). The problem is even more acute in higher eukaryotes whose genomes, typically containing 6×10^9 bp, would be 2 m long! To avoid complete chaos, most organisms have proteins to organize DNA into structures both more complex and more compact than the Watson–Crick double helix (Section 2.6).

Packaging DNA: the eukaryotic nucleosome

When a metaphase chromosome is suspended in a medium of low ionic strength, it extrudes individual chromatin strands which, in an electron micrograph, look like beads on a string. The beads are nucleosomes: protein–DNA complexes which can be isolated by cleaving the intervening DNA (the string) with micrococcal nuclease. The nucleosome core particle thus produced consists of about 146 bp of DNA and eight small, highly-basic proteins called histones: two each of H2A, H2B, H3, and H4. All of them have a highly-conserved globular domain. In the low (7 Å) resolution structure of the nucleosome core particle, the heart-shaped octameric protein core is in the centre with B-form DNA wrapped around it to form 1.8 to 1.9 turns of a left-handed, negatively supercoiled superhelix (Fig. 9.4a). The superhelix appears to be formed by a series of kinks rather than a smooth curve. In the centre of the octamer, the two H3 and two H4 histones associate to form a tetramer which by itself binds DNA. H2A and H2B associate as dimers on the top and bottom of the core particle and interact with DNA at the ends of the core particle and beyond.

The next 22 bp of DNA interact with histone H1 (a functional subtype is called H5 in birds). The conserved globular domain of histone H5, whose structure has recently been solved, is remarkably similar to CAP (Section 9.3.2). It contains a modified helix–turn–helix motif (Fig. 9.19) which suggests how it might interact with DNA. Histone H1 (H5) helps form higher-order structures which package the DNA even more compactly: the 10 nm filaments have a zig-zag arrangement of nucleosomes and the 30 nm solenoid, a helical arrangement (Fig. 9.4b). The more tightly-packaged structures are transcriptionally inactive and eukaryotic transcriptional regulation may involve competition for the DNA between H1 (H5) and RNA polymerase and its associated transcription factors (Sections 9.3.3 and 9.3.5).

Packaging DNA in prokaryotes

Although prokaryotes do not have any direct equivalent of the histone proteins, several of their proteins have somewhat similar roles: they are small, basic, and rich in lysine and alanine, and induce higher-order structure in B-DNA. They do not, however, form particles like the nucleosome. The only protein in this class whose structure is known is *E. coli* DNA binding protein II (also called HU). There are two types of HU subunits, α and β, which have more than 70 per cent sequence identity. HU is active as a dimer, but the dimer can be the $\alpha\beta$ heterodimer, the α_2 homodimer, or the β_2 homodimer.

HU is a wedge-shaped molecule that probably binds DNA by encircling it with two extended β-sheet arms that make direct contacts with the target DNA (Fig. 9.5); *met* repressor also uses a β-sheet motif to bind DNA (Section 9.2.5). Site-directed mutagenesis studies of the β-sheet arms show that DNA-binding ability increases with increasing basic residues. Furthermore, the short β-strand that connects the arm region with the N-terminal helical region (Fig. 9.5) may also take part in DNA binding. However, it is not known to which groove the arms bind, nor the precise role of the disordered loops at the ends of the β-sheet arms. A model of how the wedge-shaped HU packages DNA suggests that the protein wedges polymerize, thus inducing the DNA to form a supercoiled structure. It has been shown that HU binds both single- and double-stranded DNA: therefore the protein probably interacts with the sugar-phosphate backbone of the nucleic acid and not with the bases.

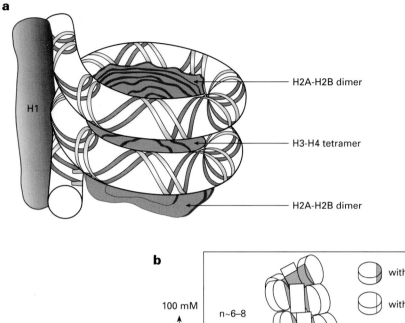

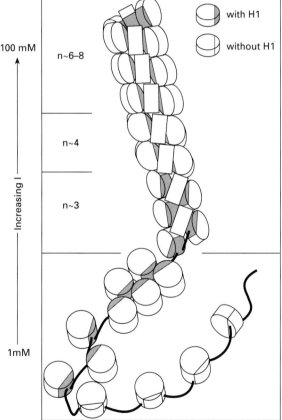

Fig. 9.4 A schematic view of the nucleosome and the higher-order structures that it can form. (**a**) The nucleosome consists of an H3–H4 histone tetramer, capped at either end with an H2A–H2B dimer (colour). The DNA winds about twice around the histone core particle in a left-handed supercoil. (**b**) Without histone H1 (H5), no regular structures form but, when it binds, the nucleosomes first form the 10 nm zigzag structure and then the 30 nm solenoid (redrawn from Thoma, F., Koller, Th., and Klug, A. (1979). *J. Cell Biol.*, **83**, 402–27. Copyright (1979) Rockefeller University Press).

Other nucleic acid packages: viruses

Viruses are particles made up of protein and nucleic acid components. The study of protein–nucleic acid interactions in viruses is not straightforward. The protein shell always contains a repeating arrangement of many copies of a few proteins; therefore, the single nucleic acid inside, is randomly oriented with respect to the protein. As a result, only the average structure of the nucleic acid bound to all possible nucleic acid binding sites is seen, even if specific interactions occur. Protein–nucleic acid interactions have been seen in the structures of the icosahedral Flock House virus, bacteriophage ϕX174, and satellite tobacco mosaic virus and in the structure of the helical tobacco mosaic virus. The nucleic acid can be either DNA or RNA and in single- or double-stranded form, depending on the virus.

Fig. 9.5 A schematic representation of DNA-binding protein II; α-helices are shown as spirals and β-strands as curved arrows. Each monomer in the dimer is shaded differently and the broken connections at the top of the figure indicate the disordered region between the two strands that should wrap around the DNA. One subunit is shown lighter, the other darker and one N- and one C-terminus are marked (redrawn from Tanaka, I., Appelt, K., Dijk, J., White, S.W., and Wilson, K.S. (1984). *Nature*, **310**, 376–81. Copyright (1984) Macmillan Magazines Limited).

Tobacco mosaic virus

Tobacco mosaic virus (TMV) has a well-ordered single-stranded RNA genome, which forms a helix inside the helical protein shell. Unlike most of the spherical viruses, TMV can pack its own or closely related RNA only, but not DNA. The specificity for RNA over DNA is because of the ribose sugars of RNA, not the difference between T and U: the ribose 2'-hydroxyl groups form contacts to the virus protein. Thus, sugar–protein interactions determine which type of nucleic acid can be incorporated into the virus.

Each protein monomer in the shell binds three bases, each in a separate pocket, but the protein–nucleic acid interactions are mostly not sequence-specific. The bases all lie flat against one of the α-helices in the coat protein, making non-specific van der Waals contacts to hydrophobic residues and non-specific hydrophilic contacts to hydrophilic residues in the protein. The binding site for base-1, however, is particularly well suited for guanine: Arg^{122} and Asp^{115} can form hydrogen bonds to a guanine O-6 and N-2, explaining why the origin of the RNA sequence that directs TMV virus assembly has a G every third residue.

The phosphate groups of the RNA are neutralized by several arginines of the protein via direct salt bridge formation and by a calcium atom that binds between a phosphate and the carboxylate of Asp^{116}. This suggests how TMV disassembly might occur: the low calcium concentration and rather high pH of a plant cell would tend to make the calcium dissociate, thus electrostatically destabilising the virus. Once some 1.5 turns of coat protein are lost, ribosomes can then attach to the now naked RNA and start translating it.

Flock House virus

Flock House virus (FHV) is a small icosahedral nodavirus that infects various animals. It is unusual in that its single-stranded RNA has an important role in the assembly of the virus capsid because, unlike many other types of viruses, empty nodavirus capsids do not occur.

The protein capsid contains 60 subunits, 20 each of three identical polypeptides, A, B, and C, each containing the eight-stranded antiparallel β-barrel motif seen in many other viruses (Fig. 9.6). However, C–C dimers associate differently from the other dimers in FHV. At each C–C interface, a peptide arm (residues 20–30) from each C monomer becomes ordered and the RNA binds to one pair of α-helices per monomer (Fig. 9.7). The ssRNA is an approximately A-form duplex, to obey the local twofold symmetry. The protein only makes non-sequence-specific electrostatic and van der Waals contacts to the sugar-phosphate backbone of the RNA (Fig. 9.7) — unlike TMV, where some sequence-specific contacts occur.

Summary

The fundamental building block of chromatin in eukaryotes is the **nucleosome**, a protein–DNA complex. The nucleosome core particle consists of 146 bp of DNA and eight small, highly basic **histone proteins**. The DNA wraps around the histone octamer to form a negative supercoil. Bacteria also use small basic proteins to package DNA, such as the dimeric HU protein from *E. coli*, whose long β-strand arms presumably wrap around a double-stranded DNA molecule.

Viruses are highly symmetric particles that can pack their nucleic acid genome efficiently inside the **protein capsid**. Protein subunits containing many basic amino acids interact with the viral nucleic acid in a non-sequence-specific manner. In the helical TMV, some sequence-specific contacts are involved in directing assembly of the virus, but there are no such contacts in the icosahedral FHV.

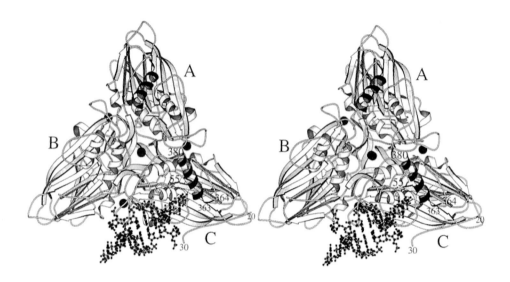

Fig. 9.6 An icosahedral asymmetric unit of the Flock House virus, showing one half of the CC dimer. The view is along a 3-fold symmetry axis from inside the virus. The RNA is shown as a ball-and-stick model, stabilizing calcium ions are shown as black spheres, and the γ-peptides (cleaved from FHV during maturation) are in black. The peptide arm (residues 20–30) that is ordered only in the CC dimers is marked, as are the two helices (αI and αIII or γ) that interact with the RNA (reprinted with permission from Fisher, A.J. and Johnson, J.E. (1993). *Nature*, **361**, 176–9. Copyright (1993) Macmillan Magazines Limited). (Stereopair diagram for parallel viewing.)

9.2.2 Single-stranded nucleic acid binding proteins

Single-stranded DNA is formed during replication and most organisms produce proteins to bind it. These proteins form an important but diverse group but, with the exception of gene-5 protein from bacteriophage fd, there is little structural information on how they interact with nucleic acids.

During the replication of phage fd, gene-5 protein seems to bind to freshly synthesized viral DNA daughter strands, thus preventing other proteins such as nucleases or polymerases from gaining access to the viral DNA. The protein is a closely associated dimer (Fig. 9.8), consisting almost entirely of β-strands which loop out from a hydrophobic core. Two such loops bind DNA. The DNA binding surface of gene-5 protein is rather unusual because many aromatic side chains are clearly exposed to solvent. A model has been suggested (Fig. 9.8) in

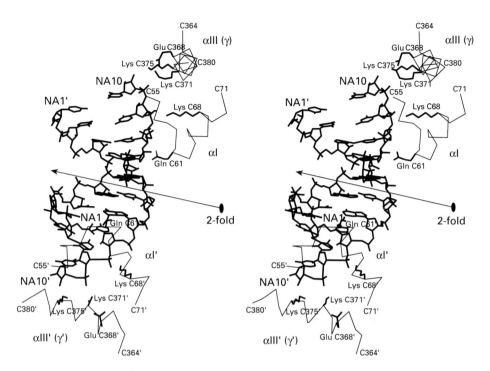

Fig. 9.7 Details of protein–RNA interactions in Flock House virus. RNA is in bold, amino acid side chains in medium, and protein backbone in thin lines. Only those side chains that interact with the RNA are shown. An icosahedral 2-fold axis is shown by an arrow pointing towards the centre of the virus. Primes specify symmetry-related atoms. γ-Peptides (Fig. 9.6) are labelled. Lys[68], Lys[371] and Lys[375] form salt bridges to the RNA, are conserved in all nodaviruses and may be important in virion assembly (reprinted with permission from Fisher, A.J. and Johnson, J.E. (1993). *Nature*, **361**, 176–9. Copyright (1993) Macmillan Magazines Limited). (Stereopair diagram for parallel viewing.)

which lysines and arginines neutralise the DNA phosphate backbone and the bases stack against aromatic amino acid side chains.

9.2.3 Non-sequence-specific nucleases

All organisms must degrade nucleic acids during their life cycle. There is no one enzyme designed for this purpose, but rather a large number of enzymes with different specificities (see also Sections 9.2.4, 9.3.6). These include exo- and endonucleases and enzymes specific for single- and double-stranded nucleic acids and for base sequences. We shall discuss only those nucleases whose structures are known in complexation with a nucleic acid.

Ribonuclease A, barnase, and binase

Ribonuclease A (RNase A) from bovine pancreas is a small, 13.7 kDa, enzyme that cleaves RNA but can also bind ssDNA, which is a competitive inhibitor. Despite its relatively small size, RNase A has an extended binding surface that can accommodate long polynucleotide

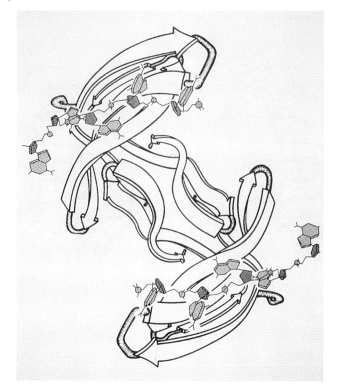

Fig. 9.8 Model of how gene-5 protein from bacteriophage fd interacts with single-stranded DNA showing one dimer. β-strands are shown as ribbons and the paths of two strands of ssDNA are shown with the bases and sugars stippled differently (reprinted with permission from Brayer, G.D. and McPherson, A. (1984). *Biochemistry*, **23**, 340–9. Copyright (1984) American Chemical Society).

chains. Indeed, 12 of the 16 nucleotides in the complex formed between four d(pA)$_4$ molecules and RNase A bind directly to the protein, forming a chain running 5′-to-3′ from a group of positively-charged residues (called the anion binding site) to the active site (Fig. 9.9). The main contacts between the protein and the DNA are electrostatic, resulting from ionic bonds between positively charged protein side chains and the polyphosphate DNA backbone. This explains why RNase A is not sequence specific. It is, however, specific for sequences that contain a pyrimidine 3′ to the site cleaved because Thr^{45} forms hydrogen bonds to that base (Fig. 9.11).

Barnase and binase are very similar microbial ribonucleases from bacilli. Both are small (12 kDa) α/β proteins. Fortuitously, crystallographic studies of inhibitor complexes of the two enzymes have allowed one to construct a precise model of their mode of action. Binase binds 3′-GMP so that Glu^{60} interacts with N-1 of the guanine: the nucleotide, a competitive inhibitor of binase, sits where the 3′-end of the cleaved RNA product would be. In contrast, d(GpC) binds non-productively in the pocket that would be occupied by the 5′-end of the cleaved RNA. By superimposing these two complexes, one can build a model of the active site during catalysis (Fig. 9.10).

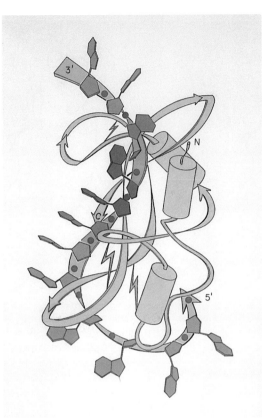

Fig. 9.9 Binding of $d(pA)_4$ to RNase A. The tetranucleotide chains (in blue) bind smoothly along the surface of the enzyme (in orange) with electrostatic interactions holding the nucleic acid in place. α-Helices are shown as cylinders, β-strands as arrows, and zig-zags show the position of the disulphides. The salt bridges are between the phosphates and six lysines and three arginines. The position of cleavage would be between the two darker blue nucleotides in the centre of the figure (reprinted with permission from McPherson, A., Brayer, G., Cascio, D., and Williams, R. (1986). *Science*, **232**, 765–8. Copyright (1986) by the AAAS).

RNase A, barnase, binase, and RNase T1 all share a common catalytic mechanism (Fig. 9.11). For RNase A, His[12] acts as a general base and activates the 2'-OH group to attack the 3'-phosphate ester and generate a 2',3'-cyclic intermediate. His[119] acts as a general acid and donates a proton to the 5'-O leaving group of the next nucleotide. In the second step, the pyrimidine ribose 3'-phosphate product is formed when a water molecule donates a proton to His[119] and a hydroxyl group to the phosphate. The reaction is stabilized by a salt bridge between Lys[41] and a phosphate ester oxygen. It can be easily seen from the scheme why DNA inhibits RNases: deoxy sugars lack the 2'-OH, making it impossible to form the cyclic intermediate.

Deoxyribonuclease I

DNase I is a non-specific nuclease that cleaves double-stranded DNA at varying rates depending on the sequence of the DNA. Although the chemical catalytic mechanism is not known,

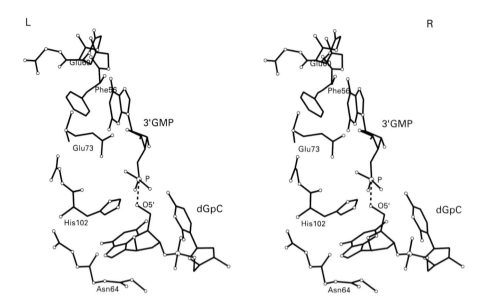

Fig. 9.10 Proposed mode of binding of the substrate GGC to barnase or binase, based on studies of different inhibitor complexes of barnase and binase. The active site region of the barnase/d(GpC) complex and the binase/3'-GMP complex are superimposed on each other. The guanine of the 3'-GMP from the binase would interact with the recognition loop of barnase (top of the figure) while its free 3'-phosphate is within bonding distance of the 5'-OH of the barnase d(GpC). Furthermore, the GMP phosphate is close to His[102] in the active site, which acts analogously to His[119] in RNase A (Fig. 9.11) (reprinted with permission from Baudet, S. and Janin, J. (1991). *J. Mol. Biol.*, **219**, 123–32. Copyright (1991) Academic Press).

the reason for sequence specificity is based on the structure of the DNase I·DNA octamer (Fig. 9.12). An exposed loop of the enzyme binds in the minor groove of a 14 bp oligonucleotide which mimics the nicked B-DNA product because it is formed from two octamer duplexes (Fig. 9.13). No specific secondary structural element binds the DNA: instead it is bound by surface loops. Tyr[76] is in van der Waals contact with the deoxyribose ring of T313 and the phosphate groups of both DNA strands form hydrogen bonds and a salt bridge to residues on both sides of the binding loop (Fig. 9.13).

As in CAP repressor (Section 9.3.2) the DNA, not the protein, undergoes a conformational change upon binding. It bends towards the major groove, thus widening the minor groove so that the loop fits better. This protein-induced DNA bend may explain the sequence dependency of DNase I. DNA strands with stretches of A and T bases form narrow minor grooves and thus will not be cleaved as efficiently as stretches of G and C bases, which induce the DNA to bend towards the major groove (Section 2.3.3). The narrow minor groove may also prevent sufficient protein-phosphate contacts from forming.

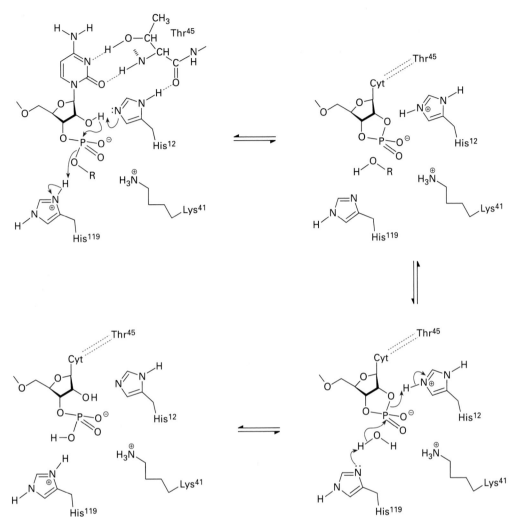

Fig. 9.11 Catalytic mechanism of Ribonuclease A. Both chemical steps in the reaction go by linear $S_N2(P)$ displacement mechanisms. In the first step, which generates the $2',3'$-cyclic phosphate intermediate, the attacking oxygen of $2'$-OH group is opposite the leaving $5'$-OH of the next nucleotide. In the second step, the attacking water is opposite the $2'$-O of the intermediate. Lys[41] stabilizes the transition state.

Summary

RNase and DNase have different reaction mechanisms because RNase uses the ribose $2'$-hydroxyl group, not present in DNA, to attack the $5'$-phosphate ester linkage. RNase A is not sequence specific because, although it can bind long nucleotide chains, it only interacts with the base at the active site: all other contacts are electrostatic ones to the sugar-phosphate backbone. DNase I cleaves different sequences with different rates because of sequence-dependent steric hindrance at the active site. G-C tracts accommodate the catalytic loop better because they have wider minor grooves than A-T tracts.

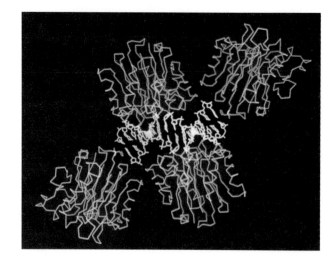

Fig. 9.12 The packing of DNA in the crystals of DNase I–DNA complex. Two DNA octamers (shown as yellow and orange stick figures) form a quasi-continuous 14-mer strand of B-DNA. The α-carbon traces of two molecules of DNase I are shown in blue. They form contacts to the minor grooves of the DNA (reprinted with permission from Suck, D., Lahm, A., and Oefner, C. (1988). *Nature*, **332**, 464–8. Copyright (1988) Macmillan Magazines Limited).

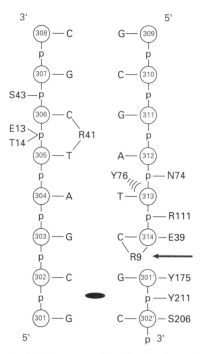

Fig. 9.13 Schematic drawing of the contacts between one DNase I molecule and a DNA duplex. An arrow indicates the nick in one of the DNA strands and a black oval indicates the crystallographic symmetry axis of the DNA (see text). Phosphates are marked 'p', deoxyriboses by circles, and the bases by sequence. The protein–DNA contacts are shown (reprinted with permission from Suck, D., Lahm, A., and Oefner, C. (1988). *Nature*, **332**, 464–8. Copyright (1988) Macmillan Magazines Limited).

9.2.4 Polynucleotide polymerases

There are four classes of template-directed polynucleotide polymerases: DNA- or RNA-dependent and DNA- or RNA-polymerizing. All add nucleotides to the 5′-end of a growing polynucleotide chain but they differ widely in how accurately they replicate the nucleic acid (their fidelity) and how many nucleotides they add before dissociating (their processivity). Many have multiple subunits, each carrying different functions. All cellular DNA-dependent DNA polymerases have a 3′-5′ proof-reading exonuclease, require a primer to begin synthesis, and replicate their own nucleic acid the most faithfully. The structures of two polymerase-DNA complexes are known: an editing complex of **E. coli DNA polymerase I (Pol I)** and a polymerizing complex of the AIDS virus polymerase, **HIV-1 reverse transcriptase (RTase)**.

DNA-dependent DNA polymerases: *E. coli* DNA polymerases I and III

Escherichia coli has three DNA-dependent DNA polymerases (Pols): I, II, and III (Sections 4.6 and 5.2.2). Pol III is the chief replicative polymerase and is much more processive than Pol I. Even without DNA, the structure of its β subunit (Fig. 9.1) neatly explains why. The β subunit is a doughnut-shaped dimer through which the DNA passes; it would have to dissociate for the DNA to get out.

Pol I is a rather small and simple DNA polymerase but contains all the functions of the larger enzymes. It repairs damaged DNA and converts the Okazaki fragments left by Pol III into complete genomic DNA (Section 5.2). Its single polypeptide chain contains three activities: the 5′-3′ DNA polymerase, the 3′-5′ (proof-reading) exonuclease, and a 5′-3′ exonuclease (to remove RNA primers). Mild proteolysis cleaves Pol I into two fragments, the smaller of which contains the 5′-3′ exonuclease. The larger fragment (the Klenow fragment) has a molecular weight of 68 kDa and contains the other two activities, each on a separate domain.

The 200-residue N-terminal domain of Klenow fragment contains the 3′-5′ proof-reading exonuclease activity. Like the class I tRNA synthetases (Section 9.3.7), it has a five-stranded variant of the Rossmann dinucleotide fold, with mostly parallel β-strands sandwiched between two layers of α-helices (Fig. 9.14). Three acidic residues create a binding pocket for an essential divalent cation; a second essential divalent cation binds when the single-stranded DNA substrate, such as dT_4, binds (Fig. 9.15). The first cation positions a hydroxide ion so that it can attack the phosphorus atom and generate the pentacoordinate transition state, while the second cation stabilizes the leaving group (the 3′-hydroxyl of the previous nucleotide). This two-metal reaction mechanism, which also occurs in the RNase H activity of HIV-1 reverse transcriptase, may well be unique to polymerases. It is not the same as the reaction mechanism of RNases (Section 9.2.3) nor of the restriction endonucleases (Section 9.3.6). The editing function of Pol I depends not on reading the sequence, but on competition between the polymerase and exonuclease sites for the newly-synthesized DNA strand. When the wrong base is incorporated, the relative rates of polymerization and hydrolysis change (Section 5.2.5). The protein makes no sequence-specific contacts; instead it makes hydrogen bonds to the sugar-phosphate backbone, and hydrophobic and base-stacking contacts to the bases.

The 400-residue C-terminal domain contains the polymerase activity (Fig. 9.14). It has a rather unusual shape which, as in HIV-1 RTase (see below), looks like a right hand ready to grip a DNA rod. The DNA, presumably, rests in the 'palm' subdomain and is gripped by the 'fingers' and 'thumb' subdomains, but quite how is still unknown. The way the DNA binds in the editing complex, however, shows unequivocally that the initial polymerase model was wrong: the primer strand approaches the polymerase from the direction of the exonuclease domain, not the other way around. In the editing complex, the 'thumb' region (Fig. 9.14) becomes more ordered and swings down towards the exonuclease domain (Fig. 9.15), creating a cleft in which the duplex DNA binds.

Three highly-conserved acidic residues (Asp^{705}, Asp^{882}, and Glu^{883}) (Fig 9.14), also seen in the structure of RTase, are essential for catalysis and may bind *two* essential divalent cations. Therefore, template-directed polymerases, their 3'-5' exonucleases, and RNase Hs may all share a common mechanism which requires two divalent cations. Such a mechanism would

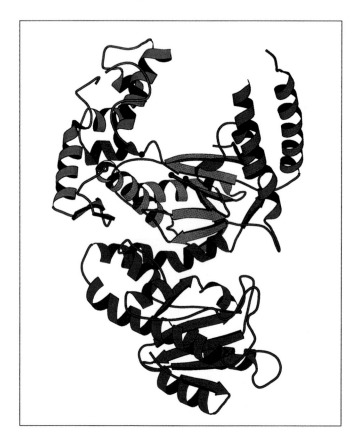

Fig. 9.14 Ribbon representation of the Klenow fragment of DNA polymerase I: the spirals are α-helices; the curved arrows β-sheets. The C-terminal polymerase domain is on the left and the N- terminal 3'–5' exonuclease domain, in magenta, on the right. The thumb subdomain is green, the fingers cyan, and the palm red and gold: gold marks the region conserved in HIV-1 RTase and Pol I. Black spheres mark the three conserved acidic residues (redrawn from Ollis, D.L., Brick, P., Hamlin, R., Xuong, N., and Steitz, T.A. (1985). *Nature*, **313**, 762–6. Copyright (1985) Macmillan Magazines Limited).

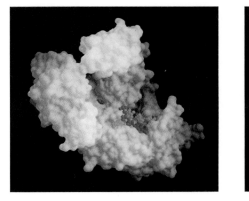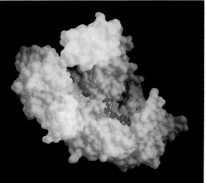

Fig. 9.15 A stereo drawing of the solvent-accessible area of the Klenow fragment of DNA polymerase I with a ball-and-stick representation of dT_4 bound in the 3′–5′ exonuclease active site cleft. The rightmost nucleotide is the one cleaved. The bases are coloured in light blue and the sugar-phosphate backbone in dark blue. In the editing complex, the DNA approaches down the exonuclease cleft and swings to the left into the polymerase active site; the thumb subdomain, on the right hand side of the polymerase domain, moves down towards the exonuclease domain (from Beese, L.S. and Steitz, T.A. (1991). *EMBO J.*, **10**, 25–33. By permission of Oxford University Press). (Stereopair diagrams for parallel viewing.)

also have existed in the ribozyme polymerases of the ancestral RNA world and, presumably, protein polymerases merely inherited it.

RNA-dependent DNA polymerases: HIV-1 reverse transcriptase

RTase is a DNA polymerase that can use either DNA or RNA as the template and so produces either RNA/DNA hybrids or duplex DNA as a product. Like other RTases, HIV-1 RTase does not have a 3′-5′ proof-reading exonuclease activity but is unusually error prone; the lack of fidelity partly explains why HIV mutates very rapidly.

RTase is produced by proteolytic cleavage from a polyprotein, first to give a polypeptide of molecular mass 66 kDa (p66). Then the 14 kDa C-terminal RNase H domain is cleaved from some of the p66 proteins, yielding a chain of molecular mass 51 kDa (p51). Together p51 and p66 form a p66/p51 heterodimer which contains only one (5′-3′) polymerase site, one RNase H site (for removing RNA in RNA · DNA heteroduplexes), and one site for binding the lysine tRNA used as a primer for DNA synthesis. The structure of the protein neatly explains why the dimer is biochemically so asymmetric (Section 6.2.1).

The p66 subunit folds into two domains: a polymerase domain and an RNase H domain. The polymerase domain is instantly recognizable as such, complete with fingers, thumb, and palm subdomains (Fig. 9.16a). Only the part of the palm subdomain containing the three conserved acidic residues (see above) is, however, truly homologous to Pol I. The thumb and fingers subdomains of the two proteins are related to each other by convergent evolution. The fourth, C-terminal 'connection' subdomain joins the polymerase domain to the RNase H. Despite having the same sequence as the polymerase domain of p66, the appearence of p51 is completely different (Fig. 9.16b). The individual subdomains have similar structures but a different overall architecture. The p51 thumb swings away from the fingers and palm to make

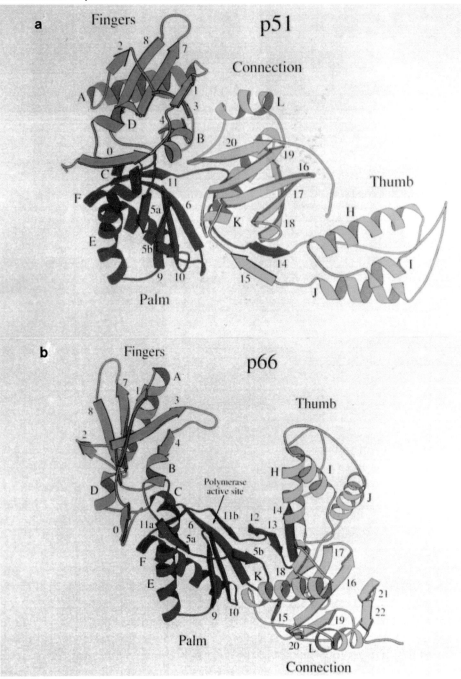

Fig. 9.16 Secondary structure drawings showing the difference between the polymerase domains of (**a**) the p66 and (**b**) the p51 subunits of HIV-1 reverse transcriptase. The palm subdomain is oriented approximately the same in both subunits. Spirals represent α-helices and arrows, β-strands. The subdomains are coloured as for the Klenow fragment (Fig. 9.14): thumb green, fingers cyan, and palm red (from Jacobo-Molina, A., Ding, J., Nanni, R.G., Clark, A.D., Jr., Lu, X., Tantillo, C., Williams, R.L., Kamer, G., Ferris, A.L., Clark, P., Hizi, A., Hughes, S.H., and Arnold, E. (1993). *Proc. Natl Acad. Sci. USA*, **90**, 6320–4).

contact with the p66 RNase H domain, thus making room for the connection subdomain, which lies roughly where the active site cleft would have been. Consequently, the heterodimer has only one polymerase site, even though there are two polymerase sequences.

The DNA in the RTase crystals, extending from the RNase H to the polymerase, is an 18 bp duplex with a single base 5′-overhang to act as a template. As RTase can use either RNA/RNA or RNA/DNA templates, both of which are A-form, it is perhaps not surprising that the DNA is somewhat A-form at the polymerase site. It is, however, more B-form at the RNase H active site and it bends by about 45° at the A–B junction (Section 2.3.1), near the p66 thumb (Fig. 9.17).

The structure of DNA and protein in the polymerase active site gives us our first clear view of a polymerase at work (Fig. 9.18). As with the Klenow editing complex (see above), the protein interacts with the phosphate backbone, not with the bases. The 3′-hydroxyl of the primer terminus is next to the three conserved acidic residues (Asp^{110}, Asp^{185}, and Asp^{186}) carried on strand 6 and in the loop between strands 9 and 10. The palm and thumb hold the nucleic

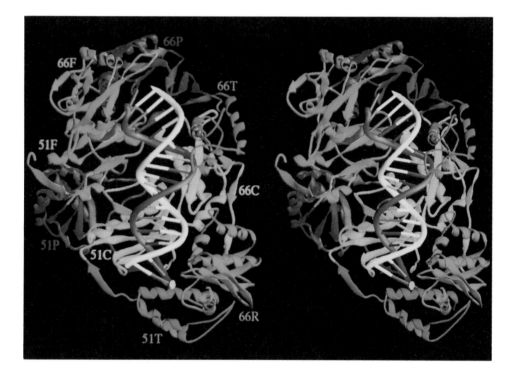

Fig. 9.17 A stereo drawing of the HIV-1 reverse transcriptase-DNA complex. Helices are represented as spirals and β-strands as arrows. The structure is oriented with the polymerase active site at the top. The thumb subdomain is green, the finger cyan, and the palm red. The one RNase H domain (marked 66R) is at the bottom in gold. Appropriately coloured letters mark each subdomain: thus the green 66T denotes the thumb subdomain of p66 and the green 51T denotes the thumb subdomain of p51. The primer strand is white and the template strand blue and the bars connecting the strands mark the positions of the bases. The DNA bends at the p66 thumb subdomain. In this view, the bulk of the p51 subdomain is to the left of the DNA; p51 may be the binding site for the lysine tRNA required for priming RTase (from Jacobo-Molina, A., Ding, J., Nanni, R.G., Clark, A.D., Jr., Lu, X., Tantillo, C., Williams, R.L., Kamer, G., Ferris, A.L., Clark, P., Hizi, A., Hughes, S.H., and Arnold, E. (1993). *Proc. Natl Acad. Sci. USA*, **90**, 6320–4). (Stereopair diagrams for parallel viewing.)

acid in the catalytic site: strands 12 and 13 of the palm (the 'primer grip') position the primer strand while strand 4 and helix B of the fingers and strand 5a of the palm (the 'template grip') position the template strand. The rest of this 'fingers subdomain' probably interacts with longer templates to keep them appropriately helical. Half a turn away from the primer terminus, two antiparallel helices in the thumb (αH and αI) interact with the phosphate backbone and helix H partially penetrates the minor groove (Fig. 9.18). They may well function like the thread on a nut to ensure that RTase spins around the nucleic acid, following the grooves correctly as it polymerizes.

Summary

Both Pol I and RTase have the same overall architecture for gripping a nucleic acid during polymerization. It is a domain that looks like a right hand, with **palm, fingers, and thumb subdomains**. Part of the palm subdomain and the direction from which the nucleic acid approaches the active site is conserved in these two polymerases. Polymerases, their 3'-5' exonucleases, and RNase Hs may all use the same mechanism, which requires **two divalent cations**, because the original all-RNA polymerases used it.

The Klenow fragment of Pol I contains two widely-separated domains, one carrying the **polymerase activity**, and the other the **3'-5' proofreading exonuclease** activity (Section 5.2.5). The DNA approaches the polymerase from the exonuclease side and bends by 90° to enter the polymerase site. The protein does not read the DNA sequence at all. Instead, when an incorrect base is added, the DNA strands separate and the daughter strand is therefore more likely to reach over to the exonuclease, which then removes the incorrect base.

RTase is a unique heterodimer because its two subunits have the same sequence yet fold differently. The p66 subunit folds into a **polymerase domain** and an **RNase H domain**. The DNA in the complex is A-form near the polymerase active site. Near the active site, the palm and thumb hold the primer strand, while the palm and fingers hold the template strand. Two helices interact with the phosphate backbone and probably ensure that RTase tracks the DNA correctly.

9.3 Specific interactions

9.3.1 The need for specificity

For a cell to function at all, proteins must distinguish one nucleic acid sequence from another very accurately. Transfer RNA synthetases must charge only their cognate tRNA, not others; transcriptional activators and repressors must turn *specific* genes on and off; restriction endonucleases must cut defined sequences, not at random; the proteins in the ribosome and snRNPs must interact with certain RNAs and not others. Proteins that bind specific nucleic acid sequences also bind non-specific ones. In some cases, like the transcriptional regulators, this binding is intrinsic to function (Section 9.1.5); in others, like the tRNAs, the binding is merely unproductive. How proteins interact in a sequence-specific manner is understood only for those systems where high-quality structures of complexes are available. These include a number of transcriptional regulators, two restriction endonucleases, and two tRNA synthetases.

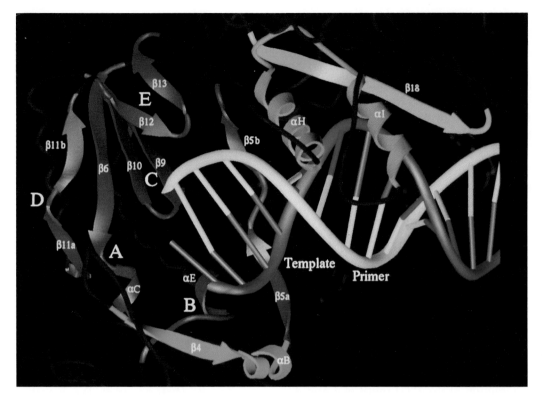

Fig. 9.18 The reverse transcriptase polymerase active site, highlighting the secondary structural elements involved in binding the nucleic acid. The representation of DNA and protein and colour-coding scheme is as in Fig. 9.17. The white capital letters A, B, C, D, and E mark regions that are conserved in all RNA-dependent polymerases. αH, which partially penetrates the minor groove, and αI keep the protein on track as it spins around the DNA. Two of the three acidic residues conserved in all polymerases are in a loop between strands 9 and 10, and the other is in the middle of strand 6 (from Jacobo-Molina, A., Ding, J., Nanni, R.G., Clark, A.D., Jr., Lu, X., Tantillo, C., Williams, R.L., Kamer, G., Ferris, A.L., Clark, P., Hizi, A., Hughes, S.H., and Arnold, E. (1993). *Proc. Natl Acad. Sci. USA*, **90**, 6320–4).

9.3.2 Transcriptional regulators: the helix–turn–helix motif

The first sequence-specific DNA binding protein structures solved were both prokaryotic: **catabolite activator protein(CAP)** from *E. coli* and cro repressor from phage λ. They used the same structure, a pair of uniquely-arranged helices, to bind DNA. This structure, the **helix–turn–helix (HTH) motif**, seems to occur only in DNA-binding proteins and is by far the most common sequence-specific DNA-binding motif in prokaryotes. It occurs in the structures of the DNA-binding domain of cI repressor from phage λ (**lambda repressor**) (Section 5.1.11), in *E. coli* **trp repressor**, **lac repressor**, and inversion stimulation factor from *E. coli*, and in phage 434 repressor (**434 repressor**) and cro protein. It also occurs, though apparently less frequently, in eukaryotic DNA binding proteins such as the homeodomain proteins that bind to homeoboxes (see below) and in histone H1(H5) (Section 9.2.1).

The canonical helix–turn–helix motif has a pair of α-helices that stack together to form a V-shape; the angle between the arms of the V is about 60°. Usually, the first helix positions the second, known as the **recognition helix**, so that it projects into the major groove and recognizes specific sequences by direct readout (Figs. 9.19; 9.20). The very high level of structural similarity engenders a lower level of sequence similarity: 6 residues out of the 20 in the motif help maintain the correct helical angle. Position-9, at the bend between the helices, is always glycine; positions -4, -8, -10, and -15 are normally hydrophobic; position -5 is almost always small (glycine or alanine). Not surprisingly, sequence comparisons have identified many other proteins that contain HTH motifs; nowadays, any protein with an HTH motif is assumed to bind DNA. There are related structures which also bind DNA: *E. coli* lexA has similarly oriented helices but with a longer connection between them.

The HTH structure, however, is a *motif*, not a *domain* and so always occurs as part of a larger structure that differs from protein to protein. In λ Cro protein and CAP, the DNA binding domain contains some β-strands but it is entirely α-helical in the λ and 434 repressors (Fig. 9.19). Secondly, the term 'recognition helix' is something of a misnomer: *both* helices in the HTH motif make contacts with the DNA and residues outside the motif itself are also often involved. All of the prokaryotic HTH proteins function as dimers and bind DNA sequences that have approximate twofold symmetry—unlike the eukaryotic homeodomain (see below) which binds as a monomer to an asymmetric DNA sequence. Many of them have more than one domain: CAP, for instance, has an N-terminal domain that binds its allosteric activator, cAMP.

The prokaryotic complexes

The interaction of the HTH motif of λ and 434 repressor with their operator DNA is rather typical of the way that prokaryotic HTH motifs interact with DNA (Fig. 9.20b). The DNA binding domain of 434 repressor binds as an approximately symmetrical dimer to an operator sequence that has approximate symmetry itself (Table 9.1, p. 405). Each monomer, therefore, recognizes a half-site. The first helix of the HTH motif (helix 2) is 'above' the major groove but its N-terminus is in contact with the DNA backbone. The repressor makes hydrogen bonds to four phosphate groups per monomer, thus clamping helix 3 (the recognition helix) in the major groove with its N-terminus closest to the edges of the base-pairs. Some of the positioning hydrogen bonds are from backbone amides, some from side chains. In 434 repressor, helix 3 then interacts with the sequence 5'-ACAA-3' in the major groove via direct readout (Fig. 9.20b). At base-pair 1, A·T, Gln[28] forms a bidentate hydrogen bond to the adenine and makes van der Waals contact with the thymine. At base-pair 2, C·G, Gln[29] forms a bidentate hydrogen bond to the guanine and Glu[32] interacts with the cytosine. At base-pair 3, A·T, the side chains of Thr [27] and Gln[29] create a pocket to bind the methyl group of the thymine; the interaction is completely non-polar. At base-pair 4, A·T, the thymine is hydrogen bonded to Gln[33] and there are hydrophobic contacts from the thymine methyl group to Gln[29] and Ser[30].

Each repressor–operator complex examined so far, however, has its own unique features. The 434 repressor uses indirect, as well as direct, readout: the DNA bends moderately toward the protein and is overwound and highly propeller-twisted in the centre and underwound at the end (Table 9.1). The DNA must be bendable in the middle of the operator; if base-pairs 6 and 7 are changed, the binding constant of 434 repressor to DNA changes even though it does not make any contact with those bases.

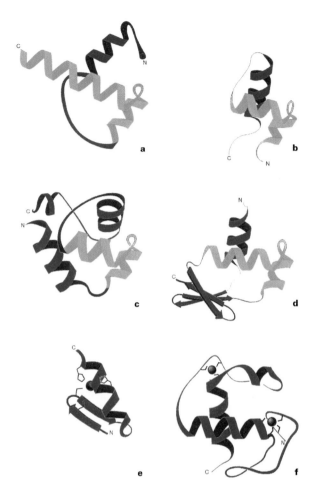

Fig. 9.19 Ribbon diagrams of some sequence-specific DNA-binding domains that use α-helices as recognition elements, drawn from the DNA looking towards the protein. α-Helices are shown as spirals and β-strands as curved arrows. The helical axis of the bound DNA is approximately vertical. In (**e**) and (**f**), the zinc ion and its ligands are in black. The first four proteins all contain HTH motifs, which are shown highlighted. (**a**) The homeodomain—a single three-helix bundle. The second HTH helix is two turns longer than in the prokaryotic HTH motifs (**b, c, d**). (**b**) The *lac* repressor, a three-helix bundle where the third helix follows the HTH motif, rather than preceding it. (**c**) The 434 repressor domain (residues 1 to 69). It is a 5-helix bundle; λ repressor is very similar but contains an N-terminal arm as an additional recognition element (Fig. 9.21). (**d**) The DNA-binding domain of CAP (and histone H1/H5), which has a three-helix bundle closed off by a 4-stranded anti-parallel sheet. (**e**) A Cys_2His_2 zinc finger. One cysteine from each β-strand and two cysteines from the α-helix bind a zinc atom. In this way the zinc atom stabilizes the structure by holding the secondary structural elements together. (**f**) The glucocorticoid receptor DNA-binding domain. Four cysteines from two helix-loop motifs bind two zinc ions. DNA-binding residues are at the beginning of the N-terminal α-helix, while residues that provide the dimerisation interface are at the beginning of the second loop (from Harrison, S.C. (1991). *Nature*, **353**, 715–19. Copyright (1991) Macmillan Magazines Limited).

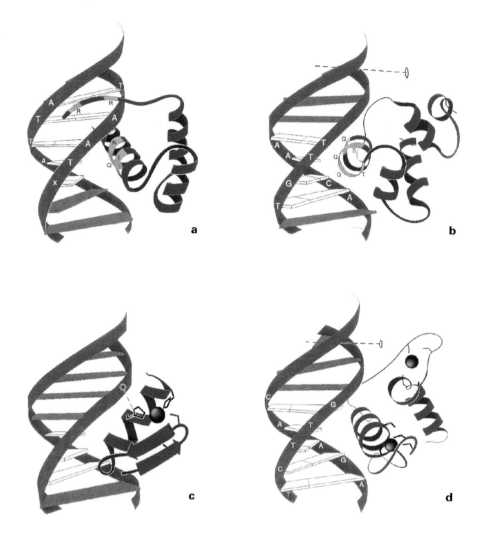

Fig. 9.20 Complexes of four of the DNA-binding domains illustrated in Fig. 9.19: the homeodomain (**9.19(a)**), 434 repressor (**9.19(c)**), the zinc finger (**9.19(e)**) and the glucocorticoid receptor (**9.19(f)**). The views in (**a**), (**b**), and (**d**) are about 90° and the view in (**c**), about 180° from the views in Fig. 9.19. The protein is blue and the DNA red; bases to which the protein binds are white. Key protein residues are shown in light blue and marked using the one-letter amino acid code. α-Helices are shown as spirals and β-strands as arrows. The little oval and dashed line in (**b**) and (**d**) represent the position of the two-fold axis of symmetry in these two complexes. In (**c**) and (**d**), the zinc ion and its ligands are in black. (**a**) Interaction of the engrailed homeodomain with the TAATX core consensus sequence of a homeobox. The residues whose position is marked are Ile^{47}, Gln^{50} and Asn^{51}, from the second and third turns of the α-helix and Arg^3 and Arg^5 on the N-terminal arm. (**b**) Interaction of 434 repressor with 434 operator. The positions of the key residues Thr^{27}, Gln^{28}, Gln^{29}, Ser^{30}, and Gln^{33} are shown. The relative position of the protein and DNA in **9.20(a)** is different than in **9.20(b)**; hence a different part of the 'recognition' helix is in contact with the DNA. (**c**) A Cys_2His_2 zinc finger with its α-helix inserted into the major groove of the DNA. Specific contacts are made between residues from the N-terminal portion of the α-helix and a three-residue subsite. The first zinc-coordinating His also binds a backbone phosphate (indicated with a broken line). (**d**) The interaction between the glucocorticoid receptor DNA-binding domain and a GRE (from Harrison, S.C. (1991). *Nature*, **353**, 715–19. Copyright (1991) Macmillan Magazines Limited).

Table 9.1 Helical parameters for the 434 operator bound to the 434 repressor fragment

Base-pair			Twist (degrees)	Propeller twist (degrees)
			35	
1R	**A·T**	1'R		9
			35	
2R	**C·G**	2'R		2
			32	
3R	**A·T**	3'R		4
			34	
4R	**A·T**	4'R		6
			38	
5R	A·T	5'R		12
			37	
6R	C·G	6'R		14
			36	
7R	T·A	7'R		28
			38	
7'L	T·A	7L		30
			40	
6'L	T·A	6L		9
			36	
5'L	C·G	5L		29
			38	
4'L	**T·A**	4L		13
			33	
3'L	**T·A**	3L		11
			32	
2'L	**G·C**	2L		3
			35	
1'L	**T·A**	1L		13
			34	

The helical twist and propeller twist observed in the complex of a synthetic oligonucleotide containing the 434 phase OR1 operator sequence with residues 1–69 of 434 repressor. The bases are highly overwound and propeller-twisted in the centre of the operator. The operator sequence is approximately two-fold symmetric: the symmetric bases are shown in bold. Adapted from Aggarwal, A.K., Rodgers, D.W., Drottar, M., Ptashne, M., and Harrison, S.C. (1988). *Science*, **242**, 899–907.

The λ repressor has an extra DNA-binding element: an N-terminal arm which, although flexible, is essential for binding DNA. It wraps around the DNA and binds on the opposite side of the DNA from the HTH motif in the major groove. The residue Lys^4 forms a bridging hydrogen bond to guanines at positions 6 and 7 in the major groove (Fig. 9.21): if this residue is mutated, λ repressor is not active. Residues Lys^3 and Lys^5 also make important contacts to the bases and the phosphate backbone.

Trp repressor, which like *met* repressor (Section 9.3.5) is a dimer of intertwined monomers, is unusual. It makes no direct hydrogen bonds to the bases in the major groove; the few that do form are water-mediated. Instead, it distorts the DNA to make 24 direct and 6 water-mediated hydrogen bonds to the phosphate backbone and the extreme flexibility of the DNA sequence is essential for recognition.

The *lac* operator system binds its repressor (*lac* repressor) and its activator (CAP) in still different ways. Nuclear magnetic resonance studies of *lac* repressor headpiece (the DNA binding domain) (Fig. 9.19b) binding to DNA show that its recognition helix binds in the major groove in the *opposite* orientation to that seen in the other repressors: instead of the C-terminus of the helix being closer to the centre of the operator site, it is further away. Consequently, even among prokaryotic HTH repressors, there is no simple code relating the identity of the side chain to the base-pair recognized.

CAP, the only prokaryotic activator solved so far, induces a remarkable kink in the DNA (Fig. 9.22). Unlike the λ and 434 repressors described above, where both protein and DNA undergo some conformational change, the structure of CAP changes not at all when it binds the DNA. Instead, the DNA bends around by about 90°, creating two 43° kinks between base-pairs 5 and 6 on either side of the twofold axis (Fig. 9.23). The bases roll by about 40° and unstack the base-pairs. Clearly, indirect readout of the DNA is important: the kink occurs at the TG step in the conserved DNA sequence GTG. Mutagenesis studies show that Glu^{181} is critical in forming the kink; it may position base-pair 5 (Fig. 9.24) so that a kink must be

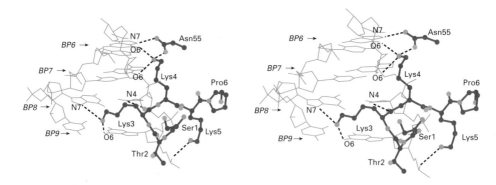

Fig. 9.21 A stereo diagram of the N-terminal arm of λ-repressor interacting with DNA. All the atoms are shown. Base-pairs 6 to 9 are in blue; residues 1 to 6 labelled. Hydrogen bonds are shown by dashed lines. The essential bridging interaction of Lys^3 with G6 and G7 and the hydrogen bond from the peptide backbone to C8 are shown (from Clarke, N.D., Beamer, L.J., Goldberg, H.R., Berkower, C., and Pabo, C.O. (1991). *Science*, **254**, 267–70. Copyright (1991) by the AAAS). (Stereopair diagrams for parallel viewing.)

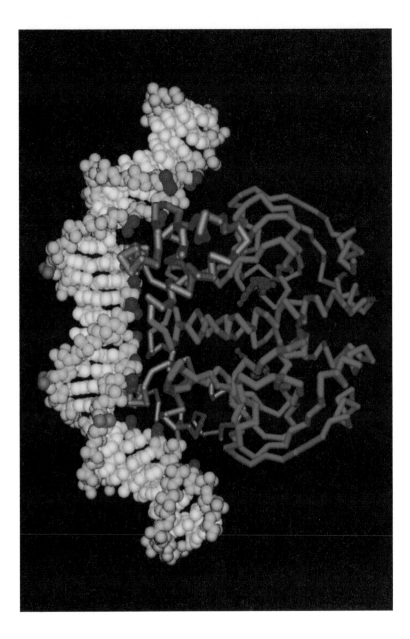

Fig. 9.22 Structure of the CAP–DNA complex. The protein is shown as an α-carbon trace. The cAMP binding domain is blue with a red ball-and-stick cAMP inside it and the DNA binding domain, purple; the DNA is shown as a space-filling model with the bases in pale-blue and the sugar-phosphate backbone yellow. Phosphates whose ethylation interferes with CAP binding are red: they all lie at the protein-DNA interface. Phosphates which are hypersensitive to DNase I are blue; they are on the outside and one is precisely at the kink in the DNA where the minor groove is about 10 Å wide (from Schultz, S.C., Shields, G.C., and Steitz, T.A. (1991). *Science*, **253**, 1001–7. Copyright (1991) by the AAAS).

formed. About 10 nucleotides from the centre of the operator, the DNA bends again by about 8°, allowing Lys^{26} in the N-terminal cAMP-binding domain to interact with a DNA phosphate. The sides of the CAP dimer have a positive electrostatic potential which supplies the free energy needed to kink the DNA by this amount.

The other interactions of CAP with its operator are more normal: three side chains from the recognition helix make direct hydrogen bonds to bases in the major groove of the DNA (Fig. 9.24); there are hydrogen bonds from the N-terminus of the first helix in the HTH motif to the phosphate backbone (Fig. 9.23), and the central 10 bp of the operator are clamped to the protein by 8 hydrogen bonds. The structure of the CAP–DNA complex explains the results from solution studies of CAP and its operator (Fig. 9.22). It also supports the model of how CAP stimulates transcription: the CAP-induced DNA bend allows RNA polymerase to interact with DNA sequences it otherwise could not reach.

Eukaryotic complexes: the homeodomain

As was first suggested by sequence comparison, the **homeodomain** proteins (Section 5.3.1), which bind the eukaryotic homeobox, have an HTH motif. Nevertheless, they are rather different from the prokaryotic HTH proteins. Firstly, the 60-residue homeodomain folds into a stable structure by itself (Fig. 9.19a); it *is* a complete domain and, secondly, it binds to the *asymmetric* homeobox as a *monomer*.

The stuctures of three homeodomain–DNA complexes are known: *Drosophila Antennapaedial* (**Antp**) and **engrailed** proteins, and yeast **MAT α2**. They all bind DNA by inserting their long third helix (the recognition helix) into the major groove of the DNA and their N-terminal

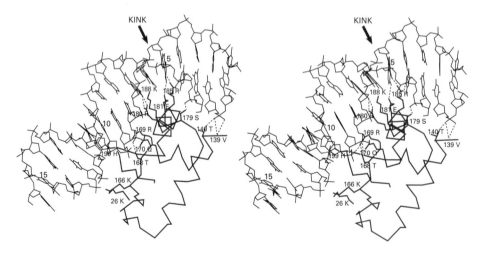

Fig. 9.23 A stereo diagram of one DNA half-site (black) bound to one CAP DNA-binding domain (red). The HTH motif is in bold; all protein side chains that interact with the DNA are shown and the hydrogen bonds between protein and DNA are shown as dotted lines. The kink that unstacks bases-5 and -6 and widens the minor groove to 10 Å is indicated, as is position-10, where the DNA bends by another 8°, allowing Lys^{26} from the larger, N-terminal cAMP-binding domain to make contact with the phosphate backbone (from Schultz, S.C., Shields, G.C., and Steitz, T.A. (1991). *Science*, **253**, 1001–7. Copyright (1991) by the AAAS). (Stereopair diagrams for parallel viewing.)

arm into the adjacent minor groove (Fig. 9.20b). The latter interaction is like that between the N-terminal arm of λ repressor and DNA (see above) but the interaction of the HTH motif with DNA is not the same as in prokaryotes. The helices in the HTH motif are much longer (compare Figs. 9.19a and b) and so helix 2 (the first helix in the HTH motif) must lie entirely above the major groove to avoid bumping into the DNA. Helix 3 is therefore positioned differently too (compare Figs. 9.20a and b): its centre, not its N-terminus, is closest to the DNA and so the residues which recognize the DNA are on turns 2 and 3 in the helix, not turn 1.

The residues which appear to be important in positioning the homeodomain correctly on the DNA are highly conserved: they include Asn[51], which forms two hydrogen bonds to an adenine in the major groove, and six others, which interact with the phosphate backbone (Figs. 9.20a; 9.25). Three of those six are at the C-terminal end of helix 3, which is kinked in the nmr structure of Antp but straight in the two crystallographic structures. Positions-47, -50 and -54 appear to be involved in discriminating one homeobox from another by direct readout (Figs. 9.20a, 9.25). The N-terminal arm may also be involved in discrimination: its sequence varies from homeodomain to homeodomain and its position in the minor groove is different in MAT $\alpha2$ and engrailed.

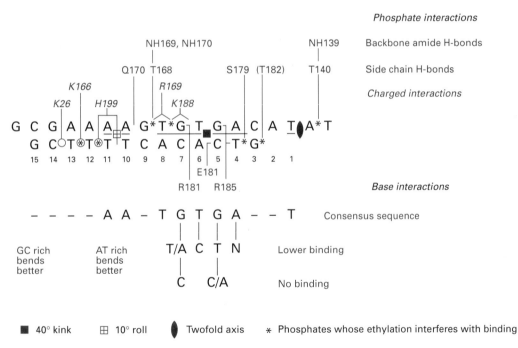

Fig. 9.24 A diagram summarizing the interactions made by CAP with a DNA half-site. NH means the backbone amide nitrogen interacts with the DNA, while the one letter code indicates which protein residues interact with DNA. Interactions with the sugar-phosphate backbone are listed above the half-site DNA sequence; interactions with the bases, below. The consensus sequence is shown below, as are mutations in the consensus sequence which reduce or eliminate CAP binding to DNA (from Schultz, S.C., Shields, G.C., and Steitz, T.A. (1991). *Science*, **253**, 1001–7. Copyright (1991) by the AAAS).

9.3.3 Exclusively eukaryotic transcriptional regulators: the zinc finger and leucine zipper

The zinc finger proteins

By far the most common eukaryotic gene regulatory proteins contain **zinc fingers**: DNA-binding domains with zinc co-ordinated to cysteines or histidines. Indeed, about 0.5 per cent of the human genome codes for such proteins. The first zinc finger discovered, transcription factor TFIIIA from *Xenopus* oocytes, appeared to have short, tandemly arranged motifs of an α-helix and an antiparallel β-sheet that point out like fingers (Fig. 9.19e). So far, three very different zinc-containing domains have been discovered: the **Cys$_2$His$_2$** finger, the **Cys$_4$** nuclear receptors, and the **GAL4** binuclear cluster. All use an α-helix inserted in the major groove to bind DNA. We discuss the one example of each whose structure is known in the presence of DNA: mouse **Zif268**, mouse **glucocorticoid receptor**, and yeast **GAL4**.

The Cys$_2$His$_2$ zinc finger

The Cys$_2$His$_2$ (or **TFIIIA-type**) zinc finger motif occurs in hundreds of eukaryotic DNA-binding proteins. All contain the characteristic sequence X$_3$-Cys-X$_{2-4}$-Cys-X$_{12}$-His-X$_{3-4}$-His-X$_4$ (where X is any amino acid). Retroviral nucleocapsid proteins that recognize signals for packaging the RNA contain a variant of the Cys$_2$His$_2$ zinc finger, where one of the histidines has been replaced by a cysteine to give a Cys$_3$His zinc finger.

The mouse Zif268 protein contains three tandemly repeated Cys$_2$His$_2$ zinc fingers: residues 5–30, 35–58, and 63–86. Each finger is less than 30 residues long, and their overall structures are very similar. Each contains one zinc ion co-ordinated by two histidines from the α-helix and two cysteines from the β-sheet (Fig. 9.20c). The motif is stabilized by the zinc ion and by interactions between hydrophobic residues on the α-helix and β-sheet. The three fingers occupy the major groove of the cognate DNA so that the N-terminal ends of the α-helices point towards and make contacts with the base-pairs of the DNA. The three fingers are separated from each other by four-residue linkers and there are no important contacts between fingers or between a finger and atoms of the DNA backbone. An exception is the first zinc co-ordinating histidine, which is bound to a DNA phosphate in Zif268 and in many other zinc finger proteins. This may be a general feature in zinc finger–DNA interactions.

Each zinc finger is oriented similarly with respect to the DNA and makes similar contacts to three base-pairs of DNA. In the case of Zif268 protein, the contacts made to the G-rich DNA strand are mainly between arginines and guanines: in fingers 1 and 3, the arginine N$^{\eta}$s donate hydrogen bonds to the N-7 and O-6 of two guanines. In finger 2, however, the second hydrogen bond donor is a histidine. Since each of the five arginines donates two hydrogen bonds and the histidine one, the three-finger domain forms 11 hydrogen bonds to the 9 bp cognate sequence.

The Cys$_4$ nuclear receptors

The Cys$_4$ zinc domains found in the nuclear receptor superfamily of eukaryotic transcriptional regulators contain two α-helix-loop motifs, each co-ordinating one zinc ion with four

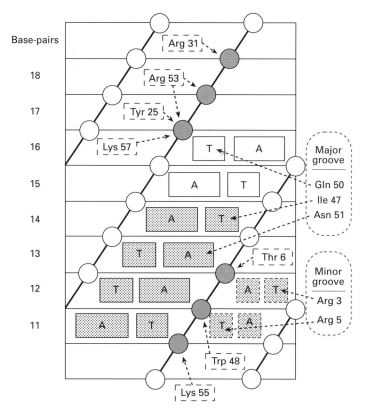

Fig. 9.25 A diagram of the interactions of an engrailed homeodomain with its homeobox. The DNA helix is shown in projection, as if it were a cylindrical pipe that had been slit open and flattened. The bases are shown as boxes and the phosphates as circles. Arrows point from a specific protein residue to the base or phosphate with which it interacts. The TAAT subsite common to all homeoboxes is shaded and the phosphates that interact with the homeodomain are hatched (from Kissinger, C.R., Liu, B., Martin- Blanco, E., Kornberg, T.B., and Pabo, C.O. (1990). *Cell*, **63**, 579–90. Copyright (1990) Cell Press).

cysteines: *two* zinc ions bind per domain (Fig. 9.19f). Furthermore they bind as a dimer to a twofold symmetric sequence. In the glucocorticoid receptor family, the DNA is called the glucocorticoid response element (**GRE**) and contains two half-sites (5′-AGAACA-3′) separated by a critical 3 bp spacing. If this spacing is too long, the second subunit binds non-specifically to a non-specific target site, while the first subunit binds correctly. Such an arrangement is seen in the crystal structure of the DNA-binding domain of the mouse glucocorticoid receptor (residues 440–525) bound to an 18 bp oligonucleotide.

The N-terminal α-helix makes contacts with the bases of the DNA, the N-terminal loops are in contact with the phosphate backbone and the C-terminus of the molecule forms the dimerization interface (Fig. 9.26). As in the case of the Cys$_2$His$_2$ zinc finger, hydrophobic contacts also help the domain fold correctly. Glucocorticoid receptor only functions as a dimer, and cognate DNA is an allosteric activator of dimerization. The correct dimer can only form when the monomers bind in the major groove to GRE half-sites spaced exactly 3 bp apart. There are, however, no protein-DNA contacts in the minor groove between the half-sites.

In the structure of the complex, the spacing between the half-sites is 4 bp. Consequently, although the first subunit binds in a sequence-specific manner to a cognate half-site (Fig. 9.27a) and a dimer does form (Fig. 9.26), the second subunit is forced out of register by 1 bp. It is, therefore, one of the few examples of a sequence-specific DNA-binding protein binding a non-specific DNA sequence (Fig. 9.27b). In the specific half-site, direct readout of the DNA sequence occurs at four bases: Lys^{461} makes one direct and two water-mediated hydrogen bonds to G(-7) and T(+6); Val^{462} is in van der Waals contact with T(+5); Arg^{466} makes two hydrogen bonds to G(+4) (Fig. 9.27a). All except the first interaction is lost in the non-specific half-site (Fig. 9.27b). There are also some changes, though not as many, to the pattern of phosphate contacts: instead of seven contacts to the phosphate backbone, there are only five.

The differences in binding between the non-cognate and cognate half-sites are what one might expect: the non-cognate half-site appears to be less tightly associated with the DNA and the α-helical recognition element is not buried as deeply in the major groove (Fig. 9.26). The role of residues (such as Lys^{461} and His^{451}) conserved in different members of the nuclear receptor superfamily is not clear: some are in contact with the nucleic acid in both specific and non-specific half-sites; others are not. Furthermore, the nuclear receptor proteins all bind similar response elements, and bind them rather weakly. How then do they distinguish the right target from the wrong one? In glucocorticoid receptor, as in *met* repressor (Section 9.3.5), binding to multiple correctly-spaced elements increases specificity by protein oligomerization, *not* by protein–DNA interaction.

The GAL4 zinc finger

The DNA binding domain of yeast transcription activator protein GAL4 (sometimes called a **binuclear zinc cluster** domain) contains two zinc ions separated by only about 3.5 Å. The zinc ions are co-ordinated by six cysteines. It is monomeric in solution, like the Cys_4 zinc domain, but dimerizes upon binding to a 17 bp symmetrical DNA sequence with specific CCG triplets at each end of the sequence. Each GAL4 monomer has three subdomains: the zinc cluster DNA binding subdomain, a linker region, and a region responsible for the dimerization. The DNA recognition α-helices of both GAL4 subunits lie in major grooves separated by about 1.5 turns of DNA helix. If the DNA is numbered so that base zero is in the middle of the 17 bp sequence, then hydrogen bonding by Gln^9 and Arg^{15} to phosphate-5 and by Lys^{20}, Cys^{21}, and Lys^{23} to phosphate-6 fixes the helix to the backbone of the DNA. Interestingly, hydrogen bonds between the bases and the protein come from side chain *and* main chain atoms: the main chain carbonyls of Lys^{17} and Lys^{18} form hydrogen bonds to C7 and C8 and the side chain of Lys^{18} forms a hydrogen bond to G6 and G7. Hydrogen bonds from main chain atoms may be more specific than hydrogen bonds from long side chains firstly because different DNA sequences cannot be accommodated by a mere side chain rotation and secondly because changes in the protein sequence would not change specificity unless the conformation of the protein changed radically.

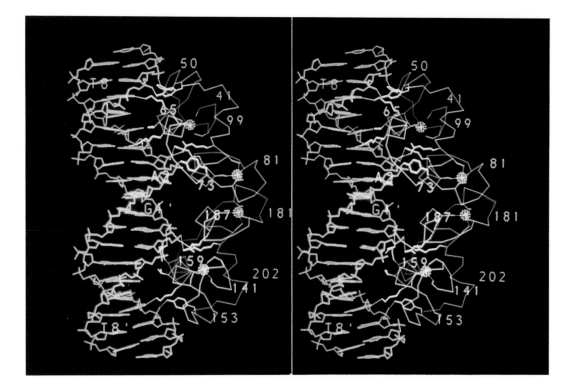

Fig. 9.26 Binding of the glucocorticoid receptor to a GRE with a four-base separation between half-sites. The subunit binding to the cognate half-site is in red and the non-cognate subunit is in blue. In each case the protein is represented by its α-carbon backbone, except for residues that are involved in contacts to the bases or phosphates of the DNA. Zinc atoms are shown as spheres (reprinted with permission from Luisi, B.F., Xu, W.X., Otwinowski, Z., Freedman, L.P., Yamamoto, K.R., and Sigler, P.B. (1991). *Nature*, **352**, 497–505. Copyright (1991) Macmillan Magazines Limited). (Stereopair figures for parallel viewing.)

The leucine zipper

The **leucine zipper** is a structural motif found in certain eukaryotic regulatory DNA-binding proteins, like the yeast transcriptional regulator **GCN4** (General Control Non-derepressable 4) and oncoproteins *fos, jun,* and *myc*. The leucine zipper motif itself does not bind DNA but dimerizes the zipper proteins so that *they* can do so. The structures of the DNA-binding domains of two leucine zipper proteins in the presence of DNA are known: GCN4 and the human transcription factor Max. Here we mainly discuss GCN4 and its interaction with one of the sequences it recognizes.

A leucine zipper (**bZIP**) DNA binding domain is typically less than 100 amino acids long and contains three subdomains: an N-terminal regulatory subdomain, a dimerization leucine zipper subdomain, and a basic DNA binding subdomain. The leucine zipper subdomain is an α-helix with a leucine every 7 residues. Because an α-helix has 3.6 residues per turn, the leucines all point in the same direction and two such helices juxtaposed in parallel wrap around each other to form a left-handed coiled coil—hence the name. The hydrophobic exclusion of

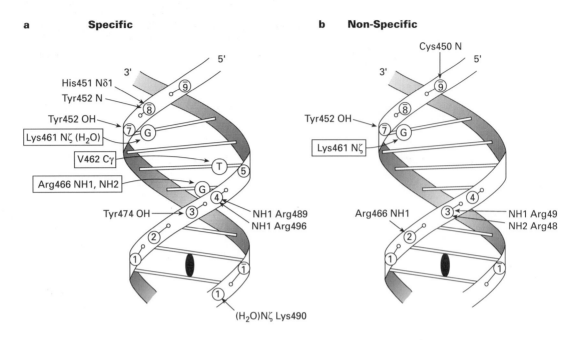

a   Specific

His451 Nδ1
Tyr452 N
Tyr452 OH
Lys461 Nζ (H₂O)
V462 Cγ
Arg466 NH1, NH2
Tyr474 OH

3'
5'

NH1 Arg489
NH1 Arg496

(H₂O)Nζ Lys490

b   Non-Specific

Cys450 N
5'
3'

Tyr452 OH
Lys461 Nζ

Arg466 NH1

NH1 Arg49
NH2 Arg48

Fig. 9.27 Diagrams showing (**a**) the specific and (**b**) the non-specific contacts between the glucocorticoid receptor and DNA. Contacts to the bases are boxed. The symmetry axis of the palindromic target sequence is indicated by the black oval (reprinted with permission from Luisi, B.F., Xu, W.X., Otwinowski, Z., Freedman, L.P., Yamamoto, K.R., and Sigler, P.B. (1991). *Nature*, **352**, 497–505. Copyright (1991) Macmillan Magazines Limited).

water from the dimer interface is essential for dimerization, but intra-coil ionic and van der Waals interactions are also important.

The basic region of GCN4, unlike that of Max, is disordered in the absence of DNA and becomes α-helical upon binding DNA. The bZIP subdomain then forms the stem of a **Y** and is joined by short loops to the basic helices, which form the arms. The DNA binds perpendicularly between the arms so that the two basic α-helices, which fit into the major groove, point in opposite directions along the DNA. This complex has twofold symmetry. In GCN4, the basic α-helices are quite straight and so extend away from the DNA after contacting three or four base-pairs: this is called an **induced helical fork**. In Max, the basic α-helices are kinked so that they follow more of the major groove: this is called a **scissors grip**.

The spacing between the leucine zipper and the basic region is functionally important in GCN4-type bZIP proteins: the spacer length must be an integral multiple of 7 amino acids for DNA binding to occur. If it is not, the basic region will be incorrectly positioned on the DNA because it is a continuation of the leucine zipper α-helix. The base-specific contacts that GCN4 makes with the DNA are typical of those found in other proteins that use α-helices as recognition elements (Table 9.2) and GCN4 also makes a number of contacts to the DNA-phosphate backbone.

Table 9.2 GCN4–DNA recognition. The oligonucleotide sequence used was based on that of the AP-1 binding site, to which the mammalian oncoproteins fos and jun also bind. The protein–DNA interactions in the complex are asymmetric because the 20 bp DNA sequence, which contains two 4 bp GCN4 half-sites, has an odd number of bases and so does not itself have perfect two-fold symmetry. Residues 226–281 are the DNA-binding bZIP domain of GCN4 and were crystallized with the oligonucleotide

DNA		GCN4	Type
Base-pair 4 (A·T)			
Weak interactions with the bases			
Base-pair 3 (T·A)			
T	O4	Asn235	H-bond from $N\Delta_1$
	Me5	Ala238	Hydrophobic contact to $C\beta$
	Me5	Ser242	Hydrophobic contact to $C\beta$
Base-pair 2 (G·C)			
C	N4	Asn235	H-bond to $O\Delta_1$
Base-pair 1 (A·T)			
T	Me5	Ala239	Hydrophobic contact to $C\beta$
Base-pair 0 (C·G)			
G	O6	Arg243	H-bond from $N\eta_1$ [a]
G	N7	Arg243	H-bond from $N\eta_2$ [a]

[a] Marks those contacts that are asymmetric in the structure of the GCN4–DNA complex.

9.3.4 Other α-helical binding motifs

The structure of bovine papilloma virus transcriptional activator **E2** bound to its specific target has revealed a new solution to the problem of placing an α-helix in the major groove of B-form DNA. The DNA binding domain of E2 (residues 326–410) is a very tightly associated dimer with a unique eight-stranded antiparallel β-barrel to which each monomer contributes four strands (Fig. 9.28). Strands β1 and β2 of each monomer are parallel to each other and are linked by a long crossover helix on the outside of the barrel. The helix, which is highly conserved among different E2s, is the recognition helix.

Unlike CAP (Fig. 9.22), E2 bends the DNA smoothly with a radius of about 45 Å: no base-pair deviates from canonical B-DNA by more than a 14° roll and both major and minor grooves are compressed when they face the protein (Fig. 9.29). As usual, the protein interacts with the phosphate backbone (10 side chains make 10 direct and 14 water-mediated contacts per half site) and it achieves specificity by interacting with the bases in the major groove (four side chains from the recognition helix make contacts with five base-pairs in each half-site) (Fig. 9.28; 9.29). Two interactions are unique: a cysteine sulfhydryl forming hydrogen bonds to base-pairs and an asparagine bridging them. Residue Cys^{340} donates a hydrogen bond to

G($+5$) and accepts one from A(-6) while Asn^{336} spans 4 bp by hydrogen bonding directly to A(-6) and C(-5) and indirectly to C($+3$) (Fig. 9.29).

9.3.5 β-Sheet binding motifs

The *met* repressor family

Most transcription factors use an α-helix to bind DNA. The *E. coli* **met repressor** was the first example of one that used a β-sheet instead. *Arc* and *Mnt* repressors also have similar structures. The *met* repressor binds co-operatively to operators which have tandem repeats of the twofold symmetric *met* box sequence, 5'-AGACGTCT-3'. It binds its co-repressor, *S*-adenosylmethionine (SAM) non-cooperatively and without a conformational change; SAM probably increases the affinity of *met* repressor for DNA simply because its sulfur atom is positively charged. The structure of *met* repressor complexed with 18 bp of DNA containing two *met* boxes neatly explains the biochemical data.

The 104-residue *met* repressor, like *trp* repressor, is a highly intertwined dimer (Figs. 9.30, 9.31); each monomer contributes one α-helix and one antiparallel β-strand to the interface.

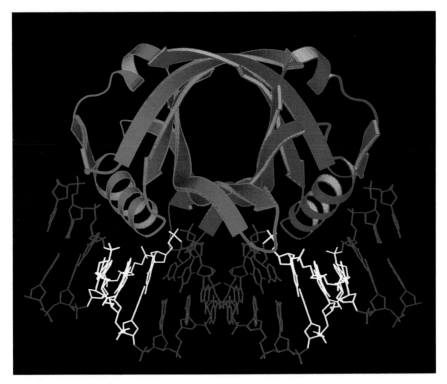

Fig. 9.28 A ribbon drawing of the E2–DNA complex looking down the barrel axis of the protein. The helices are represented by spirals, the strands by curved arrows. The two protein subunits are blue and magenta; the DNA is red, except for the nucleotides in the conserved recognition element, which are white. The two-fold axis relating the two DNA half-sites runs vertically through the drawing (reprinted with permission from Hegde, R.S., Grossman, S.R., Laimins, L.A., Sigler, P.B. (1992). *Nature*, **359**, 505–12. Copyright (1992) Macmillan Magazines Limited).

The two-stranded antiparallel β-sheet so formed binds in the major groove of the operator site (Fig. 9.31). The structure of *met* repressor is essentially unchanged on binding DNA, but the loop from residues 12–20, which is rather flexible in the unliganded structure, becomes better ordered and wraps around the phosphate backbone. Residues Lys^{23} and Thr^{25}, one from each monomer, are the major contributors to sequence specificity: their side chains form hydrogen bonds directly to the base-pairs (Fig. 9.32). The lysines also form water-mediated hydrogen bonds to the bases. The 12–20 loop and the N-terminus of helix B (residues 53–58) (Fig. 9.31) interact with the phosphate backbone: as for *trp* repressor (Section 9.3.2), there are *more* hydrogen bonds to the phosphates than to the bases. Slight increases in the roll and tilt of the bases bend the centre of each *met* box by about 25° towards the major groove (Fig. 9.30) and so increase the surface area buried on complexation from about 200 Å^2 (in the model-built complex with straight DNA) to 650 Å^2. The central T-A step between the *met* boxes is overwound (helical twist of 44°) and the flanking C-T and A-G steps are correspondingly underwound, thus pulling the phosphate of the G into position to interact with the N-terminus of helix B in the repressor. Again, subtle DNA sequence-dependent distortability increases the ability of the repressor to recognize its operator. Finally, the structure beautifully explains why *met* re-

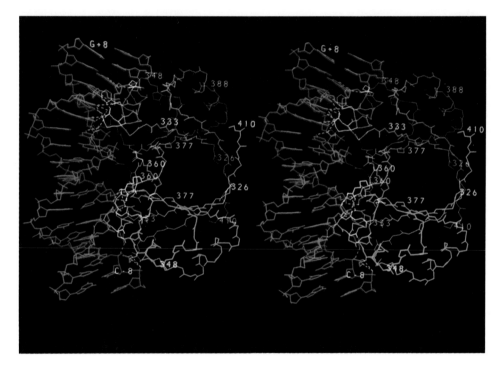

Fig. 9.29 A stereo diagram of the E2–DNA complex, turned 180° from the view in figure 9.28; the two-fold axis now runs horizontally. The DNA is coloured blue, one protein monomer is pink and the other cyan. The protein is represented as a backbone trace, except for side-chains that interact with the DNA. On the pink subunit, side-chains that interact with the bases are shown and, on the green subunit, side-chains that interact with the phosphates. The hydrogen bonds are shown as dashed green lines (reprinted with permission from Hegde, R.S., Grossman, S.R., Laimins, L.A., and Sigler, P.B. (1992). *Nature*, **359**, 505–12. Copyright (1992) Macmillan Magazines Limited). (Stereopair figures for parallel viewing.)

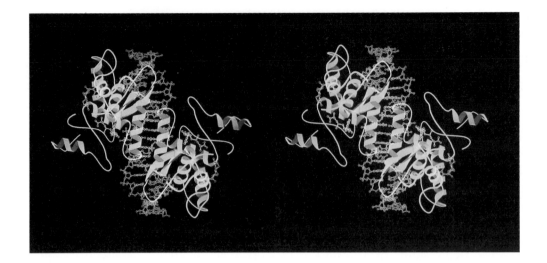

Fig. 9.30 A stereo diagram of the complex of an oligonucleotide of two *met* boxes with two repressor dimers. The repressors are represented as ribbons, with the α-helices as spirals and the β-strands as curved arrows, while the DNA and the activator SAM are represented as ball-and-stick models. The repressors are gold, with a pale green facing to the β-strands, the DNA is blue and the four SAM molecules are green. The two-fold axis relating the two *met* boxes to each other and the two repressor dimer to each other passes through the middle of the DNA and is perpendicular to the plane of the figure. Each repressor dimer is highly intertwined. The DNA curves slightly towards the repressor dimer at the bottom, then back towards the repressor dimer at the top. The major contact between the two repressors comes from the two helices, one from each dimer, that lie above the minor groove in the centre of the figure. The glycosylic bond in the adenosine residues is constrained in the *syn*-conformation (reprinted with permission from Somers, W.S. and Phillips, S.E.V. (1992). *Nature*, **359**, 387–93. Copyright (1992) Macmillan Magazines Limited). (Stereopair figures for parallel viewing.)

pressor binds poorly to a single *met* box but well to tandem boxes: a pair of antiparallel helices, one from each dimer, has an extensive hydrophobic and hydrogen bonded interface (Fig. 9.30). Thus the surface area buried on binding DNA is not twice the dimer surface area buried (1300 Å^2) but 1730 Å^2: the whole binds more tightly than the sum of its parts.

The TFIID TATA-box binding protein

To initiate transcription, all three eukaryotic RNA polymerases require the TATA-box binding polypeptide (TBP or TFIIDτ), which binds in the minor groove of DNA and recognizes the sequence TATA (Section 5.1.4). The structure of TBP-2 from *Arabidopsis thaliana* is unique. The monomeric molecule contains two 88-residue, superimposable, structural domains related by an intramolecular twofold axis. (The protein sequence is tandemly repeated but, as always, the structure is *much* more conserved than the sequence.) Each domain contains a five-stranded all-antiparallel β-sheet and the domains are connected by a seven-residue linker. The first strand in domain-1 is hydrogen bonded to its counterpart in domain-2 across the intramolecular twofold axis (Fig. 9.33); the resulting saddle-shaped structure (Fig. 9.34) is obviously complementary to B-DNA.

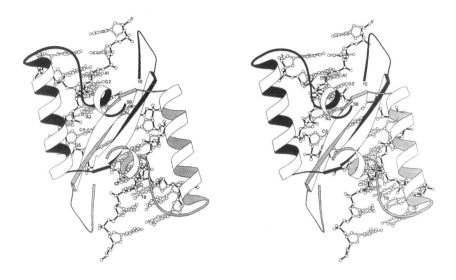

Fig. 9.31 A stereo drawing showing residues 10–58 from one repressor dimer interacting with one *met* box. The protein is represented as a ribbon, with α-helices as spirals and β-sheets as curved arrows. The view is about 90° from that in Fig. 9.30. The DNA is shown as a ball-and-stick model, with filled bonds for the sugar-phosphate backbone. The bases of one strand of a *met* box are labelled. The β-ribbon in the major groove, the N-terminus of helix B, and the 12-20 loop can be seen (reprinted with permission from Somers, W.S. and Phillips, S.E.V. (1992). *Nature*, **359**, 387–93. Copyright (1992) Macmillan Magazines Limited). (Stereopair figures for parallel viewing.)

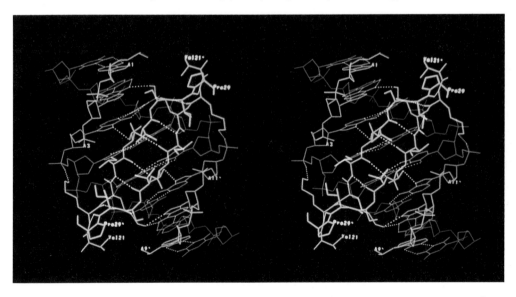

Fig. 9.32 A more detailed stereo drawing of the *met* repressor-operator interaction, showing the two-stranded β-ribbon in the major groove of the DNA. The protein is yellow, the DNA blue and hydrogen bonds are shown as dashed lines. The view is the same as in Fig. 9.31. Lys[23] forms a bifurcated hydrogen bond to base-G2 (and G10′), while Thr[25] hydrogen bonds with base-A3 (and A11′). The plane of the bases is perpendicular to the plane of the Figure in the centre of the *met* box but not on either side because of a slight increase in their roll and tilt. This bends the centre of the *met* box by about 25° (reprinted with permission from Somers, W.S. and Phillips, S.E.V. (1992). *Nature*, **359**, 387–93). (Stereopair figures for parallel viewing.)

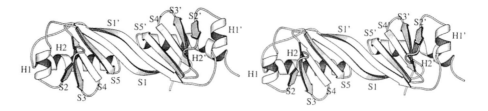

Fig. 9.33 A stereo ribbon drawing of the TATA -box binding protein. α-Helices are shown as spirals, β-strands as curved arrows. The view is looking down the *intra*molecular two-fold axis that relates the two domains. Protein secondary structural elements are labelled to emphasize how they are related: strand S1′ to strand S1 and so on. The ten-stranded curved β-sheet can be seen (reprinted with permission from Nikolov, D.B., Hu, S.-H., Lin, J., Gasch, A., Hoffmann, A., Horikoshi, M., Chua, N.-H., Roeder, R.G., and Burley, S.K. (1992). *Nature*, **360**, 40–46. Copyright (1992) Macmillan Magazines Limited). (Stereopair figures for parallel viewing.)

Not surprisingly, site-directed mutants affecting DNA-binding affinity map to the concave surface and those affecting TBP's ability to interact with other components of the pre-initiation complex map either to the convex surface or to one or other end. Side chains related by the structural two-fold are not necessarily the same because the structure is formed from a single polypeptide chain rather than from a dimer. Therefore TBP, unlike a prokaryotic transcription factor dimer, can bind DNA directionally, form an asymmetric pre-initiation complex, and help RNA polymerase transcribe DNA in the correct direction.

Summary

The placement of an α-helix in the major groove appears to be the most common way of recognizing a specific DNA sequence. It is *not* correct, however, to describe one single structural element as *the* recognition element. Other parts of the protein, which form hydrogen bonds and salt bridges to the DNA backbone, position that element on the DNA so that it can achieve recognition; there are usually as many hydrogen bonds to the backbone as to the bases.

Direct readout of the DNA sequence, most often in the major groove, is an important part of sequence-specific binding but is by no means the only component. The direct readout can involve hydrogen bonds (1) directly to side chains, (2) to the polypeptide backbone, or (3) through water molecules, or depend on hydrophobic interactions. **Indirect readout** is also very important: the correct DNA sequence may differ from canonical B-DNA in a way that increases the surface area buried, the electrostatic attraction, or the number of hydrogen bonds formed.

The protein can also change conformation upon binding, affecting the overall stability of the complex or the ability of the protein to recognize a specific sequence. Oligomerization upon binding the correct sequence, as in *met* repressor, GCN4, and glucocorticoid receptor, often increases affinity and specificity. Consequently, there is no universal code by which proteins recognize DNA sequences. Even in a single family, such as the HTH proteins, the recognition helix is presented to the DNA in many ways.

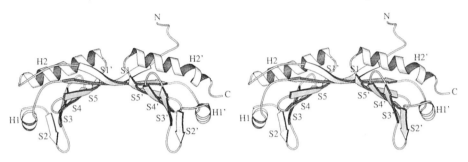

Fig. 9.34 The TATA-box binding protein, viewed perpendicularly to Fig. 9.33. α-Helices are shown as spirals, β-strands as curved arrows. The intramolecular two-fold axis is vertical and in the plane of the figure. The DNA presumably binds on the inside of the saddle and proteins on the outside (reprinted with permission from Nikolov, D.B., Hu, S.-H., Lin, J., Gasch, A., Hoffmann, A., Horikoshi, M., Chua, N.-H., Roeder, R.G., and Burley, S.K. (1992). *Nature*, **360**, 40–46. Copyright (1992) Macmillan Magazines Limited). (Stereopair figures for parallel viewing.)

9.3.6 Restriction endonucleases: *Eco*RI and *Eco*RV

Bacteria require a system to prevent foreign DNA from being replicated. This is provided by restriction endonucleases which recognize specific DNA sequences, bind to them, and cleave both DNA strands. Each bacterial strain has its own restriction enzyme or enzymes and each enzyme has its own specific recognition sequence. Bacteria protect their own DNA from being cut by specifically modifying the recognition site using a restriction methylase (Section 9.3.7).

There are three very different types of restriction and modification systems (I, II, and III) (Section 10.7.2) but structural information is only available for the type II endonucleases *Eco*RI and *Eco*RV. Type II endonucleases recognize a sequence of 4–8 bp which usually has a twofold axis of symmetry. They are usually dimeric, require magnesium, and cleave the DNA one strand at a time. The cleavage sites are usually close to or within the recognition sites; some enzymes cut symmetrically to leave a blunt (flush) end, while others cut asymmetrically to leave staggered (sticky) ends with either 5'- or 3'- overhangs. The enzymes have little sequence similarity (even those called isoschizomers that recognize the same DNA sequence) possibly because how they recognize DNA may depend on where they cut (Fig. 9.35).

The structure of the complexes

The 30 kDa *Eco*RI binds the sequence d(GAATTC) as a dimer and hydrolyses the 3'-GA phosphodiester bond in the major groove (Fig. 9.35, Table 9.4), thus leaving a four-base 5'-overhang. The major groove of the DNA is embedded into the *Eco*RI dimer, leaving the minor groove exposed. Each *Eco*RI monomer contains a central, mostly parallel β-sheet sandwiched between α-helices (Fig. 9.36). The β-sheet can be thought of as having two functional regions, joined at the only antiparallel strand ($\beta2$). Strands 1–3 are all antiparallel and contain the essential Glu[111] at the end of strand 3. Strands 3–5 are all parallel, like a Rossmann nucleotide-binding fold (Fig. 9.41); they and the two connecting crossover helices provide most of the site-specific interactions with the base-pairs (Figs 9.37 and 9.38). The four crossover helices (two from each monomer) form a parallel bundle called the **four-barrelled motif**. The connection between strand 3 and helix 4 includes a five-residue stretch of extended chain that runs parallel to the major groove, making further site-specific contacts with the pyrimidine bases on the DNA (Fig. 9.38). Although the overall path of the DNA is straight, it is kinked and unwound by about 28° in the middle of the recognition sequence, thus widening both grooves by about 3.5 Å (Fig. 9.39). The kink allows the four helices of the four-barralled motif to penetrate the major groove (Fig. 9.38) and changes the stacking of the bases in the centre of the recognition site. These protein-dependent changes in DNA conformation are almost certainly important in recognition.

*Eco*RI makes 16 hydrogen bonds to the bases in the major groove, essentially saturating all the available hydrogen bond donors and acceptors, and also makes many van der Waals contacts (Table 9.3). Every type of hydrogen bond occurs: from side chain atoms, from main chain atoms, and mediated by water. A number of interactions within base-pairs also help stabilize the kink induced in the DNA. Secondary hydrogen bonds orient the recognition side chains so that they form the correct hydrogen bonds to the bases: such interactions are as important

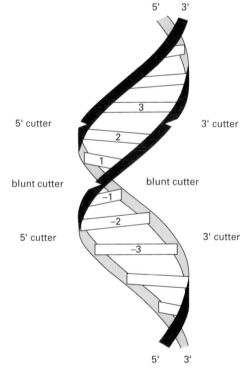

Fig. 9.35 A schematic representation of B-DNA showing where restriction enzymes could cut the DNA. The bases are shown as boxes on the DNA and are numbered outwards from the two-fold centre of symmetry. The 3′- and 5′-positions of cleavage shown in the Figure leave 4-base overhangs. Blunt cutters can approach from either side but must envelop the DNA because the cleavage sites are on opposite sides of the double helix. The cleavage sites for 5′-cutters (leaving a 3′-overhang) are above each other and on the minor groove side and the cleavage sites for 3′-cutters (leaving a 5′-overhang) are the opposite.

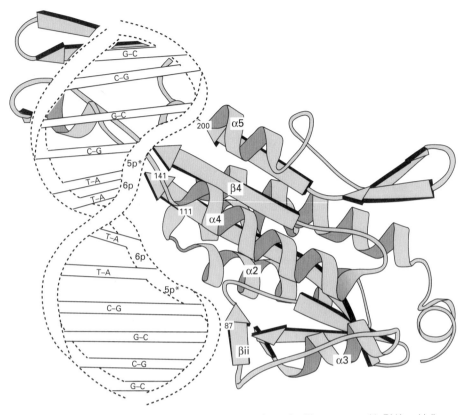

Fig. 9.36 A secondary structure diagram showing the interaction of one *Eco*RI monomer with DNA. α-Helices are shown as spirals, β-strands as curved arrows and the secondary structural elements are labelled. The approximate position of the essential Glu[111] is shown (from Rosenberg, J.M. (1991). *Curr. Opinion Struct. Biol.*, **1**, 104–13).

energetically as direct DNA–protein interactions. The protein–DNA interface forms co-operatively.

Possibly because it is a blunt-end cutter (Fig. 9.35), *Eco*RV, with a molecular mass 28 kDa, forms a completely different symmetrical complex with its cognate DNA d(GATATC) (Table 9.4.). *Eco*RV has an unusual α/β structure and envelops the DNA, placing loops in both major and minor grooves (Fig. 9.40). The loops have to move out of the way of the DNA to enter and leave the active site. However, unlike *Eco*RI, *Eco*RV does not saturate the hydrogen bond sites of the major groove: only the outer two bases in the half-site form hydrogen bonds to a short loop (residues 182–186) on the protein. The protein does not form hydrogen bonds to the inner two base-pairs at all, but does interact with bases in the minor groove and with six phosphates per half-site. Residues 182–186 are well ordered only in the complex with the cognate DNA, suggesting that *Eco*RV binds its cognate DNA co-operatively. Both endonucleases drive their cognate DNAs into unfavourable conformations, unstacking the central base-pairs and unwinding the DNA, but *Eco*RV, unlike *Eco*RI, makes the major groove narrower and deeper so that the overall path of the DNA, rather than being almost straight, is quite sharply curved (Fig. 9.39).

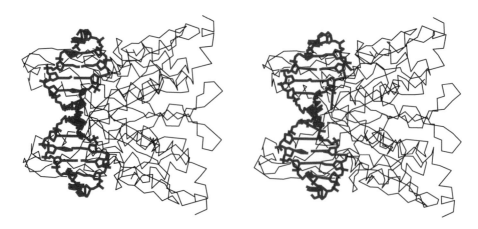

Fig. 9.37 A stereo drawing of the interaction of *Eco*RI with DNA. The α-carbon atoms of both monomers are shown in black and all the atoms of the DNA are shown in colour. The twofold axis relating the two monomers and the two DNA half-sites runs horizontally through the centre of the DNA (from Rosenberg, J.M. (1991). *Curr. Opinion Struct. Biol.*, **1**, 104–13). (Stereopair figures for parallel viewing.)

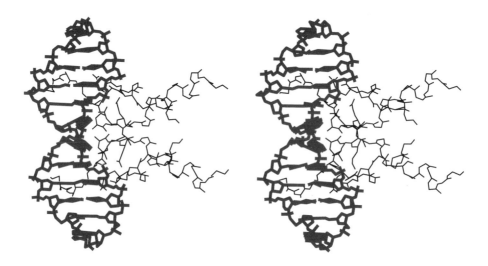

Fig. 9.38 A stereo drawing of the 'recognition motifs' of *Eco*RI. The DNA is in colour and all the atoms of the 'four-barreled motif' and the extended chain are shown. The view is as in Fig. 9.37. The protein-induced kink in the centre of the restriction site is clearly visible (from Rosenberg, J.M. (1991). *Curr. Opinion Struct. Biol.*, **1**, 104–13). (Stereopair figures for parallel viewing.)

The catalytic mechanism and sequence discrimination

Surprisingly, both enzymes have a similar structural motif that contains the catalytic machinery. One basic and two acidic side chains are near the scissile phosphodiester bond and there is even some sequence similarity. The *Eco*RI sequence is Pro-Asp-X_{19}-Glu-X-Lys and the *Eco*RV sequence, Pro-Asp-X_{15}-Asp-X-Lys (X is any amino acid). As with other nu-

Table 9.3 EcoRI–DNA recognition
A. Protein–DNA interactions

DNA		EcoRI	Type of interaction
G·C base-pair			
G	N7	Arg200 & Arg203	H-bond *via* bound water
	O6	Arg200 & Arg203	H-bond *via* bound water
C	N4	Ala138	H-bond to main chain C=O
	ring	Met137 & Ile197	Hydrophobic contact
Outer A·T base-pair			
A	N7	Arg145	H-bond from side chain Nζ
	N6	Asn141	H-bond to side chain Oδ
T	Me5	Gly140	Hydrophobic contact
Inner A·T base-pair			
A	N7	Arg145	H-bond from side chain Nη
	N6	Asn141	H-bond to side chain Oδ
T	O4	Ala142	H-bond from main chain NH
	Me5	Ala142 & Gln115	Hydrophobic contact

B. Unusual DNA–DNA interactions

From		To	Type of interaction
Outer A·T base-pair			
A	N6	*Both* Ts	Three-centre H-bond to O4
T	O4	Outer A	Watson–Crick H-bond to N6: part of three-centre H-bond.
T	Me5	Inner T	Hydrophobic contact with Me5
Inner A·T base-pair			
T	O4	Outer A	Bridging H-bond to N6: part of three-centre H-bond.
T	Me5	Outer T	Hydrophobic contact with Me5

H-bond stands for hydrogen bond. The number of hydrogen bonds per full site is twice the number shown above, which is for a half site. The three-centre interactions between the outer adenine and the two thymines helps distort the DNA and is important for recognition. Adapted from Rosenberg, J.M. (1991). *Curr. Opinion Struct. Biol.*, **1**, 104–13.

cleases (Section 9.2.3) the mechanism probably involves S_N2 attack by an activated water molecule on the scissile phosphodiester bond, with the cleaved phosphate being stabilized by the magnesium cation. The catalytic machinery seems to be preformed: in all three *Eco*RV structures (apoenzyme, complex with non-cognate DNA, and complex with cognate DNA), the catalytic residues are in the same conformation. Nonetheless *Eco*RI and *Eco*RV recognize and cleave their cognate sequences much better than sequences (called *Eco*RI* and *Eco*RV* sites) differing at only one base-pair. The sequence specificity of a restriction enzyme is much high-

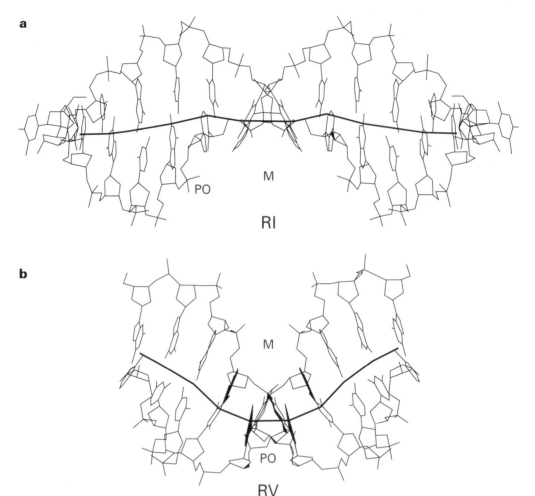

Fig. 9.39 The path of the DNA in the cognate–DNA complexes with (**a**) *Eco*RI and (**b**) *Eco*RV. The twofold axis of the complex is vertical. All the atoms of the DNA are shown and the path of the DNA is shown by the thick line: straight in (**a**) and bent in (**b**). The position of the main body of the protein is marked by RI and RV, respectively, the position of the major groove by 'M', and the position of the scissile phosphodiester bond by 'PO' (from Winkler, F.K. (1992). *Curr. Opinion Struct. Biol.*, **2**, 93–9).

er than that of a transcriptional regulator (Section 9.3.2). Were it not so, the enzyme would destroy the bacterium's own DNA. How is this specificity achieved?

As restriction enzymes cleave each DNA strand independently and can dissociate in between, discrimination can happen in three ways: (1) by not binding to the DNA, (2) by not cleaving the first half-site, and (3) by dissociating rather than cleaving the second half-site. *Eco*RI binds *Eco*RI* sites worse than the cognate site and cleaves the first *cognate* strand of an *Eco*RI* site more slowly. It is allosterically activated on binding the cognate sequence. Furthermore, the balance between cleavage of the second strand and dissociation from the DNA changes: cleavage is more probable than dissociation for the cognate site but the reverse is true for an *Eco*RI* site. Solution studies have shown that *Eco*RI — like the transcription factors — binds to DNA in two different ways. One, called **isosteric binding**, occurs with the cognate site and with sites containing base-mispairs or base analogues. The other, **adaptive binding**, occurs when *Eco*RI binds *Eco*RI* sites. The former is symmetric and the latter asymmetric.

*Eco*RV seems to discriminate between cognate and non-cognate sequences in a similar way. As suggested by solution kinetic data on *Eco*RI and *Eco*RV, the complex of *Eco*RV with non-cognate DNA is asymmetric and the complex with cognate DNA is symmetric. Furthermore, in the non-cognate complex, the protein makes few contacts to bases in either groove and the elaborate web of co-operative interactions seen in the cognate complex has disappeared. Restriction enzymes, therefore, seem to discriminate accurately because only the cognate sequence can bind in a co-operative, symmetrical manner. Much of the free energy of complexation is used to form the co-operative complex because the DNA is in a very unfavourable conformation — one which non-cognate sequences cannot achieve. Only this DNA conformation positions the scissile phosphodiester bond in the active site and completes the magnesium ion binding pocket.

Summary

*Eco*RI and *Eco*RV have very different structures and interact with DNA differently: the former only in the major groove; the latter in both grooves. However, both employ the same enzyme mechanism and catalytic residues and both achieve their high degree of sequence specificity similarly: they complex with the cognate DNA in a highly co-operative and symmetric manner. The symmetric complex cannot form with non-cognate DNA. In the complex with cognate DNA, much of the free energy of binding has been used to drive the cognate DNA into an unfavourable conformation that places the scissile phosphodiester bond in the active site and completes the binding site for the essential magnesium ion.

9.3.7 Enzymes that 'flip-out' nucleosides

The process of bacterial restriction involves an endonuclease, designed to cut alien DNA, and a modification enzyme, designed to protect the bacterium from host-dependent restriction cleavage. The modification enzyme is invariably a methyl transferase which recognizes the same nucleotide sequence as the restriction enzyme (Table 9.4). It catalyses the transfer of a methyl group from *S*-adenosyl methionine, AdoMet, to one of the bases within the recognition sequence of the endonuclease with the consequence that the cognate sequence now becomes fully resistant to cleavage by the restriction endonuclease.

Three different types of methylase transfer the methyl group from AdoMet either to cytidine C-5 or to the exocyclic amino group of cytidine N^4 or to adenosine N^6. For example, the restriction methylase from *H. haemolyticus* is a DNA **cytosine 5-methyltransferase** known as M.*Hha*I and has the recognition sequence (G^mCGC), where the symbol ^mC designates a 5-methylcytosine residue (the symbol ^{Me}C is also used). This correlates with the restriction en-

Table 9.4 Relationship between the recognition sequences for Type II restriction endonucleases and the target for methylation by the corresponding restriction methylase activity

Source of restriction Enzyme	Abbreviation	Recognition sequence	
		Endonuclease	Methylase
		(C^5-cytosine methylation)	
Bacillus subtilis R	*Bsu*RI	GG$^\downarrow$CC	GG^{m}CC
Haemophilus haemolyticus	*Hha*I	G$^\downarrow$CGC	G^{m}CGC
Haemophilus parainfluenzae	*Hpa*II	C$^\downarrow$CGG	C^{m}CGG
Xanthomonas malvacearum	*Xma*I	C$^\downarrow$CCGGG	C$_m$CCGGG
Serratia marcescens Sb	*Sma*I	CCC$^\downarrow$GGG	CC$_m$CGGG
		(N^6-adenine or N^4-cytosine methylation)	
Diplococcus pneumoniae	*Dpn*I	GA$^\downarrow$TC	G^{m}ATC
E. coli RY13	*Eco*RI	G$^\downarrow$AATTC	G^{m}AATTC
E. coli J62plg74	*Eco*RV	GAT$^\downarrow$ATC	G^{m}ATATC
Thermus aquaticus	*Taq*I	T$^\downarrow$CGA	TGC^{m}A
B. amyloliquifaciens H	*Bam*HI	G$^\downarrow$GATCC	GGATC$_m$C
Proteus vulgaris	*Pvu*II	CGAT$^\downarrow$CG	CGAT$_m$CG

^{m}C denotes a 5-methylcytidine residue; ^{m}A and N^6-methyladenosine residue; and $_m$C an N^4-cytidine residue. These methylases all use *S*-adenosyl methionine as the source of the methyl group. The cleavage point is indicated by $^\downarrow$.

donuclease from the same organism, *Hha*I, which operates on the recognition sequence (G$^\downarrow$CGC) (Section 9.3.6 and Table 9.4).

The X-ray structures of two of these methylases have shown us how the protein interacts with the DNA target. Xiadong Cheng has solved the structure of a ternary complex between the M.*Hha*I methyltransferase, a dodecanucleotide containing the recognition sequence, and AdoMet (see front cover). The target cytidine is flipped[†] completely out of the helix by torsional rotation of backbone bonds (designated α, β, ε, and ζ; Section 2.1.4 and Fig. 2.11). This places the cytosine in a cavity where the catalytic chemistry takes place. The methylation reaction involves addition of the nucleophilic thiol group of a cysteine residue to C-6 and of an electrophilic methyl group (from AdoMet) to C-5. Loss of the proton from C-5 and elimination of the cysteine thiol provides C-5-methylcytosine (Section 7.4 and Fig. 7.7b). This mechanism was

[†] We should note that this behaviour has also been described as **'base-flipping'**. However, there appears to be no involvement of rotation of the glycosylic bond, torsion angle χ. We therefore prefer to adopt the term **'nucleoside-flipping'** and conserve use of **'base-flipping'** to describe those proteins that bind a nucleoside with its glycosylic bond rotated into the *syn*-conformation, e.g. the MetJ repressor protein.

proposed by Dan Santi in line with that established by him for thymidylate synthase. As expected, 5-fluorodeoxycytidine acts as a **suicide substrate** for M.*Hha*I forming an irreversible complex with the cysteine residue because of its inability to lose a fluorine cation. This is closely analogous to the suicide inhibition of thymidylate synthase by 5-fluorouracil (Fig. 4.15 and Section 4.7.1).

The complex of M.*Hae*III with its cognate dsDNA and AdoMet also shows that the target cytidine is 'flipped-out'. In this case, there is significant distortion of the adjacent base-pairs to close the gap left by the dC residue, which is not seen in the case of M.*Hha*I. The **adenine methyltransferase** M.*Taq*I has the recognition sequence (TCG^{m}A). A model of the structure of its complex with sinefungin (an analogue of AdoMet) and the cognate DNA places the target adenine-N^6 some 15 Å distant from the methyl donor. This suggests that this adenosine has to flip out of the helix for methylation to proceed. Because there are clear similarities in key protein motifs for both the N-6 adenine and N-4 cytosine methyltransferases, it seems that these two families of enzymes, like the N-6 cytosine methylases, will use nucleoside flipping to gain access to their target residues.

Crystal structures have been solved for other enzymes which suggest that 'flipping-out' of nucleosides may be a general phenomenon for enzymes that perform chemistry on bases within a DNA duplex. Thus, uracil-DNA glycosylase, UDGase, is an enzyme that repairs cytosine→uracil lesions in dsDNA (Sections 7.4, 7.11.2 and Fig. 7.7a). Structures have been described for the UDGase from a herpes simplex virus. HSVI, and for the human enzyme. The specific binding to uracil-containing dsDNA is mediated by flipping the target nucleoside into a specific pocket in the enzyme that is precisely tailored to accommodate uracil but neither thymine nor cytosine. Three other potential examples of nucleoside-flipping are the *E.coli* photolyase that repairs thymine photodimers in dsDNA (Section 7.11.1), the Ada repair enzyme of *E. coli* that effects the suicide repair of O^6-methylguanine residues (Section 7.11.1), and the *E. coli* exonuclease III, which is a major bacterial apurinic/apyrimidinic repair enzyme (Section 7.11.2). It has been suggested that when this enzyme recognizes an abasic site, the nucleoside residue opposite the gap takes up a **flipped-out** conformation.

Finally, bacteriophage T4 employs 5-hydroxymethylcytosine residues instead of cytosine as a barrier to host restriction endonucleases. Following DNA synthesis, most of these residues are glucosylated by the transfer of glucose from UDPglucose (Fig. 3.45c) in a reaction catalysed by T4 β-glucosyl transferase. The catalytic site in the crystal structure of this enzyme has been identified. It is buried in a deep cleft in the protein where it appears to be inaccessible to the substrate when a normal B-DNA helix is modelled into the protein structure. However, it appears that nucleoside flipping of the target hydroxymethylcytosine residue would bring its nucleophilic hydroxyl group adjacent to C-1' of the UDPglucose co-factor and so facilitate glucosyl transfer.

A key feature of all of these substrate–enzyme interactions is that no external energy source is required to cause the nucleoside flipping. However, it has been observed that tighter complexes are formed between M.*Hha*I and base mismatches within the recognition sequence (GCGC), which makes the process especially appropriate for mismatch repair. In addition, Roberts has sugggested that such nucleoside flipping may emerge as a key component of other enzymes that need to open up the DNA helix, especially DNA and RNA polymerases.

Summary

Nucleoside flipping involves rotation of backbone bonds to expose an out-of-stack base in dsDNA. It can then be a substrate for an enzyme-catalysed chemical reaction. Rotations about α, β, γ, ε, and ζ torsion angles but not the glycosylic bond, χ, appear to be required. The phenomenon is fully established for two restriction methyltransferases, M.*Hha*I and M.*Hae*III. There is strong evidence for nucleoside flipping for two key DNA repair enzymes, UDGase and *E. coli* photolyases. Other examples are emerging and the phenomenon is likely to prove general for enzymes that require access to unpaired bases. It appears to need no external source of binding energy.

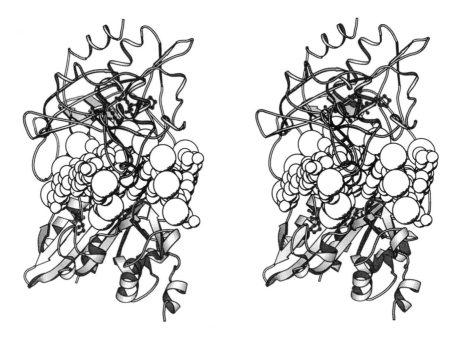

Fig. 9.40 A stereo diagram of the interaction of *Eco*RV with its cognate DNA, showing how the protein envelops the DNA. The DNA is shown as a space-filling model, with the phosphates overemphasized. The secondary structure of one monomer is shown, with α-helices as spirals and β-sheets as curved arrows, while the second monomer is represented only as a ribbon. The residues in both monomers that form the key catalytic loop are visible as ball-and-sticks in red as is the loop on the first monomer that must move out of the way for DNA to enter. (Stereopair figures for parallel viewing.)

9.3.8 RNA–protein interactions: tRNA synthetases

The aminoacyl-tRNA synthetases (**aaRS**, where aa is one of the amino acids) are essential for correctly translating a gene into a protein because they charge the tRNAs (**tRNA^{aa}**) with amino acids. There are usually 20 synthetases in every cell, one for each amino acid, and each catalyses a two-step reaction (where X is any amino acid):

$$xRS + X + ATP \longrightarrow xRS\text{-}(X\text{-}AMP) + PP_i$$

$$xRS\text{-}(X\text{-}AMP) + tRNA^x \longrightarrow X\text{-}tRNA^x + xRS + AMP$$

The amino acid is added to the ribose of the adenosine in the conserved CCA acceptor stem. (Section 6.3.2).

Synthetase classification: the catalytic motifs

Superficially, the only similarity among the synthetases is that they all charge tRNAs. They vary widely in size and oligomeric structure, in mechanism, and in method of distinguishing cognate from non-cognate tRNAs. Nevertheless, they can be grouped into two broad families of ten RSs each. All class I synthetases transfer the amino acid to the 2′-hydroxyl on the ribose of the terminal adenosine and almost all class II synthetases transfer it to the 3′-hydroxyl. Luckily, the first two structures of synthetase-tRNA complexes include one member of each class: for class I the ternary complex of the *E. coli* glutamine system, GlnRS·tRNA^{Gln}·ATP, and for class II the binary complex of the yeast aspartate system, AspRS·tRNA^{Asp}.

The difference between the two classes lies in their ATP binding domains. Both have a domain that binds all three substrates (ATP, amino acid, and acceptor adenosine) and catalyses the charging reaction, but the structure of the domain and the organization of the substrates is completely different. Class I synthetases have a five-stranded, all-parallel β-sheet domain (Fig. 9.41a)—a variant of the Rossmann dinucleotide binding domain, which normally contains six strands. In the GlnRS · tRNA^{Gln} · ATP ternary complex, the ATP binds to the first half of the sheet and the glutamine and acceptor adenosine bind to the second. Class II enzymes, in contrast, bind the same substrates using a unique domain built around an *anti*-parallel six-stranded β-sheet (Fig. 9.41b). The active site domain is N-terminal in class I aaRSs but C-terminal in class II aaRSs.

There are tiny regions of sequence similarity (and much larger regions of structural similarity) among all members of each class (Table 9.4; Fig. 9.41); each characteristic motif corresponds to a key structural or functional element. In class I synthetases, the histidines in the HIGH sequence help bind ATP and the second lysine in the KMSKS sequence is needed to activate the amino acid (Fig. 9.41a). In class II synthetases, motif 1 occurs only in the homodimeric synthetases because it is part of the monomer–monomer interface, and motifs 2 and 3 interact with the CCA acceptor stem of the tRNA and with the ATP. PheRS is considered a class II enzyme, even though it aminoacylates the 2′-hydroxyl, because it has motifs 2 and 3 of the class II aaRSs.

Sequence similarity can also be used to classify the RSs into subgroups. The grouping that emerges correlates roughly with the size and chemical properties of the amino acid charged (Table 9.4). For instance, class Ia synthetases charge tRNAs for all the large aliphatic and the two sulfur-containing amino acids, while class IIc synthetases charge the tRNAs for the two smallest amino acids. This is consistent with the shape of the active site pockets in the enzymes: GlnRS (class I) has a more open active site pocket than AspRS (class II).

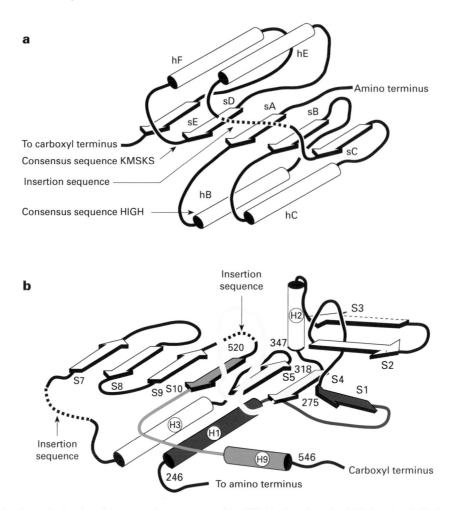

Fig. 9.41 A schematic drawing of the secondary structure of the ATP-binding domain of (**a**) class I and (**b**) class II synthetases. α-Helices are shown as cylinders, β-sheets as arrows. The positions of the conserved HIGH and KMSKS sequences
are shown for class I synthetases. The positions of the conserved structural/sequential motifs for class II synthetases are
also shown: conserved motif 1 as red, motif 2 as yellow, and motif 3 as blue. The dotted lines mark where other domains
are inserted into the ATP-binding domains (redrawn from Moras, D. (1992). *Trends Biochem. Sci.*, **17**, 159–64).

The tRNA binding motifs and the basis of tRNA discrimination

Not only do class I and class II synthetases have different catalytic modules, they also
approach the L-shaped tRNA completely differently. When the complexes are oriented so
that the tRNA is in front of the RS, GlnRS binds tRNA^{Gln} with the short arm of the L (the
acceptor stem) pointing to the right, while AspRS binds tRNA^{Asp} the other way round (Fig.
9.42). GlnRS binds its cognate tRNA from the minor groove side of the acceptor stem; AspRS,
from the major groove side.

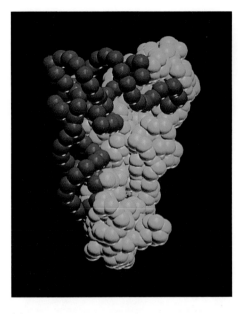
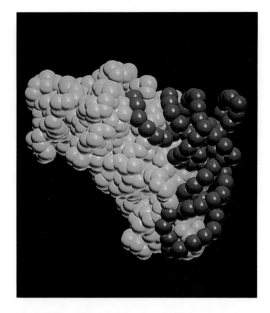

(a)   (b)

Fig. 9.42 A space-filling model of the RNA phosphate (red) and protein α-carbon (blue) backbones of (**a**) the class I GlnRs–tRNA^{Gln} complex and (**b**) the class II AspRS–tRNA^{Asp} complex. The view emphasizes that GlnRS binds tRNA^{Gln} from the opposite side to AspRS binding tRNA^{Asp} (from Ruff, M., Krishnaswamy, S., Boeglin, M., Poterszman, A., Mitschler, A., Podjarny, A., Rees, B., Thierry, J.C., and Moras, D. (1991). *Science*, **252**, 1682–9. Copyright (1991) by the AAAS).

Recognition by class I synthetases: the GlnRS·tRNA^{Gln} complex.

GlnRS has four domains: the dinucleotide fold domain containing the active site (see above) and three others involved in discriminating tRNA^{Gln}. All interact with only one side of the tRNA molecule (Fig. 9.43). Between the two halves of the N-terminal dinucleotide fold is the acceptor binding domain. It contains a five-stranded antiparallel β-sheet and distorts the acceptor by denaturing the last base-pair in the acceptor stem (U1·A72) and bending the CCA acceptor end back in a hairpin loop towards the anticodon. As had been suggested by mutational analysis, GlnRS distinguishes the hairpin loop structure by **indirect readout** rather than by making sequence-specific contacts to U1, A72, or G73. G73 helps denature the U1·A72 base-pair because its 2-amino group forms a hydrogen bond to the phosphate oxygen of A72, thus stabilizing the hairpin conformation. There is also **direct readout** of the tRNA sequence. GlnRS penetrates the minor groove of the acceptor stem; protein main chain and side chain atoms form direct and water-mediated hydrogen bonds to G2·C71, G3·C70, and G10·C25.

GlnRS uses the anticodon to discriminate between cognate and non-cognate tRNAs. The carboxyl half of GlnRS folds into two six-stranded antiparallel β-barrel domains that interact with the seven-base anticodon loop (Fig. 9.43) and distort it. In all known free tRNA structures, the bases of the anticodon stack on top of each other, with the Watson–Crick hydrogen bond donors and acceptors facing outward. In the complex, two novel, non-Watson–Crick base-pairs form (between $^{2'm}$U32·Ψ38 and U33·^{2m}A37), thus extending the anticodon stem

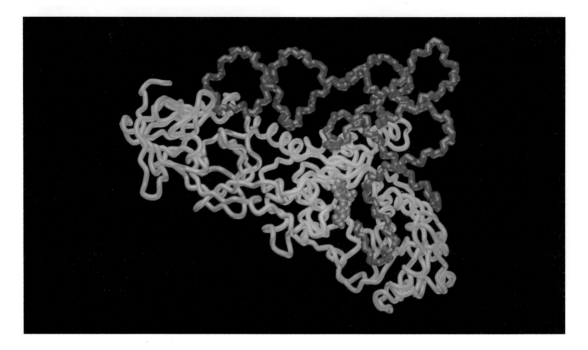

Fig. 9.43 A schematic diagram of GlnRS interacting with tRNA^{Gln} and ATP. The RNA phosphate backbone is red and, from left to right in the figure, the protein subdomains are coloured: distal β-barrel, cyan; proximal β-barrel, blue; ATP-binding domain, gold; acceptor binding domain, green. A ball-and-stick model of the ATP is in dark blue in the centre of the ATP binding domain and adjacent to the acceptor end of the tRNA (from co-ordinates provided by Rould, M. A. and Steitz, T.A. *cf.* Rould, M.A., Perona, J.J., Söll, D., and Steitz, T.A. (1989). *Science*, **246**, 1135–42).

and splaying the bases of the anticodon apart (Fig. 9.44). Each splayed-out base in the anticodon has a separate recognition pocket on the protein (Fig. 9.45) which forms tight networks of hydrogen bonds directly to the Watson–Crick donors and acceptors of the bases (Section 2.1.2). These hydrogen bonds often come from charged side chains that are buried when the complex forms. At each of the three positions, the phosphate backbone of the tRNA forms a salt bridge with a lysine or an arginine and the hydrophobic portion of that side chain packs against the hydrophobic underside of the ribose sugar.

There are two tRNA isoacceptors for glutamines: one with a CUG anticodon and the other with a UUG anticodon (Section 6.6.8). How does GlnRS bind both and yet discriminate against non-cognate tRNAs, which have purines or pyrimidines with bulky modifications? In the complex (with the CUG isoacceptor tRNA), all three Watson–Crick hydrogen bond donors and acceptors on C34 are, of course, satisfied. The key difference between uridine and cytidine is the exocyclic substituent at position-4: an amino group in cytosine and a keto group in uridine. The 4-amino group of C34 hydrogen bonds with the backbone carbonyl of Ala^{414}, which obviously could not hydrogen bond with another carbonyl group. However, if the base moves slightly, the substituent at position-4 can then hydrogen bond to the guanidinium group of Arg^{410}, thus allowing both U and C to fit into the binding pocket. The binding pocket

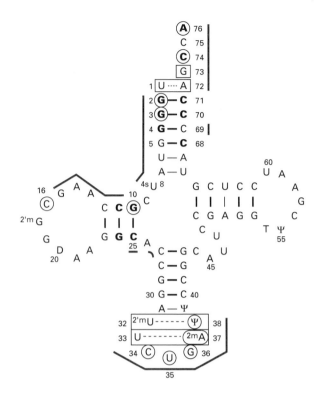

Fig. 9.44 A diagram summarising the protein–RNA and RNA–RNA interactions in the GlnRS–tRNA^{Gln} complex. The tRNA is drawn as a 'cloverleaf' with the D stem to the left, the T stem to the right and the seven-base anticodon loop at the bottom. Bases that interact with GlnRS directly are circled, bases that interact via a water molecule are in bold, and bases that help the tRNA assume the conformation necessary to bind GlnRS are boxed. Thick red lines indicate where the enzyme interacts with the phosphate backbone. The two extra base-pairs created in the anticodon loop are shown by dashed lines. In AspRS, a 10·25·45 base-triple forms; the bases at these positions are G, U, and G respectively (reprinted with permssion from Rould, M.A., Perona, J.J., and Steitz, T.A. (1991). *Nature*, **352**, 213–18. Copyright (1991) Macmillan Magazines Limited).

is too small for purines and also seems to have a floppy lid which may help exclude bulkier modified pyrimidines, thus allowing GlnRS to discriminate cognate from non-cognate tRNAs.

Recognition by class II synthetases: the AspRS·tRNA^{Asp} complex

The AspRS monomer has two domains: the multifunctional C-terminal domain (Fig. 9.41b) and an N-terminal domain. The latter contains a five-stranded antiparallel β-barrel which has been seen in other enzymes that bind nucleotides. It binds the anticodon loop of tRNA^{Asp} from the major groove side. AspRS is a symmetrical dimer and each tRNA makes almost all of its contacts with only one of the monomers.

The C-terminal domain uses the loop between the strands of motif 2 (Table 9.4; Fig. 9.41b) to form sequence-specific hydrogen bonds to the acceptor stem of the tRNA. The loop interacts with G73 and with the U1·A72 base-pair via the major groove (Fig. 9.46). Even though the acceptor stems of tRNA^{Asp} and tRNA^{Gln} have the same sequence, their RSs bind the stems

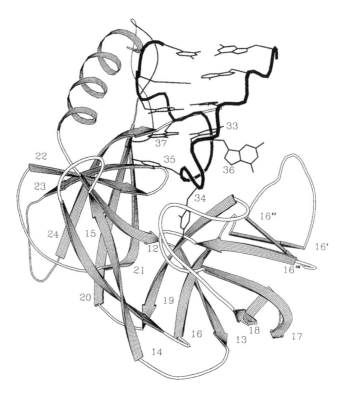

Fig. 9.45 A schematic drawing of the interaction of the two β-barrel domains of GlnRS with the anticodon stem and loop. α-Helices are represented by spirals, β-strands by curved arrows. The secondary structural elements are labelled and numbered as in Fig. 9.43. The sugar-phosphate backbone is represented by a thick line and bases 30 to 40 are shown as sticks, with the bases 33–37 numbered. The protein binds the three anticodon bases thus: C34 by the loop between strands 16‴ and 17; U35 by strands 22 and 12; G36 by a loop between strands 16′ and 16″. The loop between strands 20 and 21 packs against the ATP-binding domain (reprinted with permission from Rould, M.A., Perona, J.J., and Steitz, T.A. (1991). *Nature*, **352**, 213–18. Copyright (1991) Macmillan Magazines Limited).

completely differently. Rather than a denatured U1·A72 base-pair and a stem bent back in a hairpin, the base-pair remains and the single-stranded GCCA acceptor end continues as a helix directly into the active site.

The hinge between the C- and N-terminal domains in AspRS discriminates cognate from non-cognate tRNAs by *indirect* readout. The G10·U25 base-pair in the D-stem (Fig. 9.44) forms a base-triple with G45; when G10 or U25 is mutated, the binding affinity of AspRS for tRNA[Asp] drops. AspRS, however, does *not* interact directly with these bases. Instead, the hinge region interacts with the phosphates of U11 and U12: mutations of G10 and U25 would disrupt the base-triple, change the conformation of the tRNA, and so decrease the binding constant of AspRS for the tRNA.

One of the few similarities between AspRS and GlnRS is that both interact with the anti-codon loop of the tRNA and both splay the three bases of the anticodon apart. They do so in different ways. AspRS does not extend the anticodon stem. Instead, its N-terminal domain

Table 9.5 Classification of aminoacyl-tRNA synthetases

	Class I	*Class II*
Aminoacylation site on terminal ribose	2′-OH	3′-OH (except Phe)
Sequence motifs	..**HIGH**..	(1) ..**P**..
	..KM**SKS**..	(2) ..**FRXE/D**.. ..R/HXXX**F**..
		(3) ..**GXGXGXER**..
Subclass members		
(a)	Leu, Ile, Val, Cys, **Met**	**Ser,** Thr, Pro, His
(b)	**Tyr,** Trp	**Asp***, Asn, Lys
(c)	Glu, **Gln***, Arg	Gly, Ala
(d)		Phe

In the sequence motifs section, residues in bold type are fully conserved, residues in normal type are consensus residues and X means that the residue is not conserved. In the subclass members section, bold indicates that the structure of the corresponding RS is known, and the '*' indicates the two RS–tRNA complex structures. Adapted from Moras, D. (1992). *Trends Bioch. Sci.*, **17**, 159–64.

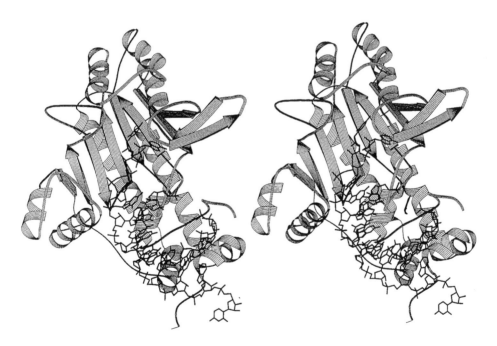

Fig. 9.46 A stereo ribbon drawing of the ATP-binding domain of AspRS and its interaction with the acceptor stem and CCA end of tRNA^Asp. α-Helices are shown as spirals, β-strands as curved arrows, and the tRNA acceptor end is shown as a stick model. It continues as a smooth helix into the active site, on the convex surface of the six-stranded antiparallel sheet (from Ruff, M., Krishnaswamy, S., Boeglin, M., Poterszman, A., Mitschler, A., Podjamy, A., Rees, B., Thierry, J.C., and Moras, D. (1991). *Science*, **252**, 1682–9. Copyright (1991) by the AAAS). (Stereopair figures for parallel viewing.)

gently bends the stem inward towards the acceptor, unravels the anticodon loop from ψ32 to C38, and so leaves the five bases in between completely unstacked and accessible. Two hydrogen bonds, between C38 and the phosphate backbone at C36 and between the 1-methylguanine-37 (^{1m}G37) and the phosphate backbone at U25, stabilize a bulge in the anticodon loop at ^{1m}G37. This is somewhat like the G73–A72 phosphate interaction in the acceptor stem of tRNA^{Gln}. AspRS recognizes the GUC anticodon bases by forming direct hydrogen bonds to them but it interacts with G34 the least. There are only two interactions with G34 but four each with U35 and C36, including a stacking interaction between Phe^{127} and U35.

Allosteric activation in the synthetases

As with the restriction endonucleases (Section 9.3.6), synthetases are allosterically activated by binding their cognate tRNA: binding a non-cognate tRNA affects not only K_m but also k_{cat}. Allosteric activation (as in haemoglobin) is normally thought to require more than one protein monomer to allow sufficient mobility to accommodate two different states. Such a model fits the dimeric AspRS more naturally. Indeed, in the Asp system, the U1 from the first tRNA^{Asp}—the **allosteric activator**—makes contacts with residues in the active site domain of the second AspRS. Furthermore, the conserved KMSKS sequence (Table 9.4), which interacts with ATP in the active site, is disordered in the absence of tRNA. In the Gln system, allosteric activation happens without a protein–protein interface and so the allosteric signal must be transmitted from the anticodon loop to the acceptor CCA by a single polypeptide chain, by the nucleic acid, or by both. Binding the correct tRNA may cause the two β-barrel domains (Fig. 9.43) to assume the conformation seen in the GlnRS·tRNA^{Gln}·ATP complex. Without this conformation, the active site may not be correctly assembled, ATP may not be able to bind productively, and the first part of the charging reaction may not be able to occur.

Summary

There are two classes of tRNA-synthetase: **class I synthetases** charge the 2′-hydroxyl of the adenosine ribose and **class II synthetases** charge the 3′-hydroxyl. Their active site domains, which are completely different, both bind the CCA acceptor stem, ATP, and the amino acid. The two synthetase–tRNA complex structures are very dissimilar: AspRS binds from the major groove side of the acceptor stem, GlnRS from the minor groove side. GlnRS breaks the hydrogen bonds between the last base-pair (A · U) in the acceptor stem and bends the CCA acceptor end back in a hairpin loop into the active site. In the AspRS system, these bases remain paired and the GCCA acceptor end continues as a helix into the active site. The conserved structural and sequence motifs occur in the acceptor-binding domains and so these modes of binding apply to other members of each class of synthetase.

Unlike the active site domain, the discrimination motifs seen in GlnRS may not apply to other class I synthetases, and the discrimination motifs seen in AspRS may not apply to other class II synthetases. Both GlnRS and AspRS discriminate **cognate** from **non-cognate tRNAs** by reading the anticodon loop but use completely different ways of splaying the anticodon bases apart. GlnRS does so by extending the anticodon stem, while AspRS does so by bending it and unravelling the anticodon loop.

As RNAs can be macromolecular catalysts, it is not surprising that sequence-specific tRNA recognition, even more than sequence-specific DNA recognition, resembles protein-protein interaction. Every kind of direct and indirect readout occurs in the mutual recognition of a synthetase and its cognate tRNA.

Further Reading

The following general reviews provide excellent starting points for further reading.

Harrison, S. C. and Saeur, R. T. (1994). Protein–nucleic acid interactions. *Curr. Opinion Struct. Biol.*, **4**, 1–66.

Phillips, S. E. V. (ed). (1992). Protein–nucleic acid interactions. *Curr. Opinion Struct. Biol.*, **2**, 69–149.

Phillips, S. E. V. and Moras, D. (ed). (1993). Protein–nucleic acid interations. *Curr. Opinion Struct. Biol.*, **3**, 1–56.

Sigler, P. B. (ed). (1991). Protein–nucleic acid interations. *Curr. Opinion Struct. Biol.*, **1**, 61–143.

9.1

Church, G. M., Sussman, J. L., and Kim, S.-H. (1977). Secondary structural complementarity between DNA and proteins. *Proc. Natl. Acad. Sci. USA*, **74**, 1458–62.

Churchill, M. E. A. and Travers, A. A. (1991). Protein motifs that recognize structural features of DNA. *Trends Biochem. Sci.*, **16**, 92–7.

Gabler, R. (1978). *Electrical interactions in molecular biophysics*. Academic Press, New York.

Hélène, C. and Lancelot, G. (1982). Interactions between functional groups in protein–nucleic acid associations. *Progr. Biophys. Molec. Biol.*, **39**, 1–68.

Saenger, W. and Heinemann, U. (ed). (1989). *Protein–nucleic acid interaction*. Macmillan Press, London.

Seeman, N. C., Rosenberg, J. M and Rich A. (1976). Sequence-specific recognition of double helical nucleic acids by proteins. *Proc. Natl. Acad. Sci. USA*, **73**, 804–8.

Tanford, C. (1973). *The hydrophobic effect — formation of micelles and biological membranes*. Wiley, New York.

Travers, A. A. (1989). DNA conformation and protein binding. *Annu. Rev. Biochem.* **58**, 427–52.

9.2

Baldwin, J. P. (1992). Protein–nucleic acid interactions in nucleosomes. *Curr. Opinion Struct. Biol.*, **2**, 78–83.

Baudet, S. and Janin J. (1991). Crystal structure of a barnase-d(GpC) complex at 1.9 Å resolution. *J. Mol. Biol.*, **219**, 123–32.

Brayer, G. D. and McPherson, A. (1984). Mechanism of DNA binding to the gene 5 protein of bacteriophage fd. *Biochemistry* **23**, 340–9.

Beese, L. S. and Steitz, T. A. (1991). Structural basis for the 3'-5' exonuclease activity of *Escherichia coli* DNA polymerase I: a two metal ion mechanism. *EMBO J.*, **10**, 25–33.

Fisher, A. J. and Johnson, J. E. (1993). Ordered duplex RNA controls capsid architecture in an icosahedral animal virus. *Nature*, **361**, 176–9.

Grunstein, M. (1992). Histones as regulators of genes. *Sci. Amer.*, Oct. 40–7.

Jacobo-Molina, A., Ding, J., Nanni, R. G., Clark, A. D. Jr., Lu, X., Tantillo, C., Williams, R. L., Kamer, G., Ferris, A. L., Clark, P., Hizi, A., Hughes, S. H. and Arnold, E. (1993). Crystal structure of human immunodeficiency virus type I reverse transcriptase complex with double-stranded DNA at 3.0 Å resolution shows bent DNA. *Proc. Natl. Acad. Sci. USA*, **90**, 6320–4.

Lahm, A. and Suck, D. (1991). DNase I-induced DNA conformation: 2 Å structure of a DNase I-octamer complex. *J. Mol. Biol.*, **221**, 646–67.

Larson, S. B., Koszelak S., Day, J., Greenwood, A., Dodds, J. A. and McPherson, A. (1993). Double-helical RNA in satellite tobacco mosaic virus. *Nature*, **361**, 179–82.

McPherson, A., Brayer, G., Cascio, D. and Williams, R. (1986). The mechanism of binding of a polynucleotide chain to pancreatic ribonuclease. *Science*, **232**, 765–8.

Ollis, D. L., Brick, P., Hamlin, R., Xuong, N. and

Steitz, T. A. (1985). Structure of the large fragment of *Escherichia coli* DNA polymerase I complexed with dTMP. *Nature*, **313**, 762–6.

Ramakrishnan, V., Finch, J. T., Graziano, V., Lee, P. L. and Sweet, R. M. (1993). Crystal structure of globular domain of histone H5 and its implications for nucleosome binding. *Nature*, **362**, 219–23.

Steitz, T. A. (1993). DNA- and RNA-dependent DNA-polymerases. *Curr. Opinion Struct. Biol.*, **3**, 31–8.

Suck, D. (1992). Nuclease structure and catalytic function. *Curr. Opinion Struct. Biol.*, **2**, 84–92.

Tanaka, I., Appelt, K., Dijk, J., White, S. W. and Wilson, K. S. (1984). 3 Å resolution structure of a protein with histone-like properties in prokaryotes. *Nature*, **310**, 376–81.

9.3

Aggarwal, A. K., Rodgers, D. W., Drottar, M., Ptashne, M. and Harrison, S. C. (1988). Recognition of a DNA operator by the repressor of phage 434: a view at high resolution. *Science*, **242**, 899–907.

Berg, J. M. (1993). Zinc-finger proteins. *Curr. Opinion Struct. Biol.*, **3**, 11–16.

Clarke, N. D., Beamer, L. J., Goldberg, H. R., Berkower, C. and Pabo, C. O. (1991) The DNA binding arm of λ repressor: critical contacts from a flexible region. *Science*, **254**, 267–70.

Ellenberger, T. E., Brandl, C. J., Struhl, K. and Harrison S. C. (1992). The GCN4 basic region leucine zipper binds DNA as a dimer of uninterrupted α-helices: crystal structure of the protein–DNA complex. *Cell*, **71**, 1223–37.

Ferr-D'Amar, A. R., Prendergast, G. C., Ziff, E. B. and Burley, S. K. (1993). Recognition by Max of its cognate DNA through a dimeric b/HLH/Z domain. *Nature*, **363**, 38–45.

Harrison, S. C. (1991). A structural taxonomy of DNA-binding domains. *Nature*, **353**, 715–19.

Hegde, R. S., Grossman, S. R., Laimins, L. A. and Sigler, P. B. (1992). Crystal structure at 1.7 Å of the bovine papillomavirus-1 E2 DNA-binding domain bound to its DNA target. *Nature*, **359**, 505–12.

Koudelka, G. B. and Carlson, P. (1992). DNA twisting and the effects of non-contacted bases on affinity of 434 operator for 434 repressor. *Nature*, **355**, 89–91.

Luisi, B. F., Xu, W. X., Otwinowski, Z., Freeman, L. P., Yamamoto, K. R. and Sigler, P. B. (1991). Crystallographic analysis of the interation of the glucocorticoid receptor with DNA. *Nature*, **352**, 497–505.

Marmorstein, R., Carey, M., Ptashne, M. and Harrison S. C. (1992). DNA recognition by GAL4: structure of a protein–DNA complex. *Nature*, **356**, 408–14.

Nikolov, D. B., Hu, S.-H., Lin, J., Gasch, A., Hoffmann, A., Horikoshi, M., Chua, N.-H., Roeder, R. G. and Burley, S. K. (1992). Crystal structure of TFIID TATA-box binding protein. *Nature*, **360**, 40–46.

Otting, G., Qin, Y. Q., Billeter, M., Müller, M., Affolter, M., Gehring, W. J. and Wüthrich, K. (1990). Protein–DNA contacts in the structure of a homeodomain–DNA complex determined by nuclear magnetic resonance spectroscopy in solution. *EMBO J.*, **9**, 3085–92.

Otwinowski, Z., Schevitz, R. W., Zhang, R.-G., Lawson, C. L., Joachimiak, A., Marmorstein, R. Q., Luisi, B. F. and Sigler, P. B. (1988). Crystal structure of *trp* repressor/operator complex at atomic resolution. *Nature*, **335**, 321–9.

Pabo, C. O. and Sauer, R. T. (1992). Transcription factors: structural families and principles of DNA recognition. *Annu. Rev. Biochem.*, **61**, 1053–95.

Pabo, C. O., Aggarwal, A. K., Jordan, S. R., Beamer, L. J., Obeysekare, U. R. and Harrison S. C. (1990). Conserved residues make similar contacts in two repressor–operator complexes. *Science*, **247**, 1210–13.

Pathak, D. and Sigler, P. B. (1992). Updating structure–function relationships in the bZip family of transcription factors. *Curr. Opinion Struct. Biol.*, **2**, 116–23.

Pavletich, N. P. and Pabo, C. O. (1991). Zinc finger–DNA recognition: crystal structure of a Zif268–DNA complex at 2.1 Å. *Science*, **252**, 809–17.

Schultz, S. C., Shields, G. C. and Steitz, T. A. (1991). Crystal structure of a CAP–DNA complex: the DNA is bent by 90°. *Science*, **253**, 1001–7.

Somers, W. S. and Phillips, S. E. V. (1992). Crystal structure of the *met* repressor–operator complex at 2.8 Å resolution reveals DNA recognition by β-strands. *Nature*, **359**, 387–93.

9.3.6: Restriction enzymes

Anderson, J. E. (1993). Restriction endonucleases and modification methylases. *Curr. Opinion Struct. Biol.*, **3**, 24–30.

Lesser, D. R., Kurpiewski, M. R. and Jen-Jacobson, L. (1990). The energetic basis of specificity in the *Eco* RI endonuclease–DNA interaction. *Science*, **350**, 776–86.

Rosenberg, J. M. (1991). Structure and function of restriction endonucleases. *Curr. Opinion Struct. Biol.*, **1**, 104–13.

Winkler, F. K. (1992). Structure and function of restriction endonucleases. *Curr. Opinion Struct. Biol.*, **2**, 93–9.

9.3.7: Enzymes that 'flip-out' nucleosides

Cheng, X. (1995). DNA modification by methyltransferases. *Curr. Opinion Struct. Biol.*, **5**, 4–10.

Klimasauskas, S., Kumar, S., Roberts, R.J., and Cheng, X. (1994). *Hhal* methyltransferase flips its target base out of the DNA helix. *Cell*, **76**, 357–69.

Mol, C. D., Arvai, A. S., Slupphaug, G., Kavli, B., Alseth, I., Krokan, H. E., and Tainer, J. A. (1995). Crystal structure and mutational analysis of human uracil DNA glycosylase: structural basis for specificity and catalysis. *Cell*, **80**, 869–78.

Park, H-W., Kim, S-T., Sancar, R. A., and Deisenhofer, J. (1995). Crystal structure of DNA photolyase from *E. coli Science*, **286**, 1866–72.

Roberts, R.J. (1995). On base flipping. *Cell*, **82**, 9–12.

9.3.8: tRNA synthetases

Cavarelli, J., Rees, B., Ruff, M., Thierry, J.-C. and Moras, D. (1993). Yeast tRNA[Asp] recognition by its cognate class II aminoacyl-tRNA synthetase. *Nature*, **362**, 181–4.

Moras, D. (1992). Structural and functional relationships between aminoacyl-tRNA synthetases. *Trends Biochem. Sci.*, **17**, 159–64.

Rould, M. A., Perona, J. J., Söll, D. and Steitz, T. A. (1989). Structure of *E. coli* glutaminyl-tRNA synthetase complexed with tRNA[Gln] and ATP at 2.8 Å resolution. *Science*, **246**, 1135–42.

Rould, M. A., Perona, J. J. Steitz, T. A. (1991). Structural basis of anticodon loop recognition by glutaminyl-tRNA synthetase. *Nature*, **352**, 213–18.

Acknowledgement

Some of the figures in Chapter 9 were produced using the program MOLSCRIPT (Kraulis, P. J. (1991). MOLSCRIPT: a program to produce both detailed and schematic plots of protein structures. *J. Appl. Crystallogr.*, **24**, 946–50).

TECHNIQUES APPLIED
TO NUCLEIC ACIDS

10.1 Isolation of DNA from cells

Although historically the isolation of polymeric DNA from living organisms was problematical (Section 1.2), nowadays the procedure can be carried out easily. This is mainly due to the chemical stability of DNA and the ease with which most deoxyribonucleases can be activated. By contrast, the isolation of intact RNA poses a much greater problem (Section 6.1).

DNA can be isolated from prokaryotic or eukaryotic whole cells or from purified nuclei. Usually the use of purified nuclei is unnecessary. DNA isolations can be essentially divided into four steps: (1) isolation of cells; (2) cell lysis; (3) removal of protein; and (4) isolation of pure DNA.

Two essential requirements during DNA isolation are the exclusion of free magnesium ions (and any other divalent cation if possible) and the avoidance of rough physical treatment. Magnesium ions are a cofactor of all common deoxyribonucleases and can normally be excluded by addition of 10 mM EDTA as a chelating agent. To ensure that no magnesium-independent nucleases cause any degradation, an anionic detergent such as sodium dodecyl sulfate is usually also added. Sodium dodecyl sulfate also serves to lyse eukaryotic cells and nuclear membranes. To lyse prokaryotic cells, lysozyme can be used in most cases but sometimes other methods, such as a French press, may need to be employed. The second important point to note is that genomic DNA, when released from any cell, is of extremely high molecular weight and is acutely sensitive to physical shearing. Over-zealous pipetting, vortexing, or shaking should be avoided if DNA larger than 10 000 bp is required. The general procedures are as follows.

1. *Isolation of cells.* For prokaryotes, monocellular eukaryotes, and tissue culture cells, this simply involves harvesting by low-speed centrifugation at 4°C followed by resuspension of the cells in isotonic neutral buffered solution. Tissues will need to be disrupted by physical or enzymatic means.

2. *Cell lysis.* To avoid clumping, lysis of eukaryotic cells is best achieved by addition of a concentrated lysis buffer to a suspension of cells. A typical lysis buffer is: 1 per cent sodium dodecyl sulfate, 10 mM Tris–HCl (pH 7.5), 10 mM EDTA. Prokaryotic cells can be pretreated with lysozyme and then suspended in a lysis buffer containing sodium dodecyl sulfate. Alternatively they can be physically disrupted in a French press or by use of a sonicator.

3. *Removal of protein.* A general protease, such as fungal proteinase K, is added to lysed cells to degrade protein, thereby allowing easy removal by repeated extraction from aqueous solution into phenol. Typically, a 30 minute digestion with 100 μg ml^{-1} proteinase K is followed by four successive extractions with redistilled phenol (phenol oxidizes readily and must be redistilled before use). Prior to use it should be equilibrated to pH 7–8 with 1 M Tris base). After deproteinization, nucleic acids and polysaccharides are precipitated by addition of sodium acetate (pH 7) to 0.3 M followed by two volumes of 95 per cent ethanol. Genomic DNA can be spooled from the water–ethanol interface using a glass pipette or precipitated by gentle vortexing. Low molecular weight plasmid DNA forms a colloidal suspension and is not separable from RNA by this method.

4. *Isolation of pure DNA by CsCl density ultracentrifugation.* The best way to remove con-

taminating RNA and polysaccharides from crude DNA preparations is by equilibrium density gradient ultracentrifugation. This method makes use of the characteristic density of DNA in aqueous solution (1.7 g mL^{-1}). Caesium chloride is added to the crude preparation to a final concentration of 1.7 g mL^{-1} and the solution is centrifuged until a density gradient of caesium chloride is formed in the tube (typically 48 h at 80 000 g in an angle rotor or overnight in a vertical rotor). If plasmid DNA is to be isolated, ethidium bromide is added prior to centrifugation. This compound intercalates between the nucleotide pairs in any DNA duplex without denaturing it (Section 8.4.1). This intercalation causes a slight unwinding of the helix and a reduction in the density of the DNA. Because plasmids are closed circles this unwinding is compensated by the formation of supercoils (Section 2.3.5). Eventually, the increased free energy thus generated prevents further intercalation at an ethidium : DNA ratio which is significantly below that for a topologically relaxed duplex. The net result is that supercoiled plasmid DNA is less dense than nicked plasmid or genomic DNA and therefore migrates to a different region of the CsCl gradient.

Summary

DNA is isolated by harvesting and lysis of cells, removal of protein, and separation from RNA and polysaccharides. Care is necessary to avoid enzymatic or mechanical breakage of the high molecular weight DNA. DNAs are usually purified according to their buoyant density in caesium chloride solutions by equilibrium ultracentrifugation.

10.2 Spectroscopic techniques

10.2.1 UV absorption

The light absorption characteristics of nucleic acids result from the combination of the strong ultraviolet absorption of the purine and pyrimidine bases in the 240–280 nm range modulated by the stereochemistry and conformational influences of a ribose-phosphate backbone that is essentially transparent to light of that wavelength (Section 2.1.3).

Oligonucleotides exhibit a strong UV absorption maxima, λ_{max}, at approximately 260 nm and a molar extinction coefficient, ε, of the order of 10^4 (dm^3 mol^{-1} cm^{-1}) (Table 2.2). This absorption arises almost entirely from complex electronic transitions in the purine and pyrimidine components. The intensity and exact position of the λ_{max} is a function not only of the base composition of the nucleic acid but also of the state of the base-pairing interactions present, the salt concentration of the solution, and its pH. Most importantly, base-base stacking results in a **decrease** in ε known as **hypochromicity**. This arises from dipole–dipole interactions that depend on the three-dimensional structure of an oligonucleotide and ranges in magnitude from 1–11 per cent for deoxyribonucleoside phosphates to 30 per cent for most helical polynucleotides. While some degree of stacking is apparent for all dimers, it appears that UpU, UpA, UpC, GpU, and UpU are generally less stacked than other ribodinucleoside

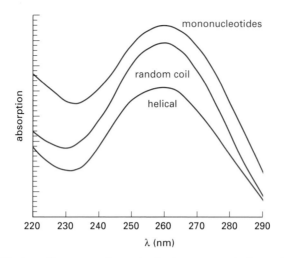

absorption

λ (nm)

Fig. 10.1 Typical UV absorption curves for equimolar base concentrations of mononucleotides, of single stranded (random coil) oligonucleotide, and of double-helical DNA.

phosphates at pH 7. In practice, the effects of structure on the UV absorption of oligonucleotides are so complex that only the most basic interpretations can be made (Fig. 10.1).

UV absorption is a sensitive and convenient way to monitor the 'melting behaviour' of DNA and RNA. When the UV absorption of a nucleic acid sample is measured as a function of temperature, the resulting plot is known as a **melting curve**. The midpoint in the increase in absorption with increasing temperature is known as the **melting temperature**, T_m. This is dependent on the base-composition of the sample, the salt concentration of its solution, and even the types of counter-ion present (Section 10.10). Such melting is a co-operative phenomenon and the observed melting curves become progressively steeper with increasing length of the oligonucleotide. Simple sigmoid melting curves are observed for many DNA samples, but it is also possible to observe more complex, multi-phasic melting in some cases (Fig. 10.2). The deconvolution of such multi-phase melting curves makes it possible to examine the effects of base-modification on the stability and nature of nucleic acid secondary structure in more detail.

10.2.2 Fluorescence

Fluorescence is defined as the emission of radiation given out as a molecule returns to its ground state from an excited electronic state. In addition to excitation and emission spectra, one can also measure the lifetime, intensity, and polarization of the fluorescence. Excitation and emission occur in about 10^{-15} s, while the lifetime of the excited state has a duration of the order of 10^{-9} s. It follows that physical processes which take place on the same time-scale as the lifetime of fluorescence can be investigated by the analysis of changes in the emission.

The emission from monomers and dinucleoside phosphates is very weak at room temperature (and is usually studied in frozen solutions at low temperatures, 77–80 K). As a result, considerable use has been made of one of the few naturally occurring fluorophores, the wyo-

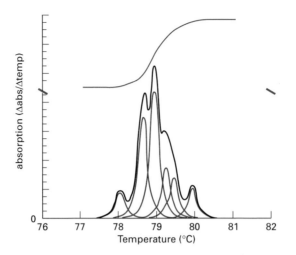

Fig. 10.2 The oligonucleotide melting curve, the derivative of the curve (colour), and the deconvolution of the derivative into its composite components.

sine base found in transfer RNA (Section 3.1.2). In the absence of natural nucleic acid fluorophores, synthetic ones can be introduced earlier into an oligonucleotide or into a natural nucleic acid by specific covalent or non-covalent addition or by total synthesis, which has become very important in the application of fluorescence energy transfer, FRET, to the study of nucleic acids.

FRET is a spectroscopic process by which energy is passed non-radiatively between molecules over long distances (10–100 Å). The donor molecule, D, is a fluorophore which has an absorption maximum at a shorter wavelength, can be excited selectively, and transfers the energy of an adsorbed photon non-radiatively to an acceptor molecule, A. The acceptor species usually has an excitation $\lambda_{\max}$ at longer wavelength and with significant absorption overlap with the fluorescence emission envelope of D. The result is that the energy of a photon absorbed by D can be lost by fluorescent photon emission from D or by radiationless transition to A leading to photon emission from A at a longer wavelength. (Other quenching processes are also possible.)

Such FRET can be measured by the efficiency of depopulation (E_T) and this is related to the fluorescent lifetimes in the presence (τ_T) and absence (τ) of resonance energy transfer (as described in equation 10.2.1).

$$E_T = 1 - \frac{\tau_T}{\tau} \tag{10.2.1}$$

The donor–acceptor distance, R, is related to E_T by the system constant R_0, which is the distance at which there would be 50 per cent of maximal energy transfer (shown in equations 10.2.2 and 10.2.3):

$$E_T = \frac{R_o^6}{R^6 + R_o^6} \tag{10.2.2}$$

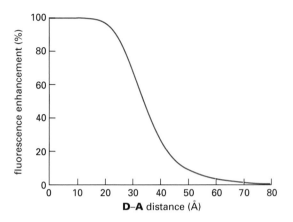

Fig. 10.3 A typical relationship between fluorescence enhancement, resulting from FRET, and the distance between donor and acceptor fluorophores.

$$R = R_0 \left(\frac{1 - E_T}{E_T} \right)^{1/6} \qquad (10.2.3)$$

The theory of fluorescence transfer predicts that fluorescence transfer depends on the sixth power of the distance. In practice, the distances over which FRET can be measured vary between about 40 and 100 Å (Fig. 10.3). As there are several factors that can influence the extent of fluorescence transfer, some care has to be taken to control and account for all possible pathways for fluorescence transfer or damping for FRET measurements to be interpreted quantitatively.

In general, the donor–acceptor pairs can be free in solution, one or both bound (covalently or non-covalently) to the nucleic acid, or be an inherent part of the structure. The type and range of information available from oligonucleotides using FRET is ideal for investigating many structural and symmetry features concerning oligonucleotides that range over distances up to 100 Å. A particularly good example has been Lilley's application of FRET to resolve the problem of helical four-way junctions in DNA topology (Holliday junctions, see cf. Section 2.3.3). FRET measurements were carried out on four-way junctions constructed from four synthetic oligonucleotides, where one end of two different strands had been covalently labelled with a fluorescent donor, fluorescein, and an acceptor species, tetramethylrhodamine. The experiments nicely confirmed earlier predictions of junction chirality and crossing angles that had been deduced from gel mobility studies (Fig. 10.4). Moreover, by fixing the location of the fluorescein donor on one arm and 'walking' the tetramethylrhodamine acceptor around a 9 bp region of a second arm, it was possible to establish that the four-way junction has a right-handed non-crossed structure.

10.2.3 Circular dichroism

Circular dichroism (CD) measurements have been in use for over 25 years to study the conformations of nucleic acids in solution. The technique depends on the differential absorption

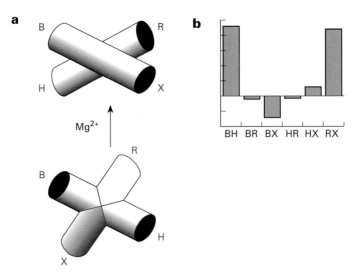

Fig. 10.4 (a) Schematic representation of the nature of topological changes proposed for a DNA four-way junction on addition of magnesium ions. (b) The relative change in fluorescence enhancement as a result of adding magnesium to the DNA observed between fluorophore pairs situated at the ends of junction arms, as performed for each of the six different possible combinations.

of left- and right-handed circularly polarized light as a result of the chirality (or 'handedness') of the chromophoric species under investigation. Data are presented as the difference in the left- and right-handed molecular extinction coefficients ($\varepsilon_L - \varepsilon_R$) as a function of wavelength.

The nucleic acid bases themselves are planar and so are intrinsically optically inactive, and thus show no circular dichroism. However, in nucleosides and nucleotides the glycosylic linkage to C-1' of the pentofuranose imparts a chiral perturbation to their UV absorption (Section 2.1.4). The CD of polynucleotides is very much larger as a result of the co-operativity of the chiral interactions of contiguous bases. This is a result both of sequence effects of nearest-neighbouring bases and of gross secondary structure.

The information from CD spectra is largely complementary to that from UV absorption spectra. In neither case is it possible to obtain detailed structural information, such as that obtained from X-ray crystallography or NMR spectroscopy. Rather, nucleic acid CD spectra provide a reliable determination of the overall conformational state by comparison with the CD spectra of reference samples. It is thus used for studies on nucleic acids that are difficult to crystallize or available in only very small amounts.

The response of CD to changes in secondary structure is well illustrated by the effects of humidity on DNA. It is known from early DNA fibre diffraction studies that DNA can undergo a co-operative structural transition when dehydrated or treated with ethanol (Section 2.2.2). The transition from the normal, hydrated B-form to a relatively dehydrated and RNA-like A-form is accompanied by marked changes in the CD spectrum (Fig. 10.5a). These changes in the profile of the CD spectrum reflect changes in helical twist and tilt between A-form and B-form DNA (Section 2.3.1).

Circular dichroism has been used to characterize many of the more unusual conformations and arrangements in DNA. One of its early successes was the identification by Pohl and Jovin of Z-DNA and the characterization of its dependency on environment. When the sodium chloride concentration is increased to about 4 M, poly(dG·dC) undergoes a major structural transformation. This B–Z conformational change involves a switch from the *anti-* to the *syn-* conformation at the purine glycosylic linkage with a consequent change from a right-handed to a left-handed helix (Section 2.2.5). As a result, the CD spectrum undergoes an almost complete inversion in character (Fig. 10.5b).

Other areas of investigation by CD have included the effect of interactions of drugs and carcinogens with DNA, but such spectra remain rather difficult to interpret with molecular precision.

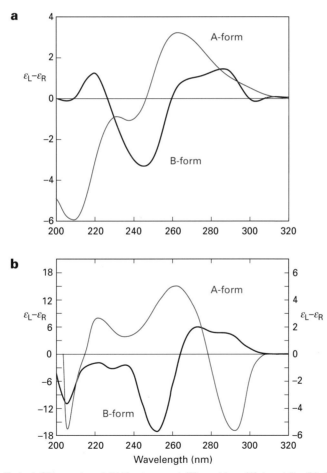

Fig. 10.5 (**a**) Typical CD spectra of DNA at normal (B) and low (A) humidity. (**b**) CD spectra for poly[d(G-C) · (G-C)] in 10 mM salt at pH 7 before (black) and after (colour) adding NaCl to 3.9 M.

10.2.4 Infrared and Raman spectroscopy

Infrared (IR) and Raman spectroscopy are often regarded as closely related techniques. Both techniques observe the vibrational frequencies of localized parts of the sample molecule. Both techniques are largely non-destructive and can be used on microscopically small samples. A major advantage is that DNA can be analysed in crystals, gels, or fibres, as well as in solution, and this has supported direct correlations between the observed vibrational frequencies and three-dimensional structures derived by X-ray diffraction.

IR absorptions from nucleic acids are observed in the frequency range between 1800 and 700 cm^{-1}. The problem of strong IR absorption by water near 1600 cm^{-1} and also below 1000 cm^{-1} can be circumvented using D_2O as solvent when the water signal at 1600 cm^{-1} is shifted to about 1200 cm^{-1} and the absorption at 1000 cm^{-1} is shifted by an almost equivalent amount (Fig. 10.6). So, measuring IR spectra in both solvents makes it possible to observe the full spectral range of interest. The use of D_2O also causes small but significant shifts in nucleic acid absorptions resulting from deuterium exchange, and these can be used to monitor H–D exchange processes.

Fourier transform infrared absorption, FTIR, is so sensitive that it is possible to make measurements on very small crystals of nucleic acids, and needs about 100 μg of material.

IR spectra are largely unaffected by the external environment and so FTIR has supported the identification of structurally significant bands by calibration with X-ray crystallography of nucleic acid samples. Such 'marker bands' are sensitive to helical type and can be used to show the existence of a conformational transition when different factors, such as temperature, hydration, concentration, or the nature and amount of cations are varied (B → A, B → C, B → Z form, helix → coil, etc.). FTIR is also used to support studies on the recognition of DNA sequences by a wide variety of molecules, such as oligonucleotides (triple-stranded structures), drugs, and proteins.

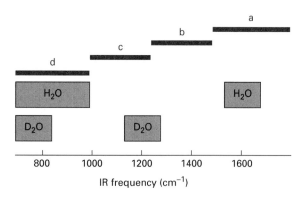

Fig. 10.6 Schematic distribution of IR absorption bands of DNA and solvent. (**a**) 1800–1500 cm^{-1} corresponding to stretching of C=X double bonds; (**b**) 1500–1250 cm^{-1} corresponding to base-sugar entities (including glycosylic torsion angle effects); (**c**) 1250–1000 cm^{-1} corresponding to phosphate and sugar absorptions; (**d**) below 1000 cm^{-1} associated with phosphodiester chain coupled with the sugar vibrations.

Raman spectroscopy also depends on the vibrational frequencies of groups within the molecule and, like IR, provides information concerning vibrational modes of nucleic acid components that are conformationally sensitive. The Raman technique has some useful advantages over IR. First, the incident radiation in Raman work is not strongly absorbed and so does little damage to the sample. Secondly, water has weak scattering properties and lacks absorption at the irradiation frequencies used for sample irradiation, so its presence in the sample is not a problem. Thirdly, unlike IR, the intensity of the Raman bands is proportional to the concentration of the target species. As with IR, the accurate assignment of resonance lines can be simplified by calibration using samples of known X-ray structure.

Raman spectroscopy has been used to examine nucleic acids in a wide variety of situations, including microcrystals and even within living cells.

10.2.5 Nuclear magnetic resonance

Since about 1978, nuclear magnetic resonance, NMR, has been used extensively in the study of structure and dynamics of nucleic acids in solution. The technique has become a viable alternative and complement to single-crystal X-ray diffraction for the solid state.

NMR spectroscopy is based on the fact that certain atomic nuclei when oriented by a strong magnetic field (2–14 Tesla) absorb radiation at characteristic frequencies, typically a few hundred MHz (megahertz) in the radiofrequency part of the spectrum. For the chemist and biologist, the value of NMR is that nuclei of the same element, particularly ^{1}H, give rise to different spectral lines according to their environment. As a result, it is possible to observe discrete signals from individual atoms even in large macromolecules, up to some 30 kDa in mass for proteins and approaching some 20 bp for nucleic acids. In addition, the NMR response can be transferred between nuclei, thereby providing information on internuclear distances and on the relative orientation of nuclei. The NMR parameters that can be derived from the resulting spectra give accurate information about molecular structure, conformation, and dynamics of nucleic acids.

The NMR of nucleic acids makes use of atomic species ^{1}H, ^{13}C, ^{15}N, and ^{31}P. Of these, hydrogen (^{1}H, the proton) is much the most important on account of its high sensitivity to NMR detection and its near 100 per cent natural isotopic abundance. Modern instrumentation usually detects other nuclei, for example ^{13}C and ^{15}N, through their attached protons in order to maximize the advantage of this sensitivity of protons. Lastly, the fact that isotopes of hydrogen, carbon, and nitrogen have very different NMR characteristics means that isotopic labelling (^{1}H $\rightarrow$ ^{2}H, ^{12}C $\rightarrow$ ^{13}C, or ^{14}N $\rightarrow$ ^{15}N) can be used to change the NMR behaviour of a molecule.

From its early applications to biological problems in the 1950s, NMR has leaped forward as a result of three main developments. First, superconducting magnets cooled by liquid helium have taken field strengths from 1 to about 20 T. Secondly, samples are now subjected to irradiation at radiofrequency by pulse Fourier transform methods, in which the radiation is applied in the form of a more or less complex sequence of computer-generated pulses while the spectrum is obtained by Fourier transformation of the response of the various nuclear spins to these pulses. Thirdly, multidimensional NMR involves the use of two, three, or even four frequency variables in the recording of resonance intensities. In one-dimensional NMR, the ^{1}H spectrum has one resonance signal for each chemically distinct species of proton, and this is measured as the **chemical shift**.

Computer-based manipulation of nuclear spins forms the basis of modern multi-dimensional NMR. In two dimensions, or higher, the diagonal of the spectrum corresponds to a conventional one-dimensional NMR spectrum while the cross-peaks, which occur off the diagonal, contain information about the *connections* between the resonances on the diagonal. This is illustrated by a section of the two-dimensional NOESY (nuclear Overhauser spectroscopy) spectrum for the triple helix formed between a hairpin d5′(TCTCTCTTTGAGAGA) and the oligonucleotide d5′(TCTCTCT) (Fig. 10.7), showing that part of the proton spectrum with resonances from the imino protons involved in the base-triples (Section 2.3.6).

There are three types of such connectivities: *through-bond* connections relate nuclei that are separated by two, three, or even four bonds (COSY or correlation spectroscopy); *through-space connections*, between the resonances of nuclei that are spatially closer than 5 Å, and *exchange connections*, that involve the same nucleus in two different environments, typically that of a ligand in fast equilibrium between free solution and bound to the nucleic. In three-dimensional and four-dimensional NMR, the two-dimensional ^{1}H NMR spectra are effectively spread out into a third or fourth dimension, usually by means of the ^{13}C and/or ^{15}N chemical shift.

The first step in any NMR study is the assignment of resonances to individual nuclei, and this is usually the rate-limiting step of the operation. The selective substitution of protons by deuterium can greatly accelerate the assignment of a ^{1}H spectrum by simplifying the problems of overlapping resonances, and this approach is now gaining strength both for DNA and for RNA structures. It is then possible to proceed to study through-space connectivities by NOESY and ROESY (rotating frame Overhauser spectroscopy) methods.

Most of the information about solution structure comes from the analysis of internuclear distance geometry NOESY, where the intensity of the NOE signal is proportional to the inverse sixth power of the distance between two hydrogen nuclei. Typically, proton NOE intensities are sorted into 'strong', 'medium', and 'weak' for distance ranges of 1.8–2.5 Å, 1.8– 3.5 Å, and 1.8–5 Å. The three-dimensional structure that is derived from such an analysis is essentially a 'protonic' picture of the molecule. This has to be converted into a full three-dimensional structure by adding in knowledge of standard chemical bond lengths and bond angles.

The through-bond features of NMR spectra provide good data about torsional bond angles in oligonucleotides. This was initially used to measure the sugar pucker by application of the Karplus equation (Fig. 10.8) to the three-bond coupling constants involving C1′, C2′, C3′ and C4′ of the furanose ring (Section 2.1.4). The conformation of the glycosylic bond, whose torsion angle is χ, can be analysed from the H6/H8–H1′ connectivities while some of the backbone torsion angles, β, γ, δ, and ε can be determined by appropriate three-bond connectivities using ^{1}H–^{1}H, ^{1}H–^{13}C, and ^{1}H–^{31}P interactions.

One of the solutions to the problem of peak assignment in two-dimensional NMR spectra of nucleic acids involves reference to crystal structures and the short proton–proton distances observed in the different types of structure. The lack of unambiguous through-bond proton–

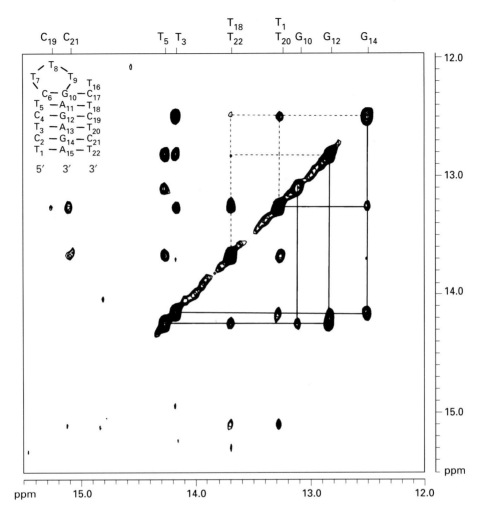

Fig. 10.7 Demonstration of the assignment procedure in a NOESY spectrum of a DNA triplex formed by the hairpin d5'($T_1C_2T_3C_4T_5C_6$-$T_7T_8T_9$-$G_{10}A_{11}G_{12}A_{13}G_{14}A_{15}$)3' and the oligonucleotide d5'($T_{16}C_{17}T_{18}C_{19}T_{20}C_{21}T_{22}$)3'. A sequential walk can be made between the imino protons involved in Watson–Crick hydrogen bonding in the hairpin as is shown below the diagonal, starting at the imino proton resonance position of residue T_1, via the imino proton of G_{14}, T_3, G_{12}, and T_5, ending at the imino proton resonance position of residue G_{10}. Note that the incompletely resolved $T_3 \cdot G_{12}$ and $G_{12} \cdot T_5$ cross-peaks (because of limited digital resolution in F1) are resolved above the diagonal. Above the diagonal the sequential walk is shown between the imino protons of the Hoogsteen base-paired residues. The latter involve the purine strands of the hairpin stem and the pyrimidine oligonucleotide. Starting at the resonance position of the imino proton of residue G_{12} the NOE walk proceeds to the resonance position of the imino proton of residue G_{14}.

proton connectivities in the backbone of oligonucleotides has meant that the assignment has relied on NOEs between contiguous nucleotide residues. From measurements made on crystal structures, strategies have evolved for using an 'NOE walk' procedure. It is possible to 'walk' up

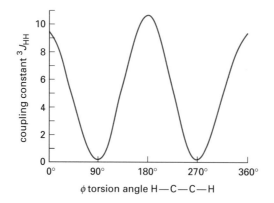

Fig. 10.8 The dependence of the ^{1}H–^{1}H coupling constant through three bonds (H–C–C–H) is a function of the dihedral torsion angle, ϕ. Its magnitude is well fitted by the **Karplus equation**. This has been developed to take account of the number and type of substituents on the C–C bond, but in a simplified form expresses the vicinal coupling constant as: $^3J_{HH} = 13.7 \cos^2\phi - 0.73 \cos\phi$ (Hz). For the backbone torsion angles, β and ε, the corresponding equation for P–O–C–H coupling is
$$^3J_{PH} = 15.3 \cos^2\phi - 6.2 \cos\phi + 1.5 \text{ (Hz)}.$$

the hydrogen-bonded protons in the centre of a double helix and 'walk' along the sugar–base backbone using connectivities that can be observed between non-exchanging base protons (H-8 or H-6) and the H-1′ and H-2′/H-2″ of the sugars (Fig. 10.9).

Even when the full range of multi-nuclear and multi-dimensional NMR data has been assigned and analysed and the sugar pucker, backbone, and glycosylic torsion angles measured, the determination of the structure of an oligonucleotide still calls for heavy computing using restrained molecular dynamics (Section 10.5.1) and various algorithms that work in distance and torsion angle space. While these are generally beyond the scope of this discussion, one of the many examples of structure analysis by NMR serves as an illustration of the power of the method. Other examples are to be found in Chapters 7, 8, and 9.

The solution structure of the DNA·RNA hybrid duplex, d(GTCACATG)·r(caugugac) has been determined by two-dimensional NOESY and analysis of the two-dimensional NOE spectra using restrained molecular dynamics. It emerges that the duplex has neither an A-form nor a B-form in solution but an intermediate duplex structure. The RNA strand has a normal C3′-*endo* sugar pucker but the DNA strand furanose rings have an unexpected intermediate O4′-*endo* conformation and the minor groove width is intermediate (8.6–9 Å) between A-form (11 Å) and B-form (7.4 Å) duplexes (Section 2.2.4). The detailed three-dimensional structure of this hybrid duplex has permitted its interaction with RNase H to be modelled in a fashion that appears to explain why RNase H can discriminate between a DNA·RNA hybrid duplex and a pure RNA·RNA duplex (Section 2.4.3).

In addition to the considerable progress achieved in the structure determination of DNA and RNA, studies on the structures of DNA complexed with groove binding or intercalating compounds have been especially successful (Sections 8.3.1 and 8.4.1) although there still remain severe restrictions on the size of an oligonucleotide that can be assigned fully.

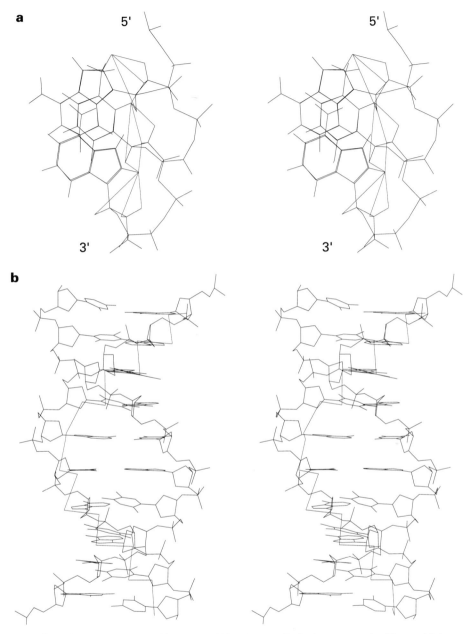

Fig. 10.9 (**a**) Some of the typical NOE connectivities observed in the base–sugar backbone of B-form DNA (diagram in stereopair for parallel viewing). (**b**) A trace of the base (H-8, H-6) to sugar (H-1′) NOE connectivities observed in a self-complementary 12 bp, B-like DNA oligomer (diagram in stereo for parallel viewing, terminal base-pairs not shown). (**c**) The one-dimensional proton NMR spectrum of the dodecamer d(CGAATATATTCG)$_2$ (next page). The regions marked D1 and D2 represent mainly the signals from the base and sugar H-1′ protons. (**d**) A part of the two-dimensional NOESY spectrum showing the NOEs resulting from interactions between these two regions, and a line showing a sequential walk through the structure.

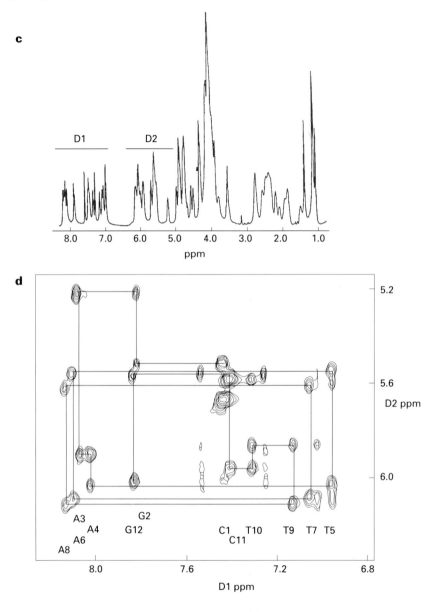

Fig. 10.9 (c) and 10.9 (d)

As a method used largely for analysis of materials in solution, NMR has certain advantages over X-ray crystallography in that a wide range of conditions of temperature, pH, and salt concentration can be studied. The information available from dynamic NMR includes proton exchange rate constants and dynamic properties of oligonucleotides on the long and medium time scale.

While DNA duplexes are not too well defined by NMR, except at the level of A-, B-, and Z-forms, this technique has been very useful in determining the structures of three- and four-stranded complexes for DNA and for RNA, of DNA–drug complexes (Section 8.3.2), and DNA–protein complexes. The DNA is generally assumed to be the standard B-DNA in these complexes, but NMR analysis can show how the ligands fit round the DNA. NMR has already done much to define the structure of DNA hairpins (Section 2.3.3 and Fig. 10.7) and much more can be expected as methods for the isotopic labelling of DNA become more readily available. Rather recently, RNA has become a challenging target for NMR study because it forms a range of interesting structures and because it can be labelled with isotopes, particularly ^{15}N and ^{13}C.

10.2.6 X-ray diffraction

X-ray diffraction has played a pivotal role in our understanding of nucleic acid structure and function. While the information available from early fibre work was quite crude by today's standards, it provided enough information totally to transform many of the theories in molecular genetics (Section 1.4). Through later refinement, it has been able to define many of the available types of DNA structure, not least in the field of oligodeoxynucleotides.

Early work relied on the diffraction pattern from DNA fibres, which contain information regarding the repeat spacings in the vertical dimension (Fig. 10.10). The resolution is rarely better than 3 Å and there is no phase information. However, while such fibre diffraction data are adequate for the construction of quite detailed models in some cases, it is inevitable that confidence in such models is rather limited.

The first step in a crystallographic study is the production of large, well-diffracting crystals. Whilst DNA oligonucleotides are relatively easy to crystallize, crystals with good diffraction properties are less common. In general, in order to achieve crystals with the necessary diffraction characteristics, it has proved desirable to consider the specific properties of a DNA molecule, to determine the physicochemical parameters which favour three-dimensional assembly, and to analyse the intermolecular contacts of the packing. What seems to be clear is that, in contrast to proteins, the packing environment can stabilize or induce important DNA alterations, and the conditions used for crystallization can induce structural transitions.

Single crystal diffraction data contain all the information needed to reconstruct the three-

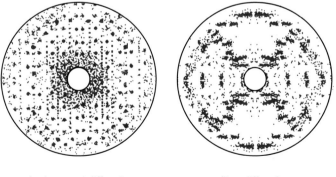

single crystal diffraction   fibre diffraction

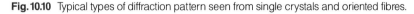

Fig. 10.10 Typical types of diffraction pattern seen from single crystals and oriented fibres.

dimensional structure of the unit cell and the molecules in that structure excepting phase information (Fig. 10.10). The phase problem has been the object of much investigation and still presents a formidable hurdle to solving some structures. The most basic approach to determining phase information is the use of multiple isomorphous replacement (MIR). This requires the incorporation of so-called heavy atoms or heavy atom-containing derivatives into specific positions within the molecular lattice. Such incorporation of heavy atoms can be done either by soaking the crystals in a solution containing the heavy atom, or by crystallizing the compound in the presence of the heavy atom, or by incorporation of a heavy atom into the compound by chemical synthesis or modification, as in the case of 5-bromodeoxyuridine. It is an absolute requirement that the crystals containing the heavy atom are of the same form as the original so that their X-ray diffraction pattern can be used together with the diffraction pattern from the non-derivatized crystal to identify the positions of the heavy atoms. The consequent phase and amplitude can then be used to calculate the phases of all other reflections. In many DNA structure determinations, this phase problem can be circumvented by molecular replacement techniques, where a related structure or model structure is used in searching for solutions to the phase problem.

It is then possible to calculate the electron density within all points in the unit cell by Fourier transform using all the intensities and phase values from the diffraction pattern. The quality of resolution of the electron density map depends on the number of reflections that can be included in the Fourier transform and on the accuracy of measurement of their intensities and phases.

A major difficulty in the interpretation of oligonucleotide crystal structures is the unknown extent to which crystallization conditions and crystal packing may have affected the structure. Crystallization of nucleic acids usually requires a precipitant solvent, and this often decreases the effective hydration of the molecule. It has been known for some time that dehydration of DNA will cause structural change from B-type to an A-type structure (Section 2.2).

The effect of crystal packing can be a more serious source of structural distortion. Oligonucleotides in crystals often form extremely close contacts with neighbouring molecules and their interactions can even include limited base–base hydrogen bonding. The extent to which these intermolecular interactions could influence structure was only fully appreciated when an oligonucleotide was crystallized with both A- and B-forms in the same crystal. The octamer d(GGG^{Br}UA^{Br}UACC)$_2$ crystallized in the form of hexameric rings of duplex DNA in the A-form, with an apparently solvent-filled channel running down their centre. On close inspection of the diffraction pattern, it became apparent that underlying the well-defined diffraction spots there was a pattern indicative of partially ordered B-form DNA. The interpretation of the disordered B-type diffraction was that the central solvent-filled channel in fact contained one-dimensionally oriented DNA in the B-form. At least one other example has been reported where a crystal is made up of predominantly A-form DNA with a significant amount of B-form present.

The possibility that many crystal structures determined are unrepresentative of the solution structure of the species, at least to some degree, does not diminish the important contribution that high-resolution crystal structures have made to our understanding of nucleic acid geometries.

Well over one hundred structures have been determined with DNA bound to a protein (Chapter 9). The DNA structure within these complexes is mostly of the B-form, with some degree of variation. Protein structures seem to be far less susceptible to distortion by crystallization than do nucleic acids and it is generally believed that the structure of the DNA in the protein complex is also largely undistorted.

Summary

Progress in a number of areas has led to the current view of DNA as a structurally adventurous molecule. Following the advent of methods for the chemical synthesis of oligonucleotides of defined sequence and high purity in milligram amounts, analysis of many DNA structures at near atomic resolution by X-ray crystallography has identified the sequence dependence of DNA structure and given details of many protein–nucleic acid complexes.

Following the first nuclear magnetic resonance structures of DNA in 1982 using two-dimensional spectroscopic methods, rapid advances in instrumentation, in computational techniques, and in isotopic labelling have placed the solution structures of DNA and RNA molecular species of over 15 kDa within the range of complete analysis. In addition, complexes of nucleic acids with proteins and drugs are viable targets for structure determination.

The early use of ultraviolet spectroscopy for the study of nucleic acids remains invaluable for melting temperature analysis and has developed with useful applications of circular dichroism, fluorescence energy transfer, Fourier transform infrared, and laser Raman spectroscopic methods. Many of the results of these newer techniques rely on calibration against standard structures pre-established by X-ray analysis.

10.3 Hydrodynamic and separation methods

10.3.1 Centrifugation

Centrifugation has a long history as a useful tool in both analysis and purification of nucleic acids. The behaviour of large nucleic acid molecules under the high forces that can be attained in an ultracentrifuge (Fig. 10.11a) can be used to determine molecular weights, molecular densities, and (indirectly) molecular shapes.

Equilibrium density-gradient centrifugation in caesium chloride was used in one of the milestone experiments in molecular biology. Meselson and Stahl reported an experiment in 1957 that established the **semi-conservative nature of DNA replication**. Using a growth medium containing ^{15}N ammonium chloride as the only source of nitrogen, they grew cells that contained only 'heavy' DNA. By transferring these cells to a medium containing normal ^{14}N ammonium chloride and then isolating DNA after discrete numbers of replication cycles, Meselson and Stahl were able to observe the distribution of $^{15}N/^{14}N$ labelled DNA by caesium chloride density gradient centrifugation (Section 5.2.1). After one replication, the dense DNA had been diluted with 50 per cent ^{14}N. After a second generation, two bands were seen: one

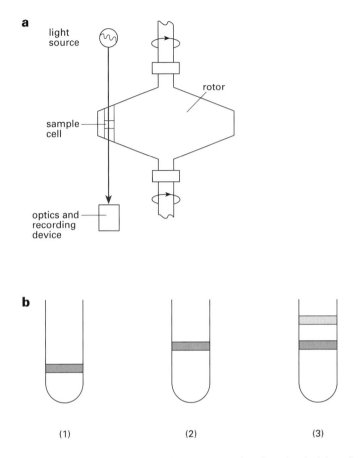

Fig.10.11 (a) Diagram showing the arrangement of the rotor, sample cell, and optical detection system, used in ultracentrifugation. **(b)** Illustration of the Meselson-Stahl experiment. **(1)** parental DNA containing 100 per cent ^{15}N; **(2)** first-generation progeny DNA containing 50 per cent ^{15}N; **(3)** second-generation progeny DNA containing 25 per cent ^{15}N DNA sedimenting as two bands: one of 50 per cent ^{15}N and one of 100 per cent ^{14}N DNA.

with 50 per cent ^{15}N and the second with only ^{14}N. Subsequent generations showed a diminishing intensity of the band containing 50 per cent ^{15}N. The conclusion is clear: the parental duplex containing ^{15}N is not scrambled among the progeny but its two strands remain intact, to be complemented by ^{14}N progeny strands in all subsequent generations (Fig. 10.11b).

In isopycnic ultracentrifugation, a caesium chloride density gradient is established by CsCl diffusion under the influence of the centrifugal force to give a range of densities which encompass the buoyant densities of the molecular species to be separated. A sample of DNA or any other macromolecule will migrate when centrifuged in this gradient until it reaches the point at which its own buoyant density equals that of the CsCl gradient. The method is thus sometimes known as **equilibrium ultracentrifugation.**

A number of factors affect the observed buoyant density of DNA. The (G + C) mole fraction, degree of hydration, the presence of other ions, especially silver and mercury(II), pH, or the presence of DNA binding or intercalating agents (Section 8.4) all significantly alter the buoyant density.

In those cases where caesium chloride is not soluble enough to generate a sufficiently high density or is unsuitable for some other reason, caesium sulfate can be used, as in the generation of gradients suitable for the separation of RNA.

The fact that intercalating dyes alter the buoyant density of DNA (Section 2.3.5) has been exploited to provide a technique for the preparative purification of DNA and also different forms of DNA. The concentration of ethidium bromide bound to DNA is dependent on the superhelical state of the DNA. Closed circular plasmid DNA bands at higher density than either nicked circular or linear DNA. Other cellular contents are removed either as a pellet at the bottom of the tube (RNA) or as a surface layer (protein).

Measurement of molecular weight can be achieved by application of the Svedberg equation which requires a knowledge of the sedimentation coefficient, s, and the diffusion coefficient, D. The term k includes the specific volume of the sample, the density of water, and the Gas Constant.

$$M = k\frac{s}{D}$$

Svedberg equation

An alternative to equilibrium centrifugation is the use of viscosity. This employs sucrose density gradients in a process known as **rate-zonal ultracentrifugation**. This method is much used for RNA species, where separation depends largely on size of the RNA, and it can also be employed in the separation of different classes of ribosome (Section 6.6). As these species all have a higher buoyant density than the sucrose gradient, equilibrium is never attained and separation depends on the different rates of migration of molecules through the sucrose density gradient.

$$M = c\left(s^0_{20,w}[\eta]^{1/3}\right)^{3/2}$$

Mandelkern–Flory equation

The analytical relationship used is based on the Mandelkern–Flory equation and has the form, where M is the molecular mass, $s^0_{20,w}$ is the sedimentation coefficient in Svedberg units, η is the viscosity, and c is a constant which includes values for the viscosity and density of water, the partial specific volume of the nucleic acid, and Avogadro's number. Solutions of high molecular weight DNA are non-Newtonian fluids whose viscosity is dependent on the shear stress and the rate of change of shear stress. The measurement of the required parameters for the computation of viscosity is relatively straightforward if the apparatus used has been adapted to provide data under conditions of low shear stress. Thus the estimation of molecular weight from either s or η alone is possible by empirical comparisons in cases where the species is of known overall structure.

10.3.2 Light scattering

When light is scattered from a solution of macromolecules it shows a variation in scattering intensity that is related to fluctuations in local polarizability. This fluctuation in polarizability is caused by Brownian motion and is therefore related to the viscosity, η, of the solution and the diffusion coefficient, D, of the macromolecule. The diffusion coefficient is directly related to the hydrodynamic radius, R_h, as shown in equation (10.3.1) where k_B is the Boltzmann constant and T is the temperature.

$$R_h = \frac{k_B T}{6\pi\eta D} \tag{10.3.1}$$

In solutions of large DNA molecules one can observe the translational diffusion coefficient and also the rotational coefficient of the species. Also, because large DNA molecules show a certain degree of flexibility, it is possible to determine the motions of small internal segments of these species.

The measurement of **dynamic light scattering**, DLS, can be accomplished by use of equipment shown schematically in Fig. 10.12. The photomultiplier and optics assembly is mounted on a goniometer to enable rotation to be centred on the sample cell. The parameter measured is the autocorrelation function, $G_2(\tau)$, of the scattered intensity. This autocorrelation function is related to the diffusion coefficient and the scattering vector, K, as shown in equation (10.3.2). K is described by equation (10.3.3), where θ is the scattering angle and λ is the wavelength of the light source.

$$G_2(\tau) = 1 + e^{-2K^2 D\tau} \tag{10.3.2}$$

$$K = \left(\frac{4\pi}{\lambda}\right)\sin\left(\frac{\theta}{2}\right) \tag{10.3.3}$$

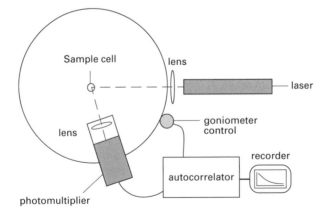

Fig. 10.12 Arrangement of equipment used for the measurement of dynamic light scattering.

10.3.3 Electrophoresis

In free solution, DNA moves under the influence of an electric field in a way that is independent of its shape and molecular mass but dependent on charge. When this movement takes place in a gel, then the speed of migration becomes dependent on both size and shape as well as charge.

The gels commonly used for nucleic acid electrophoresis are made of agarose and polyacrylamide. Both types of gel consist of three-dimensional networks of cross-linked polymer strands which contain pores whose size varies according to the concentration of the polymer. The mobility of DNA in such gels is dependent on its size and shape, since the charge per unit length of DNA is effectively constant.

For linear DNA, there is an inverse relationship between size and rate of migration while the mobility is approximately inversely proportional to $\log_{10}$ of the molecular mass. Horizontal **agarose gel electrophoresis** is useful for separation of linear DNA molecules up to 2000 kbp. Usually, the position of DNA on the plate is revealed by the use of ethidium bromide. Polyacrylamide gel electrophoresis, PAGE, is used for smaller DNA species: it can be employed preparatively for species up to 1 kbp and it forms the basis of rapid nucleic acid sequencing methods (Section 10.6), when 8 M urea is added to denature the oligonucleotides.

The theoretical basis for size selection in gel electrophoresis is not by any means exact. The simplest explanation proposes that the mobility is proportional to the volume fraction of pores that can be entered. This would predict that mobility would *decrease* with increasing gel concentration and also with *increasing molecular mass*. This agrees with the observation that very small DNA fragments move independently of molecular mass. A more complex theory assumes **reptation**, an end-to-end movement of the DNA through the pores of the gel. This theory explains the observed mobility of DNA more accurately, especially at high voltages where the mobility becomes independent of molecular mass. In practice, the analysis of observed mobility in gel electrophoresis is made by comparison with and extrapolation from known standard samples.

Topological conformation can have a marked effect on the rate of migration and the response of DNA to gel concentration and voltage. Relaxed circular DNA and supercoiled DNA migrate anomalously in agarose gel electrophoresis relative to linear DNA of the same molecular weight. The mobility of supercoiled DNA is also dependent on the linking number: as the linking number increases, so does the mobility of the supercoiled DNA (Section 2.3.5). In addition, its mobility changes with the concentration of ethidium bromide because increasing intercalation of ethidium changes the number of superhelical turns (Section 8.4).

The shape of linear DNA also has a strong effect on its electrophoretic mobility. Bends in the DNA helix have the effect of slowing the migration relative to non-bent DNA (Section 2.3.3). The extent of retardation depends on the end-to-end distance of the DNA and so is dependent on the position of the bend within a piece of DNA (Fig. 10.13). A qualitative theory to account for the mobility of bent DNA was derived by Lumpkin and Zim, which supports the general rule that mobility is related to the mean square end-to-end distance. In addition to the above examples, many elegant experiments in electrophoresis have been performed to examine the properties of polyA tracts, extra-base bulges (Section 2.4.4), protein binding, and Holliday junctions (Section 5.4.2).

The fact that DNA of very high molecular mass does not separate well under normal electrophoretic conditions has led to the development of **pulsed field electrophoresis.** There are several variations of technique in this general area. The use of a voltage pulse with a resting period between pulses changes the mobility of large DNA fragments, and can be used to separate DNA species of $\geq$1000 kbp. The use of an alternating polarity field, with the positive pulse being longer than the negative pulse, enables the efficient resolution of very large DNA fragments, while the application of inhomogeneous, perpendicular electric fields has been used to separate whole yeast chromosomes. In all of these methods, the size and duration of pulses can tuned to obtain the specific resolution required.

10.3.4 Microcalorimetry

There are several types of microcalorimetry which operate on quite different physical principles and provide significantly different types of information. Two of these have provided most of the applications of microcalorimetry to the study of nucleic acids. They will be described briefly in turn and then examples of their application to the investigation of nucleic acid behaviour will be illustrated.

The objective of **differential scanning calorimetry,** DSC, is to measure the heat capacity of a system as a continuous function of temperature. Two matched cells, one for reaction and one for reference, are filled completely (about 1 mL sample volume) and their temperature is then raised from below zero to about 90°C by electrical heaters in an adiabatic enclosure. When a thermally-induced endothermic transition occurs in the sample solution, the temperature of this cell lags behind that of the solvent reference cell. This temperature difference is sensed by a thermopile and is used to cause a boost to the electrical energy supplied to the reaction cell to maintain it at the same temperature as the reference cell. The extra energy input to the reaction cell is recorded as a function of temperature—hence the term 'differential scanning'. Analysis of this differential energy provides the excess heat capacity, C_p^{ex}, of the DNA solution relative to the reference solution. One single DSC profile can provide a wealth of thermody-

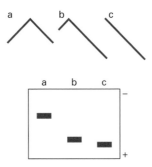

Fig. 10.13 The exact location of a bend (a, b, or c) will alter the electrophoretic mobility of that segment of DNA.

namic information, which includes ΔG^0, ΔH^0, ΔS^0, and ΔC_p for the transition taking place in the sample.

The DSC method has been used to characterise the helix–coil transition for oligomeric and polymeric DNA species leading to a number of practical applications. These include:

- predicting the stability of probe–gene complexes
- selecting optimal conditions for hybridization experiments
- choosing the minimum length of a probe required for hybridization
- predicting the influence of a specific transversion or transition on the stability of an affected DNA region.

In addition, DSC has been used to characterize the B–Z transition (Section 2.3.4), the 'un-bending' of bent DNA, and the melting of a DNA duplex containing an abasic site. Recent studies have investigated the stability of triplexes as a means of improving the rational design of third-strand oligonucleotides (Sections 2.3.6 and 2.4.5).

In **batch isothermal mixing calorimetry**, two solutions are mixed and the resulting heat change is measured. Many of the shortcomings of conventional instrumentation have been overcome by the recent development of **stopped-flow microcalorimetry**, which can be nearly two orders of magnitude more sensitive, with significant reductions in volumes of reactants (160 μL), reduced concentration of nucleic acid species (20 μM), and total run time (200 s). Moreover, such equipment can support many more experiments a day!

The study of small ligand DNA interactions has particularly benefited from this technique. For example, the interaction of the anti-cancer drug, daunomycin (Section 8.4.2) with DNA results in aggregation of the drug under the conditions needed for batch calorimetry. However, by changing to stopped-flow equipment, Breslauer has been able to study the enthalpy of interactive binding of daunomycin to a variety of synthetic and natural DNA partners, showing a dependence on base composition and base sequence.

Because the information provided by these two microcalorimetric techniques is complementary, they can effectively be used in combination to provide thermodynamic characterization of the forces that control DNA structures. For example, DSC can be used to analyse the *disruption* of a DNA duplex, while isothermal mixing calorimetry can be used to study the *formation* of the same duplex from its component strands. It has emerged from such analysis that the order in single-stranded DNA can contribute significantly to the thermodynamics of duplex formation, and this information can be used to guide selective DNA hybridization experiments and the analysis of data on the formation of higher-order DNA structures (such as triplexes and tetraplexes).

Summary

Hydrodynamic methods were among the first techniques to be used in the investigation of structures of nucleic acids. Equilibrium ultracentrifugation of nucleic acids is carried out isopycnically in density gradients of caesium chloride. Rate-zonal ultracentrifugation relies on the viscosity of migration of nucleic acids through a sucrose gradient. The separation and purification of nucleic acids by electrophoresis is a widely used application of hydrodynamic principles. Polyacrylamide gels are used for identification and purification of smaller species and in DNA sequencing while agarose gels are employed for molecules up to thousands of kbp in size. The detailed understanding of the fundamental mechanisms of gel electrophoresis is still incomplete.

The *dissociation* of complexes involving nucleic acids can be analysed thermodynamically by differential scanning microcalorimetry, while their *association* can be studied by isothermal mixing calorimetry.

10.4 Microscopy

10.4.1 Electron microscopy

Electron microscopy (EM) is an invaluable tool for gaining structural information on systems that are too large for the atomic resolution methods of X-ray diffraction or nuclear magnetic resonance and yet too small for optical microscopy. EM uses beams of electrons with wavelengths of between 0.001 and 0.01 nm. The resolving power of EM can be below 2 nm but is largely restricted by factors such as radiation damage and the process of sample preparation. The information obtainable from EM is determined largely by the methods of sample preparation.

The lenses in an electron microscope consist of shaped electric or magnetic fields. Electrons from a heated tungsten filament are accelerated across a voltage difference of up to 100 kV and the sample under study is examined in a vacuum. In **transmission EM** the electron beam is passed through the sample then through a suitable lens system and is detected on a fluorescent plate or photographic film. In **scanning EM**, the electron beam is focused down to a point and scanned across the specimen. Secondary electrons are produced when electrons interact with the target specimen and these secondary electrons are detected and used to build up a raster image of the sample.

Both transmission and scanning EM require samples to be dehydrated and often coated by deposition of a film of either carbon or a metal such as tungsten. One of the oldest means of sample preparation involves combining the nucleic acid with a thin film of denatured protein floating on water. The nucleic acid and protein film can be lifted off the surface onto a grid and then, after drying, coated with a fine layer of palladium or uranium. The process of coating, or shadow casting, involves depositing the metal from a heated filament that is oriented at an acute angle to the supporting grid (Fig. 10.14a). The effect of the metal deposition is to highlight regions that are raised above the grid. In order to achieve good contrast, either the samples can be shadowed from several angles or the sample can be rotated during the coating procedure. The protein film has two benefits: one is that it supports and protects the DNA during preparation; the other is it causes thickening of the DNA strands. Nucleic acids will also bind directly to carbon supports that have been physically and chemically pre-treated in a suitable fashion. They can then be dried and shadowed in the same manner as in film deposition. An example of the use of this technique by the author is shown (Fig. 10.14b).

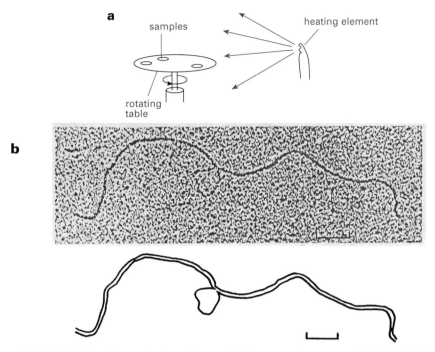

Fig. 10.14 (**a**) Dry, grid-mounted, samples under high vacuum are exposed while rotating to evaporated metal from a heating element. (**b**) Transmission electron micrograph of a heteroduplex between two cloned extrachromosomal circular *copia* transposon DNAs that have been linearized by HindIII restriction digestion. The loop is the result of an 800 base sequence missing from one strand and present in the complement. The bar indicates 0.1 μm. (Reprinted from Flavell, A. J. and Ish-Horowicz, D. (1981), *Nature*, **292**, 591–5, with permission.)

The technique of **cryo EM** uses samples that have been flash frozen into a thin film of vitreous water. The sample is prepared as an aqueous solution and applied to a carbon grid with holes of about 3 μm. Blotting of the grid followed by quick freezing in ethane at near its freezing temperature produced vitrified samples which can be directly imaged by EM.

While the contrast in cryo EM is relatively poor, it has the great advantage that a biological specimen is preserved in its solution conformation without many of the possible artefacts that can result from the dehydration and coating processes. Thus, the sample appears to be preserved in its native solution conformation. Two images are recorded on the same sample, which is tilted between recordings, and can be used to reconstruct the full three- dimensional shape of the sample.

10.4.2 Scanning tunnelling microscopy

Electron tunnelling was proposed in 1928 to explain the possibility of electrons passing through a potential barrier when the distance of travel is small. The **scanning tunnelling microscope** (STM), was invented in the early 1980s to exploit that property.

The basic technique uses an atomically sharp probe that scans across the sample under study. The height of the probe from the sample is maintained by the action of a piezoelectric crystal which, in the reverse of the normally observed phenomenon, bends in response to an applied voltage. A constant flow of electrons is maintained between the probe tip and the sample in order to keep the probe at a constant potential distance from the sample (Fig. 10.15). By scanning the sample with the probe, a raster image of the surface of the sample can be obtained, which is ostensibly topological. The image produced is in reality not strictly distance-based but results from a combination of the distance and the electronic nature of the substrate under study.

One of the main requirements for STM is an atomically flat substrate that extends across several hundred square nanometres. A wide variety of substrates has been used, including heat-treated gold foil, gold films deposited epitaxially on mica, platinum–iridium–carbon films on mica, and highly oriented pyrolytic graphite.

The simplest way of mounting and examining a sample is to spread a solution of the species under study onto the substrate and to evaporate the solvent. A wide variety of different deposition techniques has been described: they include spraying on to the substrate and forcing droplets of solution from the probe tip by application of a negative potential pulse. Scanning TM can then be carried out either in air or in vacuum.

The process of building up an image by STM is slow because of to the need to equilibrate the detection system and because of the scanning speeds used. Choice of scan speed and the mode of scanning are dictated by the sample. Samples that are not bound tightly to the substrate can be moved as the probe is swept over them. The need to sweep the sample repeatedly can lead to blurring of the image. Probe speeds of around 250 nm s^{-1} seem to be generally required to avoid disturbing the sample.

Although DNA is thought of as being an insulator, what is seen in STM is the change in work function as the probe tip moves over the surface of the molecule. There is enough variation in DNA to provide sufficient contrast for imaging at near atomic resolution and it is believed that it is the sugar-phosphate backbone, together with bound water and ions, that gives the strongest signal. The interpretation of STM images of DNA is somewhat open to question.

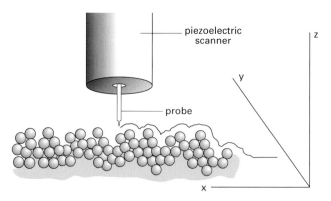

Fig. 10.15 In scanning tunnelling microscopy, an atomically sharp probe is swept across the sample following the contours of the surface-mounted sample.

The results are often susceptible to different interpretations and until some of the apparent major inconsistencies in the method are resolved it seems unlikely to contribute significantly to a new understanding DNA structure.

Summary

Electron microscopy techniques enable nucleic acid structures to be studied on a scale that lies between light microscopy and the atomic resolution methods such as NMR and X-ray diffraction but is inaccessible to them. The images produced from either the more traditional surface mounted samples or those in vitrified ice confirm model structures developed from less direct methods.

A fundamental problem in all techniques of microscopy is the discernment of the real from the artefactual. Cryo electron microscopy is an attempt to visualize samples as a snapshot of their native solution condition but is limited by the problem of low resolution with vitrified samples. Scanning tunnelling microscopy is a new addition to the methods of direct observation of macromolecular species but is still in the developmental stage and is of uncertain promise for the future.

10.5 Molecular modelling

10.5.1 Mechanics and dynamics

Molecular modelling in its simplest and least assuming form is the three-dimensional visualization of structure based on indirect structural information. As an aid to comprehension and stimulation to thought, modelled structures can be invaluable. In spite of an increasing array of sophisticated methods for the elucidation of macromolecular structure, there will always be structures which cannot be observed directly. An unstable or transient structure is often of more importance than a stable long-lasting one. Much of the modelling carried out is done as a way of testing the understanding of the underlying physical processes. By constructing a predictive model and then testing it against direct observation, the hope is to increase the accuracy of the model building procedure.

The first three-dimensional structure of DNA constructed by Crick and Watson was largely a model that fitted a collection of disparate and in some cases very indirect data. The success of the double helix model came not from its accuracy but because it supported the development of many different ideas and observations to form a holistic view of DNA structure.

Early DNA modelling was done using wired-up physical shapes to represent the bonding between atomic components of the helices. At a later stage, space-filling model kits were developed, with Corey–Pauling–Koltun models being widely employed to make DNA helices. Computers are now universally used both to construct and to analyse nucleic acid models.

A prerequisite to modelling of a molecule is the definition of a force field. In a solid model, the force field has an obvious physical basis but is often limited in its accuracy and inflexibility. Computer models are able to use quite complex descriptions of the forces acting on the molecule. The empirical force field is defined as the sum over all atom pairs of all of the energy terms. The energy terms defining bond angle, dihedral-torsion, improper-torsion, hydrogen bond, van der Waals, and electrostatic interactions are generally defined (Fig. 10.16). The actual energy potentials used in an individual modelling exercise may include only a subset of these energy terms or include other special terms such as distance restraint, dihedral restraint, symmetry restraint, or more esoteric terms describing hydrophobicity, *etc.*

Given a suitable force field, there are several ways in which it can be used. Molecular mechanics consists largely of search mechanisms for pathways to conformations with lower overall energy as defined by the force field. A large variety of different, sometimes elaborate, algorithms have been developed. All molecular mechanics suffers from the problem of multiple minima, in so far as the number of minima that exist with energies above the global minimum is so large as to be virtually unsearchable in anything but extremely small molecules. Molecular mechanics can be used in conjunction with the force field to find local minima for structures that have been created by other means, as for instance the energy minimization of models created using standard helical parameters. This type of use will allow the creation of energetically reasonable structures and permit the direct comparison of different models.

The Metropolis, or Monte Carlo, method is one approach to searching conformational space in such a way as to be able to overcome energy barriers. This algorithm simply moves the starting structure in a random way and then compares the energy of the new state of the structure with the energy of the previous state. If the new energy is lower than the last move, it is accepted and the algorithm goes on to another cycle. If the energy of the new state is higher than the previous state then the last move is accepted only with a certain probability.

Newtonian molecular dynamics was developed as a way of simulating molecular properties. The structures are treated as collections of atoms with mass and velocity. Integration of Newton's equations of motion over time is employed in order to decide the size and direction of the next step. The time step for the integration must be extremely small in order to avoid instabilities arising from fast molecular vibration. The process of calculating even a few picoseconds of dynamics is extremely expensive computationally. Newtonian dynamics has been adapted by molecular modellers for use as a mechanism for searching conformational space

$$E = \sum k(r - r_0)^2 \qquad \text{bond stretch}$$

$$E = \sum k(\theta - \theta_0)^2 \qquad \text{bond angle}$$

$$E = \sum k(1 + \cos(n\phi - \delta)) \qquad \text{torsion angle}$$

$$E = \sum \varepsilon_{ij} \left[\left(\frac{r'_{ij}}{r_{ij}} \right)^{12} - 2 \left(\frac{r'_{ij}}{r_{ij}} \right)^6 \right] \qquad \text{van der Waals}$$

$$E = \sum \frac{Q_i Q_j}{\varepsilon_{ij} \cdot r_{ij}} \qquad \text{electrostatic}$$

$$E = \sum \left(\frac{A}{r^i_{AD}} - \frac{B}{r^j_{AD}} \right) \cos^m(\theta_{A-H-D}) \qquad \text{hydrogen bond}$$

Fig. 10.16 The commonly used force field equations used for describing the potential energy of a molecular system.

to find structures compatible with some external experimental data. The use of 'unnatural' force constants and potentials enables a considerable increase in searching speed. The present application of modelling techniques to the refinement of structures that fit very detailed experimental information, such as crystallography or nuclear magnetic resonance data, is producing extremely good results.

The application of molecular dynamics and mechanics to problems with much more limited information is harder to assess. The criteria used for judging success will be different for different circumstances. A recent modelling study of the carcinogenic (+) *anti*-benzo[*a*]pyrene-diol-epoxide (BPDE) and its non-carcinogenic enantiomer, (−) BPDE, has been tested by NMR structure determination. Benzo[*a*]pyrene is a chemical present in car exhaust gas which, following metabolic activation in the liver, can bind covalently to the amino group of guanine in DNA *in vivo* (Section 7.6.3). Molecular mechanics was used produce wide conformational searches for the two adducts formed from the (+) and (−) enantiomeric dihydrodiolepoxides (Fig. 10.17). The conclusion from the modelling exercise was that the two mirror image isomers would be oriented in opposite directions along the minor groove of B-DNA. The different orientations of the BPDE adduct with DNA then result in a differential response by the enzymes responsible for DNA replication and repair (Section 7.11). The modelling results agreed with the structures of the two isomeric adducts derived from NMR analysis.

As modelling techniques improve it can be expected that quantifiably accurate predictions will become more common.

Summary

Models have been used as visual and conceptual aids from the earliest investigations of molecular biology. The development and availability of powerful computers and of suitable algorithms has enabled the building of ever more accurate molecular models. The simulation of molecular dynamics is one of the most important developments in modelling. The testing of model-derived predictions by experimental structure determination is increasingly proving the techniques. Many of the techniques originally developed for modelling are being used in structure refinement from nuclear magnetic resonance data and X-ray crystallography. The inclusion of experimental data into a model begins to blur the boundary between modelling and structure refinement.

10.6 DNA sequence determination

The methods for determination of the sequence of DNA are now so rapid that they have super-seded protein sequencing as the major method of determining polypeptide sequences. Rather than attempting the direct sequence analysis of a protein, it is usually simpler to clone the corresponding gene and determine its DNA sequence and hence the amino acid sequence of the protein.

There are two major ways of determining the sequence of a DNA molecule. These methods were developed in the laboratories of Gilbert and of Sanger for which each received a Nobel

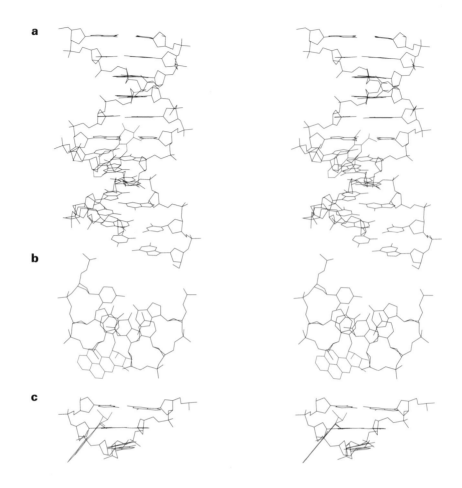

Fig. 10.17 Stereo pair diagram (for parallel viewing) showing three views of the structure of the (+) *anti-* benzo[*a*]pyrene-diol-epoxide DNA adduct predicted by modelling. The adduct formed with the guanine-N^2 in a B-type DNA dodecamer lies in the minor groove, directed towards the adducted guanine: (**a**) view normal to the helix axis, (**b**) view down the helix axis, and (**c**) view orthogonal to (**a**).

prize. Both are dependent upon the partial fragmentation of DNA at one of the four bases followed by an accurate determination of the size of each fragment. Both methods rely upon sequencing only one strand at a time. For example, let us consider how we can determine the sequence of a 20 nucleotide fragment. If we could somehow create one specific cleavage in the top strand at any of the phosphodiester bonds following a C-residue, we would generate a mixture of shorter sequences (Fig. 10.18).

Maxam and Gilbert sequencing relies upon radioactive labelling of only one end of the DNA. If in the above example the 5′-end is labelled, then only the left-hand sequence of each of the pairs of fragments would be radioactive (Fig. 10.19). These three oligonucleotides can be separated from one another by virtue of their different sizes. In practice, this is achieved by

```
5'  AGAATTCTACAGTAAATGCT  3'
3'  TCTTAAGATGTCATTTACGA  5'


5'  AGAATTC + TACAGTAAATGCT   3'

5'  AGAATTCTAC + AGTAAATGCT   3'

5'  AGAATTCTACAGTAAATGC + T   3'
```

Fig. 10.18 Cleavage 3′ to C-residues.

```
5'  AGAATTC  3'

5'  AGAATTCTAC  3'

5'  AGAATTCTACAGTAAATGC  3'
```

Fig. 10.19 DNA fragments radioactively labelled at their 5′-ends. (Radiolabelled residues in colour.)

polyacrylamide gel electrophoresis (Section 10.3.3). Here, the mixture of oligonucleotides generated by C-cleavage is subjected to an electric field in a polymeric gel matrix such as polyacrylamide. Oligonucleotides migrate towards the anode at a rate which is inversely proportional to their size and thus three radioactive bands are generated in the gel (Fig. 10.20). If the experiment is repeated, but this time with cleavage only after G bases, another set of bands is generated (Fig. 10.21). Parallel sets of such reactions can be carried out which are specific for each of the four bases and electrophoresis of these reaction products generates a sequencing **ladder** from which the complete sequence can be read upwards from the bottom of the gel (Fig. 10.22), corresponding to reading $5' \rightarrow 3'$ on the DNA.

In practice, the types of partial cleavage most commonly used are shown in Table 10.1. Each is a two-stage process. First, a small fraction of the bases is chemically modified (or removed in the case of G- and A-specific reactions), then the cleavage 3′- to these bases is achieved by alkaline hydrolysis (Sections 7.1, 7.4, and 7.5.3).

The **Sanger DNA sequencing** method shares the features of base-specific discontinuities that are resolved by gel electrophoresis, but employs an alternative procedure to generate the ladder. A new strand of DNA is synthesized enzymatically (usually using either the Kle-

Table 10.1. Base-selective cleavages for sequencing DNA.

3′- Cleavage adjacent to	Modification	Reagent	Strand breakage
G	Methylation	Dimethyl sulfate	1 M piperidine 90°C 30 min
G and A	Depurination	88% Formic acid	1 M piperidine 90°C 30 min
T and C	Base ring-opening	Hydrazine	1 M piperidine 90°C 30 min
C	Base ring-opening	Hydrazine, high salt	1 M piperidine 90°C 30 min

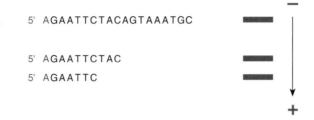

Fig.10.20 Electrophoretic separation of end-labelled fragments. (Radiolabelled residues in colour.)

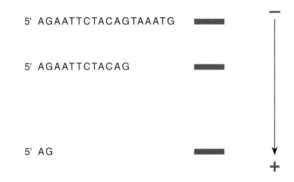

Fig.10.21 Separation of DNA fragments with 3′-terminal G-residues.

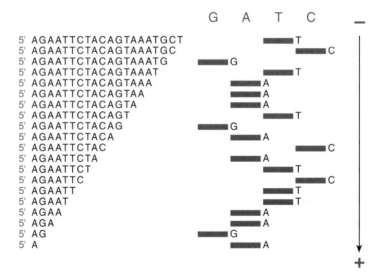

Fig.10.22 Generation of a DNA sequence ladder.

now fragment of DNA polymerase I (Section 9.2.4) or the DNA polymerase from bacterioph-age T7) with the DNA to be sequenced acting as a template. Radioactive label is incorporated into the growing strand. In the example already considered, a new DNA strand is synthesized

5' A G A A T T C T A C A G T A A A T G C T 3'   Newly synthesized strand

3' T C T T A A G A T G T C A T T T A C G A 5'

Fig. 10.23 Dideoxy DNA sequencing involving the synthesis of a new DNA strand.

on the single-stranded complementary template (Fig. 10.23). Termination of this reaction 3'- to dC residues generates the same three fragments as shown in Fig. 5.3 and termination following each of the four bases in turn gives the same pattern as seen in Fig. 10.22.

How is the polymerization terminated at a specific point? 2',3'-Dideoxynucleoside 5'-triphosphates (Fig. 10.24 and Section 7.2) can be incorporated into a growing DNA strand, but, since they possess no 3'-hydroxyl group, they are unable to accept the addition of any extra bases. They are thus **chain terminators**. The addition of a small amount of one of these, together with all four of the normal 2'-deoxynucleoside 5'-triphosphates (one of which is radio-labelled), to a polymerization reaction gives rise to a series of oligonucleotides, each terminated by a dideoxynucleotide. By carrying out four reactions, each with a different dideoxynucleoside triphosphate, and electrophoresing all four in parallel, a sequencing ladder is generated.

In practice, it is necessary to elongate a short primer which has already been annealed to the template, since DNA polymerases can only elongate existing hybrids (Fig. 10.25). For this purpose one usually subclones the DNA fragment to be sequenced into a vector (Section 10.7.1) with a known sequence flanking the insertion site. Chemically synthesized oligonucleotides (typically 17–25 nucleotides in length) which correspond to one or the other side of the insert are annealed to the subclone of DNA and the dideoxy-sequencing reactions are carried out on these templates. The polymerization reaction can proceed on double-stranded templates, with one strand being displaced by the elongated primer. More usually, single-

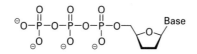

Fig. 10.24 A 2',3'-dideoxynucleoside 5'-triphosphate.

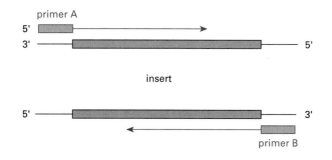

Fig. 10.25 Use of oligonucleotide primers in dideoxy-sequence analysis.

stranded templates are used, such as the viral DNA from bacteriophage M13-derived recombinants (Section 5.2.8).

Sanger dideoxy-sequencing has several powerful advantages over the Maxam–Gilbert method. First, a set of reactions can be carried out in only an hour or two as opposed to approximately a day for the Maxam–Gilbert reactions. Secondly, the reactions are much cleaner, with few contaminants that might lead to poor quality gel electrophoretograms. This means that 300 nucleotides can be sequenced routinely by this approach for each set of reactions, as opposed to approximately half this number with the Maxam–Gilbert method. Maxam–Gilbert sequencing does have the advantage that one need only purify a restriction fragment in order to sequence it whereas dideoxy-sequencing requires a sub-cloning step into a vector containing the appropriate oligonucleotide primer site. There are also several artefacts generated during both methods that render unambiguous determination difficult. For this reason it is essential to determine the sequences of both strands of a duplex and it is often advantageous to adopt the alternative method when an artefact renders difficult the interpretation of a particularly tricky sequence.

Machines have now been developed to separate and identify the products of dideoxy-sequencing reactions. Here, instead of radiolabelling, a fluorescent molecule is attached to the 5'-end of the primer or is built into an alternative chain terminator (Section 7.5.4). Fluorescently labelled oligonucleotides travelling through a polyacrylamide gel are detected by a laser. Such machines are proving valuable in attempts at large-scale sequencing of genomes.

The long-term aim of such projects is the complete sequence determination of entire genomes. The major candidates include the yeast *Saccharomyces cerevisiae* (several chromosomes, of which have already been sequenced in their entirety), the fruitfly *Drosophila melanogaster*, and *Homo sapiens*. Such huge projects demand automated sequence determination, powerful computer support for data analysis, and multiple inter-laboratory collaborations.

Summary

There are two major ways of determining the sequence of a DNA molecule. Both rely on the exact measurement of the size of fragments with one end fixed and the other being at one of the four nucleoside bases. Maxam–Gilbert sequence analysis uses base-selective partial chemical cleavage. Sanger dideoxy-sequence analysis relies upon the use of 2',3'-dideoxynucleoside 5'-triphosphates as base-specific chain terminators of the synthesis of a complementary DNA copy by a DNA polymerase.

Automated DNA sequence determination now renders the dream of sequencing entire genomes a feasible goal.

10.7 Gene cloning

It is not the object of the chapter to teach cloning and recombinant DNA methodology, which are covered in depth in other reference works (see Further reading; page 498). Instead, we shall survey the sorts of manipulation that are possible using cloning and recombinant DNA techniques.

Cloning is the technique of growing large quantities of genetically identical cells or organisms which are derived from a single ancestor (clones). Gene cloning is an extension of this whereby a particular gene of interest (or group of genes) is amplified to a huge extent. The 'classical' way of achieving this is by the insertion of the desired DNA segment into a carrier or **vector** DNA and the introduction of the hybrid or **recombinant DNA** into cells by transformation. The cells containing the recombinant DNA are propagated and each cell in a colony contains an exact copy (or copies) of the gene 'cloned' in the vector.

Nowadays, a separate, complementary approach using the **polymerase chain reaction** (PCR) can achieve the same objective of amplifying specifically particular DNA segments without involving any living cells.

Gene cloning is carried out by many people in different areas of chemistry and biology. A chemist may be interested in structural studies of defined DNA sequences which are too large to be synthesized economically by chemical means. For biochemists, who spend much time trying to isolate tiny amounts of a protein from an inaccessible or expensive source, cloning can allow production of far greater amounts of a protein with greater convenience. A molecular biologist is interested in the sequence of a gene, its transcription, and the factors controlling its expression, all of which are extremely arduous without cloning. A molecular geneticist may be interested in a gene which causes an important effect, such as cancer or a developmental or neurological fault, and cloning enables the elucidation of the information that gene encodes. Lastly, a population geneticist is interested in the sequence and evolution of DNA itself and the cloning of equivalent genes from different organisms allows a comparison of their sequences.

10.7.1 Vectors

Several classes of vector exist into which foreign DNA can be inserted and amplified. The three major classes are **plasmids**, **bacteriophage**, and **cosmids**. Prokaryotic plasmids are almost always circular double-stranded DNAs containing antibiotic resistance genes as markers and a variety of restriction sites which can be used for insertion of the foreign DNA. A large number of plasmids are available for use in *E. coli*. The most useful bacteriophage is λ. Bacteriophage λ has been engineered in many different ways to accept inserts of many different sizes and types. Because the transformation frequency is high, screening is easy, and it can accept large inserts, bacteriophage λ is often the vector of choice for an initial cloning step, especially for cloning eukaryotic genes. Cosmids are large plasmids which contain the packaging site for bacteriophage λ DNA. They can therefore either be packaged into phage particles or they can be replicated as plasmids. Since the amount of DNA which can be packaged into a λ phage particle is 50 000 bp, the potential size of cosmid inserts is very large. Cosmids are often used in **chromosome walking** (see Section 10.7.4), but they are somewhat more difficult to manipulate than is λ. Yeast artificial chromosomes (**YACs**) are vectors containing the constituents of natural yeast chromosomes, namely a **centromere** (the region required for correct segregation of daughter chromosomes during mitosis and meiosis) and

Table 10.2. The choice of vector for gene cloning

	Plasmid	λ Phage	Cosmid	YAC
Ease of screening	++	++++	++	+
Transformation efficiency (recombinants per microgram)	10^7	10^8	10^8	500
Insert size (nucleotide-pairs)	≤15 000	≤24 000	≤45 000	≤100 000
Ease of making DNA	+++	+	+	+/−

two **telomeres** (the ends of the chromosome). YACs can accept inserts of many hundreds of thousands of base-pairs and are nowadays the vectors of choice for large-scale analysis of eukaryotic genomes. They replicate inside yeast cells in exactly the same way as a natural yeast chromosome. This carries the disadvantage that only one YAC molecule can exist per yeast cell.

The choice of vector is dependent on four parameters (Table 10.2). Where the DNA of interest is a very small part of a mixture of DNAs that is being used for cloning, then the ease of screening for the correct clone is often the deciding factor. A second important parameter is transformation efficiency, especially if one needs to generate a large number of recombinants in order to ensure the presence of a correct one. However, if the average insert size is large, then less recombinants are needed in order to obtain a representative recombinant **library**. Finally, and least important, is the ease of making DNA from the recombinant. The use of a vector renders DNA isolation relatively simple and one can always subclone from the original vector into a more amenable one once the initial cloning is achieved.

10.7.2 Enzymes useful in gene manipulation

Gene cloning would not have become possible without the discovery and isolation of a range of enzymes which act on DNA to enable manipulation of particular sections. One important class of enzymes is the DNA polymerases (see Sections 9.2.4 and 10.7.3).

Polynucleotide kinase

A polynucleotide kinase isolated from bacteriophage T4 catalyses the transfer of the γ-phosphate of ATP to the 5′-hydroxyl terminus of DNA, RNA, or an oligonucleotide in a reaction that requires magnesium ions. The enzyme is particularly useful for introducing a radioactive label on to the end of a polynucleotide, where the phosphate donor is γ-[^{32}P]-ATP. Both single- and double-stranded polynucleotides can be phosphorylated although recessed 5′-hydroxyl groups in double-stranded DNA, such as those obtained by cleavage with certain restriction enzymes, are poorly phosphorylated. This sort of polynucleotide kinase activity, though not found in bacteria, has been found in some mammalian cells. The T4 enzyme is the only well-characterized kinase which has polynucleotides as substrates. This T4 protein also has a 3'-phosphatase activity which is unusually specific for a 3'-phosphate of a nucleoside or polynucleotide.

Alkaline phosphatase

Phosphatases catalyse the hydrolysis of phosphomonoesters to produce inorganic phosphate and the corresponding alcohol. Most phosphatases are non-specific. Alkaline phosphatases are found in bacteria, fungi, and higher animals (but not plants) and will remove terminal phosphates from polynucleotides, carbohydrates, and phospholipids. The *E. coli* enzyme is a dimer of molecular weight about 89 kDa, requires a zinc (II) ion, and is allosterically activated by magnesium ions. During dephosphorylation of the substrate, its phosphate is transferred to a serine residue on the enzyme located in the sequence Asp-Ser-Ala. This same sequence is found in mammalian alkaline phosphatases (the calf intestinal enzyme is particularly well characterized) and it is similar to the active centre of serine proteases. Acidic phosphatases are also common, but these do not usually operate on polynucleotides as substrates.

DNA ligase

A ligase is an enzyme that catalyses the formation of a phosphodiester linkage between two polynucleotide chains. In the case of DNA ligases, a 5'-phosphate group is esterified by an adjacent 3'-hydroxyl group and there is concomitant hydrolysis of pyrophosphate in NAD^+ (bacterial enzymes) or ATP (phage and eukaryotic enzymes). Particularly efficient joining takes place when the phosphate and hydroxyl groups are held close together within a double helix, typically where the joining process seals a 'nick' and creates a perfect duplex (Fig. 10.26). This situation occurs both in gene synthesis (Section 3.4.5) and in recombinant DNA technology in ligating identical 'sticky ends' formed by cleavage with a restriction of endonuclease.

E. coli and phage T4 DNA ligases are well- characterized enzymes which have an important role in DNA replication (Section 5.4.3). T4 DNA ligase will join blunt DNA duplex ends when

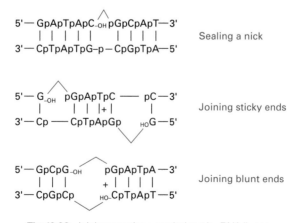

Fig. 10.26 Joining reactions carried out by DNA ligase.

used at high concentrations and it will also catalyse the joining of two oligoribonucleotides in the presence of a complementary splint oligodeoxyribonucleotide (Section 3.5.4).

Restriction endonucleases

Restriction endonucleases recognize and bind to DNA sequences at specific sites and make a double-stranded cleavage (Section 9.3.6). There are three types of restriction and modification systems, termed I, II, and III in order of their discovery. Type I consists of a large enzyme complex containing subunits encoding endonuclease, methylase, and several other activities. The recognition sequence comprises a trinucleotide and a tetranucleotide separated by about six non-specific base-pairs (Table 10.3), but the endonucleolytic cleavage site can be up to 7000 base-pairs distant. Type II systems have independent endonucleases and methylases (Section 9.3.7) that act on the same DNA sequence. These sequences are generally palindromic (i.e. they have a twofold axis of symmetry) and the cleavage sites are usually within or very close to the recognition sites. A range of enzymes with different specificities has been isolated from a wide variety of organisms and type II restriction enzymes are now highly useful tools in recombinant DNA research. The type III system shares features in common with both type I and type II. There are two independent polypeptides, one of which acts independently as a methylase, but both are required for specific endonucleolytic activity. In the case of *Eco*PI, for example, the recognition sequence is an asymmetric pentanucleotide and the cleavage site is 25 bp downstream.

Other nucleases

Almost every organism contains a wide variety of nucleases, of which some are involved in the salvage of nucleotides and some feature as intrinsic activities of proteins used in replication and repair processes. Apart from non-specific nucleases, such as DNase I, and ribonucleases (Section 9.2.3), there are several other nucleases which are used in the manipulation of DNA and RNA (Table 10.4).

10.7.3 Cloning techniques

In virtually all cases, a fragment of DNA to be cloned is inserted as a duplex into a restriction site of the vector by use of the enzyme DNA ligase (Fig. 10.27; see Section 10.7.2). Usually the insert is a restriction fragment with termini compatible with the vector ends (Fig. 10.26). Sometimes oligonucleotide **linkers** need to be joined to the insert. These are self-complementary synthetic duplex oligonucleotides that specify a recognition site for a restriction enzyme. Linkers are ligated to the fragment to be cloned and then treated with the restriction enzyme, thus generating new termini, which are now identical or compatible for joining to the cleaved vector. Often the insert is a cDNA copy of mRNA which has been generated using the enzyme reverse transcriptase. The insert DNA can also be synthesized chemically (see Section 3.4.5).

Another way of molecularly cloning DNA is by means of the **polymerase chain reaction** (PCR). This method involves use of a pair of synthetic oligonucleotide primers flanking the region to be amplified, each complementary to a different DNA strand (see Fig. 10.34, Section

Table 10.3. Some restriction endonucleases and their recognition sequences

Type	Enzyme	Recognition site
Type I	EcoK	A A C (N)$_6$ G T G C T T G (N)$_6$ C A C G
	EcoB	T G A (N)$_8$ T G C T A C T (N)$_8$ A C G A
Type II	EcoRI	G▾A A T T C C T T A A▴G
	SmaI	C C C▾G G G G G G▴C C C
	PstI	C T G C A▾G G▴A C G T C
	Sau3AI	▾G A T C C T A G▴
	NotI	G C▾G G C C G C C G C C G G▴C G
	MnlI	C C T C (N)$_7$▾ G G A G (N)$_7$▴
Type III	EcoPI	A G A C C T C T G G

Cleavage sites indicated ▾ ▴

10.8.1). The target duplex DNA is denatured by heat and annealed to the primers, which are in vast excess to prevent the target DNA strands renaturing with each other. The two primers are next elongated on the separated target DNA template strands, using a heat stable DNA polymerase (for example, Taq DNA polymerase from the thermophilic bacterium *Thermophilus aquaticus*) to give a two-fold amplification. Since the primers are derived from different DNA strands, each newly synthesized strand now contains a binding site for the primer used for copying of the *other* strand. A second round of denaturation by heat, annealing, and primer extension results in a four-fold amplification. After 20 rounds of amplification, 2^{20} copies of the original target DNA are formed. Such a powerful technique can produce as much DNA as can be made by cloning methods.

10.7.4 Assaying for the required clone

The rate-limiting step in cloning is almost always the identification of the correct clone from a huge excess of unrequired molecules. Typically, a vast number of visually identical bacterial

Table 10.4. Some exonucleases and their activities

Nuclease	Origin	Activities
Exonuclease III	*E. coli*	(1) ss *exo*-cleavage from 3′-ends of dsDNA
		(2) *endo*-cleavage for apurinic DNA
		(3) RNase H
		(4) 3′- phosphatase
Exonuclease VII	*E. coli*	ss *exo*-cleavage from 5′- or 3′-end of ssDNA
Bal31	*Alteromonas espejiana*	(1) ss *exo*- and *endo*-cleavage from 5′- or 3′-end of dsDNA
		(2) ssDNA *endo*-cleavage
S1	*Aspergillus oryzae*	ssDNA or RNA *exo*- and *endo*-cleavage
Lambda exonuclease	Infected *E. coli*	SS *exo*-cleavage from 5′-end of dsDNA
Phosphodiesterase I	Bovine spleen	ss *exo*-cleavage from 5′-end of ssDNA or RNA
Phosphodiesterase II	*Crotalus adamanteus* (or other snakes)	ss *exo*-cleavage from 3′-end of ssDNA or RNA

colonies or bacteriophage plaques are generated on the surface of agar in Petri dishes. How do we identify the very few clones of interest against the background of millions of uninteresting ones? Several approaches can be used, but there are some basic rules to be considered. In almost all cases, a copy of the entire set of recombinants is made by touching a nitrocellulose or nylon filter membrane to the agar surface. The 'master' copy agar plate is stored away until the location of the required clone has been established on the filter copy. How is this achieved?

In some cases, nucleic acid **hybridization** (annealing of complementary strands, Section 2.5.1) is useful. This is particularly true if a related sequence, such as that from another species, has already been cloned. Alternatively, one can deduce the DNA sequence from the cor-

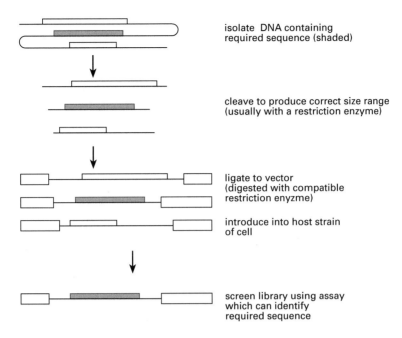

isolate DNA containing
required sequence (shaded)

cleave to produce correct size range
(usually with a restriction enzyme)

ligate to vector
(digested with compatible
restriction enyzme)

introduce into host strain
of cell

screen library using assay
which can identify
required sequence

Fig. 10.27 Basic cloning procedure.

responding protein sequence (if available) and chemically synthesize oligonucleotide **probes** complementary to part of the target DNA for screening of clones. A complication here is that most amino acids are encoded by more than one nucleotide triplet. The result is that many different oligonucleotides need to be made to be sure of using the correct one. The number can be reduced by choosing a region of protein sequence containing the less ambiguous amino acids such as methionine and tryptophan which are specified by a single codon (TGG and ATG respectively). It is also possible to synthesize a mixture of oligonucleotides with two, three, or four bases at the points where ambiguity is present, since the first two bases are often invariant for a particular amino acid (Fig. 10.28; see Fig. 6.24). Lastly, several different regions of a protein can be used to derive a battery of probes all of which can be used to screen the library. In this way artefacts can be discounted.

The probe is labelled (either by radioactivity or by use of a non-radioactive reporter molecule) and a solution of the probe is incubated with the DNA clones on the filter. After careful washing of the filter, only that probe which is exactly complementary to the desired sequence is left attached to the filter and positive clones can be identified by autoradiography or by vi-

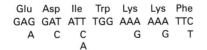

Glu	Asp	Ile	Trp	Lys	Lys	Phe
GAG	GAT	ATT	TGG	AAA	AAA	TTC
A	C	C		G	G	T
		A				

Fig. 10.28 Mixed sequence oligonucleotides in gene cloning. A mixed oligonucleotide incorporating each alternative base can be used to probe for the gene encoding this oligopeptide.

sualization of the reporter molecule. The hybridization conditions used in such experiments are often crucial to a successful outcome.

In other cases, antibody screening of the filter copies can be used to detect the required clone. Of course this can only succeed if an antibody to the polypeptide product of the required gene is available and the vector into which the gene has been cloned contains appropriate transcriptional and translational regulatory sequences for the expression of the cloned gene as protein.

Several shortcuts in molecular cloning can be considered.

Transposon tagging

A previously cloned transposon (see Section 5.4.7) is used to create mutations in the required gene (Fig. 10.29). The transposon can then itself be used as a molecular 'tag' to isolate the gene by hybridization (the transposon and its surrounding DNA must both be isolated by this method). Note that the detection method is based entirely upon the mutant phenotype and therefore no knowledge of the structure or biochemical function of the gene or gene product is needed.

Microdissection

It is possible physically to dissect and clone the required part of the chromosome (provided the chromosomal location of the gene of interest is known). Chromosomes may be separated from one another by pulsed-field gel electrophoresis or by fluorescence-activated sorting.

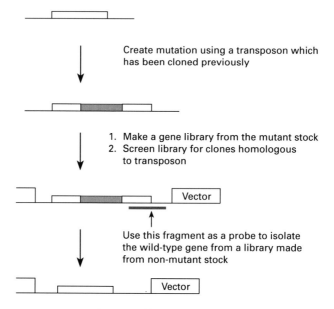

Fig. 10.29 Transposon tagging.

Chromosome walking

If an overlapping series of clones can be isolated, it is possible to use one clone to isolate the next in line and thus 'walk' along the DNA to the required sequence (Fig. 10.30). This is a very time-consuming process, but nevertheless has been used frequently.

Chromosome jumping

This is an extension of chromosome walking that proceeds by larger steps and ignores the large DNA stretches in the middle of each step (hence the word 'jumping').

Summary

Gene cloning is the technique of isolation and amplification of a gene by insertion into a vector DNA followed by the transformation of cells in order to obtain colonies, each containing identical copies of the gene. Plasmids, bacteriophage, cosmids, and yeast artificial chromosomes are the main types of vector. Each has its own advantages and disadvantages depending on the circumstances. The polymerase chain reaction is an entirely different way of cloning nucleic acids which avoids the use of living organisms. Often, assaying for the correct clone is the rate-limiting step in cloning. Transposon tagging, microdissection, chromosome walking, and chromosome jumping are all possible shortcuts in cloning.

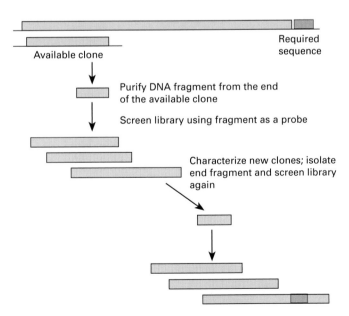

Fig. 10.30 Chromosome walking.

10.8 *In vitro* mutagenesis

The process of engineering specific changes in a DNA sequence in a test tube in order to assay the effect or function of that sequence is termed *in vitro* **mutagenesis**. It is an invaluable procedure since it enables a region of interest in a DNA to be chosen and altered in a premeditated manner. In classical mutagenesis, alterations are created randomly and the effects of each mutation need to be screened in turn.

Before mutagenesis of a sequence, the DNA must first be cloned so that it can easily be manipulated. Three types of alteration can now be made: deletions, insertions, and replacements.

10.8.1 Deletions

Deletions can be created at restriction sites by cleavage with the corresponding enzyme and then by treatment for a short period with the enzyme Bal 31 exonuclease, which removes both double- and single-stranded DNA from both ends (Fig. 10.31). Alternatively, if the restriction enzyme cut leaves overhanging single-stranded ends, these may be trimmed prior to re-ligation by use of a nuclease that is specific for single-strands, such as S1 nuclease (Tables 10.3 and 10.4).

Deletions of regions of DNA that are not near useful restriction sites may be achieved by use of synthetic oligonucleotides. In this procedure, an oligonucleotide which flanks the required deletion, but does not contain it, is used as a primer to create a complete complementary strand of the template DNA (Fig. 10.32). In the process of cloning, mutant DNA segregates from wild type DNA and clones containing mutant DNA (i.e. carrying the deletion) can be selected.

One problem encountered with the above technique is that the bacteria often prefer to repair the mutagenized strand because the *in vivo*-generated DNA strand is methylated. This results in very low efficiencies of isolation of the mutated sequence. There are several ways

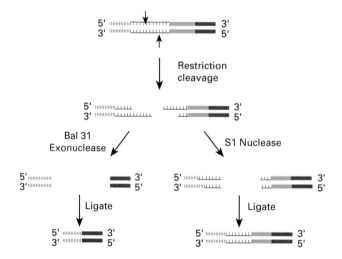

Fig. 10.31 Generation of DNA deletions.

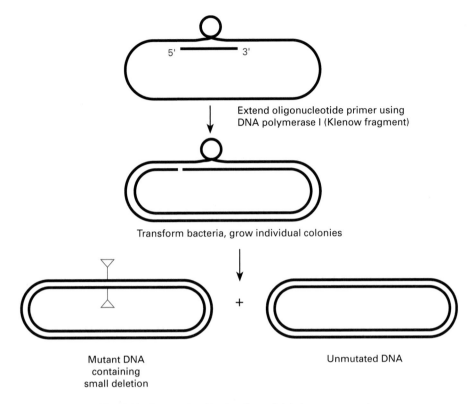

5'   3'

Extend oligonucleotide primer using
DNA polymerase I (Klenow fragment)

Transform bacteria, grow individual colonies

+

Mutant DNA
containing
small deletion

Unmutated DNA

Fig. 10.32 Oligonucleotide site-directed deletion mutagenesis.

around this problem. Eckstein has developed a reliable method which incorporates **phosphor-othiate**-modified nucleotides (Section 3.4.6) into the *in vitro*-generated strand. Such nucleotides are more resistant to nuclease degradation with the result that the above procedure is possible (Fig. 10.33).

PCR can also be used to create DNA deletions. This technique also uses oligonucleotides which flank the required deletion (Fig. 10.34).

10.8.2 Insertions

Insertions at restriction sites may be made by ligation of oligonucleotide linkers into restriction sites after cutting with the appropriate enzyme. Sequence additions at other sites can be achieved by means of site-directed mutagenesis using oligonucleotides in an analogous way to that described for deletions (Fig. 10.32). Here, however, the oligonucleotide contains the DNA sequence specifying the insert and flanked by sequences exactly complementary to those in the target DNA on either side of the desired point of insertion.

10.8.3 Replacements

Sometimes it is desired to keep the same number of nucleotides in a sequence but to alter the sequence at a specific point. This can be achieved by making a small deletion at a restriction

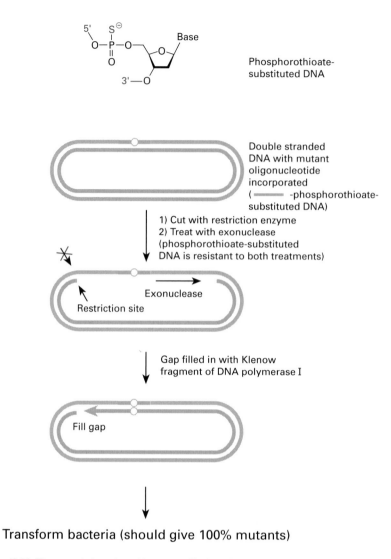

Fig. 10.33 The use of phosphorothioate-modified nucleotides for *in vitro* mutagenesis.

site and inserting into the gap an oligonucleotide linker of the same size but of different sequence.

A more general approach involves use of a synthetic oligonucleotide in an analogous way to the introduction of deletions and insertions, but with the same number of nucleotides in the mutant strand as wild type (Fig 10.32). This procedure works particularly well for single-base alterations and is widely used to change the sequence of a cloned gene that codes for a protein. Expression of the mutated gene leads to the production of a protein with a single amino acid alteration (**protein engineering**).

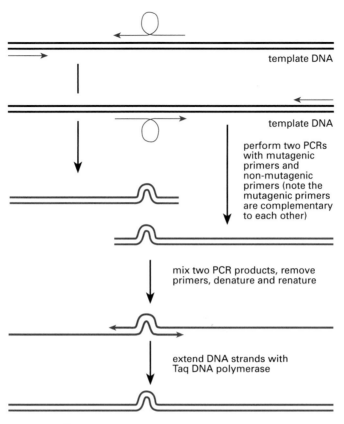

Fig. 10.34 The creation of DNA deletions by PCR.

Summary

In vitro mutagenesis is the process of engineering specific changes in a DNA sequence in a test tube and is now a major way of probing gene function. Deletions, insertions, and replacements of sequence can be achieved at any position, using synthetic oligonucleotides.

10.9 Gene replacement in organisms

Often one wishes to determine the effect of a mutation synthesized *in vivo*. For this purpose the normal gene must be replaced by the mutated gene. It is frequently impossible to remove the former and then add the latter because the first step is often a lethal event. Therefore, one needs to be able to substitute the mutated gene for the normal gene in two steps by keeping an extra copy of the original until the mutant gene is in place. This has been achieved in the yeast *Saccharomyces cerevisiae* by means of **homologous recombination** (Fig. 10.35). The mutated

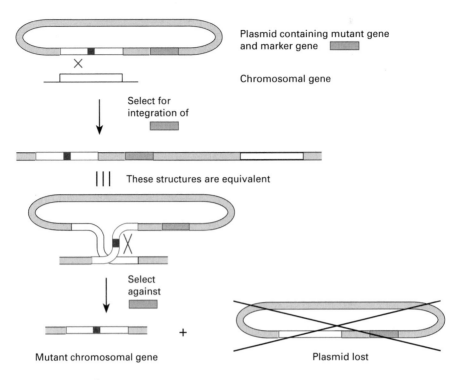

Fig. 10.35 Gene replacement in *Saccharomyces cerevisiae.*

gene is ligated next to a marker gene which can be easily selected *for* under restrictive conditions (often a gene encoding a protein necessary for amino acid or nucleic acid biosynthesis). The construct is inserted into cells which lack the marker gene and these cells are placed in a medium lacking the specific metabolic precursor synthesized by the marker gene product. This procedure kills all cells except those in which the mutated gene has been exchanged for the regular one by a normal homologous recombination event. Subsequent removal of the normal gene is achieved by selection *against* the marker gene by use of a substrate analogue which is converted into a toxic metabolite by the marker gene product.

The functioning of the mutated gene in its correct chromosomal location can now be scrutinized. This method has been used to replace a normal yeast actin gene by a duplicate of the gene that lacks any intron (see Section 5.3.2). The mutated gene functioned perfectly well, showing that an intron can be dispensable to the gene carrying it.

10.9.1 Gene therapy and genetic engineering in mammals

The ability to introduce new or altered genes into a mammalian genome has tremendous implications. For example, it may prove possible to cure some genetic diseases by introducing a healthy copy of a gene into an afflicted individual. It is already possible to introduce into mammals genes which encode economically or medically important polypeptides such as insulin, growth hormones, and interferon. The intention is that the animal either grows faster or produces large amounts of protein which can be harvested. We will address the ways in which this can be carried out, leaving the ethical questions raised by this issue to others.

There are three major ways of introducing DNA into mammalian germ tissue such that the progeny of the recipient will carry the gene. The first involves microinjection of DNA solutions into the nucleus of an egg by means of an extremely fine capillary. Such a technique works very well with a mouse egg but is more difficult with other mammals, such as sheep, where it is extremely hard to see the nucleus. In this way **transgenic** animals have been created which carry functioning genes from another organism.

The second method involves the use of retrovirus-based vectors (Fig. 10.36). As described in Section 5.4.7, retroviruses can infect a cell and then insert their DNA into its chromosomes. The gene to be introduced into the host is ligated into the genome of the retrovirus. The retro-

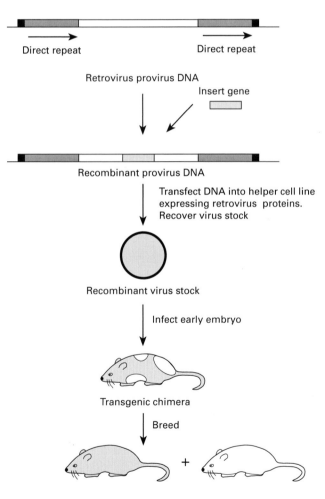

Fig. 10.36 Creation of transgenic animals by use of retroviral vectors.

viral DNA is then introduced into a cultured cell line which is capable of producing all of the components of a retrovirus except for the viral RNA (such a cell culture is called a **helper cell line**). This cell line will then package the recombinant virus stock into virus particles which can be harvested from the culture medium. Helper cells are necessary because the presence of the insert in the retroviral genome disrupts some of the normal retroviral genes needed for viral production. The harvested recombinant virus stock is then used to infect an early embryo which is then replaced into a donor mother. During growth, some cells of the embryo become infected by the virus and the retroviral gene, including the gene insert, becomes stably inserted into the DNA of these cells. Because not all cells become infected, the animal is a **chimera**. However, if the germ cells of this animal contain proviral DNA then its offspring will retain the recombinant in every cell of its body.

The third method for introducing DNA into the mammalian germ line relies upon the existence of cultured cell lines which can become germ cells if injected into early embryos. This approach is particularly useful in the mouse, where such cells, **embryonal carcinoma cells**, can be grown in dishes. The gene of interest can be introduced into these cells, which are then injected into embryos. This approach is particularly useful for gene replacement, when the same basic approach is used as that employed in yeast (Section 10.9).

Summary

Complete genes can be replaced in organisms by homologous recombination. Genes can be introduced into some mammals by DNA microinjection, retrovirus-based vectors, or DNA transfection into cultured cells which can become incorporated into the germ line.

10.10 The detection of specific nucleic acid sequences by hybridization

Molecular cloning is only the beginning of the study of a gene. Often it is important to study the same gene from a variety of different individuals. For example, much can be learned from structural analysis of a series of mutants in the gene. While it is possible to molecularly clone the gene from each mutant individual, it is often much easier simply to analyse the uncloned nucleic acid. It is also important to be able to detect the RNA encoded by a gene and to determine its levels of transcription and tissue specificity. Methods exist for both of these purposes and both depend upon the ability of a single-stranded nucleic acid to pair specifically with its complementary strand.

10.10.1 Parameters affecting nucleic acid hybridization

Temperature (T)

The rate of association of single-stranded DNA into a duplex varies markedly with temperature (Fig. 10.37; Section 2.5.1). The shape of this curve is governed by two factors. At low temperatures, the reassociation rate is determined by the difference in free energy between the unassociated and the transition state:

$$k = Ze^{-E_a/RT}$$

where k is the reassociation rate constant, E_a is the activation free energy, R is the gas constant, and T is the absolute temperature. At higher temperatures, the stability of the duplex is markedly reduced until eventually it is unstable and the hybrid melts. Thus we see a fall off in reassociation rate as this point is approached.

Monovalent cation concentration (*M*)

The **melting temperature** of a hybrid is reduced at lower salt concentration because cations stabilize the DNA duplex. Divalent cations such as magnesium are much more effective in stabilizing hybrids, but are rarely used in hybridization studies (Section 8.2.1).

Base composition (*%GC*)

G·C base-pairs are stronger than A·T, giving a higher melting temperature.

Duplex length (*L*)

Hybrids shorter than a few hundred base-pairs have significantly lower melting temperatures.

In practice, these four factors can be combined into an empirical equation giving the melting temperature T_m of a hybrid DNA.

$$T_m = 69.3 + 0.41(\%GC) + 18.5 \log_{10} M - 500L^{-1} \qquad /^{\circ}C.$$

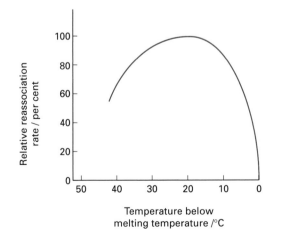

Fig. 10.37 Dependence of reassociation rate of DNA upon temperature.

Use of hybridization temperatures from $10°$ to $20°C$ below the calculated T_m of the hybrid is optimal in practice. For synthetic oligonucleotide probes of 15–20 residues and under standard hybridization conditions, the calculation of T_m is simplified to $2°C$ per $dA \cdot dT$ and $4°C$ per $dG \cdot dC$ base-pair in approximately 1 M sodium chloride solution. This is known as the **Wallace rule**.

10.10.2 Southern and Northern blot analysis

It is possible to use nucleic acid hybridization to detect uncloned genomic DNA. Genomic DNA is immobilized on a nitrocellulose or nylon filter, in basically the same way as described for gene cloning (Section 10.7.3). The gene of interest is detected on the filter by hybridizing a complementary nucleic acid strand labelled either with radioactivity or an affinity label such as biotin or digoxygenin which can be detected with great sensitivity.

Of course, if the DNA is just spotted onto the filter, all that is seen is a spot whose intensity reflects the concentration of the corresponding gene in the sample. This technique is called **dot blotting** and is very useful in this limited respect. However, if the DNA is fractionated before transfer, much more information is acquired. **Southern blot analysis** (named after its inventor, Ed Southern) involves fractionation of DNA by gel electrophoresis, followed by transfer of the DNA out of the gel onto a filter (Fig. 10.38). The filter is then probed for the gene of interest as before. In the commonest case, the DNA has been digested with restriction enzymes and the result is the detection of those restriction fragments which are homologous to the gene probe. In this way, restriction maps of genes can be derived from genomic DNA without resort to cloning.

This technology can be extended to the study of RNA (**Northern blot analysis**). RNA can be electrophoresed in gels and immobilized on filters, provided it is denatured by treatment with formaldehyde (see Section 7.5.3). It can then be detected in the same way as DNA. Unfortunately, RNA cannot be cut into large defined fragments with the same ease as DNA, so such an approach is more limited. It is particularly useful for the determination of the sizes of RNAs and their tissue specificities, the latter approach relying upon the isolation of RNAs from different tissues. Northern blot analysis is often used to determine the transcribed regions in a stretch of DNA. By this approach, a battery of different restriction fragments, which together span the DNA of interest, are separately used to probe a Northern blot. Those DNA fragments

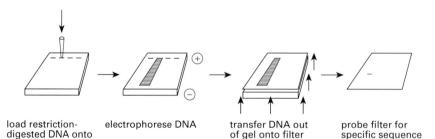

load restriction-   electrophorese DNA   transfer DNA out   probe filter for
digested DNA onto   of gel onto filter   specific sequence
agarose gel   of interest

Fig. 10.38 Southern blot analysis.

which are transcribed detect bands in the Northern blot. The complementary approach (using radioactive RNA to probe restriction-digested DNA) is only possible if the transcripts arising from the DNA are particularly abundant.

10.10.3 *In situ* analysis of RNA in whole organisms

Hybridization can be used to detect transcripts in a cell or organism. Cells and organisms smaller than about 1 mm are fixed (the macromolecules are immobilized) by treatment with a fixative, such as formaldehyde, glutaraldehyde, or methanol/acetic acid. Larger organisms are normally sliced into thin sections before fixation. The fixed specimens are then probed with labelled nucleic acid in the same way as for a Southern blot. In this way, the location of RNAs can be determined at the cellular or even the sub-cellular level.

Summary

Hybridization of a nucleic acid strand to its complementary strand is used to detect nucleic acids, either in solution, immobilized on solid supports, or in fixed tissue specimens. The factors which affect the rate of re-association of a DNA strand with its homologous strand include temperature, monovalent cation concentration, base composition, and duplex length. Nucleic acids can be detected with very high sensitivity using Southern or Northern blot analysis for detecting DNA and RNA respectively.

Further reading

10.1

Arrand, J. E. (1985). Preparation of nucleic acid probes. In *Nucleic acid hybridization: a practical approach* (ed. B. D. Hames and S. J. Higgins), pp. 17–44. IRL Press, Oxford.

Sambrook, J., Fritsch, E. F. and Maniatis, T. (eds) (1989). *Molecular cloning, a laboratory manual*, 2nd edn. Cold Spring Harbor Laboratory Press, New York.

10.2

Bloomfield, V. A., Crothers, D., and Tinoco, I. (1974). *Physical chemistry of nucleic acids.* Harper and Row, New York.

Cantor, C. R. and Schimmel, P. R. (1980). Techniques for the study of biological structure and function.

In *Biophysical chemistry,* Part 2. W. H. Freeman, San Francisco.

Clegg, R. M. (1992). Fluorescence resonance energy transfer and nucleic acids. *Methods in Enzymology,* **211**, 353–88.

Fedoroff, O. Y., Salazar, M., and Reid, B. R. (1993). Structure of a DNA : RNA hybrid duplex: why RNase H does not cleave pure DNA. *Journal of Molecular Biology,* **233**, 509–23.

Feigon, J., Sklenar, V., Wang, E., Gilbert, D. A., Macaya, R. F., and Schultze, P. (1992). ^{1}H NMR spectroscopy of DNA. *Methods in Enzymology,* **211**, 235–53.

Gorenstein, D. (1992). ^{31}P NMR of DNA. *Methods in Enzymology,* **211**, 254–86.

Gray, D. M., Ratcliff, R. L., and Vaughan, M. R. (1992). Circular dichroism spectroscopy of DNA. *Methods in Enzymology,* **211**, 389–405.

Lilley, D. J. M. (1990). The structure of the helical four-way junction in DNA, and its role in genetic recombination. In *Nucleic acids and molecular biology* (ed. D. J. M. Lilley and F. Eckstein), Vol. 4, pp. 55–77, Springer-Verlag, Berlin.

Taillander, E. and Liquier, J. (1992). Infrared spectroscopy of DNA. *Methods in Enzymology*, **211**, 307–35.

Thomas, G. J. and Wang, A. H.-J. (1988). Laser Raman spectroscopy of nucleic acids. In *Nucleic acids and molecular biology* (ed. D. J. M. Lilley and F. Eckstein), Vol. 2, pp. 1–30, Springer-Verlag, Berlin.

Timsit, Y. and Moras, D. (1992). Crystallization of DNA. *Methods in Enzymology*, **211**, 409–29.

Wijmenga, S. S., Mooren, M. M. W., and Hilbers, C. W. (1994). NMR of nucleic acids, from spectrum to structure. In *NMR of macromolecules* (ed. G. C. K. Roberts), pp. 217–88, IRL Press, Oxford.

Wüthrich, K. (1986). *NMR of proteins and nucleic acids*. Wiley, New York.

10.3

Andrews, A. T. (1986). *Electrophoresis: theory, techniques and biochemical and clinical applications*, 2nd edn. Clarendon Press, Oxford.

Beme, B. and Pecors, R. (1976). *Dynamic light scattering with applications to biology, chemistry and physics*. Wiley, New York.

Breslauer, K. J., Friere, E., and Straume, M. (1992). Calorimetry: a tool for DNA and ligand-DNA studies. *Methods in Enzymology*, **211**, 533–67.

Lilley, D. M. J. and Dahlberg, J. E. (eds) (1992). Structural methods of DNA analysis, pp. 409–567. *Methods in Enzymology*, **211A**, Section IV. Academic Press, New York.

Schachman, K. H. (1959). *Ultracentrifugation in biochemistry*. Academic Press, New York.

10.4

Arscott, P. G. and Blomfield, V. A. (1992). Scanning tunnelling microscopy of nucleic acids. *Methods in Enzymology*, **211**, 490–506.

Fisher, H. W. and Williams, R. C. (1979). Electron microscopic visualization of nucleic acids and of their complexes with proteins. *Annual Review of Biochemistry*, **48**, 649–79.

Purrugganan, M. D., Kumar, C. V., Turro, N. J., and Barton, J. K. (1988). Accelerated electron transfer between metal complexes mediated by DNA. *Science*, **241**, 1645–9.

Thresher, R. and Griffith, J. (1992). Electron microscopic visualization of DNA and DNA-protein complexes as adjunct to biochemical studies. *Methods in Enzymology*, **211**, 481–9.

10.5

von Kitzing, E. (1992). Modelling DNA structures: molecular mechanics and molecular dynamics. *Methods in Enzymology*, **211**, 449–66.

McCammon, J. A. and Havey, S. C. (1987). *Dynamics of proteins and nucleic acids*. Cambridge University Press.

Vologodskii, A. V. and Frank-Kamenetskii, M. D. (1992). Modelling supercoiled DNA. *Methods in Enzymology*, **211**, 467–80.

10.6

Maxam, A. M. and Gilbert, W. (1980). Sequencing end-labelled DNA with base-specific chemical cleavages. *Methods in Enzymology*, **65**, 499–560.

Sanger, F. (1981). Determination of nucleotide sequences in DNA. *Science*, **214**, 1205–10.

Sanger, F. (1988). Sequences, sequences and sequences. *Annual Review of Biochemistry*, **57**, 1–28.

10.7

Sambrook, J., Fritsch, E. F. and Maniatas, T. (eds) (1989). *Molecular cloning a laboratory manual*, 2nd edn, Chapters 1–4 and 12. Cold Spring Harbor Laboratory Press, New York.

10.8

Higuchi, R. (1989). In *PCR technology*, ed. H. A. Erlich, pp. 61–70.

Lathe, R. F., Lecocq, J. P. and Everett, R. (1983). DNA engineering: the use of enzymes, chemicals and oligonucleotides to restructure DNA sequences *in vitro*. In *Genetic engineering* (ed. R. Williamson, Vol. 4, pp. 2–57, Academic Press, London.

Sayers, J. R. and Eckstein, F. (1988). Phosphorothioate-based oligonucleotide-directed mutagenesis. In *Genetic engineering: principles and methods*, Vol. 10, pp. 109–22, Plenum Press, New York.

10.9

Carter, B. L. A., Irani, M., Mackay, V., Searle, R. L., Sledziewski, A., and Smith, R. (1987). Expression and sectreation of foreign genes in yeast. In *DNA cloning* (ed. D. M. Glover), Vol. III, pp. 141–62, IRL Press, Oxford.

Murphy, D. and Hansen, J. (1987). The production of transgenic mice by the microinjection of cloned DNA into fertilized one-cell eggs. In *DNA cloning* (ed. D. M. Glover), Vol. III, pp. 213–48, IRL Press, Oxford.

10.10

Bentley Lawrence, J., Singer, R. H., and Marselle, S. M. (1989). Highly localized tracks of specific transcripts within interphase nuclei visualized by *in situ* hybridization. *Cell*, **57**, 493–502.

GLOSSARY

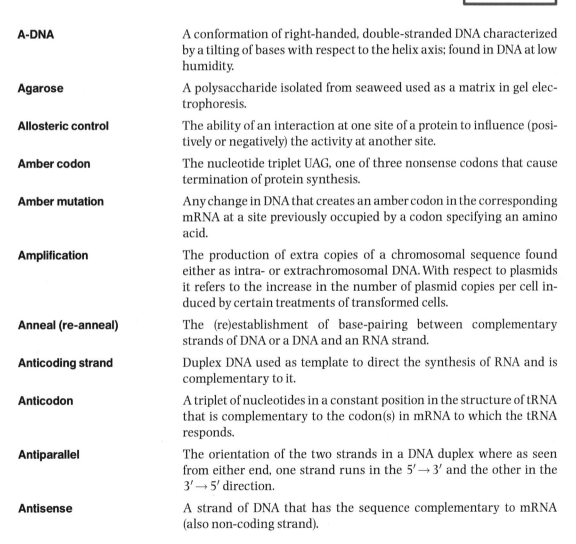

A-DNA
A conformation of right-handed, double-stranded DNA characterized by a tilting of bases with respect to the helix axis; found in DNA at low humidity.

Agarose
A polysaccharide isolated from seaweed used as a matrix in gel electrophoresis.

Allosteric control
The ability of an interaction at one site of a protein to influence (positively or negatively) the activity at another site.

Amber codon
The nucleotide triplet UAG, one of three nonsense codons that cause termination of protein synthesis.

Amber mutation
Any change in DNA that creates an amber codon in the corresponding mRNA at a site previously occupied by a codon specifying an amino acid.

Amplification
The production of extra copies of a chromosomal sequence found either as intra- or extrachromosomal DNA. With respect to plasmids it refers to the increase in the number of plasmid copies per cell induced by certain treatments of transformed cells.

Anneal (re-anneal)
The (re)establishment of base-pairing between complementary strands of DNA or a DNA and an RNA strand.

Anticoding strand
Duplex DNA used as template to direct the synthesis of RNA and is complementary to it.

Anticodon
A triplet of nucleotides in a constant position in the structure of tRNA that is complementary to the codon(s) in mRNA to which the tRNA responds.

Antiparallel
The orientation of the two strands in a DNA duplex where as seen from either end, one strand runs in the $5' \rightarrow 3'$ and the other in the $3' \rightarrow 5'$ direction.

Antisense
A strand of DNA that has the sequence complementary to mRNA (also non-coding strand).

Antisense oligomers	Single-stranded oligodeoxynucleotides or modified deoxynucleotides that bind either to a single-stranded target nucleic acid (RNA or single-stranded DNA) by Watson–Crick base-pairing or to double-stranded target (i.e. DNA) by Hoogsteen or reverse Hoogsteen base-triples.
Anti-terminator proteins	Allow RNA polymerase to transcribe through certain terminator sites.
Attenuation	The regulation of premature termination of transcription involved in controlling expression of some bacterial operons.
Autoradiography	The detection of radioactively labelled molecules present, for example, in a gel or on a filter by exposing an X-ray film to it.
Bacteriophages	Viruses that infect bacteria; often abbreviated as phages.
Base-pair (bp)	A partnership of A with T or of C with G in a DNA double helix; other pairs are possible in RNA under some circumstances.
B-DNA	A conformation of right-handed, double-stranded DNA characterized by a perpendicular arrangement of bases with respect to the helix axis; thought to be the form of DNA commonly found in nature.
Bidirectional replication	Is accomplished when two replication forks move away from the same origin in opposite directions.
Blunt-end ligation	The covalent attachment of two duplex DNA molecules having no single-stranded extensions (sticky ends).
Buoyant density	The density of a particle or molecule when suspended in an aqueous salt or sugar solution.
Capping	The post-transcriptional attachment of a cap to the 5′-terminus of most eukaryotic mRNAs.
Catenane	Interlinked double-stranded DNA circles, as in the links of a chain.
cDNA	A single-stranded DNA complementary to an RNA, synthesized from it by *in vitro* reverse transcription.
cDNA clone	A duplex DNA sequence representing an RNA, carried in a cloning vector.
Cell cycle	The period from one division to the next.
Centromere	A constricted region of a chromosome that includes the site of attachment to the mitotic or meiotic spindle.
Chromatids	Copies of a chromosome produced by replication.
Chromatin	The complex of DNA and proteins (mostly histones) present in the nucleus of a eukaryotic cell.
Chromosome	A discrete unit of the genome carrying many genes, consisting of a very long molecule of DNA, complexed with a large number of different proteins (mostly histones). Chromosomes are visible as a morphological entity only during the act of cell division.

Chromosome walking   Sequential isolation of clones carrying overlapping sequences of DNA to span large regions of the chromosome (often in order to reach a particular locus).

Cistron   The genetic unit defined by the *cis/trans* test; equivalent to gene in comprising a unit of DNA representing a protein.

Clone   A large number of cells or molecules.

Closed circular DNA   Double helical DNA in which both strands are closed circles, i.e. there are no free ends.

Coding strand (of DNA)   Has the same sequence as mRNA.

Codon   A triplet of nucleotides that represents an amino acid or a termination signal.

Cohesive ends   See **Sticky ends**.

Concatenated (circles of) DNA Interlocked-like rings on a chain.

Consensus sequence   An idealized sequence in which each position represents the base most often found when many actual sequences are compared.

Constitutive genes   Genes that are expressed without additional regulation.

Copy number   The average number of copies of a particular (recombinant) plasmid present in a single host cell. Also used for individual genes.

Cordycepin $3'$-Deoxyadenosine, an inhibitor of polyadenylation of RNA.

Core DNA   The 146 bp of DNA contained on a core particle.

Core particle   A digestion product of the nucleosome that retains the histone octamer and has 146 bp of DNA; its structure appears similar to that of the nucleosome itself.

Cosmids   Plasmids into which phage λ cos sites have been inserted; as a result, the plasmid DNA can be packaged *in vitro* into the phage coat.

Covalently-closed-circular (CCC) DNA   A completely double-stranded circular DNA molecule without nicks or other discontinuities, usually in a supercoiled conformation.

Crossing-over   The reciprocal exchange of material between chromosomes that occurs during meiosis and is responsible for genetic recombination.

Cross-linking   Introduction of covalent intra- or intermolecular bonds between groups that are normally not covalently linked. Used to detect proximity of (parts of) (macro)molecules.

Cruciform   The structure produced at inverted repeats of DNA if the repeated sequence pairs with its complement on the same strand (instead of with its regular partner in the other strand of the duplex).

Curved DNA   DNA in which the preferred conformation has a curved major axis.

Cyclic AMP   A molecule of AMP in which the phosphate group is joined to both the $3'$- and $5'$-positions of the ribose; one of its functions is the activation of CAP, a positive regulator of prokaryotic transcription.

Deletions   Constitute the removal of a sequence of DNA, the regions on either side being joined together.

Denaturation (of DNA or RNA) Conversion from the double-stranded into the single-stranded state; separation of the strands is most often accomplished by heating.

Density gradient centrifugation The separation of particles or molecules on the basis of differences in their buoyant density, usually in concentrated solutions of sucrose or caesium salts.

Diploid (set of chromosomes) Contains two copies of each autosome and two sex chromosomes.

Discontinuous replication Refers to the synthesis of DNA in short (Okazaki) fragments that are later joined into a continuous strand.

DNA sequencing The determination of the order of nucleotides in the strands of a DNA molecule.

Downstream Identifies sequences proceeding farther in the direction of expression; for example, the coding region is downstream from the initiation codon.

Duplex The double-stranded structure of DNA.

End labelling The addition of a radioactively labelled group to one end ($5'$ or $3'$) of a DNA or RNA strand.

Endonucleases Cleave bonds within a nucleic acid chain; they may be specific for RNA or for single- or double-stranded DNA.

Enhancer element A DNA sequence that increases the utilization of (some) eukaryotic promoters in *cis* configuration, but can function in any location, upstream or downstream, relative to the promoter.

Ethidium bromide A chemical that, upon intercalation between base-pairs of single-stranded DNA or RNA, fluoresces under ultraviolet light. Used in the detection of nucleic acid in density gradients or on gels.

Eukaryotic Organisms containing a nucleus.

Excision-repair Systems remove a single-stranded sequence of DNA containing damaged or mispaired bases and replace it in the duplex by synthesizing a sequence complementary to the remaining strand.

Exon Any segment of an interrupted gene that is represented in the mature RNA product.

Exonucleases Cleave nucleotides one at a time from the end of a polynucleotide chain; they may be specific for either the $5'$- or $3'$-end of DNA or RNA.

Exonuclease III An enzyme from *E. coli* which catalyses the stepwise removal of mononucleotides from double-stranded DNA carrying a $3'$-OH terminus in the $3' \rightarrow 5'$ direction.

Fingerprint The characteristic array of oligopeptides or oligonucleotides obtained upon two-dimensional electrophoresis of a protein digested with a specific endopeptidase or an RNA digested with a specific endonuclease.

Flexible DNA DNA which can be deformed more easily than average, particularly by bending.

Footprinting In this context is a technique for identifying the site on DNA bound by some protein by virtue of the protection of bonds in this region against attack by nucleases.

Gap in DNA The absence of one or more nucleotides in one strand of the duplex.

Gel electrophoresis Electrophoresis performed in a gel matrix (usually agarose or poly-acrylamide) which allows separation of molecules of similar electric charge density on the basis of their difference in molecular weight.

Gene A DNA sequence involved in the production of an RNA or protein molecule as the final product. Includes both the transcribed region and any sequences upstream and/or downstream responsible for its correct and regulated expression (e.g. promoter and operator sequences).

Genetic code The complete set of codons specifying the various amino acids, including the nonsense codons. The code is usually written in the form in which it occurs in mRNA.

Genome The entire genetic material of a cell.

Genotype The genetic constitution of an organism.

GT–AG rule Describes the presence of these constant dinucleotides at the first two (GT) and the last two (AT) positions of introns of nuclear genes.

Gyrase A type II topoisomerase of *E. coli* with the ability to introduce negative supercoils into DNA.

Hairpin The double-stranded region formed by base-pairing of adjacent complementary sequences in the same DNA or RNA strand.

Haploid (set of chromosomes) Contains one copy of each autosome and one sex chromosome; the haploid number *n* is characteristic of gametes.

H-DNA A triple-stranded form of DNA comprising a polypyrimidine strand lying in the major groove of a polypyrimidine:polypurine duplex.

Heteroduplex (hybrid) DNA DNA generated by base-pairing between partly non-complementary single strands derived from the different parental duplex molecules; it occurs during genetic recombination.

Helical repeat (*h*) The number of base-pairs per turn of a double–stranded DNA helix. It is measured relative to the helix axis (*twist-related h*) or relative to a surface on which the DNA lies (*surface-related h*).

Highly repetitive DNA DNA sequences represented more than 100 000 times per genome; satellite DNA is one family thereof, as are Alu-sequences.

Histones Conserved DNA binding proteins of eukaryotes that form the nucleosome, the basic subunit of chromatin.

Holliday junction An intermediate in recombination formed by strand-exchange between two homologous duplexes. It is structurally related to a cruciform.

Homology The degree of identity existing between the nucleotide sequences of two related but not complementary DNA or RNA molecules. Seventy per cent homology means that on the average 70 out of every 100 nucleotides are identical. The same term is used in comparing the amino acid sequences of related proteins.

Hybridization The pairing of complementary RNA and DNA strands to give an RNA–DNA hybrid, and is also used to describe the pairing of two single-stranded DNA molecules.

Hyperchromicity The increase in optical density that occurs when DNA is denatured.

Induced mutations Result from the treatment of cells with a mutagen.

Induction The expression of a gene in response to an external stimulus.

Initiation codon (AUG; sometimes GUG) Codes for the first amino acid in protein sequences, which is formyl-methionine in prokaryotes; fMet is often removed post-translationally.

Initiation factors (IF) Proteins that associate with the small subunit of the ribosome specifically to render initiation of synthesis.

Insertions Additional (sequences of) nucleotides in DNA.

Integration (of viral or another DNA sequence) Insertion into a host genome as a region covalently linked on either side to the host sequences.

Intercistronic region The region between the termination codon of one gene and the initiation codon of the next gene in a polycistronic transcription unit.

Intervening sequence An intron.

Intron A segment of DNA that is transcribed, but is removed from within the transcript by splicing together the sequences (exons) on either side of it.

Inverted repeats Two copies of the same sequence of DNA repeated in opposite orientation on the same molecule. Adjacent inverted repeats constitute a palindrome.

In vitro (lit. 'in glass') Refers to any (biological) process occurring outside the living cell.

In vivo Refers to any biological process occurring within the living cell.

Isoschizomers Two restriction enzymes that recognize the same DNA sequence and cut it differently.

Klenow fragment A piece obtained from DNA polymerase I by proteolytic cleavage; it lacks the 5′-to-3′ exonuclease activity.

Knot A knotted length of linear DNA closed into a circle.

Lariat The branched intermediate of eukaryotic mRNA formed during splicing.

Library A set of cloned fragments together representing the entire genome.

Ligase (DNA ligase) Catalyses the formation of a phosphodiester bond at the site of a single-strand break in duplex DNA. Some DNA ligases can also ligate blunt-end DNA molecules. RNA ligase covalently links separate RNA molecules.

Ligation The formation of a phosphodiester bond to link two adjacent bases separated by a nick in one strand of a double helix of DNA. (The term can also be applied to blunt end ligation and to joining of RNA.)

Linkage

The tendency of genes to be inherited together as a result of their location on the same chromosome; measured by percentage recombination between loci.

Linking number (*Lk*)

The number of times the two strands of closed-circular DNA are linked. It is strictly defined as half the sum of the inter-strand nodes in a two-dimensional projection of the DNA. Linking number is distributed between the two geometric parameters *twist* (*Tw*) and writhe (*Wr*) as

$$Lk = Tw + Wr.$$

Locus

The position of a chromosome at which the gene for a particular trait resides; locus may be occupied by any one of the alleles for the gene.

Loop

A single-stranded region at the end of a hairpin in RNA (or single-stranded DNA) or a non-paired segment in duplex DNA.

Map distance

Distance between genes (or rather mutations in genes) on the chromosome. Is measured as cM (centriMorgans) = percentage recombination (sometimes subject to adjustments).

Maxam–Gilbert sequencing

A DNA sequencing technique based on specific chemical modification of each of the four bases.

Melting (of DNA)

Results in denaturation.

Melting temperature (*T*$_m$)

The temperature where hyperchromicity is half-maximal.

Micrococcal nuclease

An endonuclease that requires Ca^{2+} ions for activity and thus can easily be inactivated by chelating these ions with EGTA. Used among other things in degrading the endogenous mRNA in cell-free translation systems. Chromatin DNA is cleaved preferentially between nucleosomes.

Missense mutation

Any mutation that changes a codon specifying one amino acid to one coding for another amino acid.

Modification (of DNA or RNA)

Includes all changes made to the nucleotides after their initial incorporation into the polynucleotide chain.

Modified bases

All those except the usual four from which DNA (T, C, A, G) or RNA (U, C, A, G) are synthesized; they result from postsynthetic changes in the nucleic acid.

Multigene family

A set of identical or related genes present in the same organism, usually coding for a family of related proteins.

Mutagens

Increase the rate of mutation by causing changes in DNA.

Mutation

Any change in the sequence of genomic DNA.

Negative supercoiling

The twisting of a duplex of DNA in space in the opposite sense to the turns of the strands on the double helix.

Nick

In duplex DNA is the absence of a phosphodiester bond between two adjacent nucleotides on one strand.

Nick translation

The ability of *E. coli* DNA polymerase I to nick as a starting point from which one strand of a duplex DNA can be degraded and replaced by resynthesis of new material; is used to introduce radioactively labelled nucleotides into DNA *in vitro*.

Nitrocellulose
A type of paper which binds nucleic acids. Used in Southern, Northern (and also Western) blotting as well as other filter hybridization techniques.

Node
A cross-over point of two single or double strands in a two-dimensional projection. The sign of the node is determined by the directions assigned to the DNA strands.

Nonsense codon
Any one of three triplets (UAG, UAA, UGA) that cause termination of protein synthesis. (UAG is known as *amber*; UAA as *ochre*; UGA as *opal*.)

Nonsense mutation
Any mutation that changes a normal codon into a nonsense codon.

Non-Watson–Crick base-pair
Any base-pair other than the standard G·G and A·T (U in RNA) pairs. Occur mainly in intrastrand pairing of RNA.

Northern blotting
A technique for transferring RNA from an agarose gel to a nitrocellulose filter on which it can be hybridized to a complementary DNA.

Nucleosome
The basic structural subunit of chromatin, consisting of about 200 bp of DNA and an octamer of histone proteins.

Ochre codon
The triplet UAA, one of three nonsense codons that cause termination of protein synthesis.

Ochre suppressor
A gene coding for a mutant tRNA able to respond to the UAA codon to allow continuation of protein synthesis.

Okazaki fragments
The short stretches of 1000–2000 (in eukaryotes 100–200) nucleotides produced during replication of the lagging strand of the DNA duplex which are subsequently covalently linked.

Oncogene
A retroviral gene that causes transformation of the mammalian infected cell. Oncogenes are slightly changed equivalents of normal cellular genes called proto-oncogenes. The viral version is designated by the prefix **v**, the cellular version by the prefix **c**.

Open circular DNA
A double-stranded DNA circle containing a broken phosphate diester bond in one strand.

Operator
The site on DNA at which a repressor protein binds to prevent transcription from initiating at the adjacent promoter.

Operon
A complete unit of bacterial gene expression and regulation, including structural genes, regulator gene(s), and control elements in DNA recognized by regulator gene product(s).

Origin (ori)
A sequence of DNA at which replication is initiated.

Overwinding (of DNA)
Caused by positive supercoiling (by applying further tension in the direction of winding of the two strands about each other in the duplex).

Packing ratio
The ratio between the length of the DNA double helix and the length of the fibre containing the DNA.

Palindrome	A sequence of DNA that is the same when one strand is read left to right or the other is read right to left; consists of adjacent inverted repeats.
PCR (polymerase chain reaction)	The use of oligonucleotide primers and a DNA polymerase to amplify a section of DNA.
Periodicity (of DNA)	The number of base-pairs per turn of the double helix.
Phenotype	In contrast to genotype, it is the appearance of an organism.
Plasmid	An autonomous self-replicating extrachromosomal circular DNA.
Polyacrylamide gel electrophoresis	See **Gel electrophoresis**.
Polyadenylation	The post-transcriptional attachment of up to 200 AMP residues to the 3′-terminus of most eukaryotic mRNAs.
Polycistronic mRNA	Includes coding regions representing more than one gene.
Polymerase	An enzyme that catalyses the assembly of nucleotides into RNA or of deoxynucleotides into DNA; usually the enzyme requires single-stranded DNA (sometimes RNA) as a template.
Polymorphic DNA	The property of DNA to adopt a number of helical forms, A, B, H, or Z, depending on the base sequence and on the conditions in the environment both in fibres and in crystals.
Polynucleotide kinase	An enzyme which transfers the γ-phosphate group from ATP on to the 5′-OH terminus of a DNA or RNA molecule. Used in end-labelling DNA and RNA for sequencing.
Polysome (polyribosome)	A mRNA associated with several ribosomes engaged in translation.
Positive supercoiling	When the double helix coils itself in the same direction as the winding of the two strands of the double helix itself.
Primary transcript	The original unmodified RNA product corresponding to a transcription unit.
Primer	A short sequence (of DNA or RNA) that is paired with one strand of DNA and provides a free 3′-OH end at which a DNA polymerase starts synthesis of a deoxyribonucleotide chain.
Probe (hybridization)	A labelled DNA or RNA molecule used to detect a complementary sequence by molecular hybridization.
Prokaryotic	Organisms lacking membrane-enclosed nuclei.
Promoter (in bacteria)	The region of the gene involved in binding of the RNA polymerase. (In eukaryotes) usually all regions of the gene required for maximum expression (excluding enhancer sequences).
Proofreading	Refers to any mechanism for correcting errors in protein or nucleic acid synthesis that involves scrutiny.
Propeller twist	The dihedral angle formed between individual base-planes when viewed along the C^6–C^8 vector of a base pair in a double helix.
Pseudogene	A DNA sequence that shows a large degree of homology to the normal

	(expressed) gene but is itself not expressed.
Reading frame	One of three possible ways of reading a nucleotide sequence as a series of triplets.
RecA protein	The product of the *rec*A locus of *E. coli* with dual activities, acting as a protease in the presence of single-stranded DNA (resulting from strong mutagenesis) and also able to exchange single strands of DNA molecules in co-operation with RecB and RecC proteins. The protease activity controls the SOS response; the nucleic acid handling facility is involved in general homologous recombination.
Recombinant DNA	Any DNA molecule created by ligating pieces of DNA that normally are not contiguous.
Relaxed DNA	Closed-circular DNA formed without any constraint of the DNA helix. It normally consists of a distribution of topoisomers.
Renaturation (of DNA or RNA)	The re-establishment of the DNA duplex or instrastrand hairpin structures in an RNA molecule after denaturation. (Of a protein): the conversion from an inactive into a biologically active conformation.
Replication eye	A region in which DNA has been replicated within a longer, unreplicated region.
Replication fork	The point at which strands of parental duplex DNA are separated so that replication can proceed.
Replicon	The regulatory unit of an origin and proteins necessary for initiation of replication (specific for this origin).
Repression	The blocking of the synthesis of certain enzymes when their products are present; more generally, refers to inhibition of transcription (or translation) by binding of repressor protein to specific site on DNA (or mRNA).
Restriction enzymes	Recognize specific short sequences of (usually) unmethylated DNA and cleave the respective DNA molecule (sometimes at target site, sometimes elsewhere, depending on type).
Retrovirus	A virus containing a single-stranded RNA genome that propagates via conversion into double-stranded DNA by reverse transcription.
Reverse transcriptase	RNA-dependent DNA polymerase. Originally detected in retroviruses. It is, however, also present in normal eukaryotic cells and even in *E. coli*.
Reverse transcription	Synthesis of DNA on a template of RNA; accomplished by reverse transcriptase enzyme.
Reversion (of mutation)	A change in DNA that either reverses the original alteration (true reversion) or compensates for it (second site reversion in the same gene).
Ribosomes	Subcellular particles consisting of several RNA and numerous protein molecules. Involved in translating the genetic code on mRNA into the amino acid sequence of the corresponding protein.
Ribosome binding site	See **Shine–Dalgarno sequence**.

Rifamycins (**including rifampicin**)	Antibiotic inhibiting transcription in bacteria.
rRNA	Ribosomal RNA. Forms part of the ribosome.
Sanger–Coulson sequencing	DNA sequencing technique based on transcription of single-stranded DNA by Klenow polymerase in the presence of dideoxynucleotides. The same technique can also be used for sequencing of RNA.
Satellite DNA	DNA consisting of many tandem repeats (identical or related) of a short basic repeating unit.
SDS (sodium dodecyl sulfate)	A detergent.
SDS gel electrophoresis	Gel electrophoresis of proteins in polyacrylamide gels in the presence of SDS. Molecules of SDS associate with the protein molecules giving them all a similar electric charge density and thus allowing separation on the basis of differences in molecular weight.
Semiconservative relication	Separation of the strands of parental duplex, each then acting as a template for synthesis of a complementary strand.
Semidiscontinuous replication	Mode in which one new strand is synthesized continuously while the other is synthesized discontinuously.
Shine–Dalgarno sequence	Part or all of the polypurine sequence AGGAGG located on bacterial mRNA just prior to an AUG initiation codon; is complementary to the sequence at the 3'-end of 16S rRNA; involved in binding of ribosome to mRNA.
Silent mutations	Do not change the product of a gene.
Site-directed mutagenesis	Introduction in the test tube of a specific mutation(s) into a DNA molecule at a predetermined site.
Site-specific recombination	Occurs between two specific (not necessarily homologous) sequences, as in phage integration/excision or resolution of cointegrate structures during transposition.
Southern blotting	A procedure for transferring denatured DNA from an agarose gel to a nitrocellulose filter where it can be hybridized with a complementary nucleic acid.
Spliceosome	The ribonucleoprotein particle containing precursor mRNA and the various splicing factors on which splicing takes place.
Splicing	Describes the removal of introns and joining of exons in RNA; thus introns are spliced out, while exons are spliced together.
SSB (single-strand binding)	Protein of *E. coli*, a protein that binds to single-stranded DNA.
Sticky ends	Complementary single-stranded extensions at the ends of a DNA fragment or two different fragments resulting from a staggered cut (or introduced by tailing).
Stop codon	Same as termination codon.
Strand displacement	A mode of replication of mitochondrial and some viral DNA in which a new DNA strand grows by displacing the previous (homologous) strand of the duplex.

Structural gene — Gene coding for any RNA or protein product other than a regulator.

Supercoiled DNA — Closed-circular DNA formed under torsional stress. A vernacular term for DNA with a non-zero linking difference. Supercoiling can be manifest in a change in twist, or in writhe, or in both.

Suppressor tRNA — A minor tRNA species that responds to a termination codon (e.g. a minor tyrosine tRNA with an AUC anticodon).

T_m — The abbreviation for melting temperature.

Template — Portion of single-stranded DNA or RNA used to direct the synthesis of a complementary polynucleotide.

Termination codon — One of three triplet sequences, UAG (amber), UAA (ochre), or UGA, that cause termination of protein synthesis; they are also called nonsense codons.

Terminator — A sequence of DNA that, after being transcribed, causes RNA polymerase to terminate transcription.

Topoisomerase — An enzyme which catalyses changes in linking number of closed-circular DNA.

Topological isomers — Molecules of DNA that are identical except for a difference in linking number.

Transcription — Usually the synthesis of RNA on a DNA template. Also used to describe the synthesis of DNA on an RNA template by reverse transcriptase, the copying of a (primed) single-stranded DNA by DNA polymerase, and the copying of RNA by (viral) RNA polymerase.

Transformation — The acquisition by a cell of new genetic markers by incorporation of added DNA. In eukaryotic cells it also refers to conversion into a state of unrestrained growth in culture resembling or identical to the tumourigenic condition.

Transition — A mutation in which one pyrimidine is substituted by the other or in which one purine is substituted for the other.

Translation — Synthesis of protein on the mRNA template.

Translocation (of a chromosome) — A rearrangement in which part of a chromosome is detached by breakage and then becomes attached to some other chromosome.

Translocation (of the ribosome) — Moving one codon along mRNA after the addition of each amino acid to the polypeptide chain.

Transposition — The movement of part of the DNA to another location within the genome.

Transposon — A DNA sequence able to replicate and insert one copy at a new location in the genome, carrying genetic information additional to that necessary for transposition mechanism.

Transversion — A mutation in which a purine is replaced by a pyrimidine or vice versa.

Triplet — A sequence of three nucleotides in DNA or RNA. Usually means the same as codon.

Twist (Tw)	Corresponds to the number of double-helical turns in a given length of DNA as measured relative to the helix axis. (Twist cannot normally be measured simply by counting and is in reality a more complex function of the path of a DNA strand about the helix axis).
Underwinding (of DNA)	Produced by negative supercoiling (because the double helix is itself coiled in the opposite sense from the intertwining of the strands).
Upstream	Identifies sequences proceeding in the opposite direction from expression; for example, the bacterial promoter is upstream from the transcription unit, the initiation codon is upstream from the coding region.
Vector (cloning)	Any plasmid or phage into which a foreign DNA may be inserted to be cloned.
W-DNA	A left-handed, zig-zag DNA duplex created by modelling that has the same sense of direction for the backbones as does B-DNA but has other features in common with Z-DNA.
Watson–Crick rules	The base-pairing rules that underly gene structure and expression. G pairs with C and A with T (U in RNA).
Wild type	The genotype or phenotype commonly encountered in the natural population or laboratory stock of a given organism.
Wobble hypothesis	Accounts for the ability of a tRNA to recognize more than one codon by unusual (non-G·C; A·T) pairing with the third base of a codon.
Writhe (W_r)	A parameter describing the path of a DNA helix axis in space. Planar DNA and DNA whose axis can be laid on the surface of a sphere without crossing itself have $W_r = 0$.
Writhing number (W)	The amount of supercoiling of a DNA molecule (see linking number).
Z-DNA	A left handed form of Watson–Crick double helix.

INDEX